# ASTRONOMY AND
# ASTROPHYSICS LIBRARY

**Series Editors:**  I. Appenzeller · M. Harwit · R. Kippenhahn
P. A. Strittmatter · V. Trimble

T0178150

# ASTRONOMY AND
# ASTROPHYSICS LIBRARY

**Tools of Radio Astronomy**
By K. Rohlfs

**Physics of the Galaxy and Interstellar Matter**
By H. Scheffler and H. Elsässer

**Galactic and Extragalactic Radio Astronomy** 2nd Edition
Editors: G. L. Verschuur and K. I. Kellermann

**Observational Astrophysics**
By P. Léna

**Astrophysical Concepts** 2nd Edition
By M. Harwit

**The Sun** An Introduction
By M. Stix

**Stellar Structure and Evolution**
By R. Kippenhahn and A. Weigert

**Relativity in Astrometry, Celestial Mechanics
and Geodesy**
By M. H. Soffel

**General Relativity, Astrophysics, and Cosmology**
By A. K. Raychaudhuri, S. Banerji and A. Banerjee

**The Solar System**
By T. Encrenaz and J.-P. Bibring

**Physics and Chemistry of Comets**
Editor: W. F. Huebner

**Supernovae**
Editor: A. Petschek

**Astrophysics of Neutron Stars**
By V. M. Lipunov

**Gravitational Lenses**
By P. Schneider, J. Ehlers and E. E. Falco

**The Stars**
By E. L. Schatzman and F. Praderie

# R. Kippenhahn   A. Weigert

# Stellar Structure and Evolution

With 192 Figures

Springer-Verlag
Berlin Heidelberg New York
London Paris Tokyo
Hong Kong Barcelona
Budapest

Professor Dr. Rudolf Kippenhahn

Rautenbreite 2, D-37077 Göttingen, Germany

Professor Dr. Alfred Weigert †

*Series Editors*

Immo Appenzeller

Landessternwarte, Königstuhl
D-69117 Heidelberg
Germany

Martin Harwit

The National Air and Space Museum
Smithsonian Institution
7th St. and Independence Ave. S. W.
Washington, DC 20560, USA

Rudolf Kippenhahn

Rautenbreite 2
D-37077 Göttingen
Germany

Peter A. Strittmatter

Steward Observatory
The University of Arizona
Tucson, AZ 85721, USA

Virginia Trimble

Astronomy Program
University of Maryland
College Park, MD 20742, USA
and Department of Physics
University of California
Irvine, CA 92717, USA

1st Edition 1990
Corrected 3rd Printing 1994

ISBN 3-540-58013-1 Springer-Verlag Berlin Heidelberg New York
ISBN 0-387-58013-1 Springer-Verlag New York Berlin Heidelberg

Springer-Verlag Berlin Heidelberg New York
a member of Springer Science+Business Media

© Springer-Verlag Berlin Heidelberg 1990
Printed in Germany

The use of general descriptive names, registered names, trademarks, etc. in this publication does not imply, even in the absence of a specific statement, that such names are exempt from the relevant protective laws and regulations and therefore free for general use.

SPIN: 10983287        55/3111 - 5 - Printed on acid-free paper

*To Our Wives*

# Preface

The attempt to understand the physics of the structure of stars and their change in time – their evolution – has been bothering many physicists and astronomers ever since the last century. This long chain of successful research is well documented not only by numerous papers in the corresponding journals but also by a series of books. Some of them are so excellently written that despite their age they can still be recommended, and not only as documents of the state of the art at that time. A few outstanding examples are the books of R. Emden (1907), A. S. Eddington (1926), S. Chandrasekhar (1939), and M. Schwarzschild (1958). But our science has rapidly expanded in the last few decades, and new aspects have emerged which could not even be anticipated, say, 30 years ago and which today have to be carefully explored.

This does not mean, however, that our ambition is to present a complete account of the latest and most refined numerical results. This can well be left to the large and growing number of excellent review articles. The present book is intended rather to be a textbook that will help students and teachers to understand these results as far as possible and present them in a simple and clear manner. We know how difficult this is since we ourselves have tried for the largest part of our scientific career to understand "how the stars work" – and then to make others believe it. In these attempts we have found that often enough a simplified analytical example can be more helpful than the discussion of an exceptionally beautiful numerical solution. Therefore we do not hesitate to include many simple considerations and estimates, if necessary, even at the expense of rigour and the latest results. The reader should also note that the list of references given in this book is not intended to represent a table of honour for the (known and unknown) heroes of the theory of stellar structure; it is merely designed to help the beginner to find a few first paths in the literature jungle and gives those papers from which we have more or less randomly chosen the numbers for figures and numerical examples. (There are others of at least the same quality!)

The choice of topics for a book such as this is difficult and certainly subject to personal preferences. Completeness is neither possible nor desirable. Still, one may wonder why we did not include, for example, binary stars, although we are obviously interested in their evolution. The reason is that here one would have had to include the physics of essentially non-spherical objects (such as disks), while we concentrate mainly on spherical configurations; even in the brief description of rotation the emphasis is on small deviations from spherical symmetry.

This book would never have been completed without the kind and competent help of many friends and colleagues. We mention particularly Wolfgang Duschl and Peter Schneider who read critically through the whole manuscript; Norman Baker,

Gerhard Börner, Mounib El Eid, Wolfgang Hillebrandt, Helmuth Kähler, Ewald Müller, Henk Spruit, Joachim Wambsganß, and many others read through particular chapters and gave us their valuable advice. In fact it would probably be simpler to give a complete list of those of our colleagues who have *not* contributed than of those who helped us.

In addition we have to thank many secretaries at our institutes; several have left their jobs (for other reasons!) during the five years in which we kept them busy. Most of this work was done by Cornelia Rickl and Petra Berkemeyer in Munich and Christa Leppien and Heinke Heise in Hamburg, while Gisela Wimmersberger prepared all the graphs. We are grateful to them all.

Finally we wish to thank Springer-Verlag for their enthusiastic cooperation.

Munich and Hamburg                                            *Rudolf Kippenhahn*
December 1989                                                   *Alfred Weigert*

# Contents

## Part I  The Basic Equations

**1. Coordinates, Mass Distribution, and Gravitational Field in Spherical Stars** ......................................... 2
   1.1  Eulerian Description ................................. 2
   1.2  Lagrangian Description ................................. 3
   1.3  The Gravitational Field ................................. 4

**2. Conservation of Momentum** ................................. 6
   2.1  Hydrostatic Equilibrium ................................. 6
   2.2  The Role of Density and Simple Solutions ................. 7
   2.3  Simple Estimates of Central Values $P_c$, $T_c$ ................. 8
   2.4  The Equation of Motion for Spherical Symmetry ............. 9
   2.5  The Non-spherical Case ................................. 11
   2.6  Hydrostatic Equilibrium in General Relativity ................. 12
   2.7  The Piston Model ................................. 13

**3. The Virial Theorem** ................................. 15
   3.1  Stars in Hydrostatic Equilibrium ................................. 15
   3.2  The Virial Theorem of the Piston Model ................. 17
   3.3  The Kelvin–Helmholtz Time-scale ................. 18
   3.4.  The Virial Theorem for Non-vanishing Surface Pressure ....... 18

**4. Conservation of Energy** ................................. 19
   4.1  Thermodynamic Relations ................................. 19
   4.2  Energy Conservation in Stars ................................. 21
   4.3  Global and Local Energy Conservation ................. 23
   4.4  Time-scales ................................. 25

**5. Transport of Energy by Radiation and Conduction** ............. 27
   5.1  Radiative Transport of Energy ................................. 27
       5.1.1  Basic Estimates ................................. 27
       5.1.2  Diffusion of Radiative Energy ................. 28
       5.1.3  The Rosseland Mean for $\kappa_\nu$ ................. 29
   5.2  Conductive Transport of Energy ................................. 31
   5.3  The Thermal Adjustment Time of a Star ................. 33
   5.4  Thermal Properties of the Piston Model ................. 34

6. **Stability Against Local, Non-spherical Perturbations** ............ 36
   6.1 Dynamical Instability ....................................... 36
   6.2 Oscillation of a Displaced Element ......................... 41
   6.3 Vibrational Stability ....................................... 42
   6.4 The Thermal Adjustment Time ............................ 43
   6.5 Secular Instability ......................................... 44
   6.6 The Stability of the Piston Model ......................... 45

7. **Transport of Energy by Convection** ......................... 48
   7.1 The Basic Picture .......................................... 49
   7.2 Dimensionless Equations ................................... 51
   7.3 Limiting Cases, Solutions, Discussion ..................... 52

8. **The Chemical Composition** ................................... 56
   8.1 Relative Mass Abundances ................................. 56
   8.2 Variation of Composition with Time ....................... 57
       8.2.1 Radiative Regions ................................... 57
       8.2.2 Diffusion ........................................... 58
       8.2.3 Convective Regions ................................. 61

**Part II The Overall Problem**

9. **The Differential Equations of Stellar Evolution** ............... 64
   9.1 The Full Set of Equations ................................. 64
   9.2 Time-scales and Simplifications ............................ 66

10. **Boundary Conditions** ....................................... 68
    10.1 Central Conditions ....................................... 68
    10.2 Surface Conditions ....................................... 69
    10.3 Influence of the Surface Conditions and Properties
         of Envelope Solutions .................................. 72
         10.3.1 Radiative Envelopes ............................. 72
         10.3.2 Convective Envelopes ............................ 75
         10.3.3 Summary ........................................ 75
         10.3.4 The $T-r$ Stratification ......................... 76

11. **Numerical Procedure** ...................................... 77
    11.1 The Shooting Method ..................................... 77
    11.2 The Henyey Method ...................................... 78
    11.3 Treatment of the First- and Second-Order Time Derivatives .... 83

12. **Existence and Uniqueness of Solutions** ..................... 85
    12.1 Notation and Outline of the Procedure ................... 86
    12.2 Models in Complete Equilibrium ......................... 87
         12.2.1 Fitting Conditions in the $P_c - T_c$ Plane ............. 87

     12.2.2 Local Uniqueness ................................. 88
     12.2.3 Variation of Parameters .......................... 89
  12.3 Hydrostatic Models without Thermal Equilibrium ........... 91
     12.3.1 Degrees of Freedom and Fitting Conditions .......... 91
     12.3.2 Local Uniqueness ................................. 93
     12.3.3 Variation of Parameters .......................... 93
  12.4 Connection with Stability Problems ....................... 95
  12.5 Non-local Properties of Equilibrium Models ............... 97

## Part III Properties of Stellar Matter

**13. The Ideal Gas with Radiation** ............................... 102
  13.1 Mean Molecular Weight and Radiation Pressure ............. 102
  13.2 Thermodynamic Quantities ............................... 104

**14. Ionization** ................................................ 107
  14.1 The Boltzmann and Saha Formulae ....................... 107
  14.2 Ionization of Hydrogen ................................. 110
  14.3 Thermodynamical Quantities for a Pure Hydrogen Gas ....... 111
  14.4 Hydrogen–Helium Mixtures ............................. 112
  14.5 The General Case ...................................... 114
  14.6 Limitation of the Saha Formula ......................... 115

**15. The Degenerate Electron Gas** ............................... 118
  15.1 Consequences of the Pauli Principle ..................... 118
  15.2 The Completely Degenerate Electron Gas ................. 119
  15.3 Limiting Cases ........................................ 122
  15.4 Partial Degeneracy of the Electron Gas .................. 123

**16. The Equation of State of Stellar Matter** .................... 129
  16.1 The Ion Gas .......................................... 129
  16.2 The Equation of State ................................. 130
  16.3 Thermodynamic Quantities ............................. 132
  16.4 Crystallization ....................................... 134
  16.5 Neutronization ....................................... 135

**17. Opacity** .................................................. 137
  17.1 Electron Scattering .................................... 137
  17.2 Absorption Due to Free–Free Transitions ................. 138
  17.3 Bound–Free Transitions ................................ 139
  17.4 Bound–Bound Transitions .............................. 140
  17.5 The Negative Hydrogen Ion ............................ 141
  17.6 Conduction ........................................... 142
  17.7 Opacity Tables ........................................ 143

**18. Nuclear Energy Production** .................................... 146
18.1 Basic Considerations .......................................... 146
18.2 Nuclear Cross-sections ......................................... 150
18.3 Thermonuclear Reaction Rates ................................. 152
18.4 Electron Shielding ............................................. 157
18.5 The Major Nuclear Burnings ................................... 161
    18.5.1 Hydrogen Burning .................................... 162
    18.5.2 Helium Burning ...................................... 165
    18.5.3 Carbon Burning etc. .................................. 167
18.6 Neutrinos ..................................................... 169

**Part IV Simple Stellar Models**

**19. Polytropic Gaseous Spheres** .................................. 174
19.1 Polytropic Relations ........................................... 174
19.2 Polytropic Stellar Models ...................................... 175
19.3 Properties of the Solutions ..................................... 177
19.4 Application to Stars ............................................ 178
19.5 Radiation Pressure and the Polytrope $n = 3$ ................. 180
19.6 Polytropic Stellar Models with Fixed $K$ ..................... 180
19.7 Chandrasekhar's Limiting Mass ................................ 181
19.8 Isothermal Spheres of an Ideal Gas ........................... 183
19.9 Gravitational and Total Energy for Polytropes ................ 184
19.10 Supermassive Stars ........................................... 186
19.11 A Collapsing Polytrope ....................................... 187

**20. Homology Relations** ......................................... 191
20.1 Definitions and Basic Relations ............................... 191
20.2 Applications to Simple Material Functions .................... 194
    20.2.1 The Case $\delta = 0$ ................................. 194
    20.2.2 The Case $\alpha = \delta = \varphi = 1, a = b = 0$ .......... 194
    20.2.3 The Role of the Equation of State .................... 197
20.3 Homologous Contraction ...................................... 198

**21. Simple Models in the $U$–$V$ Plane** .......................... 200
21.1 The $U$–$V$ Plane ............................................. 200
21.2 Radiative Envelope Solutions .................................. 203
21.3 Fitting of a Convective Core .................................. 205
21.4 Fitting of an Isothermal Core ................................. 206

**22. The Main Sequence** .......................................... 207
22.1 Surface Values ................................................ 207
22.2 Interior Solutions ............................................. 210
22.3 Convective Regions ........................................... 212
22.4 Extreme Values of $M$ ........................................ 215

**23. Other Main Sequences** ....................................... 216
  23.1 The Helium Main Sequence ............................. 216
  23.2 The Carbon Main Sequence ............................. 218
  23.3 Main Sequences as Linear Series of Stellar Models .......... 219
  23.4 Generalized Main Sequences ............................ 221

**24. The Hayashi Line** ........................................... 224
  24.1 Luminosity of Fully Convective Models ................... 224
  24.2 A Simple Description of the Hayashi Line ................. 226
  24.3 The Neighbourhood of the Hayashi Line and the Forbidden Region 229
  24.4 Numerical Results .................................... 231
  24.5 Limitations for Fully Convective Models ................. 232

**25. Stability Considerations** ................................... 234
  25.1 General Remarks ..................................... 234
  25.2 Stability of the Piston Model .......................... 235
    25.2.1 Dynamical Stability ............................ 236
    25.2.2 Inclusion of Non-adiabatic Effects ................. 236
  25.3 Stellar Stability ...................................... 238
    25.3.1 Perturbation Equations .......................... 239
    25.3.2 Dynamical Stability ............................ 240
    25.3.3 Non-adiabatic Effects ........................... 241
    25.3.4 The Gravothermal Specific Heat ................... 242
    25.3.5 Secular Stability Behaviour of Nuclear Burning ........ 243

**Part V  Early Stellar Evolution**

**26. The Onset of Star Formation** ............................... 248
  26.1 The Jeans Criterion .................................. 248
    26.1.1 An Infinite Homogeneous Medium ................. 248
    26.1.2 A Plane Parallel Layer in Hydrostatic Equilibrium ..... 250
  26.2 Instability in the Spherical Case ........................ 252
  26.3 Fragmentation ....................................... 253

**27. The Formation of Protostars** ............................... 256
  27.1 Free-Fall Collapse of a Homogeneous Sphere .............. 256
  27.2 Collapse onto a Condensed Object ...................... 258
  27.3 A Collapse Calculation ............................... 259
  27.4 The Optically Thin Phase and the Formation of a Hydrostatic Core 260
  27.5 Core Collapse ....................................... 262
  27.6 Evolution in the Hertzsprung–Russell Diagram ............. 264

**28. Pre-Main-Sequence Contraction** ............................. 266
  28.1 Homologous Contraction of a Gaseous Sphere ............. 266
  28.2 Approach to the Zero-Age Main Sequence ................ 269

**29. From the Initial to the Present Sun** .......................... 271
  29.1  Choosing the Initial Model .............................. 271
  29.2  Solar Neutrinos ...................................... 275

**30. Chemical Evolution on the Main Sequence** ................... 277
  30.1  Change in the Hydrogen Content ......................... 277
  30.2  Evolution in the Hertzsprung–Russell Diagram .............. 278
  30.3  Time-scales for Central Hydrogen Burning ................. 280
  30.4  Complications Connected with Convection ................. 280
      30.4.1  Convective Overshooting .......................... 281
      30.4.2  Semiconvection ................................. 284
  30.5  The Schönberg–Chandrasekhar Limit ...................... 285
      30.5.1  A Simple Approach – The Virial Theorem and Homology 286
      30.5.2  Integrations for Core and Envelope .................. 288
      30.5.3  Complete Solutions for Stars with Isothermal Cores .... 289

**Part VI  Post-Main-Sequence Evolution**

**31. Evolution Through Helium Burning – Massive Stars** ........... 292
  31.1  Crossing the Hertzsprung Gap ........................... 292
  31.2  Central Helium Burning ............................... 296
  31.3  The Cepheid Phase ................................... 300
  31.4  To Loop or Not to Loop ... ............................ 301
  31.5  After Central Helium Burning ........................... 306

**32. Evolution Through Helium Burning – Low-Mass Stars** .......... 308
  32.1  Post-Main-Sequence Evolution .......................... 308
  32.2  Shell-Source Homology ............................... 309
  32.3  Evolution to the Helium Flash .......................... 313
  32.4  The Helium Flash .................................... 316
  32.5  Numerical Results for the Helium Flash ................... 317
  32.6  Evolution after the Helium Flash ........................ 320
  32.7  Evolution from the Zero-Age Horizontal Branch ............. 321
  32.8  Equilibrium Models with Helium Cores – Continued ......... 324

**33. Later Phases** ....................................... 328
  33.1  Nuclear Cycles ...................................... 328
  33.2  Shell Sources and Their Stability ........................ 330
  33.3  Thermal Pulses of a Shell Source ........................ 333
  33.4  Evolution of the Central Region ......................... 336
  33.5  The Core-Mass–Luminosity Relation for Large Core Masses .... 342

**34. Final Explosions and Collapse** ......................... 344
  34.1  The Evolution of the C–O Core ......................... 344
  34.2  Carbon Burning in Degenerate Cores ..................... 348

34.2.1 The Carbon Flash ............................. 348
34.2.2 Nuclear Statistical Equilibrium .................... 349
34.2.3 Hydrostatic and Convective Adjustment ............. 351
34.2.4 Combustion Fronts ............................. 352
34.2.5 Numerical Solutions ............................ 354
34.2.6 Carbon Burning in Accreting White Dwarfs .......... 356
34.3 Collapse of Cores of Massive Stars .................... 356
34.3.1 Simple Collapse Solutions ........................ 357
34.3.2 The Reflection of the Infall ...................... 359
34.3.3 Effects of Neutrinos ............................ 360
34.3.4 Numerical Results .............................. 362
34.3.5 Pair-Creation Instability ......................... 362

**Part VII Compact Objects**

**35. White Dwarfs** ............................................ 366
35.1 Chandrasekhar's Theory ............................. 366
35.2 The Corrected Mechanical Structure .................... 370
35.3 Thermal Properties and Evolution of White Dwarfs .......... 374

**36. Neutron Stars** .......................................... 380
36.1 Cold Matter Beyond Neutron Drip ...................... 380
36.2 Models of Neutron Stars ............................. 383

**37. Black Holes** ........................................... 390

**Part VIII Pulsating Stars**

**38. Adiabatic Spherical Pulsations** ........................... 398
38.1 The Eigenvalue Problem ............................. 398
38.2 The Homogeneous Sphere ............................ 402
38.3 Pulsating Polytropes ................................ 403

**39. Non-adiabatic Spherical Pulsations** ...................... 407
39.1 Vibrational Instability of the Piston Model ................ 407
39.2 The Quasi-adiabatic Approximation ..................... 408
39.3 The Energy Integral ................................ 409
39.3.1 The $\kappa$ Mechanism ............................. 411
39.3.2 The $\varepsilon$ Mechanism ............................. 412
39.4 Stars Driven by the $\kappa$ Mechanism – The Instability Strip ....... 412
39.5 Stars Driven by the $\varepsilon$ Mechanism ......................... 417

**40. Non-radial Stellar Oscillations** .......................... 418
40.1 Perturbations of the Equilibrium Model ................... 418

40.2  Normal Modes and Dimensionless Variables ................. 420
40.3  The Eigenspectra ........................................ 422
40.4  Stars Showing Non-radial Oscillations ..................... 425

## Part IX  Stellar Rotation

**41. The Mechanics of Rotating Stellar Models** ..................... 428
41.1  Uniformly Rotating Liquid Bodies ........................ 428
41.2  The Roche Model ....................................... 431
41.3  Slowly Rotating Polytropes ............................. 433

**42. The Thermodynamics of Rotating Stellar Models** .............. 435
42.1  Conservative Rotation ................................... 435
42.2  Von Zeipel's Theorem .................................. 436
42.3  Meridional Circulation .................................. 437
42.4  The Non-conservative Case .............................. 438
42.5  The Eddington–Sweet Time-scale ........................ 439
42.6  Meridional Circulation in Inhomogeneous Stars ............. 442

**43. The Angular-Velocity Distribution in Stars** ..................... 444
43.1  Viscosity ............................................... 444
43.2  Dynamical Stability ..................................... 446
43.3  Secular Stability ....................................... 451

**References** ...................................................... 455

**Subject Index** ................................................... 461

# I  The Basic Equations

# §1 Coordinates, Mass Distribution, and Gravitational Field in Spherical Stars

## 1.1 Eulerian Description

For gaseous, non-rotating, single stars without strong magnetic fields, the only forces acting on a mass element come from pressure and gravity. This results in a spherically symmetric configuration. All functions will then be constant on concentric spheres, and we need only one spatial variable to describe them. It seems natural to use the distance $r$ from the stellar centre as the spatial coordinate, which varies from $r = 0$ at the centre to the total radius $r = R$ at the surface of the star. In addition, the evolution in time $t$ requires a dependence of all functions on $t$. If we thus take $r$ and $t$ as independent variables, we have a typical "Eulerian" treatment in the sense of classical hydrodynamics. Then all other variables are considered to depend on these two, for example the density $\varrho = \varrho(r, t)$.

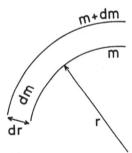

Fig. 1.1. The variation of $m$ with $r$ at a fixed moment $t = t_0$. The quantities $dm$ and $dr$ are connected by (1.2)

In order to provide a convenient description of the mass distribution inside the star, in particular of its effect on the gravitational field, we define the function[1] $m(r, t)$ as the mass contained in a sphere of radius $r$ at the time $t$ (Fig. 1.1). Then $m$ varies with respect to $r$ and $t$ according to

$$dm = 4\pi r^2 \varrho \, dr - 4\pi r^2 \varrho v \, dt \quad . \tag{1.1}$$

The first term on the right is obviously the mass contained in the spherical shell of thickness $dr$ (Fig. 1.1), and it gives the variation of $m(r, t)$ due to a variation of $r$ at constant $t$, i.e.

---

[1]    In most textbooks our function $m(r, t)$ is denoted by $M_r$.

$$\frac{\partial m}{\partial r} = 4\pi r^2 \varrho \quad . \tag{1.2}$$

Since it is preferable to describe the mass distribution in the star by $m(r, t)$ (instead of $\varrho$), (1.2) will be taken as the first of our basic equations in the Eulerian description.

The last term in (1.1) gives the (spherically symmetric) mass flow out of the sphere of (constant) radius $r$ due to a radial velocity $v$ in the outward direction in the time interval $dt$:

$$\frac{\partial m}{\partial t} = -4\pi r^2 \varrho v \quad . \tag{1.3}$$

The partial derivatives in the last two equations indicate as usual that the other independent variable ($t$ or $r$) is held constant.

Differentiating (1.2) with respect to $t$ and (1.3) with respect to $r$ and equating the two resulting expressions gives

$$\frac{\partial \varrho}{\partial t} = -\frac{1}{r^2} \frac{\partial(\varrho r^2 v)}{\partial r} \quad . \tag{1.4}$$

This is the well-known continuity equation of hydrodynamics, $\partial \varrho / \partial t = -\nabla \cdot (\varrho v)$, for the special case of spherical symmetry.

## 1.2 Lagrangian Description

It will turn out that, in the spherically symmetric case, it is often more useful to take a Lagrangian coordinate instead of $r$, i.e. one which is connected to the mass elements. The spatial coordinate of a given mass element then does not vary in time. We choose for this coordinate the above defined $m$: to any mass element the value $m$ (which is the mass contained in a concentric sphere *at a given moment* $t_0$) is assigned once and for all (see Fig. 1.1).

The new independent variables are then $m$ and $t$, and all other variables are considered to depend on them, for example $\varrho(m, t)$. This also includes the radial distance $r$ of our mass element from the centre, which is now described by the function $r = r(m, t)$. Since there is certainly no singularity of $\varrho$ at the centre, we have here $m = 0$, while the total mass $m = M$ is reached at the surface (i.e. where $r = R$). This already shows one advantage of the new description for the (normal) case of stars with constant total mass: while the radius $R$ varies strongly in time, the star always extends over the same interval of the independent variable $m$: $0 \le m \le M$.

As just indicated, there will certainly be no problem concerning a unique one-to-one transformation between the two coordinates $r$ and $m$. We then easily find the connection between the partial derivatives in the two cases from well-known formulae. For any function depending on two variables one of which is substituted by a new one $(r, t \rightarrow m, t)$, the partial derivatives with respect to the new variables are given by

$$\frac{\partial}{\partial m} = \frac{\partial}{\partial r} \cdot \frac{\partial r}{\partial m} \; ,$$

$$\left(\frac{\partial}{\partial t}\right)_m = \frac{\partial}{\partial r} \cdot \left(\frac{\partial r}{\partial t}\right)_m + \left(\frac{\partial}{\partial t}\right)_r \; . \tag{1.5}$$

Subscripts indicate which of the spatial variables ($m$ or $r$) is considered constant.

Let us apply the first of (1.5) to $m$. The left-hand side is then simply $\partial m / \partial m = 1$, and the first factor on the right-hand side is equal to $4\pi r^2 \varrho$, according to (1.2). So we can solve for the last factor and obtain

$$\frac{\partial r}{\partial m} = \frac{1}{4\pi r^2 \varrho} \; . \tag{1.6}$$

This is a differential equation describing the spatial behaviour of the function $r(m,t)$. It replaces (1.2) in the Lagrangian description and shall be the new first basic equation of our problem.

Introducing (1.6) into the first equation (1.5) gives the general recipe for the transformation between the two operators:

$$\frac{\partial}{\partial m} = \frac{1}{4\pi r^2 \varrho} \frac{\partial}{\partial r} \; . \tag{1.7}$$

The second equation (1.5) reveals the main reason for the choice of the Lagrangian description. Its left-hand side gives the so-called substantial time derivative of hydrodynamics. It describes the change of a function in time when following a given mass element, for example the change of a physical property of this mass element. The conservation laws for time-dependent spherical stars give very simple equations only in terms of this substantial time derivative. In terms of a *local* time derivative, $(\partial / \partial t)_r$, the description would become much more complicated, since the "convective" terms with the velocity $(\partial r / \partial t)_m$ [corresponding to the first term on the right-hand side of the second equation (1.5)] would appear explicitly.

## 1.3 The Gravitational Field

It follows from elementary potential theory that, inside a spherically symmetric body, the absolute value $g$ of the gravitational acceleration at a given distance $r$ from the centre does not depend on the mass elements outside of $r$. It depends only on $r$ and the mass within the concentric sphere of radius $r$, which we have called $m$:

$$g = \frac{Gm}{r^2} \; , \tag{1.8}$$

where $G = 6.673 \times 10^{-8}$ dyn cm$^2$ g$^{-2}$ is the gravitational constant. So the gravitating mass appears only in the form of our variable $m$.

Generally, the gravitational field inside the star can be described by a gravitational potential $\Phi$, which is a solution of the *Poisson equation*

$$\nabla^2 \Phi = 4\pi G \varrho \; , \tag{1.9}$$

where $\nabla^2$ denotes the Laplace operator. For spherical symmetry this reduces to

$$\frac{1}{r^2}\frac{\partial}{\partial r}\left(r^2\frac{\partial\Phi}{\partial r}\right) = 4\pi G\varrho \quad . \tag{1.10}$$

The vector of the gravitational acceleration points towards the stellar centre and may in spherical coordinates be written as $\boldsymbol{g} = (-g, 0, 0)$ with $0 < g = |\boldsymbol{g}|$. It is obtained from $\Phi$ by the vector relation $\boldsymbol{g} = -\nabla\Phi$, where in our spherically symmetric case only the radial component is non-vanishing:

$$g = \frac{\partial\Phi}{\partial r} \quad . \tag{1.11}$$

With (1.8), (1.11) becomes

$$\frac{\partial\Phi}{\partial r} = \frac{Gm}{r^2} \quad , \tag{1.12}$$

which is indeed a solution of (1.10), as is easily verified by substitution. The potential then becomes

$$\Phi = \int_0^r \frac{Gm}{r^2}dr + \text{constant} \quad . \tag{1.13}$$

Unless otherwise mentioned we will fix the free constant of integration in such a way that $\Phi$ vanishes for $r \to \infty$. $\Phi$ has a minimum at the stellar centre. Figure 1.2 shows schematically the function $\Phi(r, t)$ at a given time.

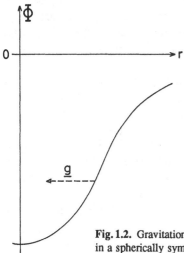

Fig. 1.2. Gravitational potential and vector of gravitational acceleration (*dashed*) in a spherically symmetric star

5

# §2 Conservation of Momentum

Conservation of momentum provides the next basic differential equation of the stellar-structure problem. We will derive this in several steps of gradually increasing generality. The first assumes mechanical equilibrium (§2.1), the equation of motion for spherical symmetry follows in §2.4, while in §2.5 even the assumption of spherical symmetry is dropped. In §2.6 we briefly discuss general relativistic effects in the case of hydrostatic equilibrium.

## 2.1 Hydrostatic Equilibrium

Most stars are obviously in such long-lasting phases of their evolution that no changes can be observed at all. Then the stellar matter cannot be accelerated noticibly, which means that all forces acting on a given mass element of the star compensate each other. This mechanical equilibrium in a star is called "hydrostatic equilibrium", since the same condition also governs the pressure stratification, say, in a basin of water. With our assumptions (gaseous stars without rotation, magnetic fields, or close companions) the only forces are due to gravity and to the pressure gradient.

For a given moment of time, we consider a thin spherical mass shell with (an infinitesimal) thickness $dr$ at a radius $r$ inside the star. Per unit area of the shell, the mass is $\varrho\, dr$, and the weight of the shell is $-g\varrho\, dr$. This weight is the gravitational force acting towards the centre (as indicated by the minus sign).

In order to prevent the mass elements of the shell from being accelerated in this direction, they must experience a net force due to pressure of the same absolute value, but acting outwards. This means that the shell must feel a larger pressure $P_i$ at its interior (lower) boundary than the pressure $P_e$ at its outer (upper) boundary (see Fig. 2.1). The total net force per unit area acting on the shell due to this pressure difference is

$$P_i - P_e = -\frac{\partial P}{\partial r}\, dr \quad . \tag{2.1}$$

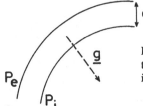

Fig. 2.1. Pressure at the upper and lower border of a mass shell of thickness $dr$, and the vector of gravitational acceleration (*dashed*) acting at one point on the shell

(The right-hand side of this equation is in fact a positive quantity, since $P$ decreases with increasing $r$.) The sum of the forces arising from pressure and gravity has to be zero,

$$\frac{\partial P}{\partial r} + g\varrho = 0 \quad , \tag{2.2}$$

which gives the condition of hydrostatic equilibrium as

$$\frac{\partial P}{\partial r} = -g\varrho \quad . \tag{2.3}$$

This shows the balance of the forces from pressure (left-hand side) and gravity (right-hand side), both per unit volume of the thin shell. Equation (1.8) gives $g = Gm/r^2$, so that (2.3) finally becomes

$$\frac{\partial P}{\partial r} = -\frac{Gm}{r^2}\varrho \quad . \tag{2.4}$$

This hydrostatic equation is the second basic equation describing the stellar-structure problem in the Eulerian form ($r$ as an independent variable).

If we take $m$ as the independent variable instead of $r$, we obtain the hydrostatic condition by multiplying (2.4) with $\partial r/\partial m = (4\pi r^2\varrho)^{-1}$, according to (1.5, 6):

$$\frac{\partial P}{\partial m} = -\frac{Gm}{4\pi r^4} \quad . \tag{2.5}$$

This is the second of our basic equations in the Lagrangian form.


## 2.2 The Role of Density and Simple Solutions

We have dealt up to now with the distribution of matter, the gravitational field and the pressure stratification in the star. This purely mechanical problem yielded two differential equations, for example, with $m$ as independent variable (a choice not affecting the discussion),

$$\frac{\partial r}{\partial m} = \frac{1}{4\pi r^2\varrho} \ , \qquad \frac{\partial P}{\partial m} = -\frac{Gm}{4\pi r^4} \quad . \tag{2.6}$$

Let us see whether solutions can be obtained at this stage for the problem as stated so far.

We have only 2 differential equations for 3 unknown functions, namely $r$, $P$, and $\varrho$. Obviously we can solve this mechanical problem only if we can express one of them in terms of the others, for example the density $\varrho$ as a function of $P$. In general, this will not be the case. But there are some exceptional situations where $\varrho$ is a well-known function of $P$ and $r$ or $P$ and $m$. We can then treat the equations as ordinary differential equations, since they do not contain the time explicitly.

If such integrations are to be carried out starting from the centre, the difficulty occurs that (2.6) are singular there, since $r \to 0$ for $m \to 0$, though one can easily overcome this problem by the standard procedure of expansion in powers of $m$, as given later in (10.3,6).

A rather artificial example that can be solved by quadrature is $\varrho = \varrho(m)$, in particular $\varrho$ = constant in the homogeneous gaseous sphere.

Physically more realistic are solutions obtained for the so-called *barotropic* case, for which the density is a function of the gas pressure only: $\varrho = \varrho(P)$. A simple example would be a perfect gas at constant temperature. After assuming a value $P_c$ for the central pressure, both equations (2.6) have to be solved simultaneously, since $\varrho(P)$ in the first of them is not known before $P$ is evaluated.

As we will see later (for instance in § 19.3,8) there are also cases for which no choice of $P_c$ yields a surface of zero pressure at finite values of $r$. In the theory of stellar structure there is even a use for these types of solution.

Among the barotropic solutions is a wide class of models for gaseous spheres called *polytropes*. These important solutions will later be discussed extensively (§ 19). Barotropic solutions also describe white dwarfs, i.e. stars that really exist (§ 35.1).

But in general the density is not a function of pressure only but depends also on the temperature $T$. For a given chemical composition of the gas, its thermodynamic behaviour yields an equation of state of the form $\varrho = \varrho(P, T)$. A well-known case is that of an ideal gas, where

$$\varrho = \frac{\mu}{\Re} \frac{P}{T} \tag{2.7}$$

with the gas constant $\Re = 8.315 \times 10^7$ erg $K^{-1}g^{-1}$ (which we define per g instead of per mole), while $\mu$ is the (dimensionless) mean molecular weight, i.e. the average number of atomic mass units per particle; in the case of ionized hydrogen $\mu = 0.5$ (see § 13.1).

Once the temperature appears in the equation of state and cannot be eliminated by means of additional conditions, it then becomes much more difficult to determine the internal structure of a self-gravitating gaseous sphere. The mechanical structure is then also determined by the temperature distribution, which in turn is coupled to the transport and generation of energy in the star. This requires new equations, with which we shall deal in § 4 and § 5.


## 2.3 Simple Estimates of Central Values $P_c$, $T_c$

The hydrostatic condition (2.5) together with an equation of state for an ideal gas (2.7) enable us to estimate the pressure and the temperature in the interior of a star of given mass and radius.

Let us replace the left-hand side of (2.5) by an average pressure gradient $(P_0 - P_c)/M$, where $P_0(= 0)$ and $P_c$ are the pressures at the surface and at the centre. On the right-hand side of (2.5) we replace $m$ and $r$ by rough mean values $M/2$ and $R/2$, and we obtain

$$P_c \approx \frac{2GM^2}{\pi R^4} \ . \tag{2.8}$$

From the equation of state for an ideal gas, and with the mean density

$$\bar{\varrho} = \frac{3M}{4\pi R^3} \quad , \tag{2.9}$$

we find with (2.8) that

$$T_c = \frac{P_c}{\varrho_c} \frac{\mu}{\Re} = P_c \frac{\mu}{\Re} \frac{\bar{\varrho}}{\varrho_c} \frac{4\pi R^3}{3M}$$

$$\approx \frac{8}{3} \frac{\mu}{\Re} \frac{GM}{R} \frac{\bar{\varrho}}{\varrho_c} \quad . \tag{2.10}$$

Since in most stars the density increases monotonically from the surface to the centre, we have $\bar{\varrho}/\varrho_c < 1$. (Numerical solutions show that $\bar{\varrho}/\varrho_c \approx 0.03 \ldots 0.01$.) Therefore (2.10) yields

$$T_c \lesssim \frac{8}{3} \frac{G\mu}{\Re} \frac{M}{R} \quad . \tag{2.11}$$

With the mass and the radius of the sun ($M_\odot = 1.989 \times 10^{33}$g, $R_\odot = 6.96 \times 10^{10}$ cm) and with $\mu = 0.5$ we find that

$$P_c \approx 7 \times 10^{15} \text{dyn/cm}^2, \quad T_c < 3 \times 10^7 \text{K} \quad . \tag{2.12}$$

Modern numerical solutions (§ 29) give $P_c = 2.7 \times 10^{17}$ dyn/cm$^2$, $T_c = 1.6 \times 10^7$K.

So we can expect to encounter enormous pressures and very high temperatures in the central regions of the stars. Moreover, our assumption of an ideal gas turns out to be fully justified for these values of $P$ and $T$.

## 2.4. The Equation of Motion for Spherical Symmetry

Our equation of hydrostatic equilibrium (2.5) is a special case of conservation of momentum. If the (spherical) star undergoes accelerated radial motions, we have to consider the inertia of the mass elements, which introduces an additional term. We confine ourselves here to the Lagrangian description ($m$, $t$ as independent variables), which is especially convenient for spherical symmetry.

We go back to the derivation of the hydrostatic equation in § 2.1 and again consider a thin shell of mass $dm$ at the distance $r$ from the centre (Fig. 1.1). Owing to the pressure gradient, this shell experiences a force per unit area $f_P$ given by (2.1), the right-hand side of which is easily rewritten in terms of $\partial P/\partial m$ according to (1.7):

$$f_P = -\frac{\partial P}{\partial m} dm \quad . \tag{2.13}$$

The gravitational force per unit area acting on the mass shell is, with the use of (1.8),

$$f_g = -\frac{g \, dm}{4\pi r^2} = -\frac{Gm}{r^2} \frac{dm}{4\pi r^2} \quad . \tag{2.14}$$

If the sum of the two forces is not equal to zero, the mass shell will be accelerated according to

$$\frac{dm}{4\pi r^2} \frac{\partial^2 r}{\partial t^2} = f_P + f_g \quad . \tag{2.15}$$

This gives with (2.13) and (2.14) the equation of motion as

$$\frac{1}{4\pi r^2} \frac{\partial^2 r}{\partial t^2} = -\frac{\partial P}{\partial m} - \frac{Gm}{4\pi r^4} \quad . \tag{2.16}$$

The signs in (2.16) are such that the pressure gradient alone would produce an outward acceleration (since $\partial P/\partial m < 0$), while the gravity alone would produce an inward acceleration.

Equation (2.16) would give exactly the equation of hydrostatic equilibrium (2.5) if the second time derivative of $r$ vanished, i.e. if all mass elements were at rest or moved radially at constant velocity. Moreover, the term on the left-hand side is certainly unimportant if its absolute value is small compared to the absolute values of any term on the right, – that is, if the two terms on the right-hand side cancel each other nearly to zero. Then the hydrostatic condition is a very good approximation and the configuration moves through neighbouring near-equilibrium states. In this sense we are allowed to apply the simpler hydrostatic equation to a much wider class of solutions than those fulfilling the strict requirement $\partial^2 r/\partial t^2 = 0$. To illustrate this further we assume a deviation from hydrostatic equilibrium such that, for example, in (2.16) the pressure term suddenly "disappears". The inertial term on the left would then have to compensate the gravitational term on the right. We now define a characteristic time-scale $\tau_{\mathrm{ff}}$ for the ensuing collapse of the star by setting $|\partial^2 r/\partial t^2| = R/\tau_{\mathrm{ff}}^2$. Then we obtain from (2.16) $R/\tau_{\mathrm{ff}}^2 \approx g$, or

$$\tau_{\mathrm{ff}} \approx \left(\frac{R}{g}\right)^{1/2} \quad . \tag{2.17}$$

This is some kind of a mean value for the free fall time over a distance of order $R$ following the sudden disappearance of the pressure. We can correspondingly determine a time-scale $\tau_{\mathrm{expl}}$ for the explosion of our star for the case that gravity were suddenly to disappear: $R/\tau_{\mathrm{expl}}^2 = P/\varrho R$, where we have replaced $\partial P/\partial r$ by $P/R$ after writing $4\pi r^2(\partial P/\partial m) = (\partial P/\partial r)/\varrho$ ($P$ and $\varrho$ are here average values over the entire star). We then find that

$$\tau_{\mathrm{expl}} \approx R\left(\frac{\varrho}{P}\right)^{1/2} \quad . \tag{2.18}$$

Since $(P/\varrho)^{1/2}$ is of the order of the mean velocity of sound in the stellar interior, one can see that $\tau_{\mathrm{expl}}$ is of the order of the time a sound wave needs to travel from the centre to the surface.

If our model is near hydrostatic equilibrium, then the two terms on the right side of (2.16) have about equal absolute value and $\tau_{ff} \approx \tau_{expl}$. We then call this time-scale the *hydrostatic time-scale* $\tau_{hydr}$, since it gives the typical time in which a (dynamically stable) star reacts on a slight perturbation of hydrostatic equilibrium. With $g \approx GM/R^2$, we obtain from (2.17) up to factors of order 1 that

$$\tau_{hydr} \approx \left(\frac{R^3}{GM}\right)^{1/2} \approx \frac{1}{2}(G\bar{\varrho})^{-1/2} \quad . \tag{2.19}$$

In the case of the sun we find the surprisingly small value $\tau_{hydr} \approx 27$ minutes. Even in the case of a red giant ($M \approx M_\odot, R \approx 100 R_\odot$) one has only $\tau_{hydr} \approx 18$ days, while for a white dwarf ($M \approx M_\odot, R \approx R_\odot/50$) the hydrostatic time-scale is extremely short: $\tau_{hydr} \approx 4.5$ seconds. In most phases of their life the stars change slowly on a time-scale that is very long compared to $\tau_{hydr}$. Then they are very close to hydrostatic equilibrium and the inertial terms in (2.16) can be ignored.

## 2.5 The Non-spherical Case

Up to now we have dealt with spherically symmetric configurations only. It is easy to see how the equations would have to be modified for more general cases without this symmetry.

After rewriting (2.16) for the independent variable $r$, we easily identify it as a special case of the Eulerian equation of motion of hydrodynamics

$$\varrho \frac{dv}{dt} = -\nabla P - \varrho \nabla \Phi \quad , \tag{2.20}$$

where $v$ is the velocity vector, and the substantial time derivative on the left is defined by the operator

$$\frac{d}{dt} = \frac{\partial}{\partial t} + v \cdot \nabla \quad . \tag{2.21}$$

The general form of (1.4) has already been shown to be the continuity equation of hydrodynamics

$$\frac{\partial \varrho}{\partial t} = -\nabla \cdot (\varrho v) \quad ; \tag{2.22}$$

and, as described in § 1.3, the gravitational potential $\Phi$ is connected with an arbitrary distribution of the density by the Poisson equation (1.9):

$$\nabla^2 \Phi = 4\pi G \varrho \quad . \tag{2.23}$$

We see in fact that the stellar-structure equations discussed up to now are just special cases of normal textbook hydrodynamics.

## 2.6 Hydrostatic Equilibrium in General Relativity

To help with subsequent work (§ 36), we briefly refer to the change of the equation of hydrostatic equilibrium due to effects of general relativity. For details see, for example, ZELDOVICH, NOVIKOV (1971).

Very strong gravitational fields, as in the case of neutron stars, are described by the Einstein field equations

$$R_{ik} - \frac{1}{2} g_{ik} R = \frac{\kappa}{c^2} T_{ik} \quad , \quad \kappa = \frac{8\pi G}{c^2} \quad , \tag{2.24}$$

where $R_{ik}$ is the Ricci tensor, $g_{ik}$ is the metric tensor and the scalar $R$ is the Riemann curvature. $T_{ik}$ is the energy–momentum tensor, which for an ideal gas has as the only non-vanishing components $T_{00} = \varrho c^2$, $T_{11} = T_{22} = T_{33} = P$ ($\varrho$ includes the energy density, $P$ = pressure). We are interested in static (time-independent), spherically symmetric mass distributions. Then the line element $ds$, i.e. the distance between two neighbouring events, is given in spherical coordinates $(r, \vartheta, \varphi)$ by the general form

$$ds^2 = e^\nu c^2 dt^2 - e^\lambda dr^2 - r^2(d\vartheta^2 + \sin^2\vartheta \, d\varphi^2) \tag{2.25}$$

with $\nu = \nu(r)$, $\lambda = \lambda(r)$. With these expressions for $T_{ik}$ and $ds$, the field equations (2.24) can be reduced to 3 ordinary differential equations:

$$\frac{\kappa P}{c^2} = e^{-\lambda}\left(\frac{\nu'}{r} + \frac{1}{r^2}\right) - \frac{1}{r^2} \quad , \tag{2.26}$$

$$\frac{\kappa P}{c^2} = \frac{1}{2} e^{-\lambda}\left(\nu'' + \frac{1}{2}\nu'^2 + \frac{\nu' - \lambda'}{r} - \frac{\nu'\lambda'}{2}\right) \quad , \tag{2.27}$$

$$\kappa\varrho = e^{-\lambda}\left(\frac{\lambda'}{r} - \frac{1}{r^2}\right) + \frac{1}{r^2} \quad , \tag{2.28}$$

where primes denote derivatives with respect to $r$. After multiplication with $4\pi r^2$, (2.28) can be integrated giving

$$\kappa m = 4\pi r \left(1 - e^{-\lambda}\right) \quad . \tag{2.29}$$

Here $m$ denotes the *"gravitational mass"* inside $r$ defined by

$$m = \int_0^r 4\pi r^2 \varrho \, dr \quad . \tag{2.30}$$

For $r = R$, $m$ becomes the gravitational mass $M$ of the star. It is the mass a distant observer would measure by its gravitational effects, for example on orbiting planets. It is not, however, the mass which we naïvely identify with the baryon number times the atomic mass unit: $M$ contains not only the rest mass, but the whole energy (divided by $c^2$). This includes the internal and the gravitational energy, the latter being negative and reducing the gravitational mass (just as the binding energy of a nucleus results in a mass defect, see § 18). The seemingly familiar form of (2.30) is treacherous. First of all, $\varrho = \varrho_0 + U/c^2$ contains the whole energy density $U$ as well as the rest-mass density $\varrho_0$; and the changed metric would give the spherical volume element as $e^{\lambda/2} 4\pi r^2\, dr$ instead of the usual form $4\pi r^2\, dr$ [over which (2.30) is integrated].

Differentiation of (2.26) with respect to $r$ gives $P' = P'(\lambda, \lambda', \nu', \nu'', r)$. When $\lambda$, $\lambda'$, $\nu'$, $\nu''$ are eliminated by (2.26, 27, 29) one arrives at the *Tolman–Oppenheimer–Volkoff (TOV) equation for hydrostatic equilibrium* in general relativity:

$$\frac{dP}{dr} = -\frac{Gm}{r^2}\varrho \left(1 + \frac{P}{\varrho c^2}\right)\left(1 + \frac{4\pi r^3 P}{mc^2}\right)\left(1 - \frac{2Gm}{rc^2}\right)^{-1} \ . \qquad (2.31)$$

Obviously this reverts to the usual form (2.4) for $c^2 \to \infty$.

For gravitational fields that are not too large (small deviations from Newtonian mechanics), one can expand the product of the parentheses in (2.31) and retain only terms linear in $1/c^2$. This gives the so-called *post-Newtonian approximation*:

$$\frac{dP}{dr} = -\frac{Gm}{r^2}\varrho \left(1 + \frac{P}{\varrho c^2} + \frac{4\pi r^3 P}{mc^2} + \frac{2Gm}{rc^2}\right) \ . \qquad (2.32)$$

## 2.7 The Piston Model

From time to time we shall make use of a simple mechanical model which in some respects mimics the behaviour of stars, and which is shown in Fig. 2.2. A piston of mass $M^*$ encloses a gas of mass $m^*$ in a box. $G^* = gM^*$ is the weight of the piston in a gravitational field described by the gravitational acceleration $g$. $A$ is the cross-sectional area of the piston, and $h$ its height above the bottom. Then $V = Ah$ is the volume of the gas, while its density is $\varrho = m^*/V$.

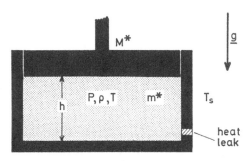

Fig. 2.2. The piston model. Gas of mass $m^*$ (with pressure $P$, density $\varrho$, temperature $T$) is held in a container with a movable piston of mass $M^*$. The gravitational acceleration $g$ acts on the piston. The container is embedded in a medium of temperature $T_s$; a possible heat leak is indicated (*dashed*) in the right wall of the container. In §2, only the mechanical properties of the model are discussed

In the case of hydrostatic equilibrium, the gas pressure $P$ adjusts in such a way that the weight per unit area is balanced by the pressure:

$$G^* = PA \ . \qquad (2.33)$$

If the forces do not compensate each other, the piston is accelerated in the vertical direction according to the equation of motion

$$M^* \frac{d^2 h}{dt^2} = -G^* + PA \ . \qquad (2.34)$$

13

In a similar manner to our considerations of §2.4, we can define two time-scales $\tau_{ff}$ and $\tau_{expl}$:

$$\tau_{ff} \approx \left(\frac{h}{g}\right)^{1/2} \quad , \tag{2.35}$$

$$\tau_{expl} \approx h \left(\frac{\varrho}{P}\right)^{1/2} \left(\frac{M^*}{m^*}\right)^{1/2} \quad . \tag{2.36}$$

In the limit of hydrostatic equilibrium both time-scales are the same, and we then call $\tau_{ff} = \tau_{expl}$ the hydrostatic time-scale $\tau_{hydr}$.

# §3 The Virial Theorem

## 3.1 Stars in Hydrostatic Equilibrium

While the virial theorem plays a relatively minor role generally in physics, it is of vital importance for the understanding of stars. It connects two important energy reservoirs of a star and allows predictions and interpretations of certain evolutionary phases.

If we multiply (2.5) by $4\pi r^3$ and integrate over $dm$ in the interval $[0, M]$, that is from centre to surface, we obtain on the left-hand side an integral which can be simplified by partial integration:

$$\int_0^M 4\pi r^3 \frac{\partial P}{\partial m} \, dm = \left[4\pi r^3 P\right]_0^M - \int_0^M 12\pi r^2 \frac{\partial r}{\partial m} P \, dm \quad , \tag{3.1}$$

where the term in brackets vanishes, since $r = 0$ at the centre and $P = 0$ at the surface. With (1.6) the integrand of the last term in (3.1) is reduced to $3P/\varrho$. Therefore, after multiplication by $4\pi r^3$ and integration, (2.5) gives

$$\int_0^M \frac{Gm}{r} \, dm = 3 \int_0^M \frac{P}{\varrho} \, dm \quad . \tag{3.2}$$

Both sides of (3.2) have the dimensions of energy and can be easily interpreted. We define the *gravitational energy* $E_g$ by

$$E_g := - \int_0^M \frac{Gm}{r} \, dm \quad . \tag{3.3}$$

Consider a unit mass at the position $r$. Its potential energy due to the gravitational field of the mass $m$ inside $r$ is $-Gm/r$. Therefore $E_g$ is the potential energy of all mass elements $dm$ of the star (normalized to zero at infinity). The energy $-E_g(> 0)$ is necessary to expand all mass shells into infinity, and it is released when the stellar configuration forms out of an infinitely distributed medium.

We see that $E_g$ varies if the configuration undergoes expansion or contraction: if all mass shells inside the configuration expand or contract simultaneously, then $E_g$ increases or decreases, respectively. And the same must be true for the integral on the right of (3.2). Note that these radial motions must be slow compared to $\tau_{\text{hydr}}$ in order that hydrostatic equilibrium is always maintained, otherwise (3.2) would not hold.

In order to understand the meaning of the term on the right of (3.2) we first assume an ideal gas. Then

$$\frac{P}{\varrho} = \frac{\Re}{\mu} T = (c_P - c_v)T = (\gamma - 1)c_v T \tag{3.4}$$

where $c_P$, $c_v$ are the specific heats per unit mass (and we make use of $\Re/\mu = c_P - c_v$ and replace $c_P/c_v$ by $\gamma$). For a monatomic gas $\gamma = 5/3$, and we have

$$\frac{P}{\varrho} = \frac{2}{3} u \quad , \tag{3.5}$$

where $u = c_v T$ is the internal energy per unit mass of the ideal gas. Therefore (3.2) can be written as

$$E_g = -2E_i \tag{3.6}$$

with the total internal energy of the star

$$E_i := \int_0^M u \, dm \quad . \tag{3.7}$$

Equation (3.6) is the *virial theorem* for an ideal monatomic gas. For a general equation of state we define a quantity $\zeta$ by

$$\zeta u := 3 \frac{P}{\varrho} \quad . \tag{3.8}$$

For an ideal gas $\zeta = 3(\gamma - 1)$, in the monatomic case $\gamma = 5/3$, and therefore $\zeta = 2$. For a pure photon gas, $P = aT^4/3$, and $u\varrho = aT^4$ ($a$ = radiation density constant), giving $\zeta = 1$. If $\zeta$ is constant throughout the star, (3.2) leads to the more general virial theorem:

$$\zeta E_i + E_g = 0 \quad . \tag{3.9}$$

We now define the *total energy* $W$ of our configuration,

$$W = E_i + E_g \quad , \tag{3.10}$$

where for a gravitationally bound system $W < 0$, and with (3.9) we find that

$$W = (1 - \zeta)E_i = \frac{\zeta - 1}{\zeta} E_g \quad . \tag{3.11}$$

In the case of $\zeta = 1$ ($\gamma = 4/3$) the total energy vanishes.

But in general $W$, $E_g$ and $E_i$ are coupled. A change of the total energy of the configuration is then connected with a change of its internal energy and with expansion or shrinking. A gas of finite temperature must radiate and $W$ must decrease. Let $L$ be the *luminosity* of the star, i.e. the total energy loss per unit time by radiation; then conservation of energy demands that $(dW/dt) + L = 0$, so that with (3.11) we obtain

$$L = (\zeta - 1)\frac{dE_i}{dt} = -\frac{\zeta - 1}{\zeta}\frac{dE_g}{dt} \quad . \tag{3.12}$$

16

We have seen that $\dot{E}_g < 0$ for contraction of all mass shells (where the dot denotes a derivative with respect to time $t$). For an ideal gas (3.12) gives $L = -\dot{E}_g/2 = \dot{E}_i$, which means that half of the energy liberated by the contraction is radiated away and the other half is used to heat the star ($L > 0$, $\dot{E}_i > 0$). The surprising fact that a star heats up while losing energy can be described by saying that the star has a negative specific heat (cf. the gravothermal specific heat defined in §25.3.4).

We have to keep in mind that it is the luminosity that causes the shrinking: a configuration in hydrostatic equilibrium has a finite temperature and therefore radiates into the (cold) universe.

## 3.2 The Virial Theorem of the Piston Model

Let us consider the situation for the piston model of §2.7 for the case of an ideal gas. Assuming $M^* \gg m^*$, we define $E_g := +G^*h$, where the free additional constant is chosen such that $E_g = 0$ for $h = 0$. Hydrostatic equilibrium (2.33) with $m^* = Ah\varrho$ and (3.4) demands that

$$hG^* = \frac{P}{\varrho} m^* = (\gamma - 1)c_v T m^* \quad . \tag{3.13}$$

The internal energy $E_i$ of the gas is $E_i = c_v T m^*$, and we find that

$$E_g = (\gamma - 1)E_i \quad , \tag{3.14}$$

which is the virial theorem for the piston model. Differentiating with respect to time, with $\gamma = 5/3$, results in

$$\frac{dE_g}{dt} = \frac{2}{3}\frac{dE_i}{dt} \quad . \tag{3.15}$$

Hence we see that in contrast to the situation in stars a reduction of $E_g$ is connected with *cooling* of the gas. Indeed the piston can only sink if the gas cools.

This different behaviour comes from the fact that the gravitational field is assumed to be constant here. In order to demonstrate this we now assume the weight $G^*$ to be a function of $h$ and differentiate (3.13) with respect to $h$:

$$G^*(1 + G_h^*) = (\gamma - 1)\frac{dE_i}{dh} \tag{3.16}$$

with $G_h^* := (d\ln G^*/d\ln h)$. Indeed, if $G_h^* = 0$ (constant gravity), we see that $E_i$ increases with $h$. If, however, $G^*$ decreases sufficiently with increasing $h$ (such that $G_h^* < -1$), then $E_i$ increases with decreasing $h$, corresponding to the behaviour of stars. In fact in an expanding star each mass shell also loses weight with increasing $r$.

## 3.3 The Kelvin–Helmholtz Time-scale

Returning now to consider stars, since according to (3.12) $L$ is of the order of $|dE_g/dt|$, we can define a characteristic time-scale

$$\tau_{KH} := \frac{|E_g|}{L} \approx \frac{E_i}{L} \qquad (3.17)$$

called the *Kelvin–Helmholtz time-scale* (after the two physicists who estimated this as the evolutionary time-scale for a contracting or cooling star).

A rough estimate for $|E_g|$ is

$$|E_g| \approx \frac{G\bar{m}^2}{\bar{r}} \approx \frac{GM^2}{2R} \quad , \qquad (3.18)$$

where quantities with a bar indicate mean values for $m$ and $r$ (which we have replaced by $M/2$ and $R/2$). Then we have

$$\tau_{KH} \approx \frac{GM^2}{2RL} \quad . \qquad (3.19)$$

For the sun, with $L = 3.827 \times 10^{33}$ erg/s, we find $\tau_{KH} \approx 1.6 \times 10^7$ years. In the early days of astrophysics the source of stellar energy was still uncertain, and it was suggested, among other proposals, that the sun "lived" from its gravitational energy $E_g$. Our estimate shows that this can work only for some $10^7$ years, after which time it would have contracted to a very condensed body. As it became obvious that the sun has been radiating in roughly the same way for some $10^9$ years, the contraction hypothesis had to be abandoned. But there are phases in a stellar life when $E_g$ is the main or even the only stellar energy source (§ 28); then the star evolves on the time-scale $\tau_{KH}$. A more detailed discussion of the evolution of a star in time appears in § 4.3.

## 3.4 The Virial Theorem for Non-vanishing Surface Pressure

One often needs the virial theorem for gaseous spheres imbedded in a medium of finite pressure. In this case, at the surface ($m = M$) $P = P_0 > 0$ instead of $P = 0$. Consequently the first term on the right of (3.1) does not vanish at the surface and (3.2) is modified to

$$\int_0^M \frac{Gm}{r}\, dm = 3 \int_0^M \frac{P}{\varrho}\, dm - 4\pi R^3 P_0 \quad . \qquad (3.20)$$

Correspondingly we find, rather than (3.9), that

$$\zeta E_i + E_g = 4\pi R^3 P_0 \quad . \qquad (3.21)$$

# §4 Conservation of Energy

Since we do not wish to interrupt the derivation of the energy equation for stars with lengthy formalisms, we first provide a few thermodynamic relations which will be used extensively later on.

## 4.1 Thermodynamic Relations

The first law of thermodynamics relates the heat $dq$ added per unit mass,

$$dq = du + P dv \quad , \tag{4.1}$$

to the internal energy $u$ and the specific volume $v = 1/\varrho$ (both also defined per unit mass).

We now assume rather general equations of state, $\varrho = \varrho(P, T)$ and $u = u(\varrho, T)$. Usually they will also depend on the chemical composition, but here this is assumed to be fixed. With the derivatives defined as

$$\alpha := \left( \frac{\partial \ln \varrho}{\partial \ln P} \right)_T = -\frac{P}{v} \left( \frac{\partial v}{\partial P} \right)_T \quad , \tag{4.2}$$

$$\delta := -\left( \frac{\partial \ln \varrho}{\partial \ln T} \right)_P = \frac{T}{v} \left( \frac{\partial v}{\partial T} \right)_P \quad , \tag{4.3}$$

the equation of state can be written in the form $d\varrho/\varrho = \alpha dP/P - \delta dT/T$.

We also need the specific heats:

$$c_P := \left( \frac{dq}{dT} \right)_P = \left( \frac{\partial u}{\partial T} \right)_P + P \left( \frac{\partial v}{\partial T} \right)_P \quad , \tag{4.4}$$

$$c_v := \left( \frac{dq}{dT} \right)_v = \left( \frac{\partial u}{\partial T} \right)_v \quad . \tag{4.5}$$

With

$$du = \left( \frac{\partial u}{\partial v} \right)_T dv + \left( \frac{\partial u}{\partial T} \right)_v dT \tag{4.6}$$

and with (4.1) we find the change $ds = dq/T$ of the specific entropy to be

$$ds = \frac{dq}{T} = \frac{1}{T} \left[ \left( \frac{\partial u}{\partial v} \right)_T + P \right] dv + \frac{1}{T} \left( \frac{\partial u}{\partial T} \right)_v dT \quad . \tag{4.7}$$

Since $ds$ is a total differential form, $\partial^2 s/\partial T \partial v = \partial^2 s/\partial v \partial T$ and

$$\frac{\partial}{\partial T}\left[\frac{1}{T}\left(\frac{\partial u}{\partial v}\right)_T + \frac{P}{T}\right] = \frac{1}{T}\frac{\partial^2 u}{\partial T \partial v} \quad , \tag{4.8}$$

which after the differentiation on the left is carried out gives

$$\left(\frac{\partial u}{\partial v}\right)_T = T\left(\frac{\partial P}{\partial T}\right)_v - P \quad . \tag{4.9}$$

Next we derive an expression for $(\partial u/\partial T)_P$, taking $P, T$ as independent variables. From (4.6) it follows that

$$\frac{du}{dT} = \left(\frac{\partial u}{\partial T}\right)_v + \left(\frac{\partial u}{\partial v}\right)_T \frac{dv}{dT} \quad , \tag{4.10}$$

and therefore

$$\begin{aligned}
\left(\frac{\partial u}{\partial T}\right)_P &= \left(\frac{\partial u}{\partial T}\right)_v + \left(\frac{\partial u}{\partial v}\right)_T \left(\frac{\partial v}{\partial T}\right)_P \\
&= \left(\frac{\partial u}{\partial T}\right)_v + \left(\frac{\partial v}{\partial T}\right)_P \left[T\left(\frac{\partial P}{\partial T}\right)_v - P\right] \quad ,
\end{aligned} \tag{4.11}$$

where we have made use of (4.9). From the definitions (4.4,5) and from (4.11) we write

$$\begin{aligned}
c_P - c_v &= P\left(\frac{\partial v}{\partial T}\right)_P + \left(\frac{\partial u}{\partial T}\right)_P - \left(\frac{\partial u}{\partial T}\right)_v \\
&= \left(\frac{\partial v}{\partial T}\right)_P \left(\frac{\partial P}{\partial T}\right)_v T \quad .
\end{aligned} \tag{4.12}$$

On the other hand, the definitions (4.2,3) for $\alpha$ and $\delta$ imply that

$$\left(\frac{\partial P}{\partial T}\right)_v = -\frac{\left(\frac{\partial v}{\partial T}\right)_P}{\left(\frac{\partial v}{\partial P}\right)_T} = \frac{P\delta}{T\alpha} \tag{4.13}$$

and therefore

$$c_P - c_v = T\left(\frac{\partial v}{\partial T}\right)_P \frac{P\delta}{T\alpha} = \frac{P\delta^2}{\varrho T\alpha} \quad , \tag{4.14}$$

where we have made use of $T(\partial v/\partial T)_P = v\delta = \delta/\varrho$; hence we arrive at the basic relation

$$c_P - c_v = \frac{P\delta^2}{\varrho T\alpha} \quad . \tag{4.15}$$

For an ideal gas this equation reduces to the well-known relation $c_P - c_v = \Re/\mu$.

We have now derived all the tools for rewriting (4.1) in terms of $T$ and $P$. The first step is to write it in the form

$$dq = du + Pdv = \left(\frac{\partial u}{\partial T}\right)_v dT + \left[\left(\frac{\partial u}{\partial v}\right)_T + P\right] dv$$

$$= \left(\frac{\partial u}{\partial T}\right)_v dT + T\left(\frac{\partial P}{\partial T}\right)_v dv \tag{4.16}$$

by making use of (4.9), and then with (4.5) and (4.13) we have

$$dq = c_v dT - \frac{T}{\varrho}\left(\frac{\partial P}{\partial T}\right)_v \frac{d\varrho}{\varrho} = c_v dT - \frac{P\delta}{\varrho\alpha}\frac{d\varrho}{\varrho}$$

$$= c_v dT - \frac{P\delta}{\varrho\alpha}\left(\alpha\frac{dP}{P} - \delta\frac{dT}{T}\right) = \left(c_v + \frac{P\delta^2}{\varrho T\alpha}\right) dT - \frac{\delta}{\varrho} dP \quad . \tag{4.17}$$

The terms in parentheses in the last expression are, according to (4.15), simply $c_P$ and therefore

$$dq = c_P dT - \frac{\delta}{\varrho} dP \quad . \tag{4.18}$$

Next we define the adiabatic temperature gradient $\nabla_{\text{ad}}$, a quantity often used in astrophysics, by

$$\nabla_{\text{ad}} := \left(\frac{\partial \ln T}{\partial \ln P}\right)_s \quad , \tag{4.19}$$

where the subscript $s$ indicates that the definition is valid for constant entropy. Since for adiabatic changes the entropy has to remain constant, that is $ds = dq/T = 0$, we can easily derive an expression for $\nabla_{\text{ad}}$ from (4.18), i.e.

$$0 = dq = c_P dT - \frac{\delta}{\varrho} dP \tag{4.20}$$

or $(dT/dP)_s = \delta/\varrho c_P$ and

$$\nabla_{\text{ad}} \equiv \left(\frac{P}{T}\frac{dT}{dP}\right)_s = \frac{P\delta}{T\varrho c_P} \quad . \tag{4.21}$$

## 4.2 Energy Conservation in Stars

By $l(r)$ we define[1] the net energy per second passing outward through a sphere of radius $r$. The function $l$ is zero at $r = 0$, since there can be no infinite energy source at the centre, while $l$ reaches the total luminosity $L$ of the star at the surface. In between, $l$ can be a complicated function, depending on the distribution of the sources and sinks of energy.

The function $l$ comprises the energies transported by radiation, conduction, and convection, transport mechanisms with which we shall deal in §5 and §7. Not

---

[1]  In many textbooks our function $l$ is denoted by $L_r$.

included is a possible energy flux by neutrinos, which normally have negligible interaction with the stellar matter (see below). Included in $l$ are only those fluxes which require a temperature gradient.

Consider a spherical mass shell of radius $r$, thickness $dr$, and mass $dm$ as indicated in Fig. 4.1. The energy per second entering the shell at the inner surface is $l$, while $l + dl$ is the energy per second leaving it through the outer surface. The surplus power $dl$ can be provided by nuclear reactions, by cooling, or by compression or expansion of the mass shell.

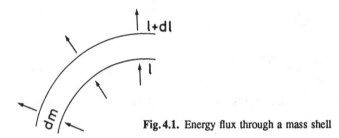

Fig. 4.1. Energy flux through a mass shell

We first consider a *stationary* case in which $dl$ is due to the release of energy from nuclear reactions only. Let $\varepsilon$ be the nuclear energy released per unit mass per second; then

$$dl = 4\pi r^2 \varrho \varepsilon \, dr = \varepsilon \, dm \quad , \qquad \text{or} \tag{4.22}$$

$$\frac{\partial l}{\partial m} = \varepsilon \quad . \tag{4.23}$$

In general, $\varepsilon$ depends on temperature and density and on the abundance of the different nuclear species that react, described in detail in § 18.

If we relax the condition of time independence, then $dl$ can become non-zero even if there are no nuclear reactions. A *non-stationary* shell can change its internal energy, and it can exchange mechanical work ($P \, dV$) with the neighbouring shells. Instead of (4.23) we write

$$dq = \left( \varepsilon - \frac{\partial l}{\partial m} \right) dt \quad , \tag{4.24}$$

where $dq$ is the heat per unit mass added to the shell in the time interval $dt$. Replacing $dq$ by the first law of thermodynamics (4.1) we obtain

$$\frac{\partial l}{\partial m} = \varepsilon - \frac{\partial u}{\partial t} - P \frac{\partial v}{\partial t}$$

$$= \varepsilon - \frac{\partial u}{\partial t} + \frac{P}{\varrho^2} \frac{\partial \varrho}{\partial t} \quad . \tag{4.25}$$

This can be rewritten in terms of $P$ and $T$, with the help of (4.18), as

$$\frac{\partial l}{\partial m} = \varepsilon - c_P \frac{\partial T}{\partial t} + \frac{\delta}{\varrho} \frac{\partial P}{\partial t} \quad , \tag{4.26}$$

where $\delta$ is defined in (4.3). This is the third of the basic equations of stellar structure. One often combines the terms containing the time derivatives in a source function $\varepsilon_g$:

$$
\begin{aligned}
\varepsilon_g : &= -T\,\frac{\partial s}{\partial t} \\
&= -c_P\,\frac{\partial T}{\partial t} + \frac{\delta}{\varrho}\,\frac{\partial P}{\partial t} \\
&= -c_P\,T\left(\frac{1}{T}\,\frac{\partial T}{\partial t} - \frac{\nabla_{ad}}{P}\,\frac{\partial P}{\partial t}\right) ,
\end{aligned}
\tag{4.27}
$$

where use is made of the fact that $ds = dq/T$ and of (4.21).

Let us now turn to the problem of *neutrino losses*. These can be formed in appreciable amounts in a star either as a by-product of nuclear energy generation or by other reactions. Stellar material is normally transparent to neutrinos and therefore they can easily "tunnel" the energy they have to the surface. This is the reason we have excluded the energy flux due to neutrinos from $l$. The only mass elements affected by the neutrinos are at the place of their creation, where they act as an energy sink; hence $\varepsilon_\nu$ is used to represent the energy taken per unit mass per second from the stellar material in the form of neutrinos. By definition, $\varepsilon_\nu > 0$. Obviously the complete energy equation is then

$$
\frac{\partial l}{\partial m} = \varepsilon - \varepsilon_\nu + \varepsilon_g \quad .
\tag{4.28}
$$

As mentioned at the beginning of §4.2, the boundary values of $l$ are $l = 0$ at the centre and $l = L$ at the surface. In between, $l$ is not necessarily monotonic, it can even become larger than $L$, or negative, since the right-hand side of (4.28) may be positive or negative. For instance, the surface luminosity $L$ of an expanding star can be smaller than the energy produced in the central core by nuclear reactions ($\varepsilon > 0$), since part of it is used to expand the star ($\varepsilon_g < 0$); and strong neutrino losses can make $l < 0$ in certain parts of the stellar interior (see §32.5).

The energy per second carried away from the star by neutrinos is often called the *neutrino luminosity*:

$$
L_\nu : = \int_0^M \varepsilon_\nu\,dm
\tag{4.29}
$$

## 4.3 Global and Local Energy Conservation

In §3 we considered gravitational energy ($E_g$) and internal energy ($E_i$), but ignored nuclear and neutrino energies, as well as the kinetic energy $E_{kin}$ of radial motion. We now define the total energy of the star as $W = E_{kin} + E_g + E_i + E_n$, where $E_n$ is the nuclear energy content of the whole star. Obviously the energy equation is

$$
\frac{d}{dt}\left(E_{kin} + E_g + E_i + E_n\right) + L + L_\nu = 0 \quad ,
\tag{4.30}
$$

and, of course, this must also be obtained from the *local* energy equation (4.28) by integration over $m$. Clearly, the integration of $\partial l/\partial m$ gives $L$, the integration of $-\varepsilon_\nu$ gives $-L_\nu$, while the integral over $\varepsilon$ gives $-dE_n/dt$. Integration over $\varepsilon_g$, however, needs some consideration.

Let us write $\varepsilon_g$ as in (4.25):

$$\varepsilon_g = -\frac{\partial u}{\partial t} + \frac{P}{\varrho^2}\frac{\partial \varrho}{\partial t} \quad . \tag{4.31}$$

Then integration over $-\partial u/\partial t$ gives $-dE_i/dt$. In order to deal with the last term in (4.31) we use (3.2,3) and find that

$$E_g = -3\int_0^M \frac{P}{\varrho}\,dm \quad , \tag{4.32}$$

which we differentiate with respect to time (indicated by dots):

$$\dot{E}_g = -3\int_0^M \frac{\dot{P}}{\varrho}\,dm + 3\int_0^M \frac{P}{\varrho^2}\dot{\varrho}\,dm \quad . \tag{4.33}$$

We first treat hydrostatic equilibrium ($dE_{\text{kin}}/dt = 0$). Then differentiation of (2.5) gives

$$\frac{\partial \dot{P}}{\partial m} = 4\frac{Gm}{4\pi r^4}\frac{\dot{r}}{r} \quad . \tag{4.34}$$

We multiply this by $4\pi r^3$ and integrate over $m$:

$$\int_0^M 4\pi r^3\,\frac{\partial \dot{P}}{\partial m}\,dm = 4\int_0^M \frac{Gm}{r}\frac{\dot{r}}{r}\,dm = 4\,\dot{E}_g \quad . \tag{4.35}$$

Partial integration of the left-hand side gives

$$\left[4\pi r^3 \dot{P}\right]_0^M - 3\int_0^M 4\pi r^2\,\frac{\partial r}{\partial m}\,\dot{P}\,dm \quad , \tag{4.36}$$

where the term in brackets vanishes at both ends of the interval, since either $r = 0$ or $P = 0$ independent of time. If we replace $\partial r/\partial m$ by $1/4\pi r^2\varrho$ we find from (4.35) that

$$-3\int_0^M \frac{\dot{P}}{\varrho}\,dm = 4\,\dot{E}_g \quad . \tag{4.37}$$

Introducing this into the right-hand side of (4.33) gives

$$\dot{E}_g = -\int_0^M \frac{P}{\varrho^2}\dot{\varrho}\,dm \quad , \tag{4.38}$$

and therefore the integration of the last term of (4.31) gives $\dot{E}_g$ so that the equation (4.30) without $\dot{E}_{\text{kin}}$ is now recovered.

If, instead of hydrostatic equilibrium, we had used the full equation of motion (2.16), after multiplication with $4\pi r^2 \dot{r}$ and integration over $m$, we would have obtained the full equation (4.30) *with* the term $\dot{E}_{kin}$.

## 4.4 Time-scales

Consider a star balancing its energy loss $L$ essentially by release of nuclear energy. If $L$ remains constant this can go on for a *nuclear time-scale* $\tau_n$ defined by

$$\tau_n := \frac{E_n}{L} \quad . \tag{4.39}$$

Note that $E_n$ means the energy reservoir from which energy can be released under the given circumstances, i.e. the corresponding reactions must be possible. The most important reaction is the fusion of $^1$H into $^4$He. This "hydrogen burning" releases $Q = 6.3 \times 10^{18}$ erg g$^{-1}$, and, if the sun consisted completely of hydrogen, $E_n$ would be $QM_\odot = 1.25 \times 10^{52}$ erg. With $L_\odot = 4 \times 10^{33}$ erg/s, (4.39) gives $\tau_n = 3 \times 10^{18}$ s, or $10^{11}$ years. A comparison with the earlier estimates of $\tau_{hydr}$ (§ 2.4) and $\tau_{KH}$ (§ 3.3) shows that

$$\tau_n \gg \tau_{KH} \gg \tau_{hydr} \quad , \tag{4.40}$$

which is not only true for the sun, but for all stars that survive by hydrogen and helium burning. We emphasize this point, since under these circumstances the equation of energy conservation (4.26) can be simplified. As an illustration, we assume that the star changes its properties considerably within the time-scale $\tau$ (which may be either small or large compared to $\tau_{KH}$). This change may, for instance, be due to exhaustion of nuclear fuel or artificial "squeezing" of the star from the exterior. We now give rough estimates for the four terms in (4.26), assuming an ideal gas:

$$\left|\frac{\partial l}{\partial m}\right| \approx \frac{L}{M} \approx \frac{E_i}{\tau_{KH}M} \quad , \tag{4.41}$$

$$\varepsilon \approx \frac{L}{M} = \frac{E_n}{M\tau_n} \approx \frac{E_i}{\tau_{KH}M} \quad , \tag{4.42}$$

$$\left|c_P \frac{\partial T}{\partial t}\right| \approx \frac{c_P T}{\tau} \approx \frac{E_i}{\tau M} \quad , \tag{4.43}$$

$$\left|\frac{\delta}{\varrho}\frac{\partial P}{\partial t}\right| \approx \frac{\Re}{\mu}\frac{T}{\tau} \approx \frac{c_P T}{\tau} \approx \frac{E_i}{\tau M} \quad . \tag{4.44}$$

In the case $\tau \gg \tau_{KH}$, the terms in (4.43,44) are small compared to those in (4.41,42); therefore the time derivatives in the energy equation (4.26) can be neglected ($|\varepsilon_g| \ll \varepsilon$) and the energy equation is $\partial l/\partial m = \varepsilon$, as in (4.23). This occurs if, for instance, the consumption of hydrogen and helium steers the evolution, i.e. $\tau = \tau_n$ ($\gg \tau_{KH}$), and represents a considerable simplification for calculating models which are said to be in *complete equilibrium* (i.e. mechanical and thermal equilibrium).

In the case $\tau \ll \tau_{KH}$, the right-hand sides of (4.43,44) are large compared to those of (4.41,42). Therefore in (4.26) the last two terms containing the time derivatives must (at least very nearly) cancel each other, which means that $dq/dt \approx 0$, or the change is nearly adiabatic. Note that a relatively small deviation from the strict adiabatic change can still be of the order $\varepsilon$, and therefore $\varepsilon_g$ cannot be neglected in the energy equation. An example for this case is a star pulsating with the time-scale $\tau = \tau_{hydr} \ll \tau_{KH}$ (see § 38, § 39). The variable luminosity of a pulsating star, for instance, is not due to changes of $\varepsilon$, but of $\varepsilon_g$.

Here we have assumed the simplest case, namely that the star changes more or less uniformly. The situation can be much more complicated if, for example, only parts of the star are affected and local time-scales have to be considered which may be quite different.

# §5 Transport of Energy by Radiation and Conduction

The energy the star radiates away so profusely from its surface is generally re-plenished from reservoirs situated in the very hot central region. This requires an effective transfer of energy through the stellar material, which is possible owing to the existence of a non-vanishing temperature gradient in the star. Depending on the local physical situation, the transfer can occur mainly via radiation, conduction, and convection. In any case, certain "particles" (photons, atoms, electrons, "blobs" of matter) are exchanged between hotter and cooler parts, and their mean free path together with the temperature gradient of the surroundings will play a decisive role. The equation for the energy transport, written as a condition for the temperature gradient necessary for the required energy flow, will supply our next basic equation for the stellar structure.

## 5.1 Radiative Transport of Energy

### 5.1.1 Basic Estimates

Rough estimates show important features of the radiative transfer in stellar interiors and justify an enormous simplification of the formalism.

Let us first estimate the mean free path $\ell_{ph}$ of a photon at an "average" point inside a star like the sun:

$$\ell_{ph} = \frac{1}{\kappa \varrho} \quad , \tag{5.1}$$

where $\kappa$ is a mean absorption coefficient, i.e. a radiative cross-section per unit mass averaged over frequency. Typical values for stellar material are of order $\kappa \approx 1\ \mathrm{cm^2\ g^{-1}}$; for the ionized hydrogen in stellar interiors, a lower limit is certainly the value for electron scattering, $\kappa \approx 0.4\ \mathrm{cm^2\ g^{-1}}$ (see §17). Using this and the mean density of matter in the sun, $\bar{\varrho}_\odot = 3M_\odot/4\pi R_\odot^3 = 1.4\ \mathrm{g\ cm^{-3}}$, we obtain a mean free path of only

$$\ell_{ph} \approx 2\ \mathrm{cm} \quad , \tag{5.2}$$

i.e. stellar matter is very opaque.

The typical temperature gradient in the star can be roughly estimated by aver-aging between centre ($T_c \approx 10^7\ \mathrm{K}$) and surface ($T_0 \approx 10^4\ \mathrm{K}$):

$$\frac{\Delta T}{\Delta r} \approx \frac{T_c - T_0}{R_\odot} \approx 1.4 \times 10^{-4}\ \mathrm{K\ cm^{-1}} \quad . \tag{5.3}$$

The radiation field at a given point is emitted from a small, nearly isothermal surrounding, the differences of temperature being only of order $\Delta T = \ell_{\text{ph}}(dT/dr) \approx 3 \times 10^{-4}$ K. Since the energy density of radiation is $u \sim T^4$, the relative anisotropy of the radiation at a point with $T = 10^7$ K is $4\Delta T/T \sim 10^{-10}$. The situation in stellar interiors must obviously be very close to thermal equilibrium, and the radiation very close to that of a black body. Nevertheless, the small remaining anisotropy can easily be the carrier of the stars' huge luminosity: this fraction of $10^{-10}$ of the flux emitted from 1 cm$^2$ of a black body of $T = 10^7$ K is still $10^3$ times larger than the flux at the solar surface ($6 \times 10^{10}$ erg cm$^{-2}$ s$^{-1}$). Radiative transport of energy occurs via the non-vanishing net flux, i.e. via the surplus of the outwards-going radiation (emitted from somewhat hotter material below) over the inwards-going radiation (emitted from less-hot material above).

### 5.1.2 Diffusion of Radiative Energy

The above estimates have shown that for radiative transport in stars the mean free path $\ell_{\text{ph}}$ of the "transporting particles" (photons) is very small compared to the characteristic length $R$ (stellar radius) over which the transport extends: $\ell_{\text{ph}}/R_\odot \approx 3 \times 10^{-11}$. In this case, the transport can be treated as a diffusion process, which yields an enormous simplification of the formalism.

The diffusive flux $j$ of particles (per unit area and time) between places of different particle density $n$ is given by

$$j = -D \, \nabla n \quad , \tag{5.4}$$

where $D$ is the coefficient of diffusion,

$$D = \frac{1}{3} v \, \ell_{\text{p}} \quad , \tag{5.5}$$

determined by the average values of mean velocity $v$ and mean free path $\ell_{\text{p}}$ of the particles.

In order to obtain the corresponding diffusive flux of radiative energy $F$, we replace $n$ by the energy density of radiation $U$,

$$U = aT^4 \quad , \tag{5.6}$$

$v$ by the velocity of light $c$, and $\ell_{\text{p}}$ by $\ell_{\text{ph}}$ according to (5.1).

In (5.6), $a = 7.57 \times 10^{-15}$ erg cm$^{-3}$ K$^{-4}$ is the *radiation-density constant*. Owing to the spherical symmetry of the problem, $F$ has only a radial component $F_r = |F| = F$ and $\nabla U$ reduces to the derivative in the radial direction

$$\frac{\partial U}{\partial r} = 4 \, a \, T^3 \, \frac{\partial T}{\partial r} \quad . \tag{5.7}$$

Then (5.4,5) give immediately that

$$F = -\frac{4ac}{3} \frac{T^3}{\kappa\varrho} \frac{\partial T}{\partial r} \quad . \tag{5.8}$$

Note that this can be interpreted formally as an equation for heat conduction by writing

$$F = -k_{\text{rad}} \, \nabla T \quad , \tag{5.9}$$

where

$$k_{\text{rad}} = \frac{4ac}{3} \frac{T^3}{\kappa \varrho} \tag{5.10}$$

represents the coefficient of conduction for this radiative transport.

We solve (5.8) for the gradient of the temperature and replace $F$ by the usual local luminosity $l = 4\pi r^2 F$; then

$$\frac{\partial T}{\partial r} = -\frac{3}{16\pi ac} \frac{\kappa \varrho l}{r^2 T^3} \quad . \tag{5.11}$$

After transformation to the independent variable $m$ (as in § 2.1) the basic equation for radiative transport of energy is obtained in the form

$$\frac{\partial T}{\partial m} = -\frac{3}{64\pi^2 ac} \frac{\kappa l}{r^4 T^3} \quad . \tag{5.12}$$

Of course, this neat and simple equation becomes invalid when one approaches the surface of the star. Because of the decreasing density, the mean free path of the photons will there become comparable with (and finally larger than) the remaining distance to the surface; hence the whole diffusion approximation breaks down, and one has to solve the far more complicated full set of transport equations for radiation in the stellar atmosphere. (These equations indeed yield our simple diffusion approximation as the proper limiting case for large optical depths.) Fortunately, however, we have then left the stellar-interior regime with which this book deals, and we happily leave the complicated remainder to those of our colleagues who feel the call to treat the problem of stellar atmospheres.

### 5.1.3 The Rosseland Mean for $\kappa_\nu$

The above equations are independent of the frequency $\nu$; $F$ and $l$ are quantities integrated over all frequencies, so that the quantity $\kappa$ must represent a "proper mean" over $\nu$. We shall now prescribe a method for this averaging.

In general the absorption coefficient depends on the frequency $\nu$. Let us denote this by adding a subscript $\nu$ to all quantities that thus become frequency dependent: $\kappa_\nu$, $\ell_\nu$, $D_\nu$, $U_\nu$, etc.

For the diffusive energy flux $F_\nu$ of radiation in the interval $[\nu, \nu + d\nu]$ we write now, as in § 5.1.2,

$$F_\nu = -D_\nu \, \nabla U_\nu \quad \text{with} \tag{5.13}$$

$$D_\nu = \frac{1}{3} c \, \ell_\nu = \frac{c}{3\kappa_\nu \varrho} \quad , \tag{5.14}$$

while the energy density in the same interval is given by

$$U_\nu = \frac{4\pi}{c} B(\nu, T) = \frac{8\pi h}{c^3} \frac{\nu^3}{e^{h\nu/kT} - 1} \quad . \tag{5.15}$$

$B(\nu, T)$ denotes here the Planck function for the *intensity* of black-body radiation (differing from the usual formula for the energy density simply by the factor $4\pi/c$). From (5.15) we have

$$\nabla U_\nu = \frac{4\pi}{c} \frac{\partial B}{\partial T} \nabla T \quad , \tag{5.16}$$

which together with (5.14) is inserted into (5.13), the latter then being integrated over all frequencies to obtain the total flux $F$:

$$F = - \left[ \frac{4\pi}{3\varrho} \int_0^\infty \frac{1}{\kappa_\nu} \frac{\partial B}{\partial T} d\nu \right] \nabla T \quad . \tag{5.17}$$

We have thus regained (5.9), but with

$$k_{\mathrm{rad}} = \frac{4\pi}{3\varrho} \int_0^\infty \frac{1}{\kappa_\nu} \frac{\partial B}{\partial T} d\nu \quad . \tag{5.18}$$

Equating this expression for $k_{\mathrm{rad}}$ with that in the averaged form of (5.10), we have immediately the proper formula for averaging the absorption coefficient:

$$\frac{1}{\kappa} = \frac{\pi}{acT^3} \int_0^\infty \frac{1}{\kappa_\nu} \frac{\partial B}{\partial T} d\nu \quad . \tag{5.19}$$

This is the so-called *Rosseland mean* (after Sven Rosseland).

Since

$$\int_0^\infty \frac{\partial B}{\partial T} d\nu = \frac{acT^3}{\pi} \quad , \tag{5.20}$$

the Rosseland mean is formally the harmonic mean of $\kappa_\nu$ with the weighting function $\partial B/\partial T$, and it can simply be calculated, once the function $\kappa_\nu$ is known from atomic physics.

In order to see the physical interpretation of the Rosseland mean, we rewrite (5.13) with the help of (5.14–16):

$$F_\nu = - \left( \frac{1}{\kappa_\nu} \frac{\partial B}{\partial T} \right) \frac{4\pi}{3\varrho} \nabla T \quad . \tag{5.21}$$

This shows that, for a given point in the star ($\varrho$ and $\nabla T$ given), the integrand in (5.19) is at all frequencies proportional to the net flux $F_\nu$ of energy. The Rosseland mean therefore favours the frequency ranges of maximum energy flux. One could say that an average *transparency* is evaluated rather than an *opacity* – which is plausible, since it is to be used in an equation describing the transfer of energy rather than its blocking.

One can also easily evaluate the frequency where the weighting function $\partial B/\partial T$ has its maximum. From (5.15) one finds that, for given a temperature, $\partial B/\partial T \sim$

$x^4 e^x (e^x - 1)^{-2}$ with the usual definition $x = h\nu/kT$. Differentiation with respect to $x$ shows that the maximum of $\partial B/\partial T$ is close to $x = 4$.

The way we have defined the Rosseland mean $\kappa$, which is a kind of weighted harmonic mean value, has the uncomfortable consequence that the opacity $\kappa$ of a mixture of two gases having the opacities $\kappa_1$, $\kappa_2$ is not the sum of the opacities: $\kappa \neq \kappa_1 + \kappa_2$.

Therefore, in order to find $\kappa$ for a mixture containing the weight fractions $X$ of hydrogen and $Y$ of helium, the mean opacities of the two single gases are of no use. Rather one has to add the frequency-dependent opacities $\kappa_\nu = X\kappa_{\nu H} + Y\kappa_{\nu He}$ before calculating the Rosseland mean. For any new abundance ratio $X/Y$ the averaging over the frequency has to be carried out separately.

In the above we have characterized the energy flux due to the diffusion of photons by $F$. Since in the following we shall encounter other mechanisms for energy transport, from now on we shall specify this radiative flux by the vector $F_{\text{rad}}$. Correspondingly we shall use $\kappa_{\text{rad}}$ instead of $\kappa$, etc.


## 5.2 Conductive Transport of Energy

In heat conduction, energy transfer occurs via collisions during the random thermal motion of the particles (electrons and nuclei in completely ionized matter, otherwise atoms or molecules). A basic estimate similar to that in §5.1.1 shows that in "ordinary" stellar matter (i.e. in a non-degenerate gas) conduction has no chance of taking over an appreciable part of the total energy transport. Although the collisional cross-sections of these charged particles are rather small at the high temperatures in stellar interiors ($10^{-18} \ldots 10^{-20}$ cm$^2$ per particle), the large density ($\bar{\varrho} = 1.4$ g cm$^{-3}$ in the sun) results in mean free paths several orders of magnitude less than those for photons; and the velocity of the particles is only a few per cent of $c$. Therefore the coefficient of diffusion (5.5) is much smaller than that for photons.

The situation becomes quite different, however, for the cores of evolved stars (see §32), where the electron gas is highly degenerate. The density can be as large as $10^6$ g cm$^{-3}$. But degeneracy makes the electrons much faster, since they are pushed up close to the Fermi energy; and degeneracy increases the mean free path considerably, since the quantum cells of phase space are filled up such that collisions in which the momentum is changed become rather improbable. Then the coefficient of diffusion (which is proportional to the product of mean free path and particle velocity) is large, and heat conduction can become so efficient that it short-circuits the radiative transfer (see §17.6).

The energy flux $F_{\text{cd}}$ due to heat conduction may be written as

$$F_{\text{cd}} = -k_{\text{cd}} \nabla T \quad . \tag{5.22}$$

The sum of the conductive flux $F_{\text{cd}}$ and the radiative flux $F_{\text{rad}}$ as defined in (5.9) is

$$F = F_{\text{rad}} + F_{\text{cd}} = -(k_{\text{rad}} + k_{\text{cd}}) \nabla T \quad , \tag{5.23}$$

which shows immediately the benefit of writing the radiative flux in (5.9) formally as an equation of heat conduction. On the other hand, we can just as well write the conductive coefficient $k_{cd}$ formally in analogy to (5.10) as

$$k_{cd} = \frac{4ac}{3} \frac{T^3}{\kappa_{cd}\varrho} \quad , \tag{5.24}$$

hence defining the "conductive opacity" $\kappa_{cd}$. Then (5.23) becomes

$$F = -\frac{4ac}{3} \frac{T^3}{\varrho} \left( \frac{1}{\kappa_{rad}} + \frac{1}{\kappa_{cd}} \right) \nabla T \quad , \tag{5.25}$$

which shows that we arrive formally at the same type of equation (5.11) as in the pure radiative case, if we replace $1/\kappa$ there by $1/\kappa_{rad} + 1/\kappa_{cd}$. Again the result is plausible, since the mechanism of transport that provides the largest flux will dominate the sum, i.e. the mechanism for which the stellar matter has the highest "transparency".

Equation (5.12), which, if we define $\kappa$ properly, holds for radiative and conductive energy transport, can be rewritten in a form which will be convenient for the following sections.

Assuming hydrostatic equilibrium, we divide (5.12) by (2.5) and obtain

$$\frac{(\partial T/\partial m)}{(\partial P/\partial m)} = \frac{3}{16\pi acG} \frac{\kappa l}{mT^3} \quad . \tag{5.26}$$

We call the ratio of the derivatives on the left $(dT/dP)_{rad}$, and we mean by this the variation of $T$ in the star with depth, where the depth is expressed by the pressure, which increases monotonically inwards. In this sense, in a star which is in hydrostatic equilibrium and transports the energy by radiation (and conduction), $(dT/dP)_{rad}$ is a gradient describing the temperature variation with depth. If we use the customary abbreviation

$$\nabla_{rad} := \left( \frac{d \ln T}{d \ln P} \right)_{rad} \quad , \tag{5.27}$$

(5.26) can be written in the form

$$\nabla_{rad} = \frac{3}{16\pi acG} \frac{\kappa l P}{mT^4} \quad , \tag{5.28}$$

in which conduction effects are now included. Note the difference in definition and meaning of $\nabla_{rad}$ and of $\nabla_{ad}$ introduced in (4.21), which concerns not only their (in general different) numerical values. As just explained, $\nabla_{rad}$ means a spatial derivative (connecting $P$ and $T$ in two neighbouring mass shells), while $\nabla_{ad}$ describes the thermal variation of one and the same mass element during its adiabatic compression. Only in special cases will they have the same value, and we then speak of an "adiabatic stratification".

We will use $\nabla_{rad}$ also in connection with more general cases (other modes of energy transport, deviation from hydrostatic equilibrium). It then means the gradient to which a radiative, hydrostatic layer would adjust at a corresponding point (same values of $P$, $T$, $l$, $m$), or simply an abbreviation for the expression on the right-hand side of (5.28).


## 5.3 The Thermal Adjustment Time of a Star

We can write (5.12), which holds for radiative and conductive energy transport, in the form

$$l = -\sigma^* \frac{\partial T}{\partial m} \quad , \qquad \sigma^* = \frac{64\pi^2 acT^3 r^4}{3\kappa} \quad . \tag{5.29}$$

Now, combining this with (4.25) and replacing the internal energy $u$ by its value $c_v T$ for the ideal gas, it follows that

$$\frac{\partial}{\partial m}\left(\sigma^* \frac{\partial T}{\partial m}\right) - c_v \frac{\partial T}{\partial t} = -\left[\varepsilon + \frac{P}{\varrho^2}\frac{\partial \varrho}{\partial t}\right] \quad . \tag{5.30}$$

If we put the right-hand side equal to zero, then (5.30) has the form of the equation of heat transfer with variable conductivity $\sigma^*$. Indeed variation of the temperature with time along a rod of conductivity $\sigma$ and specific heat $c$ is governed by the equation

$$\frac{\partial}{\partial x}\left(\sigma \frac{\partial T}{\partial x}\right) = c \frac{\partial T}{\partial t} \quad , \tag{5.31}$$

where $x$ is the spatial coordinate along the rod (see LANDAU, LIFSHITZ, vol 6, 1959). There exists a vast amount of mathematical theory associated with this equation, especially for the case where $\sigma$ is constant. For example, one can define an initial-value problem with given $T = T(x)$ at $t = 0$. How, then, does this initial temperature profile evolve in time? There are classical methods for determining $T = T(x,t)$ for $t > 0$. One of the basic results is that one can start with an exciting temperature profile $T(x)$, for instance one which resembles the skyline of Manhattan or the panorama of the Alps, and after some time the temperature profile always looks like the landscape of Nebraska: $T(x,t)$ approaches the limit solution $T = $ constant after sufficient time.

One can easily estimate the time-scale over which (5.31) demands considerable changes of an initially given temperature profile, *the time-scale of thermal adjustment*:

$$\tau_{\text{adj}} = \frac{c}{\sigma} d^2 \quad , \tag{5.32}$$

where $d$ is a characteristic length over which the (initally given) temperature variation changes. Obviously, only temperature profiles with variations over small distances can change rapidly in time.

The inhomogeneous term on the right of (5.30) is a source term. It takes into account that energy can be added everywhere by nuclear reactions or by compression. In the case of the rod it would correspond to extra heat sources adding heat at different values of $x$. Similarly to (5.32) we can derive a characteristic time for a star:

$$\tau_{\mathrm{adj}} = \frac{c_v M^2}{\overline{\sigma^*}} \quad , \tag{5.33}$$

where we have replaced the operator $\partial/\partial m$ by $1/M$ and introduced a mean value $\overline{\sigma^*}$. From (5.29) we find for the luminosity $L$ of the star $L \approx \overline{\sigma^*}\overline{T}/M$, where $\overline{T}$ is a mean temperature of the star. Therefore, for a rough estimate, we have from (5.33) that

$$\tau_{\mathrm{adj}} \approx \frac{c_v \overline{T} M}{L} = \frac{E_i}{L} = \tau_{\mathrm{KH}} \quad . \tag{5.34}$$

This means that the Kelvin–Helmholtz time-scale can be considered a characteristic time of thermal adjustment of a star or – in other words – the time it takes a thermal fluctuation to travel from centre to surface.

In spite of the indicated equivalence of $\tau_{\mathrm{adj}}$ and $\tau_{\mathrm{KH}}$ it is often advisable to consider $\tau_{\mathrm{adj}}$ separately, in particular if it is to be applied to parts of a star only. For example, we will encounter evolved stars with isothermal cores of very high conductivity (§ 32). The luminosity there is zero so that formally the corresponding $\tau_{\mathrm{KH}}$ becomes infinite. The decisive time-scale that in fact enforces the isothermal situation is the very small $\tau_{\mathrm{adj}}$. The difference can be characterized as follows: how much energy may be transported after a temperature perturbation is often much more important than how much energy is flowing in the unperturbed configuration.

## 5.4 Thermal Properties of the Piston Model

We now investigate the thermal properties of the piston model discussed in § 2.7 and § 3.2 by first assuming that the gas of mass $m^*$ in the container is thermally isolated from the surroundings. If the piston is moved, the gas changes adiabatically, i.e.

$$dQ \equiv m^* du + P dV = 0 \quad , \tag{5.35}$$

$dQ$ being the heat added to the total mass of the gas. For an ideal gas the energy per unit mass is $u = c_v T$, and for adiabatic conditions, with $V = Ah$, this leads to

$$dQ \equiv c_v m^* \, dT + PA \, dh = 0 \quad . \tag{5.36}$$

We now relax the adiabatic condition in three ways. First, we allow a small leak through which heat (but no gas) can escape from the interior (gas at temperature $T$) to the surroundings (at temperature $T_s$), see Fig. 2.2. The corresponding heat flow will be $\chi(T - T_s)$. Second, in order to make the gas more similar to stellar matter we assume the release of nuclear energy with a rate $\varepsilon$. Third, we assume that a radiative energy flux $F$ penetrates the gas and that the energy $\kappa F m^*$ is absorbed per second. The energy balance of the gas in the stationary case then can be expressed by

$$\varepsilon m^* + \kappa F m^* = \chi(T - T_s) \quad . \tag{5.37}$$

In general the heat $dQ$ added to the gas within the time interval $dt$ is

$$dQ = \left[\varepsilon m^* + \kappa F m^* - \chi(T - T_\mathrm{s})\right] dt \quad , \tag{5.38}$$

and, if we compare (5.36) and (5.38), we find that

$$c_v m^* \frac{dT}{dt} + PA \frac{dh}{dt} = \varepsilon m^* + \kappa m^* F - \chi(T - T_\mathrm{s}) \quad . \tag{5.39}$$

This is the equation of energy conservation of the gas.

If we assume $\varepsilon = \kappa = 0$, then (5.39) has only one time-independent solution: $T = T_\mathrm{s}$. What is the time-scale of this adjustment of $T$?

The two time derivatives on the left-hand side of (5.39) give the same estimate for $\tau$; indeed a change of $h$ occurs only as a consequence of, and together with, the change of $T$. For our rough estimate we can therefore replace the left-hand side of (5.39) by $c_v \Delta T m^*/\tau$ where $\Delta T = |T - T_\mathrm{s}|$:

$$c_v m^* \Delta T/\tau \approx \chi|T - T_\mathrm{s}| \quad . \tag{5.40}$$

For the time-scale by which $\Delta T$ decays we obtain

$$\tau_\mathrm{adj} \approx c_v m^*/\chi \quad , \tag{5.41}$$

which is the time it takes the gas to adjust its temperature to that of the surroundings. This time-scale for our piston model plays a role similar to the Kelvin–Helmholtz time-scale in stars. For sufficiently small $\chi$ (sufficiently large $\tau_\mathrm{adj}$) we have $\tau_\mathrm{hydr} \ll \tau_\mathrm{adj}$, similar to the situation in stars, where $\tau_\mathrm{hydr} \ll \tau_\mathrm{KH}$.

# §6 Stability Against Local, Non-spherical Perturbations

We have based our treatment on the assumption of strict spherical symmetry, meaning that all functions and variables (including velocities) are constant on concentric spheres. In reality there will arise small fluctuations on such a sphere, for example, simply from the thermal motion of the gas particles. Such local perturbations of the average state may be ignored if they do not grow. But in a star sometimes small perturbations may grow and give rise to macroscopic local (non-spherical) motions that are also statistically distributed over the sphere. In the basic equations the assumption of spherical symmetry can still be kept if we interpret the variables as proper average values over a concentric sphere.

However, these motions have to be considered carefully because they can have a strong influence on the stellar structure. They not only mix the stellar material but also transport energy: hot gas bubbles rise, while cooler material sinks down; i.e. energy transport is by convection, something which is known to play an important role in the earth's atmosphere.

Whether convection occurs in a certain region of a star obviously depends on the question whether the small perturbations always present will grow or stay small: a question of *stability*. We shall derive criteria which tell us whether stellar material at a certain depth is stable or not. Depending on the physical conditions one can make different simplifying assumptions which lead to different stability problems. The following dynamical problem covers most of the "normal" cases in stars.

## 6.1 Dynamical Instability

The kind of stability we are discussing here is based on the assumption that the moving mass elements have no time to exchange appreciable amounts of heat with the surroundings and therefore move adiabatically. This type of stability (or instability) is called *dynamical*. We will soon learn that there are other types of instability.

First we consider the possibility that the physical quantities (temperature, density, etc.) may not be exactly constant on the surface of a concentric sphere but rather may show certain fluctuations. In the global problem of stellar structure, one then has only to interpret the previously used functions as proper averages. For the local description, we shall simply represent a fluctuation by a mass "element" (subscript e) in which the functions have constant, but somewhat different, values than in the average "surroundings" (subscript s). For any quantity $A$ we define the difference $DA$ between element and surroundings as

$$DA := A_e - A_s \quad .\tag{6.1}$$

One can easily imagine an initial fluctuation of temperature, for example a slightly hotter element with $DT > 0$. Normally one could then also expect an excess of pressure $DP$. However, the element will expand immediately until pressure balance with the surroundings is restored, and since this expansion occurs with the velocity of sound, it is usually much more rapid than any other motion of the element. Therefore we can assume here (and in the following) that the element always remains in pressure balance with its surroundings:

$$DP = 0 \quad .\tag{6.2}$$

Consequently the assumed $DT > 0$ requires that, for an ideal gas with $\varrho \sim P/T$, $D\varrho < 0$, i.e. the element is lighter than the surrounding material, and the buoyancy forces will lift it upwards: temperature fluctuations are obviously accompanied by local motions of elements in a radial direction.

So, we can also take a radial shift $\Delta r > 0$ of the element as the initial perturbation for testing the stability of a layer. Consider an element that was in complete equilibrium with the surroundings at its original position $r$ but has now been lifted to $r + \Delta r$ (cf. Fig. 6.1). In general its density will differ from that of its new surroundings by

$$D\varrho = \left[ \left( \frac{d\varrho}{dr} \right)_e - \left( \frac{d\varrho}{dr} \right)_s \right] \Delta r \quad ,\tag{6.3}$$

$(d\varrho/dr)_e$ determining the change of the element's density while it rises by $dr$; the other derivative is the spatial gradient in the surroundings.

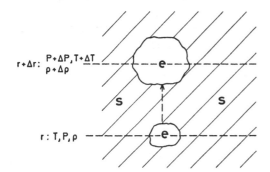

Fig. 6.1. In order to test the stability of a "surrounding" layer (s), a test "element" (e) is lifted from level $r$ to $r + \Delta r$

A finite $D\varrho$ gives the radial component $K_r = -g D\varrho$ of a buoyancy force $K$ (per unit of volume), where $g$ again is the absolute value of the acceleration of gravity. If $D\varrho < 0$, the element is lighter and $K_r > 0$, i.e. $K$ is directed upwards. This situation is obviously unstable, since the element is lifted further, the original perturbation being increased.

If on the other hand $D\varrho > 0$, then $K_r < 0$, i.e. $K$ is directed downwards. The element, which is heavier than its new surroundings, is drawn back to its original position, the perturbation is removed, and the layer is stable. As the *condition for stability* we obtain with $D\varrho > 0$ from (6.3) the result

$$\left(\frac{d\varrho}{dr}\right)_{\mathrm{e}} - \left(\frac{d\varrho}{dr}\right)_{\mathrm{s}} > 0 \quad . \tag{6.4}$$

Unfortunately this criterion is highly impractical, since it requires knowledge of density gradients that do not appear in the basic equations. It is therefore preferable to turn to temperature gradients as used in the equations of radiative and conductive transport. In order to evaluate $(d\varrho/dr)_{\mathrm{e}}$ correctly, we would have to take into account the possible energy exchange between the element and its surroundings. For simplicity let us here assume that no such exchange of energy occurs, i.e. that the element rises *adiabatically*. This is very close to reality for the deep interior of a star (see §7).

In order to transform the gradients of $\varrho$ into those of $T$, we write the equation of state $\varrho = \varrho(P, T, \mu)$ in the following differential form:

$$\frac{d\varrho}{\varrho} = \alpha \frac{dP}{P} - \delta \frac{dT}{T} + \varphi \frac{d\mu}{\mu} \quad , \tag{6.5}$$

where $\alpha$ and $\delta$ have already been defined in (4.2, 3). But here we have made allowance also for a possible variation of the chemical composition, which is characterized by the molecular weight $\mu$. We therefore have

$$\alpha := \left(\frac{\partial \ln \varrho}{\partial \ln P}\right) \quad , \qquad \delta := -\left(\frac{\partial \ln \varrho}{\partial \ln T}\right) \quad , \qquad \varphi := \left(\frac{\partial \ln \varrho}{\partial \ln \mu}\right) \quad , \tag{6.6}$$

where the three partial derivatives correspond to constant values of $T, \mu$; $P, \mu$; and $P, T$, respectively, and for an ideal gas with $\varrho \sim P\mu/T$ one has $\alpha = \delta = \varphi = 1$. In this description $d\mu$ shall represent only the change of $\mu$ due to the change of chemical composition, i.e. the variation of the concentrations of different nuclei in the deep interior. Of course, $\mu$ can also change in the outer regions for constant composition if the degree of ionization changes. This effect, however, has a well-known dependence on $P$ and $T$ and is supposed to be incorporated in $\alpha$ and $\delta$. Thus $d\mu = 0$ for the moving element that carries its composition along. But $d\mu \neq 0$ for the surroundings if the element passes through layers of different chemical composition.

We can immediately rewrite (6.4) with the help of (6.5) in the form

$$\left(\frac{\alpha}{P}\frac{dP}{dr}\right)_{\mathrm{e}} - \left(\frac{\delta}{T}\frac{dT}{dr}\right)_{\mathrm{e}} - \left(\frac{\alpha}{P}\frac{dP}{dr}\right)_{\mathrm{s}} + \left(\frac{\delta}{T}\frac{dT}{dr}\right)_{\mathrm{s}} - \left(\frac{\varphi}{\mu}\frac{d\mu}{dr}\right)_{\mathrm{s}} > 0 \quad . \tag{6.7}$$

The two terms containing the pressure gradient cancel each other owing to (6.2), and the other terms are usually multiplied by the so-called *scale height of pressure* $H_P$:

$$H_P := -\frac{dr}{d \ln P} = -P \frac{dr}{dP} \quad . \tag{6.8}$$

With (2.3), the condition for hydrostatic equilibrium, we find $H_P = P/\varrho g$, i.e. $H_P > 0$, since $P$ decreases with increasing $r$. $H_P$ has the dimension of length, being the length characteristic of the radial variation of $P$. In the solar photosphere ($g = 2.7 \times 10^4$ cm s$^{-2}$, $P = 6.8 \times 10^4$ dyn cm$^{-2}$, $\varrho = 1.8 \times 10^{-7}$ g cm$^{-3}$) one finds $H_P = 1.4 \times 10^7$ cm, while at $r = R_\odot/2$ ($g = 1.0 \times 10^5$ cm s$^{-2}$, $P = 6.7 \times 10^{14}$ dyn cm$^{-2}$, $\varrho = 1.3$ g cm$^{-3}$) $H_P$ is much bigger, at $5.2 \times 10^9$ cm. If one approaches the stellar centre – where $g = 0$, while $P$ remains finite –, then $H_P \to \infty$.

Multiplication of (6.7) by $H_P$ yields as a condition for stability

$$\left( \frac{d \ln T}{d \ln P} \right)_s < \left( \frac{d \ln T}{d \ln P} \right)_e + \frac{\varphi}{\delta} \left( \frac{d \ln \mu}{d \ln P} \right)_s \quad . \tag{6.9}$$

Similar to the previously defined quantities $\nabla_{\mathrm{rad}}$ and $\nabla_{\mathrm{ad}}$ we define three new derivatives:

$$\nabla := \left( \frac{d \ln T}{d \ln P} \right)_s, \qquad \nabla_e := \left( \frac{d \ln T}{d \ln P} \right)_e, \qquad \nabla_\mu := \left( \frac{d \ln \mu}{d \ln P} \right)_s \quad . \tag{6.10}$$

Here the subscripts s indicate that the derivatives are to be taken in the surrounding material. In both cases they are spatial derivatives in which the variations of $T$ and $\mu$ with depth are considered and $P$ is taken as a measure of depth. The quantity $\nabla_e$ describes the variation of $T$ in the element during its motion, where the position of the element is measured by $P$. In this sense $\nabla_e$ and $\nabla_{\mathrm{ad}}$ are similar, since both describe the temperature variation of a gas undergoing pressure variations; on the other hand, $\nabla_{\mathrm{rad}}$ and $\nabla_\mu$ describe the spatial variation of $T$ and $\mu$ in the surroundings.

With the definitions (6.10) the condition (6.9) for stability becomes

$$\nabla < \nabla_e + \frac{\varphi}{\delta} \nabla_\mu \quad . \tag{6.11}$$

In (5.27,28) we defined $\nabla_{\mathrm{rad}}$, which describes the temperature gradient for the case that the energy is transported by radiation (or conduction) only. Therefore in a layer that indeed transports all energy by radiation the actual gradient $\nabla$ is equal to $\nabla_{\mathrm{rad}}$. Let us test such a layer for its stability and assume the elements change adiabatically: $\nabla_e = \nabla_{\mathrm{ad}}$; the radiation layer is stable if

$$\nabla_{\mathrm{rad}} < \nabla_{\mathrm{ad}} + \frac{\varphi}{\delta} \nabla_\mu \quad , \tag{6.12}$$

a form known as the *Ledoux criterion* (named after Paul Ledoux) for dynamical stability. In a region with homogeneous chemical composition, $\nabla_\mu = 0$, and one has then simply the famous *Schwarzschild criterion* for dynamical stability (named after Karl Schwarzschild):

$$\nabla_{\mathrm{rad}} < \nabla_{\mathrm{ad}} \quad . \tag{6.13}$$

If in the criteria (6.12, 13) the left-hand side is larger than the right, the layer is dynamically unstable. If they are equal, one speaks of marginal stability. The difference between the two criteria obviously plays a role only in regions where the chemical composition varies radially. We will see that such regions occur in the interior of evolving stars, where heavier elements are usually produced below the lighter ones, such that the molecular weight $\mu$ increases inwards (as the pressure does) and $\nabla_\mu > 0$. Then the last term in inequality (6.12) obviously has a stabilizing effect ($\varphi$ and $\delta$ are both positive). This is plausible since the element carries its heavier material upwards into lighter surroundings and gravity will tend to draw it back to its original place.

If these criteria favour stability, then no convective motions will occur, and the whole flux will indeed be carried by radiation, i.e. the actual gradient at such a place is equal to the radiative one: $\nabla = \nabla_{rad}$. If they favour instability, then small perturbations will increase to finite amplitude until the whole region boils with convective motions that carry part of the flux – and the actual gradient has to be determined in a manner described in §7. This instability can be caused either by the fact that $\nabla_{rad}$ has become too high (large flux, or very opaque matter), or else by a depression of $\nabla_{ad}$; both cases occur in stars. And, finally, in a twilight zone, where one of the two criteria (6.12, 13) says stability and the other one says instability, strange things may happen (see, for instance, §6.3 and §30.4.2).

Note that (6.12) and (6.13) are strictly local criteria, which means good and bad news. They are very practical since they can be evaluated easily for any given place by using the local values of $P$, $T$, $\varrho$ only, without bothering about other parts of the star. And in most cases this will give satisfactory answers. In critical cases, however, this may not be sufficient. Strictly speaking, convective motions are not only dependent on the local forces (which are solely regarded by the criteria), but must be coupled (by momentum transfer, inertia, the equation of continuity) to their neighbouring layers. And in extreme cases the reaction of the whole star against a local perturbation should be taken into account. An obvious example is the precise determination of the border of a convective zone, where elements that were accelerated elsewhere "shoot over" until their motion is braked. We will come back later to such problems when they arise (see §30.4.1).

We can immediately derive a qualitative relation between the different gradients. They are best visualized in a diagram such as Fig. 6.2, where $\ln T$ is plotted against $\ln P$ (decreasing outwards) for an unstable layer violating the Schwarzschild criterion. In such a diagram, an adiabatic change follows a line with slope $\nabla_{ad}$, the changes in a rising element are given by a line with slope $\nabla_e$, while the stratifications in the surroundings and in a radiative layer are shown by lines with slopes $\nabla$ and $\nabla_{rad}$ respectively.

Suppose we have convection in a chemically homogeneous layer ($\nabla_\mu = 0$). The criterion (6.11) must be violated, i.e. $\nabla > \nabla_e$. If some part of the flux is carried by convection, then the actual gradient $\nabla < \nabla_{rad}$, since only a part of the total flux is left for radiative transfer. Consider a rising element that has started from a point with $P_0$, $T_0$. In Fig. 6.2 this element moves downwards to the left along the line with slope $\nabla_e$. Since $\nabla > \nabla_e$, the element (although cooling) will obviously have an increasing temperature excess over its new surroundings (the temperature of which changes with $\nabla$). Therefore it will radiate energy into its surroundings, which means that the element cools more than adiabatically: $\nabla_e > \nabla_{ad}$. Combining these inequalities, we arrive at the relation illustrated in Fig. 6.2:

$$\nabla_{rad} > \nabla > \nabla_e > \nabla_{ad} \quad . \tag{6.14}$$

The fact that $\nabla_e$ must always be between $\nabla_{ad}$ and $\nabla$ of the surroundings shows that

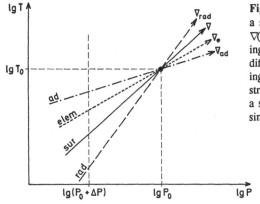

Fig. 6.2. Temperature–pressure diagram with a schematic sketch of the different gradients $\nabla (\equiv \partial \ln T / \partial \ln P)$ in a convective layer. Starting at a common point with $P_0$ and $T_0$, the different types of changes (adiabatic, in a rising element, in the surroundings, for radiative stratification) lead to different temperatures at a slightly higher point with $P_0 + \Delta P$ ($< P_0$, since $P$ decreases outwards)

the criteria (6.12,13) are also to be used in near-surface regions, where the rising elements lose much of their energy by radiation.

## 6.2 Oscillation of a Displaced Element

In a dynamically stable layer a displaced mass element is pushed back by buoyancy. When coming back to its original position, it has gained momentum and will overshoot and therefore start to oscillate. In the following we shall discuss this oscillation.

Consider a mass element lifted from its normal (equilibrium) position in the radial direction by an amount $\Delta r$ (see Fig. 6.1) There it has an excess of density $D\varrho$ over its new surroundings given by (6.3), which for balance of pressure ($DP = 0$) and with (6.5) and the definitions (6.6,8,10) can easily be written as

$$D\varrho = \frac{\varrho \delta}{H_P} \left[ \nabla_e - \nabla + \frac{\varphi}{\delta} \nabla_\mu \right] \Delta r \quad . \tag{6.15}$$

In the presence of gravity $g$, the resulting buoyancy force per unit volume is $K_r = -gD\varrho$, producing an acceleration of the element of

$$\frac{\partial^2 (\Delta r)}{\partial t^2} = -\frac{g\delta}{H_P} \left[ \nabla_e - \nabla + \frac{\varphi}{\delta} \nabla_\mu \right] \Delta r \quad . \tag{6.16}$$

Suppose now that the element, after an original displacement $\Delta r_0$, moves adiabatically ($\nabla_e = \nabla_{ad}$) through a dynamically stable layer ($D\varrho / \Delta r > 0$). The element is accelerated back towards its equilibrium position around which it then oscillates according to the solution of (6.16):

$$\Delta r = \Delta r_0 e^{i\omega t} \quad . \tag{6.17}$$

The frequency $\omega = \omega_{ad}$ of this adiabatic oscillation is the so-called *Brunt-Väisälä frequency* given by

41

$$\omega_{\text{ad}}^2 = \frac{g\delta}{H_P} \left( \nabla_{\text{ad}} - \nabla + \frac{\varphi}{\delta} \nabla_\mu \right) \quad . \tag{6.18}$$

(It plays, for example, a role in the discussion of non-radial oscillations of a star; see § 40.) The corresponding period is $\tau_{\text{ad}} = 2\pi / \omega_{\text{ad}}$.

We see immediately what happens in an unstable layer. If the Ledoux criterion (6.12) [or the Schwarzschild criterion (6.13) for $\nabla_\mu = 0$] is violated, then (6.18) gives $\omega_{\text{ad}}^2 < 0$, such that $\omega_{\text{ad}}$ is imaginary and the time dependence of $\Delta r$ is given by the factor $\exp(\sigma t)$ with a real $\sigma > 0$. Instead of oscillating, the displaced element moves away exponentially.

## 6.3 Vibrational Stability

In a dynamically stable layer an oscillating mass element has, in general, $DT \neq 0$. If $DT > 0$, it will lose heat to its surrounding by radiation; if $DT < 0$, it will gain heat. This means it will not move adiabatically. We consider the deviation from adiabaticity to be small, which means that the thermal adjustment time of the element is large compared to the period of the oscillation; then the temperature excess of the element can be written as

$$DT = \left[ \left( \frac{dT}{dr} \right)_{\text{e}} - \left( \frac{dT}{dr} \right)_{\text{s}} \right] \Delta r$$
$$= -\frac{T}{H_P} (\nabla_{\text{e}} - \nabla) \Delta r \quad . \tag{6.19}$$

Dynamical stability means that $D\varrho / \Delta r > 0$ and therefore (6.11) is fulfilled. If the layer is chemically homogeneous, then $\nabla_\mu = 0$, and (6.11) becomes $\nabla_{\text{e}} - \nabla > 0$, such that (6.19) gives $DT < 0$ for $\Delta r > 0$. Above its equilibrium position the element is cooler than the surroundings and receives energy by radiation. This reduces $\nabla_{\text{e}} - \nabla$, $D\varrho$, and the restoring force, such that the element is less accelerated back towards the equilibrium position. The result will be an oscillation with slowly decreasing amplitude. Formally this radiative damping shows up as a small positive imaginary part of $\omega$ in (6.17) after the exchange of heat with the surroundings is included in (6.16). The oscillatory part (real part of $\omega$) is still very close to the adiabatic value (6.18).

If the stable layer is inhomogeneous with $\nabla_\mu > 0$, it can be that with (6.11) $\nabla_{\text{e}} - \nabla > 0$ also (*both* criteria are fulfilled), i.e. we find again that $DT < 0$ for $\Delta r > 0$ and radiative damping as before. However, we can also imagine a situation with $\nabla_{\text{e}} - \nabla < 0$ in spite of (6.11) for large enough $\nabla_\mu$. Then $DT > 0$ for $\Delta r > 0$ according to (6.19), and the lifted element, being hotter than its surroundings, will now *lose* energy by radiation. This increases $\nabla_{\text{e}} - \nabla$, $D\varrho$, and the restoring force, and the element will oscillate with slowly increasing amplitude. This is an *over-stability*, or *vibrational instability*. The difficulties in this strange situation are obvious [it being the above mentioned twilight zone between the two criteria (6.12,13)]. The growing oscillation may lead to a chemical mixing of elements and surroundings and thus decrease, or, eventually even destroy, the stabilizing gradient $\nabla_\mu$. But then again, it is not clear whether in such critical situations a local analysis suffices at all. The reaction of other layers of the star might provide enough damping to suppress the over-stability.

With these considerations it follows that we have to distinguish between *dynamical stability* and *vibrational stability*. The first applies to purely adiabatic behaviour of the moving mass, while the second takes heat exchange into account. A layer with a temperature gradient $\nabla$ such that the Ledoux criterion is fulfilled but the Schwarzschild criterion is not, i.e.

$$\nabla_{ad} < \nabla < \nabla_{ad} + \frac{\varphi}{\delta}\,\nabla_\mu \quad , \tag{6.20}$$

is dynamically stable but vibrationally unstable.

A dynamical instability grows on a time-scale given by $(H_P/g)^{1/2}$, while in the case of a vibrational instability the growth of amplitude is governed by the time it takes a mass element to adjust thermally to its surrounding. In the following we shall estimate this time-scale $\tau_{adj}$.

## 6.4 The Thermal Adjustment Time

Let us consider a mass element with $DT > 0$, i.e. one that will radiate into the surroundings. Superposed onto the radial energy flux $F$, carrying energy from the stellar interior to the surface, there will be a local, non-radial flux $f$, carrying the surplus energy of the element to its surroundings. According to (5.9,10), the absolute value $f$ of the radiative flux from the element due to its excess temperature will be

$$f = \frac{4acT^3}{3\kappa\varrho}\left|\frac{\partial T}{\partial n}\right| \quad , \tag{6.21}$$

where $\partial/\partial n$ indicates the differentiation perpendicular to the surface of the element. Suppose our element to be a roughly spherical "blob" with diameter $d$. We will approximate the temperature gradient in the normal direction by $\partial T/\partial n \approx 2DT/d$. The radiative loss $\lambda$ per unit time from the whole surface $S$ of the blob is then

$$\lambda = Sf = \frac{8acT^3}{3\kappa\varrho}\,DT\frac{S}{d} \quad . \tag{6.22}$$

The quantity $\lambda$ is a sort of "luminosity" of the blob, and it determines the rate by which the thermal energy of the blob of volume $V$ changes:

$$\varrho V c_P \frac{\partial T_e}{\partial t} = -\lambda \quad . \tag{6.23}$$

Here we can replace $\partial T_e/\partial t$ by $\partial(DT)/\partial t$, since the temperature of the (large) surroundings scarcely changes, owing to radiative losses of the blob. Furthermore let $V/S \approx d/6$ (as for a sphere); then one obtains from (6.22, 23) that

$$\frac{\partial(DT)}{\partial t} = -\frac{DT}{\tau_{adj}} \quad , \tag{6.24}$$

with the time-scale for thermal adjustment

$$\tau_{adj} = \frac{\kappa \varrho^2 c_P d^2}{16 a c T^3} = \frac{\varrho V c_P D T}{\lambda} \quad . \tag{6.25}$$

The second equation follows from a comparison of (6.22, 23) and (6.24). We see that $\tau_{adj}$ is roughly the excess thermal energy divided by the luminosity, i.e. an equivalent to the Kelvin-Helmholtz time-scale for a star (3.17). For sufficiently large elements that are far enough from a region of marginal stability, one has $\tau_{adj} \gg 1/\omega_{ad}$, which means that the radiative losses give only a small deviation from adiabatic oscillations, as discussed in §6.2.

## 6.5 Secular Instability

Even a small exchange of heat between a displaced mass element and its surroundings can lead to another kind of instability, which is called *thermal* or *secular instability*. We first discuss this qualitatively with an experiment which can easily be carried out with water and kitchen equipment!

In a glass jar containing cold fresh water we carefully pour over a layer of warm salty water. The salt increases the specific weight of the upper layer, but the warmth shall be enough to reduce (despite the salt content) its specific weight to below that of the underlying fresh water. If, owing to a perturbation, a blob of salty water is pushed downwards, buoyancy will push it back, i.e. the two layers are then *dynamically stable*.

But the buoyancy acts as a restoring force only as long as the element stays warm during its excursion into the cold layers. On the time-scale by which it loses its excess temperature the buoyancy diminishes and the element moves downward because of its salt content. Indeed if one watches the two layers for some time, one can see (especially if the salty water is coloured) that small blobs of salty water slowly sink, a phenomenon called *salt fingers*. It is an instability controlled by the heat leakage of the element. This is *secular instability*. It can not only occur in glass jars, but also in stars!

Consider a blob of stellar matter situated in surroundings of somewhat different, but homogeneous, composition, i.e. $D\mu \neq 0$, but $\nabla_\mu = 0$. (Such a situation can occur, for example, if two homogeneous layers of different compositions are above each other and a blob from one layer is displaced into the other.) The blob is supposed to be in mechanical equilibrium with its surroundings, i.e. $DP = D\varrho = 0$. This requires, however, a temperature difference according to (6.5):

$$\delta \frac{DT}{T} = \varphi \frac{D\mu}{\mu} \quad . \tag{6.26}$$

For $D\mu > 0$, for example, the blob is hotter and therefore radiates towards the surroundings; the loss of energy under pressure balance ($DP = 0$) leads to an increased density and the blob sinks until again $D\varrho = 0$. Equation (6.26) is still valid and, since $D\mu$ is unchanged, $DT > 0$ as before, and so on. Obviously the blob will slowly sink (or rise for $D\mu < 0$) with a velocity $v_\mu$ such that $DT$ always remains constant according to (6.26).

Owing to radiation, the temperature of the blob changes at the rate $-DT/\tau_{adj}$ [see(6.24)]. While sinking or rising it changes also because of the adiabatic compression (or expansion) that occurs as a result of the change of pressure, even in the absence of energy exchange. The rate of change of $DT$ can then immediately be written as

$$\frac{1}{T}\frac{\partial}{\partial t}(DT) = \left(\nabla_{ad}\frac{\partial \ln P}{\partial t} - \frac{DT}{T\tau_{adj}}\right) - \nabla\frac{\partial \ln P}{\partial t} \quad . \tag{6.27}$$

The rate of change of pressure is simply linked to the velocity $v_\mu$ by

$$\frac{\partial \ln P}{\partial t} = -\frac{v_\mu}{H_P} \quad . \tag{6.28}$$

Using this and (6.26), together with the condition $\partial(DT)/\partial t = 0$ [which follows from (6.26), since $D\mu$ does not vary if the element moves in a chemically homogeneous region], we can solve (6.26–28) for the velocity and obtain

$$v_\mu = -\frac{H_P}{(\nabla_{ad} - \nabla)\tau_{adj}}\frac{\varphi}{\delta}\frac{D\mu}{\mu} \quad . \tag{6.29}$$

In this case of thermal instability, therefore, the blob sinks ($v_\mu < 0$ for $D\mu > 0$) through a dynamically stable surrounding ($\nabla_{ad} > \nabla$) with the adjustment time-scale for radiative losses.

The idea of blobs finding themselves in strange surroundings ($D\mu > 0$) is not far-fetched. Secular instabilities of the kind discussed here can occur in stars, for example, of about one solar mass. After hydrogen has been transformed to helium in their cores, their central region is cooled by neutrinos, which take away energy without interacting with the stellar matter. The temperature in these stars, therefore, is highest somewhere off-centre and decreases towards the stellar surface as well as towards the centre. If, then, helium "burning" is ignited in the region of maximum temperature, the newly formed carbon is in a shell surrounding the central core (§ 32.4,5). This carbon-enriched shell has a higher molecular weight than the regions below: carbon "fingers" will grow and sink inwards. In later evolutionary phases other nuclear reactions, such as neon burning, may ignite off-centre, and heavier fingers of material may sink.

## 6.6 The Stability of the Piston Model

Our piston model (§ 2.7 and 5.4) shows a stability behaviour in many respects similar to that of the blobs.

We start with the two equations that together with the equation of state describe the time dependence of the piston model. These are (2.34) and (5.39), where we assume for the sake of simplicity that $\varepsilon = \kappa = 0$. The equilibrium state is given by $T = T_s$ and $G^* = PA$.

In order to investigate the stability we denote the equilibrium values by the subscript "0" and make small perturbations of the form

$$h(t) = h_0 \left(1 + x e^{i\omega t}\right)$$

$$P(t) = P_0 \left(1 + p e^{i\omega t}\right)$$

$$T(t) = T_0 \left(1 + \vartheta e^{i\omega t}\right) \tag{6.30}$$

with $|x|, |p|, |\vartheta| \ll 1$. We therefore neglect quadratic and higher-order expressions in these quantities.

From mass conservation $\varrho h = $ constant and from the ideal gas equation $P \sim \varrho T$ we obtain

$$p = \vartheta - x \quad . \tag{6.31}$$

We now introduce (6.14) into (2.34) and obtain after linearization

$$M^* h_0 \omega^2 x + P_0 A p = 0 \quad , \tag{6.32}$$

which with $g = P_0 A / M^*$ and with (6.31) can be replaced by

$$\left(\frac{\omega^2 h_0}{g} - 1\right) x + \vartheta = 0 \quad , \tag{6.33}$$

while the corresponding perturbation and linearization of (5.39) gives

$$i\omega P_0 A h_0 x + \left(i\omega c_v m^* T_0 + \chi T_0\right) \vartheta = 0 \quad . \tag{6.34}$$

The two linear homogeneous equations (6.33, 34) for $x$ and $\vartheta$ can be solved if the determinant vanishes. This condition gives an algebraic equation of third order for the eigenvalue $\omega$.

The problem becomes simple if we assume that the trapped gas changes adiabatically, i.e. if $\chi = 0$. Then (6.34), with $m^*/(A h_0) = \varrho_0$ and with the ideal gas equation, yields

$$\frac{\Re}{\mu c_v} x + \vartheta = 0 \quad , \tag{6.35}$$

and with $\Re/\mu = c_P - c_v$ and $c_P/c_v = \gamma_{\text{ad}}$ it follows that

$$(\gamma_{\text{ad}} - 1) x + \vartheta = 0 \quad . \tag{6.36}$$

Setting the determinant of the equations (6.33, 36) to zero gives the eigenvalue for the adiabatic motion:

$$\omega = \pm \omega_{\text{ad}} , \qquad \omega_{\text{ad}} = \left(\gamma_{\text{ad}} g / h_0\right)^{1/2} \quad . \tag{6.37}$$

Since $\omega$ is real, the adiabatic motion is an oscillation with frequency $\omega$ and constant amplitude. Therefore in the language of §6.1 our ideal gas piston model is *dynamically stable*. Note that $1/\omega_{\text{ad}}$ is of the order of the hydrostatic time-scale $\tau_{\text{hydr}}$ defined in §2.7.

How do non-adiabatic effects change the picture? With the $\chi$ term in (6.34) we have, instead of (6.36),

$$(\gamma_{ad} - 1) x + \left(1 + \frac{a}{i\omega}\right) \vartheta = 0 \quad , \tag{6.38}$$

with $a = \chi/(c_v m^*)$. Setting the determinant of (6.33) and (6.38) equal to zero now gives a *cubic equation* in $\omega$. In general $\omega$ will be complex.

We assume $\chi$ to be small, so that the oscillation frequency must be close to the adiabatic value and we can put $\omega = \omega_{ad} + \xi$, with $|\xi| \ll |\omega_{ad}|$. If we neglect higher terms in $\xi$ and $\chi$, we find from the vanishing determinant of the system of homogeneous linear equations (6.33,38) and after some algebraic manipulation that

$$i\xi = -\frac{\gamma_{ad} - 1}{2\gamma_{ad}} \frac{\chi}{c_v m^*} = -\frac{\gamma_{ad} - 1}{2\gamma_{ad}} \frac{1}{\tau_{adj}} < 0 \quad , \tag{6.39}$$

where we have used (5.41). The (almost adiabatic) oscillation is therefore damped, since the exponents of (6.30), $i\omega = i\omega_{ad} + i\xi$, have a negative real part that decreases the amplitude on a time-scale $\tau_{adj}$. The piston model with a leak is *vibrationally stable*.

The cubic equation for $\omega$ must have a third root, which we find easily by assuming that it describes an evolution so slow that the inertia term in (2.34) can be neglected. (This has to be checked later.) Then (6.33) has to be replaced by

$$\vartheta - x = 0 \quad , \tag{6.40}$$

which according to (6.31) is equivalent to $p = 0$. Indeed if the evolution is so slow that there is always hydrostatic equilibrium, the pressure is given by the (constant) weight of the piston. We then have from (6.34, 40)

$$i\omega = -\frac{\chi T}{P_0 A h_0 + c_v m^* T_0} = -\frac{\chi}{cpm^*} = -\frac{1}{\gamma_{ad}\tau_{adj}} \quad . \tag{6.41}$$

For the latter equation we have used the relation $P_0 A h_0 = \Re m^* T_0/\mu$ and (5.41). The third root gives an exponential decay in time of the initial perturbation, the time-scale being comparable with $\tau_{adj}$. If $\chi$ is sufficiently small and the evolution slow, the assumption that the inertia term is negligible is justified.

Our result (6.41) means that any deviation from thermal equilibrium ($T - T_s \neq 0$) vanishes within the thermal adjustment time, i.e. the thermally adjusted piston model for $\varepsilon = \kappa = 0$ is *secularly stable*. We see that it shows the same limiting cases for the stability problem (dynamical, vibrational, and secular stability) as the blobs. In § 39.1 we will consider the influence on the stability of the piston model of the (here neglected) terms in (5.39) due to $\varepsilon$ and $\kappa$.

# § 7 Transport of Energy by Convection

Convective transport of energy means an exchange of energy between hotter and cooler layers in a dynamically unstable region through the exchange of macroscopic mass elements ("blobs", "bubbles", "convective elements"), the hotter of which move upwards while the cooler ones descend. The moving mass elements will finally dissolve in their new surroundings and thereby deliver their excess (or deficiency) of heat. Owing to the high density in stellar interiors, convective transport can be very efficient. However, this energy transfer can operate only if it finds a sufficient driving mechanism in the form of the buoyancy forces.

A thorough theoretical treatment of convective motions and transport of energy is extremely difficult. It is the prototype of the many astrophysical problems in which the bottle-neck preventing decisive progress is the difficulty involved in solving the well-known hydrodynamic equations. For simplifying assumptions solutions are available that may even give reasonable approximations for certain convective flows in the laboratory (or in the kitchen). Unfortunately convection in stars proceeds under rather malicious conditions: turbulent motion transports enormous fluxes of energy in a very compressible gas, which is stratified in density, pressure, temperature, and gravity over many powers of ten. Nevertheless large efforts have been made over many years to solve this notorious problem, and they have partly arrived at promising results (for a review of the state of art in this field see SPIEGEL, 1972). None of them, however, has reached a stage where it could provide a procedure easy enough to be handled in everyday stellar-structure calculations. Therefore we limit ourselves exclusively to the description of the old so-called "mixing-length" theory. The reason for this is not that we believe it to be sufficient; but it does provide at least a simple method for treating convection locally, at any given point of a star. And, moreover, even this poor approximation shows without any doubt that in the very deep interior of a star a detailed theory is normally not necessary.

Note that in the following we are dealing only with convection in stars that are in hydrostatic equilibrium. We furthermore assume that the convection is time independent, which means that it is fully adjusted to the present state of the star. Otherwise a convection theory for rapidly changing regions (time-dependent convection) has to be developed.

Equation (5.28) gives the gradient $\nabla_{rad}$ that would be maintained in a star if the whole luminosity $l$ had to be transported outwards by radiation only. If convection contributes to the energy transport, the actual gradient $\nabla$ will be different (namely smaller). It is the purpose of this section to estimate $\nabla$ in the case of convection.

## 7.1 The Basic Picture

The mixing-length theory goes back to Ludwig Prandtl (1925), who modelled a simple picture of convection in complete analogy to molecular heat transfer: the transporting "particles" are macroscopic mass elements ("blobs") instead of molecules; their mean free path is the "mixing length" after which the blobs dissolve in their new surroundings. Prandlt's theory was adapted for stars by L. Biermann.

The total energy flux $l/4\pi r^2$ at a given point in the star consists of the radiative flux $F_{rad}$ (in which the conductive flux may already be incorporated) plus the convective flux $F_{con}$. Their sum defines according to (5.28) the gradient $\nabla_{rad}$ that would be necessary to transport the whole flux by radiation:

$$F_{rad} + F_{con} = \frac{4acG}{3} \frac{T^4 m}{\kappa P r^2} \nabla_{rad} \quad . \tag{7.1}$$

However, part of the flux is transported by convection. If the actual gradient of the stratification is $\nabla$, then the radiative flux is obviously only

$$F_{rad} = \frac{4acG}{3} \frac{T^4 m}{\kappa P r^2} \nabla \quad . \tag{7.2}$$

Note that $\nabla$ is not yet known; in fact we hope to obtain it as the result of this consideration. The first step is to derive an expression for $F_{con}$.

Consider a convective element (a blob) with an excess temperature $DT$ over its surroundings. It moves radially with velocity $v$ and remains in complete balance of pressure, i.e. $DP = 0$ [see (6.2) and Fig. 6.1]. This gives a local flux of convective energy

$$F_{con} = \varrho v c_P DT \quad , \tag{7.3}$$

which we can take immediately as the correct equation for the average convective flux, if we consider $vDT$ replaced by the proper mean over the whole concentric sphere. One should be aware that this "proper mean" comprises most of the difficulties for a strict treatment. We adopt the following simple model.

All elements may have started their motion as very small perturbations only, i.e. with initial values that can be approximated by $DT_0 = 0$ and $v_0 = 0$. Because of differences in temperature gradients and buoyancy forces, $DT$ and $v$ increase as the element rises (or sinks) until, after moving over a distance $\ell_m$, the element mixes with the surroundings and loses its identity. $\ell_m$ is called the *mixing length*. The elements passing at a given moment through a sphere of constant $r$ will have different values of $v$ and $DT$, since they have started their motion at quite different distances, from zero to $\ell_m$. We assume, therefore, that the "average" element has moved $\ell_m/2$ when passing through the sphere. Then

$$\frac{DT}{T} = \frac{1}{T} \frac{\partial(DT)}{\partial r} \frac{\ell_m}{2}$$
$$= (\nabla - \nabla_e) \frac{\ell_m}{2} \frac{1}{H_P} \quad . \tag{7.4}$$

49

The density difference [for $DP = D\mu = 0$, see (6.3,5)] is simply $D\varrho/\varrho = -\delta\,DT/T$ and the (radial) buoyancy force (per unit mass), $k_r = -g \cdot D\varrho/\varrho$. On average half of this value may have acted on the element over the whole of its preceding motion $(\ell_m/2)$, such that the work done is

$$\frac{1}{2}\,k_r\frac{\ell_m}{2} = g\delta(\nabla - \nabla_e)\frac{\ell_m^2}{8H_P} \quad . \tag{7.5}$$

Let us suppose that half of this work goes into the kinetic energy of the element $(v^2/2$ per unit mass), while the other half is transferred to the surroundings, which have to be "pushed aside". Then we have for the average velocity $v$ of the elements passing our sphere

$$v^2 = g\delta(\nabla - \nabla_e)\frac{\ell_m^2}{8H_P} \quad . \tag{7.6}$$

Inserting this and (7.4) into (7.3), we obtain for the average convective flux

$$F_{\text{con}} = \varrho c_P T \sqrt{g\delta}\,\frac{\ell_m^2}{4\sqrt{2}}\,H_P^{-3/2}(\nabla - \nabla_e)^{3/2} \quad . \tag{7.7}$$

Finally we shall consider the change of temperature $T_e$ inside the element (diameter $d$, surface $S$, volume $V$) when it moves with velocity $v$. This change has two causes, one being the adiabatic expansion (or compression), the other being the radiative exchange of energy with the surroundings. The total energy loss $\lambda$ per unit time is given by (6.22); the corresponding temperature decrease per unit length over which the element rises is $\lambda/\varrho V c_P v$, and the total change per unit length is then

$$\left(\frac{dT}{dr}\right)_e = \left(\frac{dT}{dr}\right)_{\text{ad}} - \frac{\lambda}{\varrho V c_P v} \quad . \tag{7.8}$$

Multiplying this by $H_P/T$, we have

$$\nabla_e - \nabla_{\text{ad}} = \frac{\lambda H_P}{\varrho V c_P v T} \quad . \tag{7.9}$$

Here $\lambda$ may be replaced by (6.22), with the average $DT$ given by (7.4). The resulting equation then contains a "form factor" $\ell_m S/Vd$, which would be $6/\ell_m$ for a sphere of diameter $\ell_m$. In the literature one often finds

$$\frac{\ell_m S}{Vd} \approx \frac{9/2}{\ell_m} \quad , \tag{7.10}$$

which we will use in the following.

Equation (7.9), with the help of (6.22) and (7.10), then becomes

$$\frac{\nabla_e - \nabla_{\text{ad}}}{\nabla - \nabla_e} = \frac{6acT^3}{\kappa\varrho^2 c_P \ell_m v} \quad . \tag{7.11}$$

Let us now summarize what we have achieved and describe what is still lacking. To start with the latter, we have obviously not yet used any physics that could *determine* the mixing length $\ell_m$. Since we do not know a reasonable approach for this, we shall simply treat $\ell_m$ as a free parameter and make (more or less) plausible assumptions for its value. (This is typical for all versions of the mixing-length approach and in fact also for many others that seem to be less arbitrary at a first glance.) In any case, the heat transfer mainly operates via the largest possible elements and they can scarcely move over much more than their own diameter before differential forces destroy their identity.

Now, however, the prospect looks quite favourable: we have obtained the five equations (7.1,2,6,7,11); which we can solve for the five quantities $F_{rad}$, $F_{con}$, $v$, $\nabla_e$, and $\nabla$, if the usual local quantities ($P$, $T$, $\varrho$, $l$, $m$, $c_P$, $\nabla_{ad}$, $\nabla_{rad}$, and $g$) are given.

## 7.2 Dimensionless Equations

For a simpler treatment of the five equations obtained from the mixing-length theory we define two dimensionless quantities:

$$U := \frac{3acT^3}{c_P \varrho^2 \kappa \ell_m^2} \sqrt{\frac{8H_P}{g\delta}} \quad , \tag{7.12}$$

$$W := \nabla_{rad} - \nabla_{ad} \quad . \tag{7.13}$$

The meaning of $U$ will become clear later; that of $W$ is obvious. Note that both can be calculated immediately for any point in the star when the usual variables and the mixing length $\ell_m$ are given.

If $v$ is eliminated with the help of (7.6), then (7.11) becomes

$$\nabla_e - \nabla_{ad} = 2U\sqrt{\nabla - \nabla_e} \quad . \tag{7.14}$$

Eliminating $F_{rad}$, $F_{con}$ from (7.1, 2, 7) and using (2.4) and (6.8) we arrive at

$$(\nabla - \nabla_e)^{3/2} = \frac{8}{9}U(\nabla_{rad} - \nabla) \quad . \tag{7.15}$$

We have thus replaced the set of five equations by the two equations (7.14, 15) for $\nabla$ and $\nabla_e$; and we will now even reduce them to one final equation.

Rewriting the left-hand side of (7.14) as $(\nabla - \nabla_{ad}) - (\nabla - \nabla_e)$, one sees immediately that this is a quadratic equation for $(\nabla - \nabla_e)^{1/2}$ with the solution

$$\sqrt{\nabla - \nabla_e} = -U + \xi \quad , \tag{7.16}$$

where $\xi$ is a new variable given by the positive root of

$$\xi^2 = \nabla - \nabla_{ad} + U^2 \quad . \tag{7.17}$$

In (7.15) we insert (7.16) on the left-hand side, eliminate $\nabla$ on the right-hand side with (7.17), and obtain

$$(\xi - U)^3 + \frac{8U}{9}(\xi^2 - U^2 - W) = 0 \quad . \tag{7.18}$$

So we have arrived at a cubic equation for $\xi$ that can be solved for any given set of parameters $U$ and $W$. It turns out that (7.18) has only one real solution. The resulting $\xi$, together with (7.17), then gives the decisive quantity $\nabla$, i.e. the average temperature gradient to which the layer settles in the presence of convection.

Other characteristic quantities of the convection are then also easily calculable, for example the velocity $v$ from (7.6, 14).


## 7.3 Limiting Cases, Solutions, Discussion

For a given difference $W = \nabla_{\text{rad}} - \nabla_{\text{ad}}$, the convection depends decisively on the value of $U$. Let us write (7.2) as $F_{\text{rad}} = \sigma_{\text{rad}}\nabla$, and (7.7) as $F_{\text{con}} = \sigma_{\text{con}}(\nabla - \nabla_{\text{e}})^{3/2}$. Then $U$, defined in (7.12), is essentially the ratio of the "conductivities": $\sigma_{\text{rad}}/\sigma_{\text{con}}$.

The dimensionless quantity $U$ can also be written in terms of the time $\tau_{\text{ff}}$ it takes a mass element to fall freely over the distance $H_P$. With $\tau_{\text{ff}} = (2H_P/g)^{1/2}$ and (6.25) we have

$$U \approx \frac{\tau_{\text{ff}}}{\tau_{\text{adj}}} \frac{d^2}{\ell_{\text{m}}^2} \quad , \tag{7.19}$$

where we have ignored a factor $3/(8\delta^{1/2})$, which is of order 1. One normally assumes that $\ell_{\text{m}} \approx d$, and therefore $U \approx \tau_{\text{ff}}/\tau_{\text{adj}}$.

The quantity $U$ is also related to another dimensionless quantity $\Gamma$ defined by

$$\Gamma := \frac{(\nabla - \nabla_{\text{e}})^{1/2}}{2U} = \frac{\nabla - \nabla_{\text{e}}}{\nabla_{\text{e}} - \nabla_{\text{ad}}} \quad , \tag{7.20}$$

where we have made use of (7.14). Numerator and denominator have simple meanings as can easily be shown. For a roughly spherical convective element of radius $\ell_{\text{m}}/2$, cross-section $A$, volume $V$, lifetime $\tau_l = \ell_{\text{m}}/v$, and thermal energy $e_{\text{th}} = \varrho V c_P T$, one finds from (7.3,4) that

$$\nabla - \nabla_{\text{e}} = \frac{(F_{\text{con}}A)\tau_l}{e_{\text{th}}} \frac{4H_P}{3\ell_{\text{m}}} \tag{7.21}$$

and from (7.9) that

$$\nabla_{\text{e}} - \nabla_{\text{ad}} = \frac{\lambda \tau_l}{e_{\text{th}}} \frac{H_P}{\ell_{\text{m}}} \tag{7.22}$$

and therefore

$$\Gamma = \frac{4}{3} \frac{F_{\text{con}} A}{\lambda} \approx \frac{\text{energy transported}}{\text{energy lost}} \quad . \tag{7.23}$$

For an average element, $\Gamma$ gives the convective energy flowing through $A$ relative to the radiative energy loss per second. It is a measure for the *efficiency of convection*. Large values of $\Gamma$ (small $U$) are typical for very dense matter, where radiation losses are relatively unimportant compared to the convective flux. In regions of small density, however, the radiative losses can be so large that even very violent movements are ineffective for energy transport; the elements then lose nearly all of their excess heat through radiation to the surroundings, and cool down to $DT \approx 0$. In this case $\Gamma$ is very small (i.e. $U$ is very large). The meaning of $\Gamma$ can also be represented in terms of two typical time-scales for the elements, namely lifetime and adjustment time: in the second equation (6.25) replace $DT$ by (7.4) and solve for $\nabla - \nabla_e$. This expression is then divided by (7.22) giving

$$\Gamma = \frac{\nabla - \nabla_e}{\nabla_e - \nabla_{ad}} = 2\frac{\tau_{adj}}{\tau_l} \quad . \tag{7.24}$$

Let us consider the limiting cases for very large and very small $U$ (or $\Gamma$). One should keep in mind that all gradients are finite; except for $\nabla_{rad}$ they are all smaller than unity. And for the discussion in terms of $\Gamma$ one can easily rewrite (7.14,15) with the help of (7.20).

$\underline{U \to 0}$ (or $\underline{\Gamma \to \infty}$): Equation (7.14) gives $\nabla_e \to \nabla_{ad}$, and thus (7.15) yields $\nabla \to \nabla_{ad}$. A negligible excess of $\nabla$ over the adiabatic value is sufficient to transport the whole luminosity. This is the case in the very dense central part of a star. Here we do not need to solve the mixing-length equations ($\nabla = \nabla_{ad}$ is known), and the uncertainties of this theory do not arise.

$\underline{U \to \infty}$ (or $\underline{\Gamma \to 0}$): In (7.15), the gradients on the left-hand side must be finite, and therefore on the right-hand side $\nabla \to \nabla_{rad}$. Convection is ineffective and cannot transport a substantial fraction of the luminosity. Therefore $F \to F_{rad}$, and the gradient $\nabla$ is again known without further calculations. This is the case near the photosphere of a star.

The situation is difficult where the two limiting cases do not apply, for example in the upper part of an outer convective envelope. There the equations of the mixing-length theory have to be solved, and they will yield a value for $\nabla$ somewhere between $\nabla_{ad}$ and $\nabla_{rad}$, the convection being said to be *superadiabatic*.

The following gives a more detailed discussion of the solutions of (7.18), which depend strongly on the (given) parameters $U$ and $W$. We illustrate them in a diagram, where $\lg W$ is plotted over $\lg U$ (Fig. 7.1).

Instead of using the variable $\xi$, the solutions may be discussed in terms of the over-adiabaticity

$$x := \nabla - \nabla_{ad} = \xi^2 - U^2 \quad , \tag{7.25}$$

which describes the gradient $\nabla$ of the stratification relative to the (known) adiabatic gradient. With this definition, the cubic equation (7.18) is transformed to

$$\left[\sqrt{x + U^2} - U\right]^3 + \frac{8}{9}U(x - W) = 0 \quad . \tag{7.26}$$

53

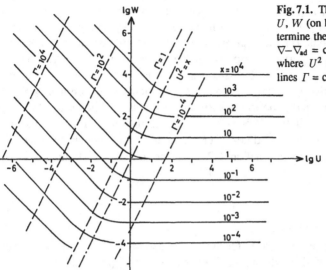

Fig. 7.1. The plane of the parameters $U$, $W$ (on logarithmic scales) that determine the convection. The lines $x = \nabla - \nabla_{ad} = $ constant are solid; the line where $U^2 = x$ is dot-dashed. Some lines $\Gamma = $ constant are dashed

(1) $\underline{\Gamma = 1}$: Let us first derive the line which separates the regimes of effective convection (at small $U$) and ineffective convection (at large $U$). Equation (7.20) for $\Gamma = 1$ is introduced into (7.16), which gives $\xi = 3U$ such that from (7.25) we have $x = 8U^2$. Inserting this into (7.26), we find the condition for $\Gamma = 1$ to be

$$W = 17 \, U^2 \quad . \tag{7.27}$$

The corresponding straight line $\lg W = 2 \, \lg U + 1.23$ is shown by dashes in Fig. 7.1. (Lines for other values of $\Gamma$ are obtained by a parallel shift.) We will now derive the lines on which $x$ is constant. This is easily done by considering the following two limiting cases.

(2) $\underline{U^2 \gg x}$: In (7.26) the term in square brackets on the left, divided by $U$, goes to zero, and one has

$$x = W \quad . \tag{7.28}$$

Therefore $x = $ constant on straight lines parallel to the abscissa (right part of Fig. 7.1).

(3) $\underline{U^2 \ll x}$: In (7.26) the term in square brackets goes to $x^{3/2} \gg Ux$, such that

$$x^{3/2} = \frac{8}{9} U W \tag{7.29}$$

and $x = $ constant on the lines $\lg W = - \lg U + \lg(9/8) + (3/2) \lg x$ (left part of Fig. 7.1)

Finally we derive the equation for the border between the regimes $U^2 \gg x$ and $U^2 \ll x$.

(4) $\underline{U^2 = x}$: With this condition (7.26) gives

$$W = U^2 \left[ \frac{9}{8} (\sqrt{2} - 1)^3 + 1 \right] \quad . \tag{7.30}$$

The corresponding straight line $\lg W = 2 \, \lg U + 0.033$ (dot-dashed line in Fig. 7.1) is below and parallel to that for $\Gamma = 1$.

The meaning of the different regions in Fig. 7.1 is now quite clear. Below and left of a line of sufficiently small $x$ (say, $x = 10^{-2}$), we have nearly $\nabla = \nabla_{ad}$; above that line the convection is superadiabatic. Not too far to the right of the line $\Gamma = 1$, the efficiency is so small that $\nabla \approx \nabla_{rad}$.

For an estimate for the interior of a star, let us assume an ideal monatomic gas with $\delta = \mu = 1$, $c_P/\Re = 5/2$, and a mixing-length $\ell_m = H_P$. For an average point in a star like the sun we may take $r = R_\odot/2$, $m = M_\odot/2$, $T = 10^7$K, $\kappa = 1$ cm$^2$g$^{-1}$, and $\varrho = 1$ g cm$^{-3}$. Then we obtain $U \approx 10^{-8}$, which is so far to the left in Fig. 7.1 that, for reasonable values of $W = \nabla_{rad} - \nabla_{ad}$ (say between 1 and $10^2$), $\nabla - \nabla_{ad} \approx 10^{-5}$ $\ldots 10^{-4}$. For the central region of the sun, $\varrho$ and $\kappa$ are larger by factors of $10^2$ and 10 respectively. Then $U \approx 10^{-13}$, and (for the same values of $W$) the difference $\nabla - \nabla_{ad}$ is even smaller by a factor $10^3$ or more, i.e. $< 10^{-7}$. The stratification of such convective zones is indeed very close to an adiabatic one and we can simply set $\nabla = \nabla_{ad}$, independent of the uncertainties of the theory. (The situation is difficult only near the interface between convective and radiative zones, where one should have a smooth transition between the two modes of transport).

Convective elements in such dense layers are so effective ($\Gamma \approx 10^6 \ldots 10^9$) that they can transport the whole luminosity with surprisingly little effort. Compared with the surroundings, they only need very small excesses of the $T$ gradient, $D(dT/dr) \approx 10^{-12} \ldots 10^{-10}$ K cm$^{-1}$, and an average temperature excess $DT \approx 10^{-2} \ldots 1$ K; their velocities are typically $v \approx 1 \ldots 100$ m s$^{-1}$ (which is $10^{-6} \ldots 10^{-4}$ times the velocity of sound), and their lifetime is between 1 and $10^2$ days.

In spite of these small velocities, the Reynolds number

$$Re = \frac{v\varrho\ell_m}{\eta} \qquad (7.31)$$

($\eta$ = viscosity) is $\gg 1$, since the flow extends over such a large distance $\ell_m$. The situation is quite different for convection near the surface of the star, where the density is low. This gives small effectivity and positive $\lg U$. Here the cubic equation for $\xi$ (or $x$) has to be solved for each point to find the proper $\nabla$ for that place, and the results are affected by the uncertainties of the theory.

In any case, we use the resulting value of $\nabla$ in the transport equation written in the form

$$\frac{dT}{dm} = -\frac{T}{P}\frac{Gm}{4\pi r^4}\nabla \quad . \qquad (7.32)$$

(Here we have replaced $dP/dm$ by the right-hand side of the hydrostatic equation, since the theory is suitable only for hydrostatic equilibrium.) For convection in the very deep interior, $\nabla = \nabla_{ad}$, where $\nabla_{ad}$ is given by (4.21), while for envelope convection we take $\nabla$ as given by the solution of the mixing-length theory. And we can even take the same equation (7.32) for transport by radiation, if we set $\nabla = \nabla_{rad}$ (compare §5.2).

Aside from the more or less effective (and more or less well-determined) transport of energy, turbulent convection, if it occurs, has a side-effect that is important for the life of the star: it mixes the stellar matter very thoroughly and rapidly compared to other relevant time-scales, and thus it contributes directly to the long-lasting chemical record of the star's history.

# §8 The Chemical Composition

## 8.1 Relative Mass Abundances

The chemical composition of stellar matter is obviously very important, since it directly influences such basic properties as absorption of radiation or generation of energy by nuclear reactions. These reactions in turn alter the chemical composition, which represents a long-lasting record of the nuclear history of the star.

The composition of stellar matter is extremely simple compared to that of terrestrial bodies. Because of the high temperatures and pressures there are no chemical compounds in the stellar interior, and the atoms are for the most part completely ionized. It suffices then to count and keep track of the different types of nuclei.

We denote by $X_i$ that fraction of a unit mass which consists of nuclei of type $i$. This requires that

$$\sum_i X_i = 1 \ . \tag{8.1}$$

The chemical composition of a star at time $t$ is then described, if for the relevant nuclei the functions $X_i = X_i(m, t)$ are given in the interval $[0, M]$ of $m$.

The commonly used particle number per volume, $n_i$, of nuclei with mass $m_i$, is related to the mass abundances by

$$X_i = \frac{m_i n_i}{\varrho} \ . \tag{8.2}$$

Usually one does not need to specify very many $X_i$, because most elements are either too rare or play no relevant role, or their abundances remain constant in time. In fact for many purposes it is even sufficient to specify only the mass fractions of hydrogen, helium, and "the rest" with the notation

$$X \equiv X_{\mathrm{H}} \ , \qquad Y \equiv X_{\mathrm{He}} \ , \qquad Z \equiv 1 - X - Y \ . \tag{8.3}$$

This requires additional conventions about the relative distribution of the elements in $Z$, in particular the amount of C, N, and O, which are important for hydrogen burning.

Young stars throughout, and most stars in their envelopes, contain an overwhelming amount of hydrogen and helium: $X = 0.6\ldots0.7$, $Y = 0.36\ldots0.3$, $Z = 0.04\ldots0.001$.

Of course, nuclear reactions will eventually change this simple picture drastically. For example, if many competing reactions occur simultaneously, or if one is interested in such aspects as isotopic ratios, one may have to specify a large number of different $X_i$. Only if inverse $\beta$ decay, the big equalizer in late stages of evolution, has destroyed all elements does the composition then return to utmost simplicity – just neutrons (§ 36).

The advantages of the use of $m$ instead of $r$ as independent variable become particularly evident when we have to describe the chemical composition. If we took $X_i(r,t)$ instead, any expansion would immediately lead to a change of all the functions $X_i$; this holds, of course, for all functions depending on the chemical composition.

## 8.2 Variation of Composition with Time

### 8.2.1 Radiative Regions

In radiative regions there is no exchange of matter between different mass shells, if we can neglect diffusion. Then the $X_i$ can change only if nuclear reactions create or destroy nuclei of type $i$ in the mass element under consideration.

The frequency of a certain reaction is described by the *reaction rate* $r_{lm}$, i.e. the number of reactions per unit volume and time that transform nuclei from type $l$ into type $m$ (see § 18). In general an element $i$ can be affected simultaneously by many reactions, some of which create it ($r_{ji}$) and some of which destroy it ($r_{ik}$). These reaction rates give directly the change per second of $n_i$. Then, with (8.2), we have

$$\frac{\partial X_i}{\partial t} = \frac{m_i}{\varrho} \left[ \sum_j r_{ji} - \sum_k r_{ik} \right] \quad , \qquad i = 1 \ldots I \tag{8.4}$$

for any of the elements $1 \ldots I$ which are involved in reactions. (If more than one nucleus of type $i$ is created or destroyed per reaction, the corresponding terms in the sums have simply to be normalized by the number of nuclei of type $i$ involved.)

The reaction $p \rightarrow q$ in which one nucleus of type $p$ is transformed may be connected with a release of energy $e_{pq}$. In the equation of energy conservation we have used the energy generation rate $\varepsilon$ per unit mass, which normally contains contributions from several different reactions. The $\varepsilon$ are simply proportional to the reaction rates:

$$\varepsilon = \sum_{p,q} \varepsilon_{pq} = \frac{1}{\varrho} \sum_{p,q} r_{pq} e_{pq} \quad . \tag{8.5}$$

Let us introduce the energy generated when one mass unit of type $p$ nuclei is transformed into type $q$:

$$q_{pq} = \frac{e_{pq}}{m_p} \quad . \tag{8.6}$$

For simple cases it is convenient to rewrite (8.4) in terms of the $\varepsilon$, which already occur in the equation of energy conservation. If all reactions give a positive contribution to $\varepsilon$, then instead of (8.4) we can write

$$\frac{\partial X_i}{\partial t} = \sum_j \frac{\varepsilon_{ji}}{q_{ji}} - \sum_k \frac{\varepsilon_{ik}}{q_{ik}} \quad . \tag{8.7}$$

If $I$ different nuclei are simultaneously subject to nuclear transformations, equations (8.4) or (8.7) form a set of $I$ differential equations. Since one of them can be replaced by the normalization (8.1), we need only $I - 1$ of them to complete the basic equations of our problem.

Note that for simple cases it may even suffice to consider just one of these equations. For example, if hydrogen burning is to be taken into account only by way of an overall generation rate $\varepsilon_H$ (giving the sum over all single reactions), then the only equation needed is

$$\frac{\partial X}{\partial t} = -\frac{\varepsilon_H}{q_H} \tag{8.8}$$

with $\partial Y/\partial t = -\partial X/\partial t$, where $q_H$ is the energy release per unit mass when hydrogen is converted into helium.

In §4.4 we defined the nuclear time-scale for a certain burning, $\tau_n = E_n/L$. One can actually define a nuclear time-scale for each type of nuclear burning, since each nuclear energy reservoir is proportional to an integral of $X_i \cdot dm$ over the whole star, where $X_i$ refers to the element consumed by the reactions; therefore $\tau_n$ is equivalent to $\tau_{X_i}$, the time-scale for the exhaustion of the element $i$.

### 8.2.2 Diffusion

Certain microscopic effects can also change the chemical composition in a star. If gradients occur in the abundances of chemical elements, then *concentration diffusion* tends to smooth out the differences. Even in chemically homogeneous stellar layers, heavier atoms can migrate towards the regions of higher temperature, owing to the effect of *temperature diffusion*. Also, the pressure gradient in a stratified layer causes the heavier particles to diffuse towards the region of higher pressure, i.e. *pressure diffusion*. The detailed statistical theory of diffusion is derived in CHAPMAN, COWLING (1970).

We start with the simplest case: concentration diffusion. Let $c$ be the concentration of particles of a certain species, i.e. the number density of particles of that type divided by the number density of all particles, and $j_D$ be the "flux of concentration"; then Fick's first law states that

$$j_D = -D\nabla c \quad , \tag{8.9}$$

where $D$ is the diffusion coefficient. (We will derive (8.9) later.) With $j_D = cv_D$, where $v_D$ is the diffusion velocity, one has

$$v_D = -\frac{D}{c}\nabla c \quad . \tag{8.10}$$

With the continuity equation

$$\frac{\partial c}{\partial t} = -\nabla \cdot \boldsymbol{j}_D \tag{8.11}$$

we find that

$$\frac{\partial c}{\partial t} = \nabla \cdot (D \nabla c) \quad , \tag{8.12}$$

and in the case of constant $D$ that

$$\frac{\partial c}{\partial t} = D \nabla^2 c \quad . \tag{8.13}$$

A rough estimate for the characteristic time-scale is given by

$$\tau_D \approx \frac{S^2}{D} \quad , \tag{8.14}$$

where $S$ is a characteristic length for the variation of $n$.

By generalizing (8.10) one can formally include the two other types of diffusion, i.e.

$$\boldsymbol{v}_D = -\frac{1}{c} D (\nabla c + k_T \nabla \ln T + k_P \nabla \ln P) \quad , \tag{8.15}$$

if the coefficients $k_T$ and $k_P$ are properly specified. In order to do that we first consider the combined effects of concentration and temperature diffusion.

We assume $\nabla T$ to be perpendicular to the $x$–$y$ plane in a cartesian coordinate system; then the flux of particles of a certain type in the $+z$ direction due to the statistical motion of the particles is determined by the density $n$ and the mean velocity $\overline{v}$, both taken at $z = -\ell$, where $\ell$ is the mean free path of the particles of this type:

$$j^+ = \frac{1}{6} c(-\ell) \overline{v}(-\ell) \quad , \tag{8.16}$$

where the numerical factor originates in averaging over $\cos^2$. This takes into account that the particles penetrating the $x$–$y$ plane had their last encounter at $z = -\ell$.

If one expands $n$ and $\overline{v}$ at $z = 0$ in (8.16) and in a corresponding expression for $j^-$, the fluxes in the $+z$ and $-z$ directions are

$$j^\pm = \frac{1}{6} \left( c(0) \mp \frac{\partial c}{\partial z} \ell \right) \left( \overline{v}(0) \mp \frac{\partial \overline{v}}{\partial z} \ell \right) \quad , \tag{8.17}$$

and therefore there is a net flux

$$j = j^+ - j^- = -\frac{1}{3} \left( \frac{\partial c}{\partial z} \ell \overline{v} + \frac{\partial \overline{v}}{\partial z} \ell c \right) \quad , \tag{8.18}$$

which in general does not vanish, i.e. we have obtained Fick's law.

We now consider the relative diffusion velocity $v_{D_1} - v_{D_2}$ resulting from the motion of two different types of particles $(1, 2)$, with fluxes $j_1, j_2$ and concentrations $c_1, c_2$:

$$v_{D_1} - v_{D_2} = \frac{j_1}{c_1} - \frac{j_2}{c_2} \quad . \tag{8.19}$$

59

With (8.18) we can replace the $j_i$ by $\ell_i$, $\bar{v}_i$, and the gradients of $c_i$, while the velocity gradient — with the help of $\bar{v}_i = (3\Re T/\mu_i)^{1/2}$ — can be replaced by the temperature gradient. Using the continuity equation (and after some algebra) an expression of the form

$$v_{D_1} - v_{D_2} = -\frac{D}{c_1 c_2}\left(\frac{\partial c_1}{\partial z} + k_T\,\frac{\partial \ln T}{\partial z}\right) \tag{8.20}$$

follows. The two terms in the brackets are responsible for concentration diffusion and temperature diffusion. In a mixture of two species ($i = 1, 2$) $D$ and $k_T$ have the form

$$D = \frac{1}{3}(c_2\ell_1\bar{v}_1 + c_1\ell_2\bar{v}_2) = \left(\frac{\Re T}{3}\right)^{1/2}(c_2\ell_1\mu_1^{-1/2} + c_1\ell_2\mu_2^{-1/2}) \quad, \tag{8.21}$$

$$k_T = \frac{1}{2}\,\frac{\ell_1\sqrt{\mu_2} - \ell_2\sqrt{\mu_1}}{\ell_1 c_2\sqrt{\mu_2} + \ell_2 c_1\sqrt{\mu_1}}\,c_1 c_2(c_2 - c_1) \quad, \tag{8.22}$$

where $\ell_1$ and $\ell_2$ are the mean free paths of the two species (LANDAU, LIFSHITZ, vol. 6, 1959). The absolute value $k_T$ is of order 1 or less, and its sign is not immediately clear, though more detailed considerations indicate that $k_T > 0$ for a typical ionized hydrogen–helium mixture in stars.

From (8.21) it is obvious that $D$ is of order

$$D \approx \left(\frac{\Re T}{3}\right)^{1/2}\ell \approx \frac{1}{3}v^*\ell \quad, \tag{8.23}$$

where $v^*$ and $\ell$ are some kind of averages of the statistical velocities and the mean free paths of both components. This expression for $D$ can be used to estimate the time-scale $\tau_D$ according to (8.14). As long as $|k_T| \approx 1$ this also gives the characteristic time-scale for temperature diffusion.

Since $D > 0$, in the case of $k_T > 0$ for pure temperature diffusion one has sign($v_D$) $= -$sign($\partial \ln T/\partial z$). Let us now consider the case of a mixture of hydrogen and helium. Here $v_D = v_H - v_{He}$ is the $z$ component of the diffusion velocity and $v_D > 0$ means that hydrogen diffuses in the direction of lower temperature, i.e. "upwards" in the star. For the central region of the sun ($T \approx 10^7$K, $\varrho \approx 100\,\mathrm{g\,cm^{-3}}$) one finds that $\ell \approx 10^{-8}$ cm and $D \approx 6\,\mathrm{cm^2 s^{-1}}$, and with a characteristic length-scale $S \approx R_\odot \approx 10^{11}$ cm the characteristic time-scale $\tau_D$ (according to (8.14)) there becomes $\tau_D \approx 10^{13}$ years. Despite the fact that $\tau_D$ is much larger than the age of the universe and that therefore the effects of concentration and temperature diffusion are astrophysically irrelevant for the sun, we will briefly discuss the situation. If a layer is homogeneous, then there is no concentration diffusion, but the hydrogen particles diffuse towards the regions of lower temperature. This causes an outward increase of $n_H$ which in turn triggers concentration diffusion acting against the temperature diffusion (sign($\partial c_H/\partial z$) $= -$sign($\partial T/\partial z$)) until both types of diffusion compensate each other.

We now turn to pressure diffusion, which is the cause of what is often called "sedimentation" or "gravitational settling". A statistical consideration similar to that used to make temperature diffusion plausible also shows that there is diffusion in isothermal layers with a non-vanishing pressure gradient. The reader is again referred to CHAPMAN, COWLING (1970). In a way similar to that for $k_T$ an expression for $k_P$ in (8.15) can also be obtained.

We here confine ourselves to the discussion of the final outcome of this process of pressure diffusion, i.e. the state of final equilibrium for an isothermal layer in hydrostatic equilibrium in a gravitational field pointing towards the $-z$ direction. Let us assume that the material consists of two components ($i = 1, 2$) of ideal gases of different molecular weights $\mu_i$ and partial pressures $P_i$. Then there exist two pressure-scale heights $H_{P_i} = -dz/d\ln P_i$ with which (6.8) can be written in the form

$$H_{P_i} = \frac{P_i}{g\varrho_i} = \frac{\Re T}{g\mu_i} \quad , \tag{8.24}$$

where $dP_i/dz = -g\varrho_i$ and $P_i = \Re\varrho_i T/\mu_i$ are used. The particle densities are proportional to the $P_i$, which are here approximately proportional to $\exp(-z/H_{P_i})$. Therefore the component with the higher $\mu_i$ falls off more sharply in the $z$ direction than that with smaller $\mu_i$, so that in a very simplified way one can say that the heavier component has "moved below" the lighter one. This is the final state, which would be brought about by pressure diffusion alone even if the species were originally in a completely mixed state. Of course, in reality the two other types of diffusion would also act and therefore influence the final state.

Estimates show that not only $|k_T|$ but also $|k_P|$ is of order one. Therefore it normally takes rather a long time before an appreciable separation occurs in stars. Although diffusion effects are relevant in certain special cases (see, for instance, ALECIAN, VAUCLAIR, 1983) we will ignore them in this book.

### 8.2.3 Convective Regions

Here we deal with the much more important effect of mixing due to turbulent convective motion, a process that is very rapid compared to the extremely slow change of the chemical composition produced by nuclear reactions. Therefore we can assume that the composition in a convective region always remains homogeneous,

$$\frac{\partial X_i}{\partial m} = 0 \quad . \tag{8.25}$$

This requires a dispersion not only of the newly created nuclei, but of all elements inside a convective zone.

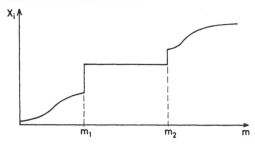

Fig. 8.1. The abundances $X_i$ are smeared out owing to rapid mixing inside a convection zone extending from $m_1$ to $m_2$. At these borders $X_i$ can be discontinuous

Suppose a convective zone extends between the mass values $m_1$ and $m_2$ (Fig. 8.1). Inside that interval all $X_i = \overline{X}_i$ are constant. At the boundaries one can generally have a discontinuity, such that the "outer" values $X_{i1}$ and $X_{i2}$ are different from the "inner" values – which are simply $X_{i1} = X_{i2} = \overline{X}_i$. But $m_1$ and $m_2$ can change in time, and hence one can easily see that the abundances in the convective zone vary with the rate

$$
\frac{\partial \overline{X}_i}{\partial t} =
$$

(8.26)

$$
\frac{1}{m_2 - m_1} \left( \int_{m_1}^{m_2} \frac{\partial X_i}{\partial t} dm + \frac{\partial m_2}{\partial t}(X_{i2} - \overline{X}_i) - \frac{\partial m_1}{\partial t}(X_{i1} - \overline{X}_i) \right) .
$$

The $X_{i1}$, $X_{i2}$ should here be taken as the value on the side that the corresponding boundary moves towards. The integral in the bracket describes the change due to nuclear reactions and can be replaced by an integral over the rates $-\varepsilon_i / q_i$, as in (8.8), where $q_i$ is the energy released if a mass unit of the nucleus $i$ is transformed. Without any nuclear reaction ($\partial X_i / \partial t = 0$) in the convective zone, its composition can still change if the boundaries move into a region of inhomogeneous composition, and this can have important consequences. For example, "ashes" of earlier nuclear burnings may be brought to the surface, fresh fuel may be carried into a zone of nuclear burning, or discontinuities can be produced that drastically influence the later evolution.

# II   The Overall Problem

# §9 The Differential Equations of Stellar Evolution

## 9.1 The Full Set of Equations

Collecting the basic differential equations for a spherically symmetric star derived in Chap. I, we are then led by (1.6), (2.16), (4.27, 28), (7.32), and (8.4) to:

$$\frac{\partial r}{\partial m} = \frac{1}{4\pi r^2 \varrho} \quad , \tag{9.1}$$

$$\frac{\partial P}{\partial m} = -\frac{Gm}{4\pi r^4} - \frac{1}{4\pi r^2} \frac{\partial^2 r}{\partial t^2} \quad , \tag{9.2}$$

$$\frac{\partial l}{\partial m} = \varepsilon_n - \varepsilon_\nu - c_P \frac{\partial T}{\partial t} + \frac{\delta}{\varrho} \frac{\partial P}{\partial t} \quad , \tag{9.3}$$

$$\frac{\partial T}{\partial m} = -\frac{GmT}{4\pi r^4 P} \nabla \quad , \tag{9.4}$$

$$\frac{\partial X_i}{\partial t} = \frac{m_i}{\varrho} \left( \sum_j r_{ji} - \sum_k r_{ik} \right) \quad , \qquad i = 1, \dots, I \quad . \tag{9.5}$$

In (9.5) we have a set of $I$ equations (one of which may be replaced by the normalization $\sum_i X_i = 1$) for the change of the mass fractions $X_i$ of the relevant nuclei $i = 1, \dots, I$ having masses $m_i$. An additional formula (8.26) regulates the mixing of the composition in convective regions. In (9.3), $\delta \equiv -(\partial \ln \varrho / \partial \ln T)_P$, and in (9.4), $\nabla \equiv d \ln T / d \ln P$. If the energy transport is due to radiation (and conduction), then $\nabla$ has to be replaced by $\nabla_{\text{rad}}$, which is given by (5.28):

$$\nabla = \nabla_{\text{rad}} = \frac{3}{16\pi acG} \frac{\kappa l P}{m T^4} \quad . \tag{9.6}$$

If the energy is carried by convection, then $\nabla$ in (9.4) has to be replaced by a value obtained from a proper theory of convection; this may be $\nabla_{\text{ad}}$ in the deep interior, or obtained from a solution of the cubic equation (7.26) for superadiabatic convection in the outer layers. Note that the expression on the right-hand side of (9.4) assumes hydrostatic equilibrium. This does not matter in the case of radiative transport, since the local adjustment time of the radiation field is very short, and the convection theory of §7 is valid only for stars in hydrostatic equilibrium. Otherwise another convection theory valid in rapidly changing regions would have to be used. Additional criteria such as (6.12, 13) distinguish between radiative and convective transport.

In the system (9.1-5) one can distinguish certain subsystems, i.e. equations (9.1,2) give the mechanical part, being coupled to the thermo-energetic part only through the density $\varrho$ – which usually also depends on $T$. If for some reason or other this dependence of $\varrho$ on $T$ is not present (or can be eliminated), then (9.1,2) can be solved regardless of the other equations to give the mechanical structure $r(m)$, $P(m)$. Equations (9.5) may be regarded as the chemical part. Under normal conditions ($\tau_n$ much larger than the other time-scales, see §9.2) they can be decoupled from the spatial parts (9.1 – 4), which describe the structure of the star for a given time and given composition $X_i(m)$. This would be questionable, of course, if the chemical composition changed as rapidly as the other variables; and for changes of $X_i(m)$ more rapid than those of $P$, $T$ one would rather assume to have an "equilibrium composition" $X_i(P, T)$ at any time (see §34).

Equations (9.1-5) contain functions which describe properties of the stellar material such as $\varrho$, $\varepsilon_n$, $\varepsilon_\nu$, $\kappa$, $c_P$, $\nabla_{ad}$, $\delta$, and the reaction rates $r_{ij}$. We shall deal with these functions in Part III. Meanwhile we assume them to be known functions of $P$, $T$, and the chemical composition described by the functions $X_i(m, t)$. We therefore have an equation of state

$$\varrho = \varrho(P, T, X_i) \tag{9.7}$$

and equations for the other thermodynamic properties of the stellar matter

$$c_P = c_P(P, T, X_i) \quad , \tag{9.8}$$

$$\delta = \delta(P, T, X_i) \quad , \tag{9.9}$$

$$\nabla_{ad} = \nabla_{ad}(P, T, X_i) \quad , \tag{9.10}$$

as well as the Rosseland mean of the opacity (including conduction)

$$\kappa = \kappa(P, T, X_i) \quad , \tag{9.11}$$

and the nuclear reaction rates and the energy production and energy loss via neutrinos:

$$r_{jk} = r_{jk}(P, T, X_i) \quad , \tag{9.12}$$

$$\varepsilon_n = \varepsilon_n(P, T, X_i) \quad , \tag{9.13}$$

$$\varepsilon_\nu = \varepsilon_\nu(P, T, X_i) \quad . \tag{9.14}$$

In these equations the arguments $X_i$ stand for *all* types of nuclei ($i = 1, \ldots, I$).

It is now time to count the equations and the unknown variables. We consider the material functions on the right-hand sides of (9.1-5) to be replaced with the help of the corresponding equations (9.7-14), i.e. by functions of $P$, $T$, $X_i$. For $I$ different types of nuclei being affected by reactions, (9.1-5) form a set of $4 + I$ differential equations for the $4 + I$ variables $r$, $P$, $T$, $l$, $X_1$, $\ldots$, $X_I$. We therefore have the same number of equations and unknown variables.

The independent variables are $m$ and $t$. If we assume that the total mass of the star does not change in time (i.e. no gain nor loss of mass) and if we define the time at which evolution starts as $t = t_0$, then we are looking for solutions in the intervals

$$0 \leq m \leq M \quad , \qquad t \geq t_0 \quad . \tag{9.15}$$

In the full problem we are confronted with a set of non-linear, partial differential equations. As usual, physically relevant solutions require the specification of boundary conditions (here at $m = 0$, $m = M$) and of initial values [for example $X_i(m, t_0)$]. The boundary conditions will be dealt with in §10. In order to see more clearly which initial values have to be specified we replace the two terms with time derivatives of $P$ and $T$ in (9.3) by one term containing the change of the entropy $s$, $-T \partial s / \partial t$, according to (4.27). Obviously the full problem requires specification of the functions $r(m, t_0)$, $\dot{r}(m, t_0)$, $s(m, t_0)$, and $X_i(m, t_0)$.

After proper initial values and boundary conditions are specified, together with the stellar mass $M$, the problem is to find solutions of the basic equations, i.e. the unknown variables as functions of $m$ and $t$. A solution $r(m)$, $P(m)$, ..., $X_i(m)$ for a given time $t$ in the interval $[0, M]$ is called a *stellar model*. But before we discuss in more detail how solutions of our set of differential equations can be obtained, we first discuss simplifications of the full problem.

## 9.2 Time-scales and Simplifications

There are three types of time derivatives in our set of equations. To each of them belongs a certain characteristic time-scale. In §2.4 the term $(\partial^2 r / \partial t^2)/4\pi r^2$ in (9.2) was used to derive $\tau_{\text{hydr}}$. From the time derivatives in (9.3) we have derived $\tau_{\text{KH}}$ in §3.3. The time derivatives in (9.5) define chemical time-scales $\tau_{Xi}$ which were shown to be equivalent to $\tau_{\text{n}}$ [see (4.39)] at the end of §8.2.1.

In §2.4 we showed that the inertia term in (9.2) can be neglected if the evolution is slow compared to $\tau_{\text{hydr}}$. Therefore, if the evolution of a star is governed by thermal adjustment or by nuclear reactions, (9.2) can be replaced by the equation of hydrostatic equilibrium

$$\frac{\partial P}{\partial m} = -\frac{Gm}{4\pi r^4} \quad , \tag{9.16}$$

since $\tau_{\text{KH}} \gg \tau_{\text{hydr}}$ and $\tau_{\text{n}} \gg \tau_{\text{hydr}}$. The star then evolves along a sequence of states of hydrostatic equilibrium. As initial conditions, the functions $s(m, t_0)$ and $X_i(m, t_0)$ have to be specified in this approximation.

If the star evolves on the time-scale $\tau_{\text{n}} \gg \tau_{\text{KH}}$, then according to the discussion in §4.4 the time derivatives in the energy equation can also be neglected and (9.3) is reduced to

$$\frac{\partial l}{\partial m} = \varepsilon_{\text{n}} - \varepsilon_{\nu} \quad . \tag{9.17}$$

The star now evolves along a sequence of states in which it is not only in hydrostatic equilibrium but also thermally adjusted. We call this *complete* (mechanical and thermal) *equilibrium*. The only initial values to be given in this case are the $X_i(m, t_0)$.

In complete equilibrium the basic equations split into two parts: the "structure equations" (9.1, 16, 17, 4) contain only spatial derivatives while the "chemical equations" (9.5) contain only time derivatives. Therefore, if at a certain time $t = t_0$ the $X_i(m, t_0)$ are given, the structure equations can be taken as a set of four *ordinary* differential equations describing the structure of the star at $t_0$.

Complete equilibrium is a good approximation for stars in many important evolutionary phases, for example the stars on the main sequence.

# § 10 Boundary Conditions

As usual in mathematical physics, the boundary conditions constitute a serious part of the whole problem, and their influence on the solutions is not easy to foresee. This is connected with the fact that the boundary conditions for the problem of stellar structure cannot be imposed at one end of the interval $[0, M]$ only, but rather are split into some that are given at the centre and some near the surface of the star. The central conditions are simple, whereas the surface conditions implicate observable quantities and a completely different, much more complicated transport equation. It is therefore advisable to get some feeling about their influence on the stellar structure. We discuss these problems for the case of complete equilibrium.

## 10.1 Central Conditions

Two boundary conditions can be immediately written down for the centre, defined by $m = 0$. Since the density $\varrho$ must go to a reasonable, finite, and non-vanishing value (there can be no singularity and no cavity in the centre), we must have $r = 0$. And since the energy sources also remain finite (positive or negative), $l$ must vanish at the centre as well:

$$m = 0 \quad : \qquad r = 0 \quad , \qquad l = 0 \quad . \tag{10.1}$$

This was the simple part. Unfortunately nothing is a priori known about the central values of pressure $P_c$ and temperature $T_c$, so the conditions (10.1) still allow a two-parameter set of solutions, obtained by outward integrations starting with arbitrary $P_c$, $T_c$, and $r = l = 0$.

It is useful to know the behaviour of the four functions $r$, $l$, $P$, $T$ in the vicinity of the centre, $m \to 0$, for a given time $t = t_0$. The equation of continuity (9.1) may be written as

$$d\left(r^3\right) = \frac{3}{4\pi\varrho}dm \quad , \tag{10.2}$$

which can be integrated for constant $\varrho = \varrho_c$, i.e. for small enough values of $m$ and $r$, giving

$$r = \left(\frac{3}{4\pi\varrho_c}\right)^{1/3} m^{1/3} \quad . \tag{10.3}$$

This can be considered the first term in a series expansion of $r$ around $m = 0$. A corresponding integration of the energy equation (9.3) yields

$$l = (\varepsilon_n - \varepsilon_\nu + \varepsilon_g)_c \, m \quad . \tag{10.4}$$

In both cases we have used the proper boundary conditions (10.1) by taking the integration constants to be zero.

Eliminating $r$ for small values of $m$ by (10.3), we obtain from the hydrostatic equation (9.16)

$$\frac{dP}{dm} = -\frac{G}{4\pi} \left(\frac{4\pi \varrho_c}{3}\right)^{4/3} m^{-1/3} \quad , \tag{10.5}$$

which can be integrated to yield

$$P - P_c = -\frac{3G}{8\pi} \left(\frac{4\pi}{3} \varrho_c\right)^{4/3} m^{2/3} \quad . \tag{10.6}$$

The pressure gradient must of course vanish at the centre, which can be seen by writing the hydrostatic equation (2.4) in the form

$$\frac{dP}{dr} \sim \frac{m}{r^2} \sim \frac{r^3}{r^2} \to 0 \tag{10.7}$$

for $r \to 0$.

The variation of temperature will first be considered in the radiative case, for which (5.12) requires that

$$\frac{dT}{dm} = -\frac{3}{64\pi^2 ac} \frac{\kappa l}{r^4 T^3} \quad . \tag{10.8}$$

With $P \to P_c$, $T \to T_c$, $\kappa$ tends to some well-defined value $\kappa_c$. Replacing $l(\sim m)$ by (10.4) and $r(\sim m^{1/3})$ by (10.3) now, we can integrate (10.8) for small values of $m$ and obtain the first equation (10.9). In the case of (adiabatic) convection we start from (7.32) with $\nabla = \nabla_{ad}$ and replace $r$ by (10.3). An integration for small values of $m$ then gives the second equation (10.9):

$$T^4 - T_c^4 = -\frac{1}{2ac}\left(\frac{3}{4\pi}\right)^{2/3} \kappa_c (\varepsilon_n - \varepsilon_\nu + \varepsilon_g)_c \, \varrho_c^{4/3} m^{2/3} \quad \text{(radiative)} \quad ,$$

$$\ln T - \ln T_c = -\left(\frac{\pi}{6}\right)^{1/3} G \frac{\nabla_{ad,c}\varrho_c^{4/3}}{P_c} m^{2/3} \quad \text{(convective)} \quad . \tag{10.9}$$

## 10.2 Surface Conditions

The strict surface conditions are rather complicated and unwieldy. For rough estimates one might therefore prefer to use a crude approximation, provided that it is simple.

An extreme step in this direction would be to take the naïve "zero-conditions"

$$m \to M \quad : \quad P \to 0 \quad , \quad T \to 0 \quad . \tag{10.10}$$

These at least reflect correctly the fact that, in the outermost region of the star, $P$ and $T$ go to very small values compared to those in the interior. But, of course, in reality there is a gradual and rather extended transition to the finite values of $P$, $T$ of the diffuse interstellar medium.

The next step is to find a sphere that we can reasonably call the "surface" of the star and that defines the total stellar radius $r = R$. The theory of stellar atmospheres suggests the use of the *photosphere*, from where the bulk of the radiation is emitted into space, and which is found where the optical depth $\tau$ of the overlying layers,

$$\tau := \int_R^\infty \kappa \varrho \, dr = \overline{\kappa} \int_R^\infty \varrho \, dr \quad , \tag{10.11}$$

is equal to 2/3. Here we have defined a mean opacity $\overline{\kappa}$, averaged over the stellar atmosphere. In hydrostatic equilibrium the pressure at this level is given by the weight of the matter above. We can well approximate the gravitational acceleration by the constant value $g_0 = GM/R^2$, since the bulk of the matter in these layers is anyway very close to the photosphere. Then

$$P_{r=R} = \int_R^\infty g \varrho \, dr = g_0 \int_R^\infty \varrho \, dr \quad , \tag{10.12}$$

and if we eliminate here the integral over $\varrho$ by that in the second equation (10.11), we find with $\tau = 2/3$ that

$$P_{r=R} = \frac{GM}{R^2} \frac{2}{3} \frac{1}{\overline{\kappa}} \quad . \tag{10.13}$$

The temperature at the photosphere is equal to the *effective temperature* $T_{r=R} = T_{\text{eff}}$ of the star defined by

$$L = 4\pi R^2 \sigma \, T_{\text{eff}}^4 \quad . \tag{10.14}$$

Here $\sigma = ac/4$ is the Stefan–Boltzmann constant of radiation. $T_{\text{eff}}$ is thus the temperature of that black body which yields the same surface flux of energy as the star.

The *photospheric conditions* (10.13, 14) represent two relations between the surface values ($m \to M$) of the functions $P$, $T$, $r$, $l$. They are certainly a better approximation for the surface conditions than (10.10). Their severest defect is that they refer to a level where the assumption made for deriving the transport equation (5.12) (small mean free path of the photons) breaks down. At this level, one should use the more complicated transport equation for stellar atmospheres.

Quite generally the correct surface conditions can be formulated as follows: the interior solution should fit smoothly to a solution of the stellar-atmosphere problem. Let us put this into a more mathematical form.

The transition between interior and outer (atmospheric) solutions is made at a certain mass value $m_F$, the "fitting mass", which should be far enough in to ensure that the interior equations are still valid there. On the other hand, $m_F$ should still be close enough to $M$ that, for simplicity, we can always use thermally adjusted outer solutions with constant $l = L$. The smaller $M - m_F$, the less energy can be stored or released in these outer layers.

For the stellar-interior problem we consider the mass $M$ and the chemical composition to be given. The theory of stellar atmospheres tells us that for given $M$ and $X_i(m)$ there is a two-parameter set of possible atmospheric solutions, the parameters being, for example, $R$ and $T_{\mathrm{eff}}$, or $R$ and $L$ [which are connected by (10.14)]. Any one of these possible atmospheric solutions can be extended by integration downwards to $m_F$ and may yield there the four "exterior" values $r = r_F^{\mathrm{ex}}$, $P = P_F^{\mathrm{ex}}$, $T = T_F^{\mathrm{ex}}$, $l = l_F^{\mathrm{ex}} = L$.

The outer boundary conditions now require for $m = m_F$ that one quartet $r_F^{\mathrm{ex}}$, ..., $l_F^{\mathrm{ex}}$ obtained from an outer solution has to match the corresponding values $r_F^{\mathrm{in}}$, ..., $l_F^{\mathrm{in}}$ of the interior solution, which extends from the centre to $m_F$:

$$r_F^{\mathrm{ex}} = r_F^{\mathrm{in}} \quad , \quad P_F^{\mathrm{ex}} = P_F^{\mathrm{in}} \quad , \quad T_F^{\mathrm{ex}} = T_F^{\mathrm{in}} \quad , \quad l_F^{\mathrm{ex}} = l_F^{\mathrm{in}} \quad . \tag{10.15}$$

These four simultaneous fits are in principle possible, since the solutions have enough degrees of freedom: the interior solution has two (we can vary the central values $P_c$ and $T_c$), and the outer solution also has two (variation of $R$ and $L$). The fact that both solutions have two degrees of freedom is reflected in the following alternative representation, which is often used in numerical computations. Imagine that many outer integrations are carried out for many pairs of parameters $R$ and $L$. At $m = m_F$ they yield the four functions $r_F^{\mathrm{ex}}(R, L)$, $P_F^{\mathrm{ex}}(R, L)$, $T_F^{\mathrm{ex}}(R, L)$, $l_F^{\mathrm{ex}}(R, L)$. The last one is very simple, namely $l_F^{\mathrm{ex}} = L$. The first one is certainly well-behaved and we can invert it without complications, obtaining $R = R(r_F^{\mathrm{ex}}, L)$. This is now used to replace the argument $R$ in the functions $P_F^{\mathrm{ex}}$ and $T_F^{\mathrm{ex}}$, which can then be considered known functions $\pi$ and $\theta$ of $r_F^{\mathrm{ex}}$ and $l_F^{\mathrm{ex}} = L$:

$$\begin{aligned} P_F^{\mathrm{ex}}\left(R\left(r_F^{\mathrm{ex}}, L\right), L\right) &:= \pi\left(r_F^{\mathrm{ex}}, L\right) \quad , \\ T_F^{\mathrm{ex}}\left(R\left(r_F^{\mathrm{ex}}, L\right), L\right) &:= \theta\left(r_F^{\mathrm{ex}}, L\right) \quad . \end{aligned} \tag{10.16}$$

For any given pair $r_F^{\mathrm{ex}}$, $L$, the $\pi$ and $\theta$ give the corresponding values of pressure and temperature for one outer solution. We now replace the variables $P_F^{\mathrm{ex}}$, ..., $l_F^{\mathrm{ex}} = L$ in (10.16) by $P_F^{\mathrm{in}}$, ..., $l_F^{\mathrm{in}}$, using the fit conditions (10.15):

$$P_F^{\mathrm{in}} = \pi\left(r_F^{\mathrm{in}}, L\right) \quad , \quad T_F^{\mathrm{in}} = \theta\left(r_F^{\mathrm{in}}, L\right) \quad . \tag{10.17}$$

These are the outer boundary conditions for the interior solution. Obviously, if these are fulfilled, there is always an outer solution that continuously matches the interior solution. We can now drop the distinction between the variables of the exterior and interior solutions at $m = m_F$ expressed in the superscripts "ex" and "in".

The fulfilment of the boundary conditions is illustrated in Fig. 10.1, where the functions $\pi$ and $\theta$ (obtained from outer solutions) are sketched over the $r_F$–$L$ plane. We have also indicated the surfaces $\tilde{\pi}(r_F, L)$ and $\tilde{\theta}(r_F, L)$, which give the corresponding functions of the *interior* solutions obtained by varying $P_c$ and $T_c$. The intersection of the surfaces ($\pi = \tilde{\pi}$ and $\theta = \tilde{\theta}$) gives the matches of $P_F$ or of $T_F$ respectively. We project the intersections into the $r_F$–$L$ plane (dot-dashed lines), and where these projections intersect we have the desired match of all four variables.

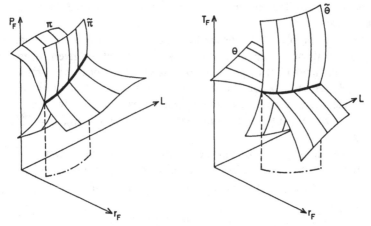

**Fig. 10.1 a, b.** The function values $P_F$ (or $T_F$) at the fitting mass $m = M_F$ are plotted over $r_F$ and $L$. The surface $\pi$ (or $\theta$) contains the values obtained by all possible integrations downwards from the photosphere. The surface $\tilde{\pi}$ (or $\tilde{\theta}$) contains the corresponding values obtained from all possible integrations outwards from the centre. The heavy line shows the intersection of $\pi$ and $\tilde{\pi}$ (or $\theta$ and $\tilde{\theta}$), the dot-dashed line the projection of this intersection into the $r_F$-$L$ plane. (All surfaces are freely invented sketches)

## 10.3 Influence of the Surface Conditions and Properties of Envelope Solutions

We confine ourselves here to "normal" stars in complete (mechanical and thermal) equilibrium. For the outer envelope of such a star, it is characteristic that $l$ and $m$ vary very little over wide ranges of $r$. (This is because $\varepsilon$ is negligible and $\varrho$ is very small; for example, only about 10% of the solar mass lies outside $r = R_\odot/2$.) This allows the derivation of approximate solutions that demonstrate the influence of the outer layers on the interior solution.

### 10.3.1 Radiative Envelopes

Since $m$ varies so little in the envelope, it seems advisable to take another independent variable, for which we may choose the pressure $P$, since it varies monotonically with $m$. The equation of radiative transport is derived from (5.12) and (2.5) as

$$\frac{\partial T}{\partial P} = \frac{3}{64\pi\sigma G} \frac{\kappa l}{T^3 m} \tag{10.18}$$

($\sigma = ac/4$). Let us approximate the dependence of $\kappa$ on $P$ and $T$ by a power law of the form

$$\kappa = \kappa_0 P^a T^b \quad , \tag{10.19}$$

with $\kappa_0 = $ constant and exponents typically $a > 0$, $b < 0$. By proper choice of $\kappa_0$, $a$, and $b$ we can represent reasonably (though of course not correctly) the run of $\kappa$ over wide ranges of the envelope. Introducing (10.19) into (10.18) results in

$$\frac{T^{3-b}}{P^a} \frac{\partial T}{\partial P} = \frac{3\kappa_0}{64\pi\sigma G} \frac{l}{m} \quad , \qquad (10.20)$$

and now we take $l \approx L$ and $m \approx M$ (this, together with the approximation of $\kappa$, determines how far inwards we are allowed to extend our solution). Then the right-hand side is constant and (10.20) can be integrated by separation of the variables:

$$T^{4-b} = B\left(P^{1+a} + C\right) \quad , \qquad (10.21)$$

where $C$ is a constant of integration, while the positive constant $B$ is given by

$$B = \frac{4-b}{1+a} \frac{3\kappa_0}{64\pi\sigma G} \frac{L}{M} \quad . \qquad (10.22)$$

For an illustrative example we now fix the exponents: $a = 1$, $b = -4.5$, which corresponds to the famous Kramers opacity for bound–free and free–free absorption in stellar material (see §17), and which is a good approximation for envelopes of moderate temperatures. Then (10.21) becomes

$$T^{8.5} = B\left(P^2 + C\right) \quad , \qquad (10.23)$$

a solution for the envelope that will now be discussed. It is illustrated in Fig. 10.2, which gives $\lg T$ against $\lg P$, so that the slope of a solution is equal to the value of $\nabla \equiv d\ln T/d\ln P$. Differentiation of (10.23) gives the slope

$$\nabla = 0.235 \frac{BP^2}{T^{8.5}} \quad . \qquad (10.24)$$

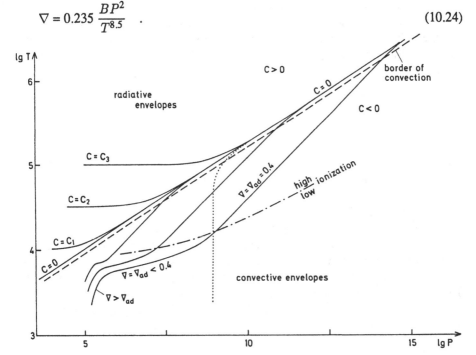

**Fig. 10.2.** A $\lg T - \lg P$ diagram for illustrating typical properties of envelope solutions as discussed in the text (see there for details)

73

The multitude of possible solutions differ by their value of the integration constant $C$.

$\underline{C = 0}$: The solution (10.23) now gives

$$\frac{T^{8.5}}{BP^2} = 1 \quad , \tag{10.25}$$

for which (10.24) yields the slope $\nabla = 2/8.5 \approx 0.235$. This is smaller than the usual value of $\nabla_{ad} = 2/5$ (see §14), and therefore the solution is consistent with (the assumed) radiative transport, shown in Fig. 10.2 as the straight solid line $\lg T = (2\lg P + \lg B)/8.5$. Obviously $T \to 0$ for $P \to 0$, and this solution would reach the zero boundary condition if we were to extend it outwards over the photosphere.

$\underline{C > 0}$: Since $B > 0$, (10.23) yields

$$\frac{T^{8.5}}{BP^2} > 1 \quad . \tag{10.26}$$

Comparing this with (10.24, 25), we see that in Fig. 10.2 the solutions with $C > 0$ lie above that with $C = 0$ and that they have a smaller slope, $\nabla < 2/8.5$. The layers are therefore all the more radiative. For $P^2 \ll C$ equation (10.23) becomes $T^{8.5} \approx BC = $ constant. This shows that towards the surface these solutions tend to a constant (and rather high) $T$. Three of them (for 3 different values $C_1 < C_2 < C_3$ of $C$) are illustrated by solid lines on the left of Fig. 10.2. On each line, one point corresponds to the photosphere with $T = T_{eff}$. Obviously we will find such radiative-envelope solutions below the photospheres with $T_{eff}$ larger than some critical value (close to $10^4$ K). Towards the interior, $P$ will finally increase so far that $P^2 \gg C$ in (10.23) and the solution approximates closely that for $C = 0$. Since all solutions with $C > 0$ asymptotically approach the solution $C = 0$, the precise starting values at the surface do not greatly influence the solution in the deep interior.

$\underline{C < 0}$: Equation (10.23) now gives

$$\frac{T^{8.5}}{BP^2} < 1 \quad , \tag{10.27}$$

which with (10.24, 25) shows that these solutions lie below the curve for $C = 0$ and that their slope is larger, $\nabla > 2/8.5$. A discussion quite analogous to that for $C > 0$ shows immediately that these solutions have the structure indicated in Fig. 10.2 by the dotted line. They bend downwards from the line $C = 0$, become gradually steeper, and tend vertically to a finite $P$ for $T \to 0$. (With a proper scaling of the coordinates the curves $C > 0$ and $C < 0$ are simply symmetric with respect to the line $C = 0$.) However, the assumption of radiative transport breaks down when convection sets in, which is the case for $\nabla = \nabla_{ad}$ (see §6.1). This is close to 0.4 in the interior of not too massive stars, while ionization effects near the surface can make it considerably smaller (see §14). This limit is derived by equating the right-hand side of (10.24) with $\nabla_{ad}$:

$$T^{8.5} = \frac{0.235}{\nabla_{ad}} BP^2 \quad . \tag{10.28}$$

For constant $\nabla_{ad}$ this corresponds to a straight line given by $\lg T = (2 \lg P + \lg B + \lg(0.235/\nabla_{ad}))/8.5$. For $\nabla_{ad} = 0.4$ this lower border for radiative solutions is plotted in Fig. 10.2 (dashed line). Near the surface, ionization effects decrease $\nabla_{ad}$ considerably below 0.4, and therefore the border line should be curved in its lowest part.

### 10.3.2 Convective Envelopes

The radiative solutions with $C < 0$ extending from the interior have to be terminated at the broken line in Fig. 10.2 given by (10.28), where convection sets in, and have to be replaced in the outer regions by solutions valid for convective transport. Three such convective solutions are shown as solid lines in the lower part of Fig. 10.2. In order to construct them we have to consider their slope $d\lg T/d\lg P$ $(= \nabla)$. As long as the solutions stay in regions of high enough density, convection is very effective (cf. §7.3) and the slope is equal to the adiabatic gradient $\nabla_{ad}$.

We can start the convective solutions near the border of convection with a slope given by $\nabla = \nabla_{ad} = 0.4$. With decreasing temperature the curves come into regions where the most abundant elements (hydrogen and helium) are no longer completely ionized (see §14). For hydrogen this occurs around $T = 10^4$ K, depending somewhat on $P$ (cf. the dependence of the Saha equation on the electron density). Partial ionization depresses $\nabla_{ad}$ appreciably below 0.4 such that the curves with a slope $\nabla = \nabla_{ad}$ are less steep and closely approach one another.

Finally the curves come into regions of such low density that convection is ineffective and the stratification is over-adiabatic, $\nabla > \nabla_{ad}$ (§7). Correspondingly the curves in Fig. 10.2 become rather steep until they reach the photospheric point. Unfortunately the precise slope $\nabla$ in the over-adiabatic part can only be calculated from a convection theory, with all its uncertainties. Anyway, convective envelopes start at cool photospheres, and with decreasing $T_{eff}$ the convection gradually reaches deeper into the interior. Small variations (due to numerical or physical uncertainties) of $T_{eff}$ or of the over-adiabatic part lead to curves that are widely separated in the interior.

### 10.3.3 Summary

Making a few simplifying assumptions, we have been able to derive convenient solutions for the temperature–pressure stratification of stellar envelopes, i.e. for the layers below the photosphere. In the case of radiative envelopes, the assumptions concerned $\kappa$, $m$, and $l$. An opacity law like (10.19) is certainly a poor approximation if one takes the same values of $a$, $b$, $\kappa_0$ for too wide a range, or for very different envelopes. The discussions can, however, be easily repeated for different values of $a$, $b$, $\kappa_0$ [for example $a = 0$, $b = 0$, $\kappa_0 = 0.2(1+X_H)$, as in the case of electron scattering, §17.1] giving essentially similar results. The assumption $l = $ constant certainly holds for $T < 10^6$ K, where nuclear burning is negligible, though the assumption $m = $ constant $= M$ breaks down much earlier. But, even if we stress these assumptions somewhat by extending the solutions too far inwards, we will still obtain the correct qualitative behaviour.

Radiative envelopes are found below all hot photospheres ($T > 9000$ K). Towards the deep interior these solutions converge rapidly to the solution with $C = 0$. The interior is therefore relatively insensitive to details of the outer boundary conditions, in particular to the photospheric details.

Below cool atmospheres there are convective envelopes, which extend farther downwards the smaller $T_{\text{eff}}$ is. This suggests that a minimum value of $T_{\text{eff}}$ might exist where the whole star has become convective (cf. the Hayashi line, § 24). The inward extension of the convective part depends rather sensitively on the precise position of the photosphere and the details of the over-adiabatic layer. Small changes in even the outer solution, which are otherwise rather unimportant, can exert a remarkable influence on the interior, and the same is true for the uncertainties in the treatment of superadiabatic convection.

### 10.3.4 The $T$–$r$ Stratification

Sometimes it is useful to know how $T = T(r)$ increases below the photosphere. From the definition of $\nabla \equiv d\ln T/d\ln P$ we have $dT = T\nabla dP/P$, where we replace $dP$ by using the hydrostatic equation in the form

$$dP = -\frac{Gm}{r^2}\varrho\, dr = Gm\varrho\, d\left(\frac{1}{r}\right) \tag{10.29}$$

and eliminate $T\varrho/P = \mu/\Re$ by means of the equation of state for an ideal gas. We then have

$$dT = \nabla\frac{G\mu}{\Re}m\, d\left(\frac{1}{r}\right) \quad . \tag{10.30}$$

For the outer envelope with low density we may approximate $m$ by the surface value $M$, so that if $\nabla$ is constant between points 1 and 2, we can integrate (10.30) to obtain

$$T_1 - T_2 = \nabla\frac{GM\mu}{\Re}\left(\frac{1}{r_1} - \frac{1}{r_2}\right) \quad . \tag{10.31}$$

Let the subscript 2 indicate the photosphere, i.e. $T_2 = T_{\text{eff}}$ and $r_2 = R$. Now at any point $r = r_1$ in the envelope we have

$$T - T_{\text{eff}} = f\left(\frac{R}{r} - 1\right) \quad , \qquad f = \nabla\frac{G\mu}{\Re}\frac{M}{R} \quad . \tag{10.32}$$

As a simple example we take $M = M_\odot$, $R = R_\odot$, and a solution with $C = 0$ (see § 10.3.1), for which we found that $\nabla = 0.235$. With $\mu = 1$ we find that $f = 5.4 \times 10^6$ K. This large value of $f$ provides for a very rapid increase of $T$ below the photosphere. Within only 2% of the radius, $T$ has reached $10^5$ K. And at $r \approx 0.8R$ (where $m \approx 0.99M$ still) the temperature exceeds $10^6$ K, which also shows that the "average" $T$ for all mass elements of the star is well above $10^6$ K.

# § 11 Numerical Procedure

For realistic material functions no analytic solutions are possible, so that one depends all the more on numerical solutions of the basic differential equations. Consequently the activity and the number of results in this field has increased with the numerical capabilities. The growth of computing facilities by leaps and bounds since the 1960s may be illustrated by a remark of M. Schwarzschild (1958): "A person can perform more than twenty integration steps per day", so that "for a typical single integration consisting of, say, forty steps, less than two days are needed". The situation has changed drastically since those days when the scientist's need for meals and sleep was an essential factor in the total computing time for one model. Nowadays one asks rather for the number of solutions produced per second. And these modern solutions are enormously more refined (numerically and physically) than those produced 30 years ago. This progress has been possible because of the introduction of large and fast electronic computers and the simultaneous development of an adequate numerical procedure connected with the name of L.G. Henyey. His method for calculating models in hydrostatic equilibrium is now generally used and will be described later. For more details and for further references see KIPPENHAHN et al. (1967). If inertia terms with $\ddot{r} \neq 0$ become important, one needs a so-called "hydrodynamic" procedure (see § 11.3).

## 11.1 The Shooting Method

It is not difficult to see that the appropriate choice of a numerical procedure is anything but a trivial matter. Consider the simplest case, the calculation of a model in complete equilibrium at a given time, for given mass $M$ and given chemical composition $X_i(m)$. The "spatial problem" can then be separated and is described by the structure equations (9.1,4,16,17). The naïve attempt simply to integrate them from one boundary to the other would encounter the difficulty that the boundary conditions are split, one pair being given at the centre, the other at the surface. Moreover, a test calculation starting with trial values $P_c$, $T_c$ at the centre has little chance of meeting the correct surface conditions. Outward integrations differing only a little near the centre have the tendency to diverge strongly when approaching the surface (see § 10.3). The reason is that for radiative transport (9.4) with (9.6) contains the factor $T^{-4}$. For inward integrations starting with trial values $R$, $L$ at the surface another divergence occurs near the centre owing to the singularity produced by the factor $r^{-4}$ in (9.16).

A compromise between these two possibilities is a fitting procedure often used in earlier, non-automized computations. Outward and inward integrations were both carried to an intermediate fitting point, where they were fitted smoothly to each other by a gradual variation of the trial values $P_c$, $T_c$ and $R$, $L$. The simultaneous fit of four variables ($r$, $P$, $T$, $l$) is, in principle, possible, since one can vary four free parameters ($P_c$, $T_c$, $R$, $L$) in the partial solutions. The fitting point is preferably chosen to be at the interface between physically different regions. For example, one takes the border between a convective central core and a radiative envelope, or between regions of different composition.

Fitting methods turned out to be unsuitable for calculating large series of complicated models. For these purposes they were generally replaced by the Henyey method. There are, however, certain applications where a fitting method is still unsurpassed, for example if one wishes to find *all* possible solutions for given core and envelope parameters.

## 11.2 The Henyey Method

This method is very practical, especially for solving boundary-value problems where the conditions are given at both ends of the interval. A trial solution for the whole interval is gradually improved upon in consecutive iterations until the required degree of accuracy is reached. In each iteration, corrections to *all* variables at *all* points are evaluated in such a way that the effect of each of them on the whole solution (including the boundaries) is taken into account. In a generalized Newton–Raphson method, corrections are obtained from linearized algebraic equations.

For spherical stars in hydrostatic equilibrium we have the partial differential equations (9.1,3,4,5,16) together with boundary conditions at the centre and at the surface. In addition the proper initial values have to be specified as well as the stellar mass $M$. The general structure of the system of equations suggests that one should treat two subsystems separately and alternately. First the system (9.1,3,4,16) is solved for given $X_i(m)$, then (9.5) is applied to a small time step $\Delta t$, after which (9.1,3,4,16) is solved for the new values of $X_i(m)$, and so on. In this way one can construct a whole evolutionary sequence of models. We now describe in detail the first of these two steps, the solution of the "spatial system".

If there is complete equilibrium ($\ddot{r} = \dot{P} = \dot{T} = 0$), the initial values to be given are the $X_i(m)$, so that we can treat them as known parameters for any point. According to (9.7–14) the material functions $\varepsilon$, $\kappa$, $\varrho$, ... on the right-hand sides of (9.1,4,16,17) can be replaced by their dependencies upon $P$ and $T$. Then we have to solve the four ordinary differential equations (9.1,4,16,17) for the four unknown variables $r$, $P$, $T$, $l$ in the interval $[0, M]$ (where $M$ is also thought to be given).

The case of hydrostatic equilibrium ($\ddot{r} = 0$) but thermal non-equilibrium ($\dot{P} \neq 0$, $\dot{T} \neq 0$) is almost equivalent, the only difference being the additional term $\varepsilon_g$ in (9.3), which contains the partial derivatives $\dot{P}$ and $\dot{T}$. This requires as initial values for the earlier time $t_0 - \Delta t$ not only the $X_i(m)$ but also $T(m)$ and $P(m)$. (See the remarks on possible initial values in §9.) Assume that we take them from a "foregoing" solution, calling these given functions $P^*(m)$, $T^*(m)$. At any point $m = m_j$, we denote the variables by $P_j$, $T_j$ and replace the time derivatives $\dot{P}_j$, $\dot{T}_j$ by

$$\dot{P}_j = \frac{1}{\Delta t}(P_j - P_j^*) \quad , \qquad \dot{T}_j = \frac{1}{\Delta t}(T_j - T_j^*) \quad . \tag{11.1}$$

The given values of $\Delta t$, $P_j^*$, $T_j^*$ can now be considered known parameters. Then $\dot{P}_j$, $\dot{T}_j$ are functions of $P_j$, $T_j$ only, as is the case with all material functions, and therefore we can also consider $\varepsilon_g$ to be replaced by the function $\varepsilon_g(P,T)$, and the situation is as before with the complete equilibrium models: we again have the 4 ordinary differential equations (9.3,4,16) for the four unknown variables $r$, $P$, $T$, $l$, but with a somewhat different right-hand side of (9.3).

Let us write these 4 differential equations briefly as

$$\frac{dy_i}{dm} = f_i(y_1, \ldots, y_4) \quad , \qquad i = 1, \ldots, 4 \quad , \tag{11.2}$$

where we have used the abbreviations $y_1 = r$, $y_2 = P$, $y_3 = T$, $y_4 = l$. The next step is discretization, i.e. we proceed from the differential equations (11.2) to corresponding difference equations for a finite mass interval $[m^j, m^{j+1}]$. Let us denote the variables at both ends of this interval by upper indices, e.g. $y_1^j$, $y_1^{j+1}$, $\ldots$, $y_4^j$, $y_4^{j+1}$. The functions $f_i$ on the right-hand sides of (11.2) have to be taken for some average arguments we call $y_i^{j+1/2}$; they are a combination of $y_i^j$ and $y_i^{j+1}$, for example the arithmetic or the geometric mean. If we define the four functions

$$A_i^j := \frac{y_i^j - y_i^{j+1}}{m^j - m^{j+1}} - f_i(y_1^{j+1/2}, \ldots, y_4^{j+1/2}) \quad , \qquad i = 1, \ldots, 4 \quad , \tag{11.3}$$

then the difference equations replacing (11.2) for the mass interval between $m_j$ and $m_{j+1}$ are

$$A_i^j = 0 \quad , \qquad i = 1, \ldots, 4 \quad . \tag{11.4}$$

It is advisable to exclude the outermost envelope of the star from the iteration procedure, since time-consuming computations may be necessary for this part (e.g. partial ionization and super-adiabatic convection). As described in §10.2 the outer boundary conditions are imposed at a fitting mass $m_F$, which may have the upper index $j = 1$, and they are formulated by the two equations (10.16) that relate the variables $y_1^1$, $\ldots$, $y_4^1$ at $m^1 = m_F$. With the definitions

$$B_1 := y_2^1 - \pi(y_1^1, y_4^1) \quad , \qquad B_2 := y_3^1 - \theta(y_1^1, y_4^1) \quad , \tag{11.5}$$

equations (10.17) become

$$B_i = 0 \quad , \qquad i = 1, 2 \quad . \tag{11.6}$$

As described in §10.2 the functions $\pi$, $\theta$ have to be derived by "downward" integrations starting with different trial values of $R$, $L$. In practice this may be greatly simplified if we content ourselves with a linear approximation for $\pi$ and $\theta$ (i.e. taking the tangential planes instead of the complicated surfaces in Fig. 10.1). Then only three trial integrations suffice to determine all coefficients in $B_1$ and $B_2$.

In the innermost interval of $m$, between the central point $m^K (= 0)$ and $m^{K-1}$, we apply series expansions for all four variables as given by (10.3,4,6,9). These four equations are written as

$$C_i(y_1^{K-1},\ldots, y_4^{K-1}, y_2^K, y_3^K) = 0 \quad , \qquad i = 1,\ldots 4 \quad , \tag{11.7}$$

which already incorporates the central boundary conditions $y_1^K = y_4^K = 0$ (i.e. $r = l = 0$ at the centre).

Consider now the whole interval of $m$, between $m^K = 0$ and the fitting mass $m^1 = m_F$, to be divided into $K - 1$ intervals (usually not equidistant) by $K$ mesh points as sketched in Fig. 11.1. At these $K$ mesh points we have $(4K - 2)$ unknown variables (since $y_1^K = y_4^K = 0$), and in order to have a solution these unknowns have to fulfil the following equations: (11.6) for the outer boundary, (11.4) for each interval except the last one $(j = 1, \ldots, K - 2)$, and (11.7) for the central boundary; thus there are $2 + 4(K - 2) + 4 = 4K - 2$ equations, which may be written:

$$\begin{aligned}
B_i &= 0 , & i &= 1,2 \quad , \\
A_i^j &= 0 , & i &= 1,\ldots,4 , & j &= 1,\ldots, K - 2 \quad , \\
C_i &= 0 , & i &= 1,\ldots,4 \quad .
\end{aligned} \tag{11.8}$$

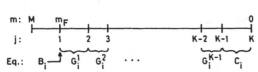

**Fig. 11.1.** Sketch of the mesh points in the interior solution, from the fitting mass $m = m_F$ to the centre ($m = 0$). It is also indicated which of the equations (11.4,6,7) have to be fulfilled at $m_F$ or between two adjacent mesh points

Suppose that we are looking for a solution for given values of $M$, $X_i(m)$, $P^*(m)$, $T^*(m)$ (which all enter into these equations as parameters). And suppose, furthermore, that we have a first approximation to this solution, say $(y_i^j)_1$ with $i = 1, \ldots, 4, j = 1, \ldots, K$. (This may be a rough first guess, for example obtained by an extrapolation of a foregoing solution or a solution for similar parameters.) Since the $(y_i^j)_1$ are only an approximation, they will not fulfil (11.8), i.e. when we use them as arguments in the functions $A_i^j$, $B_i$, and $C_i$ we find that

$$B_i(1) \neq 0 , \qquad A_i^j(1) \neq 0 , \qquad C_i(1) \neq 0 \quad , \tag{11.9}$$

where we indicate by (1) that the first approximation is used as arguments. Let us now look for corrections $\delta y_i^j$ for all variables at all mesh points such that the second approximation

$$(y_i^j)_2 = (y_i^j)_1 + \delta y_i^j \tag{11.10}$$

of the arguments makes the $B_i$, $A_i^j$, and $C_i$ vanish. The changes $\delta y_i^j$ of the arguments produce the changes $\delta B_i$, $\delta A_i^j$, and $\delta C_i$ of the functions, and we obviously have to require that

$$B_i(1) + \delta B_i = 0 , \qquad A_i^j(1) + \delta A_i^j = 0 , \qquad C_i(1) + \delta C_i = 0 . \qquad (11.11)$$

For small enough corrections, we may expand the $\delta B_i$, ... in terms of increasing powers of the corrections $\delta y_i^j$, and keep only the linear terms in this expansion; for example

$$\delta B_1 \approx \frac{\partial B_1}{\partial y_1^1} \delta y_1^1 + \frac{\partial B_1}{\partial y_2^1} \delta y_2^1 + \frac{\partial B_1}{\partial y_3^1} \delta y_3^1 + \frac{\partial B_1}{\partial y_4^1} \delta y_4^1 . \qquad (11.12)$$

With this linearization (11.11) can be written as

$$\frac{\partial B_i}{\partial y_1^1} \delta y_1^1 + \ldots + \frac{\partial B_i}{\partial y_4^1} \delta y_4^1 = -B_i , \qquad i = 1,2 ,$$

$$\frac{\partial A_i^j}{\partial y_1^j} \delta y_1^j + \ldots + \frac{\partial A_i^j}{\partial y_4^j} \delta y_4^j + \frac{\partial A_i^j}{\partial y_1^{j+1}} \delta y_1^{j+1} + \ldots + \frac{\partial A_i^j}{\partial y_4^{j+1}} \delta y_4^{j+1} = -A_i^j ,$$

$$i = 1,\ldots,4 , \qquad j = 1,\ldots,K-2 , \qquad (11.13)$$

$$\frac{\partial C_i}{\partial y_1^{K-1}} \delta y_1^{K-1} + \ldots + \frac{\partial C_i}{\partial y_4^{K-1}} \delta y_4^{K-1} + \frac{\partial C_i}{\partial y_2^K} \delta y_2^K + \frac{\partial C_i}{\partial y_3^K} \delta y_3^K = -C_i ,$$

$$i = 1,\ldots,4 .$$

(The $B_i$, $A_i^j$, $C_i$, and all derivatives have here to be evaluated using the first approximation as arguments.) This is a system of $2 + 4(K-2) + 4 = 4K - 2$ linear, inhomogeneous equations for the $4K - 2$ unknown corrections $\delta y_i^j$ ($i = 1, \ldots, 4$ and $j = 1, \ldots, K$; but $\delta y_1^K = \delta y_4^K = 0$ because of the central boundary conditions). Equation (11.13) may be written concisely in matrix form as

$$H \begin{pmatrix} \delta y_1^1 \\ . \\ . \\ . \\ \delta y_3^K \end{pmatrix} = - \begin{pmatrix} B_1 \\ . \\ . \\ . \\ C_4 \end{pmatrix} , \qquad (11.14)$$

where the matrix $H$ of the coefficients is called the *Henyey matrix*; its elements are the derivatives on the left-hand sides of (11.13).

Usually $H$ has a non-vanishing determinant, det $H \neq 0$ (see § 12.4) and we can solve these linear equations, obtaining the wanted corrections $\delta y_i^j$. These are applied as shown in (11.10) to obtain a second, better approximation $(y_i^j)_2$. When using these second approximations as arguments, we will generally still find $B_i \neq 0$, $A_i^j \neq 0$, $C_i \neq 0$, i.e. equations (11.8) are not yet fulfilled. This is because the corrections were calculated from the *linearized* equations (11.13), while equations (11.8) are non-linear. (Even if we had linear equations instead of (11.8), the solution might require several iterations, since the numerical solution of (11.13) has only limited accuracy.) Therefore in a second iteration step we calculate new corrections by the same procedure to obtain a third approximation

$$(y_i^j)_3 = (y_i^j)_2 + \delta y_i^j \quad , \tag{11.15}$$

and so on. In consecutive iterations of this type, the approximate solution can be improved until either the absolute values of all corrections $\delta y_i^j$, or the absolute values of all right-hand sides in (11.13), drop below a chosen limit. Then we have approached the solution with the required accuracy.

If a time sequence of models is to be produced, one can now change the parameters appropriately for a new small time step $\Delta t$ [by evaluating from (9.5) the change of the $X_i(m)$, and by redefining the just-calculated $P(m)$, $T(m)$ as the new $P^*(m)$, $T^*(m)$]. The new model for $t + \Delta t$ is then calculated by the Henyey method in the same manner as for the model for $t$.

Of course, there is no guarantee that the iteration procedure for improving the approximations really does converge. In fact often enough one finds divergence if the chosen approximation is too far from the solution; then the required corrections are so large that one cannot neglect the second-order terms when evaluating $\delta B_i$, $\delta A_i^j$, and $\delta C_i$ in (11.11), and the linearized equations (11.14) therefore yield wrong corrections.

What happens, on the other hand, if we take a given precise solution as the "first approximation"? It fulfils (11.8) such that the right-hand sides of (11.14) vanish. Equation (11.14) is then a system of *homogeneous* linear equations, which for det $H \neq 0$ has only the trivial solution $\delta y_i^j = 0$: in this (normal) case there is no other solution ("local uniqueness" as described in § 12.2 and § 12.4). If, however, det $H = 0$, then we obtain solutions $\delta y_i^j \neq 0$, i.e. other solutions for the same parameters. In this somewhat pathological situation the "local uniqueness" of the solution is violated.

Fig. 11.2. Mesh points in the "three-layer model"

The Henyey matrix and its determinant are obviously important quantities. This concerns also their connection with the stability properties (see § 12.4). It is worthwhile noting the general structure of $H$, which turns out to be very simple. This is most easily demonstrated by considering the simple "three-layer model", which has only 4 mesh points from centre to fitting mass (Fig. 11.2). One interval is adjacent to $m_F$, one to the centre, while the intermediate interval borders on neither of these two boundaries, so that the full generality of possible cases is exhibited. Any further mesh point will only duplicate the situation of the intermediate interval. The Henyey matrix $H$ for this three-layer star is indicated in Fig. 11.3, where a dot in a column under $y_i^j$ and in a row denoted at the left-hand side by $A_k^l$ means a matrix element $\partial A_k^l / \partial y_i^j$. Some of these derivatives will be zero, since some basic equations do not depend on all variables [for example, (9.17) does not contain $y_1 = r$]. Outside

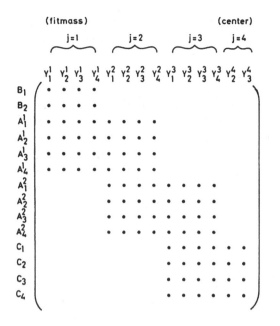

Fig. 11.3. Structure of the Henyey matrix $H$ for the three-layer star sketched in Fig. 11.2. A dot in, for example, the column $y_i^j$ and the row $A_k^l$ means the matrix element $\partial A_k^l / \partial y_i^j$. All matrix elements outside the dotted area are zero

the dotted area there are only zero elements. The Henyey matrix therefore has non-vanishing elements only in overlapping blocks along the main diagonal, so that this can be easily used for devising simple and well-behaved algorithms for computing det $H$ and inverting the matrix through elimination processes.

## 11.3 Treatment of the First- and Second-Order Time Derivatives

When devising a numerical scheme for solving our partial differential equations one can choose many details more or less arbitrarily without greatly affecting the results. This concerns questions such as the prescription for averaging between spatial mesh points, and the definition of the variables; these can be, for example, the physical quantities themselves, their logarithms, or any other functions describing them properly.

Concerning the manner in which the time derivatives are approximated, one distinguishes between explicit and implicit schemes that are known to behave differently, in particular when one is dealing with second-order time derivatives. Forward integration in time, starting from given initial values, can require time steps of various length, and the results can be unstable with respect to small numerical errors. In § 11.2 we encountered examples of both types of scheme.

An *explicit* scheme was described in the case of the chemical equations (9.5). Consider the time interval between $t^n$ (at which all variables $q^n$ are supposed to be known) and $t^{n+1}$ (for which the variables $q^{n+1}$ are to be calculated). We have used (9.5) only in order to calculate time derivatives $\dot{X}_i^n$ of the chemical composition from the known reaction rates $r_{ik}^n$ and densities $\varrho^n$. The composition for $t^{n+1}$ was then evaluated as $X_i^{n+1} = X_i^n + \Delta t \dot{X}_i^n$ before the other variables for this time were

83

derived. In fact the $X_i^{n+1}$ are used as fixed parameters when calculating the solution at $t^{n+1}$ by iteration. Such a procedure is relatively simple, and in general the results are sufficiently accurate if the time steps are kept small enough.

In the set of structure equations (9.1,3,4,16) to be solved at time $t^{n+1}$ for given $X_i^{n+1}$, the energy equation (9.3) contains the time derivatives of $P$ and $T$. With respect to these an *implicit* scheme was used in § 11.2. According to (11.1) the $\dot{P}$ and $\dot{T}$ are replaced by $(P^{n+1} - P^n)/\Delta t$ and $(T^{n+1} - T^n)/\Delta t$, respectively. These time derivatives are therefore considered to depend also on the variables at time $t^{n+1}$ and are evaluated together with them in the iteration procedure. In principle one could also have used an explicit method. For example, replace $\dot{P}$ and $\dot{T}$ in (9.3) by the time derivative of the entropy $s$ and use this equation only in order to evaluate $\dot{s}^n$ at time $t^n$. Then, as in the case of the chemical composition, the solution for $t^{n+1}$ is calculated for a given, fixed entropy $s^{n+1} = s^n + \Delta t \dot{s}^n$ from the other equations (cf. the discussion in § 12.3).

It is well known that, for differential equations that involve first-order derivatives in time and first- (or higher-) order spatial derivatives, implicit methods allow larger time steps for a given spacing in mass; for explicit difference schemes the time step has to be kept small to avoid numerical instability. (For details see, for instance, RICHTMYER, MORTON, 1967.)

Let us now turn to the so-called *hydrodynamical problem*, which arises when the inertial term in the equation of motion cannot be neglected. Then in addition to the first-order time derivatives in (9.3) there is a second-order time derivative in (9.2). One usually introduces the radial velocity

$$v = \frac{\partial r}{\partial t} \tag{11.16}$$

of the mass elements as a new variable, with which (9.2) becomes

$$\frac{\partial P}{\partial m} = -\frac{Gm}{4\pi r^4} - \frac{1}{4\pi r^2}\frac{\partial v}{\partial t} . \tag{11.17}$$

When using (11.16,17) instead of (9.2) one has again to deal with first-order time derivatives only. These can be replaced by ratios of differences, and one can use an explicit or an implicit scheme as before, the explicit being simpler but demanding smaller time steps. However, this is not the only choice to be made. For example, within the framework of an explicit method the different variables can be defined at different times (say the radius values at $t^n$, $t^{n+1}$, ..., and the velocities at the intermediate times $t^{n-1/2}$, $t^{n+1/2}$, ...). Furthermore, one may devise a scheme which treats the mechanical equations explicitly but is implicit with respect to the time derivatives in the energy equation (9.3).

The presence of the second-order time derivatives changes the properties of the equations and the behaviour of the numerical procedure considerably. Whenever an explicit scheme is used, the time steps have to be kept small in order to fulfil the Courant condition, according to which the time step $\Delta t$ must not exceed $\Delta r/v_s$, where $\Delta r$ is the thickness of the smallest mass shell and $v_s$ is the local velocity of sound.

# § 12 Existence and Uniqueness of Solutions

The purpose of the theory of stellar structure is to explain observed stars as a natural consequence of basic principles of physics. The models necessary for this, however, follow from a mathematical procedure that also produces models which are not realized in nature, for example, because of the initial conditions during star formation, or because of stability properties. The inclusion of these types of model in the discussion, even though their stellar counterparts cannot be seen through a telescope, often deepens the insight into the behaviour of real stars. We therefore devote this section to (more mathematical) problems such as those of uniqueness, the manifold of all possible solutions of the stellar-structure equations, and the stability of solutions.

An old problem is whether, for stars in complete equilibrium and of given "parameters" (stellar mass $M$ and chemical composition $X_j$), there exists one, and only one, solution of the basic equations of stellar structure. From simple considerations concerning uncomplicated cases, answers to this question were given in the 1920s by Heinrich Vogt und Henry Norris Russell; however, there is no mathematical basis for this so-called Vogt–Russell theorem, and lately – when by numerical experiments multiple solutions for the same parameters were found to exist (e.g. § 32.10) – it has had to be abandoned. It is all the more important to outline the conditions under which uniqueness is violated, and why. A linearized treatment (concerning "local" uniqueness) is easy to understand, whereas non-linear results refer to the "global" behaviour of the solutions and require a more involved mathematical apparatus; hence they are given here without proof, where we mainly follow the argumentation of KÄHLER (1972, 1975, 1978). For another representation, particularly of the linear problem, see PACZYŃSKI (1972).

Behind the questions about existence and uniqueness of solutions there is not only the mathematical interest, but also interest concerning the predicted evolution of stars. For example, after learning that often more than one solution exists, that solutions can disappear, or that new solutions appear in pairs, one might begin to wonder whether the star really "knows" how to evolve. But we should keep in mind that normally the star will be brought into one particular state (corresponding to a certain solution) according to its history. And if the equations indicate that the evolution approaches a "critical point", then this means in general only that the approximation used breaks down. For example, if an evolutionary sequence calculated for complete equilibrium comes to a critical point beyond which continuation is not possible, then the difficulties are normally removed by allowing for thermal non-equilibrium. Correspondingly if hydrostatic models that are not in thermal equilibrium evolve to a critical point, the difficulties are usually removed after the introduction of inertia terms. Nevertheless, we cannot exclude the possibility that even with the full set of general equations we might arrive at a branching point, where statistical effects finally decide the fate of the star ("statistical instability"). We will see that critical points are generally closely connected with the onset of instabilities.

## 12.1 Notation and Outline of the Procedure

In order to obtain a simple representation, we denote the dependent variables $P$, $T$, $r$, $l$ by $y_1$, $y_2$, $y_3$, $y_4$. As independent variable we take $x = m/M$, such that the total interval to be considered is always $[0, 1]$:

$$x = m/M \quad ,$$
$$y_1 = r \quad , \qquad y_2 = P \quad , \qquad y_3 = T \quad , \qquad y_4 = l \quad . \tag{12.1}$$

The left-hand sides of the four basic differential equations (9.1,2,3,4) are the derivatives $\partial y_i/\partial x$. On the right-hand sides, the material properties $\varepsilon$, $\kappa$, ... are thought to be replaced by functions of the variables $y_i$ and of the $X_j(x)$. The right-hand sides then are functions $f_1, \ldots, f_4$ of the $y_i$, possibly of the time derivatives $\dot{y}_2$ and $\dot{y}_3$ (for thermal non-equilibrium), and (for deviations from hydrostatic equilibrium) even of the second time derivative $\ddot{y}_1$. In addition there enter certain "parameters" such as $M$ and the $X_j(x)$ that may be indicated symbolically by $p$. Note that $p$ actually comprises many values; it can even be used to denote certain functions and describe their possible variations (see the example at the beginning of § 12.2.3). The basic differential equations (9.1,2,3,4) can then be written generally as

$$\frac{\partial y_i}{\partial x} = f_i(x, y_1, \ldots, y_4, \dot{y}_2, \dot{y}_3, \ddot{y}_1; p) \quad ; \qquad i = 1, \ldots, 4 \quad . \tag{12.2}$$

It is physically plausible to assume that the material functions incorporated in the right-hand sides are unique and differentiable with respect to their arguments. Therefore the functions $f_i$ are smooth, their partial derivatives with respect to their arguments being continuous functions of $x$ in the interval $0 < x \le x_F$, where $x_F$ refers to the fitting point $m_F$ at which the outer boundary conditions are given (§ 10).

Near the centre ($x = 0$), instead of (12.2) we use series expansions which start at $x = 0$ with the proper boundary conditions $y_1 = y_4 = 0$ and contain the central values of the other variables as free parameters (cf. § 10.1). Of course, these are also smooth functions of all arguments.

The outer boundary conditions at $x = x_F$ require a smooth continuation of the interior solution into a solution of the stellar-atmosphere problem, a procedure described in § 10.2. Stellar-atmosphere integrations are continued inwards to $x_F$, which represents in a certain sense a "transformation" of the surface conditions to the fitting point at $x_F$. To define one model, interior and outer solutions are fitted at $x_F$.

Most of the following discussion concerns models in complete equilibrium and the general procedure will be as follows. The first step will be to consider the infinitesimal neighbourhood of a model and see whether it is the only one existing there. This is the problem of "local uniqueness", for which the equations can be linearized. The formalism then reduces to a simple discussion of linear algebraic equations (§ 12.2). Closely connected is the question how a model changes owing to infinitesimal variations of the physical input parameters (e.g. mass and chemical composition).

In § 12.3 this local, linearized treatment will be extended to hydrostatic models without thermal equilibrium. In this case we can apply a procedure quite analogous to that of complete equilibrium if we consider models not only of given mass and chemical composition, but also of given entropy distribution. Local uniqueness ensures here the uniqueness of the thermal evolution.

It is obvious that the problem of local uniqueness must be intimately connected with the classical problem of (linear) stability of models, since both deal with infinitesimally neighbouring solutions and therefore use the linearized equations. These connections are indicated in § 12.4.

Even if a solution is locally unique, there can be several other widely separated solutions, and this raises the "global" uniqueness problem. Linear treatments do not help any more, but certain rules can be given even in this case (§ 12.5).

## 12.2 Models in Complete Equilibrium

### 12.2.1 Fitting Conditions in the $P_c$ – $T_c$ Plane

Here we consider complete (hydrostatic and thermal) equilibrium. Then there are no time derivatives in (12.2), which have become *ordinary* differential equations:

$$\frac{dy_i}{dx} = f_i(x, y_1, \ldots y_4; p) \quad , \quad i = 1, \ldots, 4 \quad . \tag{12.3}$$

Here $p$ again stands for the given parameters of the model, in particular $M$ and the $X_j(x)$.

The *interior solutions* are thought to start at the centre ($x = 0$) with $y_1 = y_4 = 0$ (boundary conditions) and with any chosen pair of central values $y_2 = P_c$, $y_3 = T_c$. They are obtained by the described series expansions (see § 10.1) and by integration of (12.3) outwards to the fitting mass $x_F$. Under the assumptions made above, a local Lipshitz condition (see e.g. INCE, 1956) holds throughout, and the solutions depend uniquely on the chosen starting values $P_c$, $T_c$. Therefore the interior solutions yield values for all variables at $x = x_F$ that are smooth functions of $P_c$ and $T_c$, say

$$
\begin{aligned}
y_{1F} &= \varrho_{in}(P_c, T_c) \quad , & y_{2F} &= \pi_{in}(P_c, T_c) \quad , \\
y_{3F} &= \theta_{in}(P_c, T_c) \quad , & y_{4F} &= \lambda_{in}(P_c, T_c) \quad .
\end{aligned} \tag{12.4}
$$

The *outer solutions* are thought to commence near the surface, with proper atmospheric solutions for assumed pairs of values $R$, $L$. When continued by integration downwards to $x = x_F$, they there give values for the variables that are functions of $R$ and $L$, denoted by

$$
\begin{aligned}
y_{1F} &= \varrho_{ex}(R, L) \quad , & y_{2F} &= \pi_{ex}(R, L) \quad , \\
y_{3F} &= \theta_{ex}(R, L) \quad , & y_{4F} &= \lambda_{ex}(R, L) \quad .
\end{aligned} \tag{12.5}
$$

In a certain sense one may regard (12.4,5) as "transformations" of the central and surface conditions to the *same* point $x = x_F$. A fit requires that the four functions in (12.4) and the corresponding ones in (12.5) give the same values. We therefore have 4 equations which we will reduce by an elimination process. We first fulfil two fitting conditions by setting $\varrho_{in} = \varrho_{ex}$ and $\lambda_{in} = \lambda_{ex}$. These two equations, solved for $R$ and $L$, give $R = R(P_c, T_c)$ and $L = L(P_c, T_c)$. (This is certainly possible if $x_F$ is not too far from the surface, where, for example, we have simply $\lambda_{ex} = L$.) These functions $R$ and $L$ are now used as arguments in $\pi_{ex}$ and $\theta_{ex}$, which thus also become functions of $P_c$ and $T_c$. Defining the two functions

$$
\begin{aligned}
g_1(P_c, T_c) &:= \pi_{in}(P_c, T_c) - \pi_{ex}(R(P_c, T_c), L(P_c, T_c)) \quad , \\
g_2(P_c, T_c) &:= \theta_{in}(P_c, T_c) - \theta_{ex}(R(P_c, T_c), L(P_c, T_c)) \quad ,
\end{aligned}
\tag{12.6}
$$

we can write the remaining two fitting conditions simply as

$$
g_1(P_c, T_c) = 0 \quad , \qquad g_2(P_c, T_c) = 0 \quad .
\tag{12.7}
$$

The fulfillment of (12.7) is the guarantee of a complete solution that satisfies the central and the surface boundary conditions. The equations $g_1 = 0$ and $g_2 = 0$ describe two curves in the $P_c - T_c$ plane (Fig. 12.1), and a fit, i.e. a model, corresponds to an intersection.

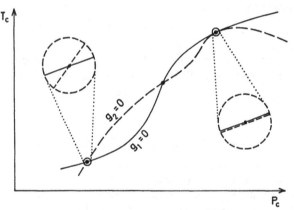

**Fig. 12.1.** Sketch of the curves $g_1 = 0$ and $g_2 = 0$ in the $P_c$–$T_c$ plane. Each intersection defines a model. Here we see three widely separated models, two of which (the left and intermediate) are locally unique, while the right one violates local uniqueness. The corresponding tangents are separately indicated in the two enlarged circles

### 12.2.2 Local Uniqueness

The question of existence and uniqueness of a solution is then equivalent to asking for the number of roots of the system (12.7) or for the number of intersections in Fig. 12.1. Over the full range of their arguments, $g_1$, $g_2$ are very complicated, non-linear functions, and therefore we should not be surprised if in certain cases no, or maybe several, roots for different pairs $P_c$, $T_c$ exist. General statements concerning this global behaviour of $g_1$, $g_2$ are certainly not easy to obtain. The prospects are much more favourable if we restrict ourselves to the simpler question of a *local uniqueness* in the following sense.

Assume that, for given stellar parameters, a solution of (12.7) exists that is an intersection in Fig. 12.1 at the arguments $P_c'$ and $T_c'$. Let us ask whether there exists another solution for the same parameters *within an infinitesimal neighbourhood* of the given solution.

Suppose there is one at $P_c' + \delta P_c$, $T_c' + \delta T_c$. These new, slightly changed arguments would then also give $g_1 = g_2 = 0$. In Fig. 12.1 this means that the two curves have a common tangent at this point. Therefore the variations $\delta g_1$ and $\delta g_2$ must vanish around $P_c'$, $T_c'$, which after linearization means

$$\frac{\partial g_i}{\partial P_c} \delta P_c + \frac{\partial g_i}{\partial T_c} \delta T_c = 0 \quad , \qquad i = 1, 2 \quad . \tag{12.8}$$

Equations (12.8) are written in matrix form as

$$G \begin{pmatrix} \delta P_c \\ \delta T_c \end{pmatrix} = 0 \tag{12.9}$$

with the $2 \times 2$ matrix

$$G = \begin{pmatrix} \dfrac{\partial g_1}{\partial P_c} & \dfrac{\partial g_1}{\partial T_c} \\ \dfrac{\partial g_2}{\partial P_c} & \dfrac{\partial g_2}{\partial T_c} \end{pmatrix} \quad . \tag{12.10}$$

Obviously the determinant of $G$ is decisive for the solutions of the linear homogeneous equations (12.9).

If $\det G \neq 0$, then we have only the trivial solution $\delta P_c = \delta T_c = 0$, i.e. no other solution exists within an infinitesimal neighbourhood of the given one. This corresponds to an intersection with different tangents in Fig. 12.1. Then we call the given solution *locally unique*. Fortunately most models describing stellar evolution have this property, though, of course, this does not say anything about other possible solutions (other intersections) far away.

If $\det G = 0$, non-vanishing $\delta P_c$, $\delta T_c$ are possible and we have neighbouring solutions. [The fact that then (12.9) yields infinitely many solutions is only due to the linearization of the $g_i$.] Then *local uniqueness is violated*. Geometrically this means that the two curves $g_1 = 0$ and $g_2 = 0$ in Fig. 12.1 intersect with coinciding tangents. Remembering that $g_1$ and $g_2$ are "transformations" of the boundary conditions to $x = x_F$, we can say that the common tangent of the two curves reflect a certain dependence between central and surface boundary conditions near this solution. Such cases can occur from time to time, for example in connection with the Schönberg–Chandrasekhar limit (§ 30.5).

### 12.2.3 Variation of Parameters

Up to now we have asked for neighbouring solutions with the *same* parameters. The next step is to ask whether one can go uniquely from a solution with given parameters $p$ to an adjacent one for a slightly changed stellar parameter $p + \delta p$. This can represent many types of changes that are discussed for purely theoretical reasons or that occur in stellar evolution. Very simple examples would be that $p$ stands directly for the total stellar mass $M$, for a core mass with special characteristics, or for a physical quantity which is not well known, so that we wish to vary it in its range of uncertainty. The parameter $p$ can also describe different chemical compositions: after defining two functions $\overline{X}_j(m)$ and $\widehat{X}_j(m)$, the present composition may be written as $X_j(m) = \overline{X}_j + p(\widehat{X}_j - \overline{X}_j)$. (Other functions can be treated analogously.) And, of course, one can define an arbitrary linear combination of such (and other) characteristic changes described by a continuous change of $p$. Following Poincaré, one speaks of *linear series* of models, if, resulting from a continuous change of $p$, there is a continuous sequence of solutions in which neighbouring ones can be derived from each other by linearized equations. Linear series are well suited for displaying uniqueness properties. Obviously, starting from a given solution, one can define many different linear series.

Suppose we have a given solution for a certain stellar parameter $p = p^*$; then this solution must satisfy (12.7). We now ask for a neighbouring solution with an infinitesimally increased parameter $p = p^* + \delta p$. This new solution would also have to fulfil (12.7), such that the resulting changes $\delta g_1 = \delta g_2 = 0$. We linearize the $g_1$ and $g_2$ as before; however, note that $g_1$ and $g_2$ in (12.7) also depend on $p$ [since the right-hand sides in (12.3) depend on $p$]. Therefore, instead of (12.8), we now have

$$\frac{\partial g_i}{\partial P_c} \delta P_c + \frac{\partial g_i}{\partial T_c} \delta T_c + \frac{\partial g_i}{\partial p} \delta p = 0 \quad , \qquad i = 1, 2 \quad , \tag{12.11}$$

or in matrix form

$$G \begin{pmatrix} \delta P_c \\ \delta T_c \end{pmatrix} = -\delta p \begin{pmatrix} \dfrac{\partial g_1}{\partial p} \\ \dfrac{\partial g_2}{\partial p} \end{pmatrix} \quad , \tag{12.12}$$

where $G$ is again the $2 \times 2$ matrix defined in (12.10). This shows that the present problem is intimately connected with that of local uniqueness. But with $\delta p \neq 0$, (12.12) is an inhomogeneous system of linear equations.

If $\det G \neq 0$ (i.e. the given solution for $p^*$ is *locally unique*), we find from (12.12) a unique pair $\delta P_c$, $\delta T_c$ (proportional to $\delta p$), i.e. one neighbouring solution exists for each value $\delta p$. In other words, when starting from a locally unique model, a continuous change of some parameter $p$ uniquely creates a continuous sequence (a linear series) of neighbouring models.

If $\det G = 0$ (i.e. the given solution for $p^*$ is *not locally unique*), (12.12) yields either no, or infinitely many, solutions for $\delta P_c$, $\delta T_c$ (depending on the rank of $G$ and the derivatives on the right-hand side). Therefore, if we start from an equilibrium model that violates local uniqueness, an infinitesimally small variation of the parameter leads either to several neighbouring equilibrium models or to none. (As before, the appearance of infinitely many solutions is only a consequence of the linearization.)

Possible examples for such cases are illustrated in Fig. 12.2, where a characteristic value $\mathcal{L}$ of a solution (for example, the radius $R$) is plotted over the parameter $p$. In the sketched linear series, the solutions are locally unique (but with two or

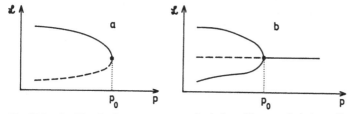

**Fig. 12.2 a, b.** Sketch of two sequences of solutions (linear series) depending on a parameter $p$. The letter $\mathcal{L}$ stands for some characteristic value of a solution. At $p_0$, the local uniqueness is violated and the linear series has a critical point; two solutions (a), or three solutions (b) merge here

three non-neighbouring solutions) for $p < p_0$, local uniqueness is violated at $p_0$, and there is either one locally unique solution for $p > p_0$ or none.

Of the many applications, we will briefly mention the most important one: *evolution of models in complete equilibrium*. A given model (time $t_0$, parameter $p^*$) evolves as the nuclear reactions change the chemical composition in the small time interval $\delta t$ from $X_j^*(m)$ to $X_j^*(m) + \dot{X}_j(m)\delta t$. The function $\dot{X}_j(m)$ is known for the given solution from the additional basic equations (9.5),

$$\dot{X}_j = F_j(x, y_1, \ldots, y_4; p) \quad , \tag{12.13}$$

where the right-hand sides are known functions containing, in particular, the reaction rates. Then we can immediately apply the above formalism if we identify $\delta p$ with the time step $\delta t$. And the statement is then simply that an *evolution of equilibrium models proceeds uniquely as long as the models remain locally unique.*

## 12.3 Hydrostatic Models without Thermal Equilibrium

### 12.3.1 Degrees of Freedom and Fitting Conditions

In (12.2), the second time derivative $\ddot{y}_1 (\equiv \ddot{r})$ is now still zero, while the first time derivatives $\dot{y}_2$, $\dot{y}_3$ have to be considered, though they appear only in the fourth equation (12.2), which is the energy equation. They can be combined as the time derivative of one function, namely the specific entropy $s$. This equation is then explicitly

$$\frac{\partial l}{\partial m} = \varepsilon - T\frac{\partial s}{\partial t} \quad . \tag{12.14}$$

The appearance of the specific entropy $s$ also suggests the introduction of $s$ in the other equations, and the elimination of another variable, say $T$, instead. This is possible, since, for given chemical composition, $s = s(P, T)$ is a well-known thermodynamic function that is monotonic with respect to its arguments $P$ and $T$. So we can invert it and obtain

$$T = T(P, s) \quad , \tag{12.15}$$

which is used to eliminate $T$ in terms of $P$ and $s$ in all basic equations (12.2). These may then be written (with $T$ replaced by $s$) as

$$\frac{\partial r}{\partial m} = \widehat{f}_1(r, P, s) \quad,$$

$$\frac{\partial P}{\partial m} = \widehat{f}_2(m, r) \quad,$$

$$\frac{\partial s}{\partial m} = \widehat{f}_3(m, r, P, s, l) \quad, \tag{12.16}$$

$$\frac{\partial l}{\partial m} = \widehat{f}_4(P, s, \dot{s}) \quad,$$

where the right-hand sides are denoted as the functions $\widehat{f}_1, \ldots, \widehat{f}_4$. Of course, they depend also on the given chemical composition.

Equations (12.16) represent an initial-value problem in time, for which we have to specify the entropy distribution $s_0(m) = s(m, t_0)$ at the starting time $t_0$. The function $s_0(m)$ plays here exactly the same role as the initial chemical composition $X_j(m, t_0)$ in the case of chemically evolving equilibrium models. It is plausible, therefore, to treat $s_0(m)$ formally in the same way as $X_j(m, t_0)$, namely also to describe it by some parameter $p$ on which the right-hand sides of (12.16) depend.

If, however, an entropy distribution $s(m)$ is given, we see immediately that $\widehat{f}_1$ and $\widehat{f}_2$ contain only $r$ and $P$ as unknown variables, i.e. the first two equations of (12.16) can be solved without considering the other two: the "mechanical" part of the system is decoupled from the thermo–energetic problem (see §9.1). The interior solution for $r$ and $P$ can be obtained as follows. We start at the centre with the boundary condition $r = 0$ and some assumed starting value $P_c$. Then we integrate the first 2 equations (12.16) outwards using the series expansion for small $m$, until we reach the chosen fitting mass $m_F$. This interior solution obviously has only one degree of freedom ($P_c$ can vary).

We now show that the (just-obtained) solution of the mechanical structure, $r(m)$ and $P(m)$, also completely fixes the thermal structure of such a star with given $s(m)$. We first obtain the temperature stratification $T(m)$ from (12.15) before considering the third of equations (12.16), the transport equation. It contains only one unknown variable, namely $l$, and may be used to obtain $l(m)$ from the temperature gradient. (This must in principle be possible even for nearly adiabatic convection.) Then the left-hand side of the fourth equation (12.16) is also known and this equation finally yields $\dot{s}(m)$ as

$$\dot{s} = \frac{1}{T} \left[ \varepsilon(P, s) - l' \right] \quad, \tag{12.17}$$

where $l' = \partial l / \partial m$ [see also (12.14)]. Obviously the *whole* interior solution is fixed merely by specifying the starting value $P_c$. Since, in contrast to the case of complete equilibrium, $s(m)$ is a given function, this reduces the number of degrees of freedom from 2 to 1.

By varying the starting values $P_c$, the interior solutions yield at the fitting mass $m_F$ the functions

$$P_F = \pi_{in}(P_c) \quad, \qquad r_F = \varrho_{in}(P_c) \quad. \tag{12.18}$$

Again, a local Lipshitz condition holds for the integrated differential equations, and the $\pi_{in}$, $\varrho_{in}$ are smooth, uniquely determined functions of $P_c$.

Consider now the outer solutions. In principle, they have to be treated quite analogously to the interior ones, i.e. as non-static solutions for given entropy distributions $s(m)$. In order to avoid unphysical situations (e.g. discontinuities in $P$ or $T$), we have to require that the given $s(m)$ is smooth at $m_F$; then, one can show that the outer solution as well is uniquely determined by one starting value only, say $R$ (for details see KÄHLER, WEIGERT, 1974), and if such outer solutions are extended to $m_F$, they yield there the function

$$P_F = \pi_{ex}(R) \quad , \qquad r_F = \varrho_{ex}(R) \quad . \tag{12.19}$$

We now require continuity of $r$ and $P$ at $m_F$ by equating (12.18) with (12.19), which also ensures continuity of $T$ and $l$. By inversion of the fitting condition $\varrho_{in}(P_c) = \varrho_{ex}(R)$ we obtain $R = R(P_c)$. This function is used to replace the argument in $\pi_{ex}$. For the remaining fitting condition we define

$$g(P_c) := \pi_{in}(P_c) - \pi_{ex}(R(P_c)) \quad , \tag{12.20}$$

and the condition for a solution fulfilling the central and surface boundary conditions can be written as

$$g(P_c) = 0 \quad . \tag{12.21}$$

### 12.3.2 Local Uniqueness

Suppose we have a solution fulfilling (12.21) for a given entropy distribution $s_0(m)$ and for given $M$ and chemical composition. We now ask for the local uniqueness of this model. [In § 12.2 we treated the corresponding problem for complete equilibrium where only $M$ and the chemical composition were given, while $s(m)$ was allowed to vary between neighbouring models.] A neighbouring solution for the same $s_0(m)$, but with a slightly changed $P_c + \delta P_c$, would also have to fulfil (12.21). After linearizing $g(P_c)$, we can write the condition for $\delta P_c$ as

$$\frac{dg}{dP_c} \delta P_c = 0 \quad . \tag{12.22}$$

[This corresponds to (12.9), and $dg/dP_c$ corresponds to $\det G$.]

For $dg/dP_c \neq 0$, the only solution of (12.22) is $\delta P_c = 0$, i.e. there exists no neighbouring solution for the same $s_0(m)$ [and the same $X_j(m)$ and $M$]. The given solution may then again be called *locally unique*.

For $dg/dP_c = 0$, there are neighbouring solutions with slightly changed $P_c$ that also fulfil (12.21) for the same $s_0(m)$, i.e., the given solution is *not locally unique*.

### 12.3.3 Variation of Parameters

The next step is to allow for a small change $\delta p$ of a parameter $p$. In particular we want to describe small changes of the given entropy distribution $s_0(m, p_0)$, say by writing

$$s(m, p) = s_0(m, p_0) + f(m)\delta p \quad, \tag{12.23}$$

where $f(m)$ is an arbitrarily chosen function of $m$. Note that $g$ as defined in (12.21) certainly depends also on $p$. A solution for the new $s(m)$ would also have to fulfil (12.21), which may be linearized to give

$$\frac{\partial g}{\partial P_c}\delta P_c + \frac{\partial g}{\partial p}\delta p = 0 \tag{12.24}$$

[corresponding to (12.11)]. For non-vanishing $\partial g/\partial p$, we obtain from (12.24) a unique, non-vanishing $\delta P_c$ only if $\partial g/\partial P_c \neq 0$. This means that, starting from a locally unique solution for $s_0(m)$, there exists one neighbouring solution for a slightly changed entropy distribution. Such a change leads to no neighbouring solutions (for $\partial g/\partial p \neq 0$), or several neighbouring solutions (for $\partial g/\partial p = 0$) if $\partial g/\partial P_c = 0$, i.e. if the given solution violates local uniqueness.

To be sure, the foregoing discussion (and its results) holds for any stellar parameter $p$ (also, for example, for one describing $M$ or the chemical composition, or any combination of them), and also for any function $f(m)$ in (12.23). But, of course, the most important application again concerns *the evolution in time*.

In this case, we start with a given solution for $s_0(m)$ at $t = t_0$. As described before, (12.17) yields uniquely $\dot{s}_0(m) \equiv \dot{s}(m, t_0)$. This function $\dot{s}_0(m)$ is now identified with $f(m)$, while we take $\delta p$ to represent an infinitesimal time step $\delta t$, so that (12.23) becomes

$$s(m, t_0 + \delta t) = s_0(m) + \dot{s}_0(m)\delta t \quad. \tag{12.25}$$

The foregoing statements then mean simply that we have a *continuous and unique thermal evolution in time, as long as the evolving model remains locally unique*. If the model arrives at a solution violating local uniqueness, there is either no continuous further evolution (which is physically unacceptable), or several possible continuations exist, i.e. there is a branching point in the evolution. In any case, this would require further considerations.

Let us briefly look back and compare the procedure used for these thermal non-equilibrium models with that derived earlier for complete equilibrium models. [We should here avoid the commonly used distinction of "time-dependent" and "time-independent" models, since *both* types depend on $t$; however, in one case it is $X_j(m)$ which changes, in the other it is $s(m)$.] The analogy between both cases is very clear: statements about local uniqueness always concern the *spatial solution at a given time* $t_0$. It is therefore a key attribute of both cases that we were able to separate the system of differential equations, one part describing the spatial problem at $t_0$ (for given parameters), and another one which can then be used to evaluate the change of the parameters [either $X_j(m)$, or $s(m)$] for a neighbouring time $t_0 + \delta t$. Thus the two equations (12.13) and (12.17) play exactly analogous roles.

## 12.4 Connection with Stability Problems

We shall here only briefly indicate the close connection which exists between the problems of local uniqueness and stellar stability owing to the fact that for both problems we ask for infinitesimally close neighbouring solutions.

For stellar-stability considerations, we start with a given solution (the "unperturbed" solution) for a certain set of parameters (mass, chemical composition, etc.). This solution is now thought to be perturbed by infinitesimal changes $\delta \overline{y}_i$ of the variables $y_i$ ($i = 1, \ldots, 4$), which means we consider a perturbed solution in which the variables have the values $y_i + \delta \overline{y}_i$. The perturbations $\delta \overline{y}_i$ are generally functions of $m$, but also of $t$, since the model will react to this perturbation (e.g. it will try to reduce it in certain cases). It is usual to separate the two dependencies by setting

$$\delta \overline{y}_i(m, t) = \delta y_i(m) e^{\sigma t} \quad , \qquad i = 1, \ldots, 4 \quad . \tag{12.26}$$

(Note that the eigenvalue $\omega$ used in §6 is related to $\sigma$ by $\sigma = i\omega$.) The requirement that the perturbed solution also fulfils the basic differential equations (which may be linearized for the small perturbations) and the proper boundary conditions leads to an eigenvalue problem for $\sigma$. Let us suppose here for simplicity that all eigenvalues are real (though complex eigenvalues also occur, see, for instance, §39). Then the "stability" of the initial solution depends only on the sign of $\sigma$.

For $\sigma < 0$, the perturbations $\delta \overline{y}_i$ decrease exponentially with time, such that the perturbed solution goes back to the initial solution, which is therefore described as being *stable* against such a perturbation. For $\sigma > 0$, the perturbation $\delta \overline{y}_i$ increases with time and the solution moves away from the initial solution, which is therefore called *unstable*. For $\sigma = 0$, the perturbations neither increase nor decrease in time and the perturbed solution remains at the same "distance" from the initial solution, which may then be called *marginally stable* (or marginally unstable, depending on the optimism or pessimism of our view).

Obviously, a zero eigenvalue is very important, since it separates the regimes of stability and instability. We will therefore now consider this case.

Let us assume that the initial (unperturbed) solution is in complete equilibrium (thermal and hydrostatic). If we have an eigenvalue $\sigma = 0$, then the exponential factor in (12.26) is equal to one, and the perturbation becomes simply $\delta y_i(m)$, i.e. independent of time. Therefore the perturbed solution $y_i(m) + \delta y_i(m)$ is also independent of time, i.e. it represents another equilibrium solution in the infinitesimal neighbourhood of the original one. This was called a violation of local uniqueness in §12.2.2 and is connected with the determinant of the matrix $G$ defined in (12.10): *a zero eigenvalue of the stability problem for an equilibrium solution occurs if (and only if) $\det G = 0$, i.e. if the solution violates local uniqueness.*

Depending on the assumptions made for the perturbation (such as adiabaticity or neglection of inertia terms) one normally distinguishes different types of stability problem (such as dynamical or secular stability, see §6, §25). Since for $\sigma \to 0$ all changes occur extremely slowly, the secular problem and the "full" problem become identical, since inertia terms play no role anyway. Dynamical instability is excluded since it would require (exact) adiabaticity, which can only be realised on short time-scales.

There must also be a close connection to the Henyey determinant $\det H$ (see §11.2), which describes a scheme of linearized equations for obtaining corrections to a given approximate solution. Suppose we take a given solution as this "approximation", then the calculated corrections would lead to a neighbouring solution of the same given parameters. If this does not exist (local uniqueness), the only corrections to be obtained must be zero, and $\det H \neq 0$. In fact one can show that

$$\det H = C \det G \quad , \tag{12.27}$$

where $C$ is a strictly positive function of certain properties of the interior solution (for this and the following see KÄHLER, 1972).

The zeros of $\det G$ (and $\det H$) coincide with $\sigma = 0$. It is thus not surprising that generally the sign of $\det G$ tells us something about the model's stability. It is certainly unstable if at least one eigenvalue is $> 0$, and if there are $k$ positive eigenvalues, one has

$$\mathrm{sign}(\det G) = \mathrm{sign}(\det H) = (-1)^k \quad . \tag{12.28}$$

This relation holds even if there are also complex eigenvalues; then $k$ is the number of unstable modes (Re $\sigma > 0$). Since complex eigenvalues always come as complex-conjugate pairs, they contribute an even number to $k$. Therefore if a model becomes unstable via a complex-conjugate pair of eigenvalues, $\det G$ does not go through zero (as it does with real $\sigma$), and the model remains locally unique. But we can certainly say that $k = 0$, and therefore sign $(\det G) = +1$ is a necessary (although not sufficient) condition for stability. On the other hand, sign $(\det G) = -1$ is sufficient for instability.

In order to find the eigenvalues, one can define a characteristic function $F = F(\sigma)$ with the properties

$$\begin{aligned} F(\sigma = 0) &= \det G \\ F(\sigma = \sigma_k) &= 0 \quad , \quad \text{for all eigenvalues } \sigma_k \quad . \end{aligned} \tag{12.29}$$

Then the problem of finding the eigenvalues is reduced to the search for the zeros of $F(\sigma)$. Correspondingly, a characteristic function for the thermal (secular) stability problem can easily be obtained from a slightly modified Henyey matrix $H'$ for non-equilibrium models; $H'$ differs from $H$ only in such a way that in the energy equation the operator $\partial/\partial t$ is replaced by a factor $\sigma$ according to (12.26).

In order to clarify further the different stability problems we have discussed, we have to go back to (12.2). These equations must hold for the unperturbed equilibrium solution given by $y_i(m)$ $(i = 1, \ldots, 4)$, as well as for the neighbouring perturbed solution $y_i(m) + \delta \overline{y}_i(m, t)$, where $\delta \overline{y}_i(m, t)$ is given by (12.26). Then the linearized version of (12.2) must also hold for the difference of both, i.e. for the perturbations $\delta \overline{y}_i$. Since they are small quantities, we can linearize the right-hand sides of (12.2) and obtain

$$\frac{\partial(\delta \overline{y}_i)}{\partial x} = \sum_{j=1}^{4} \frac{\partial f_i}{\partial y_j} \delta \overline{y}_j + \sum_{j=2}^{3} \frac{\partial f_i}{\partial \dot{y}_j} \delta \dot{\overline{y}}_j + \frac{\partial f_i}{\partial \ddot{y}_1} \delta \ddot{\overline{y}}_1 \quad , \quad i = 1, \ldots, 4 \quad . \tag{12.30}$$

From (12.26) we see that $\partial(\delta \bar{y}_i)/\partial t = \sigma \, \delta \bar{y}_i$. With this, all terms in (12.30) become proportional to $e^{\sigma t}$, and we can divide by this factor obtaining simply

$$\frac{\partial(\delta y_i)}{\partial x} = \sum_{j=1}^{4} \left[ \frac{\partial f_i}{\partial y_j} + \sigma \, \frac{\partial f_i}{\partial \dot{y}_j} + \sigma^2 \, \frac{\partial f_i}{\partial \ddot{y}_j} \right] \delta y_j \quad , \qquad i = 1, \dots, 4 \quad . \qquad (12.31)$$

The full stellar-stability problem requires the inclusion of the complete right-hand side of this equation, whereas the thermal- (or secular-) stability problem is obtained when the term proportional to $\sigma^2$ is dropped.

And what about the dynamical problem, for which the perturbation is assumed to be (completely) adiabatic? This is not included in (12.31), since all four basic equations are here perturbed, and there is no further freedom for an additional condition of adiabaticity. But it can be connected with the local uniqueness of solutions for thermal non-equilibrium (see § 12.3.2).

For the dynamical stability problem, we again take a perturbation of the form (12.26), but only for the two variables $y_1(= r)$ and $y_2(= P)$. Correspondingly, we consider only two basic equations, namely $\partial y_1/\partial x = f_1$ and $\partial y_2/\partial x = f_2$, and short-circuit the others by the assumption of adiabaticity. The second time derivative in $f_1$ introduces a factor $\sigma^2$. These eigenvalues will always be real, since they are obtained from a second-order problem (2 differential equations of first order plus boundary conditions) which is self-adjoint. Suppose there is an eigenvalue $\sigma = 0$ for a solution of given entropy distribution $s_0(m)$. The small perturbations are then independent of $t$ and lead to a neighbouring solution of the same entropy $s_0(m)$, since they are assumed to be adiabatic. Therefore a neighbouring solution of the same $s(m)$ exists and the *given solution violates local uniqueness* as defined in § 12.3.2. This means that we have $\sigma = 0$ in the dynamical problem, if $dg/dP_c = 0$ in (12.22).

Of course, one has to be careful with the terms "stability" and "instability" for solutions that depend on time. One should then always check whether the time-scale $\tau = 1/\sigma$ of a typical change of the perturbation is short compared to the time-scale with which the unperturbed solution changes. But this warning then also concerns the equilibrium models that change with the nuclear time-scale. And the distinction between thermal and dynamical stability certainly becomes questionable if both yield unstable modes of comparable positive eigenvalues (which is fortunately not usually the case). Then one has to treat the full stability problem.

## 12.5 Non-local Properties of Equilibrium Models

In § 12.2 it was shown that, for given parameters (mass, chemical composition), the existence of an equilibrium solution fulfilling the boundary conditions is equivalent to having a zero of the functions $g_1(P_c, T_c)$ and $g_2(P_c, T_c)$ defined in (12.6). "Local" statements, for an infinitesimal neighbourhood of a given solution (as discussed above), can be derived relatively easily through a linearization of the $g_i(P_c, T_c)$. It is known from numerical experiments that for given parameters several different solutions can exist that are so widely separated that the linearized analysis does not hold any more. Here one should rather have "global" statements, valid for a finite range of arguments $P_c$, $T_c$. At a first glance this might seem impossible. And it was in fact only recently that proper algebraic and topological methods allowed certain statements beyond the range of linearization. The procedure is rather involved, so that we will restrict ourselves to describing a few important results. (For more details, see KÄHLER, 1975, 1978).

For any given equilibrium solution, one can define certain characteristic properties that may be comprised in simple numbers: *the multiplicity m, and the charge c*. These are, in a certain sense, "quantum numbers" of this solution.

In order to count the total number of solutions existing for a given parameter $p$ in a mathematically relevant way, one has to see whether some of them are degenerate. This means that, in an algebraic sense, a given solution may have to be counted twice or several times if two or more solutions coincide. The *multiplicity m* is the number of coinciding solutions, i.e. there is a connection with local uniqueness: if the solution is locally unique, then $m = 1$; if local uniqueness is violated, then $m > 1$, and $m$ tells us the number of neighbouring solutions (which could not be obtained correctly in the linearized analysis). The multiplicity can be illustrated by looking at the linear series in Fig. 12.2. The multiplicity of a solution is the maximum number of single solutions (i.e. branches with $m = 1$) that emerge from the solution by a small change $\delta p$ of the parameter $p$. In Fig. 12.2, for example, the solutions have everywhere $m = 1$, except at the points $p = p_0$. At these critical points we have $m = 2$ (double solution) in Fig. 12.2(a) and a triple solution with $m = 3$ in Fig. 12.2(b).

The *charge c* of a solution contains information about the *sign* of the determinant of $G$, which was shown to be important for stability problems. The value of $c$ can be +1, −1, or 0. For $\det G \neq 0$, we have simply $c = \text{sign}(\det G)$. For $\det G = 0$, we can have $c = 0$ (if $m$ is even), or $\pm 1$ (if $m$ is odd). This quantity was called "charge" because it exhibits certain mathematical properties analogous to electric charges and their electrostatic field: a certain integral over a closed boundary in the $P_c - T_c$ plane gives uniquely the sum of the charges of all solutions contained inside this boundary.

A few applications of these definitions may be indicated. It was shown that in order to have stability of a given solution, the determinant of $G$ must be positive. A necessary condition for stability is therefore $c = m = 1$.

The total number of solutions for a given parameter $p$ may be defined as the sum $\sum m$ over the multiplicities of all existing solutions for this $p$. The total charge is correspondingly the sum $\sum c$ over all charges of these solutions. Then it can be shown that, if one changes the parameter $p$ (i.e. moves along a linear series), these two sums strictly obey certain rules (which might therefore be called the selection rules of the quantum numbers $m$ and $c$). With changing $p$, the total charge is always conserved ($\sum c = $ constant), while the number of solutions $\sum m$ is either conserved or changes by an even number. These rules can be easily checked in Fig. 12.2, where solid lines mean $c = +1$, and dashed lines $c = -1$. The requirement $\sum c = $ constant obviously means that along linear series the branches can appear or disappear only in pairs of opposite charge $c$, i.e. one knows that at least half of them must be unstable.

Suppose we start with a case that is so simple that we know there exists only one solution, and that it is a stable one. (Such simple cases can certainly be found.) Then we have $m = 1$, $c = 1$, and, of course, $\sum c = 1$. Now consider parameter variations and remember that an enormous variety of combinations of mass and chemical composition can be reached from the given simple one by simultaneously changing these parameters. In fact, we can thus reach all combinations of mass $M$ and $X_j(m)$ that occur in stars (except for those where the change of $p$ would necessarily lead through singular solutions). *All of them must therefore have the same total charge* $\sum c = +1$ according to the requirement $\sum c = $ constant. This ensures first of all the existence of at least one solution, since at least one solution with $c = 1$ must be present (existence theorem in a global sense). Then the total number $\sum m$ of solutions must be odd, since it can have changed only by an even number ($\sum m = 1 + 2n$, with $n$ an integer). And of these additional $2n$ solutions, $n$ have $c = +1$, while the other $n$ have $c = -1$. Therefore the maximum number of stable solutions is $n + 1$, while at least $n$ are necessarily unstable.

# III    Properties of Stellar Matter

In addition to the basic variables $(m, r, P, T, l)$ in terms of which we have formulated the problem, the differential equations of stellar structure (9.1–5) also contain quantities such as density, nuclear energy generation, or opacity. These describe properties of stellar matter for given values of $P$ and $T$ and for a given chemical composition as indicated in (9.7–14) and are quantities that certainly do not depend on $m$, $r$, or $l$ at the given point in the star. They could just as well describe the properties of matter in a laboratory for the same values of $P$, $T$, and chemical composition. We can therefore deal with them without specifying the star or the position in it for which we want to use them. In this chapter we shall discuss these "material functions", and we start by specifying the dependence of the density $\varrho$ on $P$, $T$, and the chemical composition. This is described by an *equation of state*, which is especially simple if we have an ideal gas.

# § 13 The Ideal Gas with Radiation

For an ideal gas consisting of $n$ particles per unit volume that all have molecular weight $\mu$, the equation of state is

$$P = nkT = \frac{\Re}{\mu} \varrho T \quad , \tag{13.1}$$

with $\varrho = n\mu m_u$ ($k = 1.38 \times 10^{-16}$ erg K$^{-1}$ = Boltzmann constant; $\Re = k/m_u = 8.31 \times 10^7$ erg K$^{-1}$g$^{-1}$ = universal gas constant; $m_u$ = 1 amu = $1.66053 \times 10^{-24}$ g = the atomic mass unit). Note that we here use the gas constant with a dimension (energy per K and per *unit mass*) different from that in thermodynamic text books (energy per K and per mole). This has the consequence that here the molecular weight $\mu$ is dimensionless (instead of having the dimension mass per mole); it is simply the particle mass divided by 1 amu.

## 13.1 Mean Molecular Weight and Radiation Pressure

In the deep interiors of stars the gases are fully ionized, i.e. for each hydrogen nucleus there also exists a free electron, while for each helium nucleus there are two free electrons. We therefore have a mixture of two gases, that of the nuclei (which in itself can consist of more than one component) and that of the free electrons. The mixture can be treated similarly to a one-component gas, if all single components obey the ideal gas equation.

We consider a mixture of fully ionized nuclei. The chemical composition can be described by specifying all $X_i$, the weight fractions of nuclei of type $i$, which have molecular weight $\mu_i$ and charge number $Z_i$. If we have $n_i$ nuclei per volume and a "partial density" $\varrho_i$, then obviously $X_i = \varrho_i/\varrho$ and

$$n_i = \frac{\varrho_i}{\mu_i m_u} = \frac{\varrho}{m_u} \frac{X_i}{\mu_i} \quad . \tag{13.2}$$

(Here and in the following, we neglect the mass of the electrons compared to that of the ions.) The total pressure $P$ of the mixture is the sum of the partial pressures

$$P = P_e + \sum_i P_i = \left( n_e + \sum_i n_i \right) kT \quad . \tag{13.3}$$

Here $P_e$ is the pressure of the free electrons, while $P_i$ is the partial pressure due to the nuclei of type $i$. The contribution of one completely ionized atom of element $i$ to the total number of particles (nucleus plus $Z_i$ free electrons) is $1 + Z_i$; therefore

$$n = n_e + \sum_i n_i = \sum_i (1 + Z_i)\, n_i \quad . \tag{13.4}$$

With this and (13.2), (13.3) becomes

$$P = nkT = \Re \sum_i \frac{X_i(1 + Z_i)}{\mu_i}\, \varrho T \quad , \tag{13.5}$$

which can be written simply in the form (13.1) with the *mean molecular weight*

$$\mu = \left( \sum_i \frac{X_i(1 + Z_i)}{\mu_i} \right)^{-1} \quad . \tag{13.6}$$

By introducing the mean molecular weight we are able to treat a mixture of ideal gases as a uniform ideal gas. We just have to replace the molecular weight in (13.1) by the mean molecular weight. In the case of pure (fully ionized) hydrogen with $X_H = 1$, $\mu_H = 1$, $Z_H = 1$ we have $\mu = 1/2$, while for a fully ionized helium gas ($X_{He} = 1$, $\mu_{He} = 4$, $Z_{He} = 2$) we find $\mu = 4/3$.

Equation (13.6) can be easily modified for the partial gas consisting of the ions only, or equivalently, for the case of a *neutral* gas where all the electrons are still in the atom. In (13.4) we just have to replace $1 + Z_i$ by 1 and we find

$$\mu_0 = \left( \sum_i \frac{X_i}{\mu_i} \right)^{-1} \quad . \tag{13.7}$$

Here we have dealt with the cases of full ionization and of no ionization at all. In § 14 we will deal with the case of partial ionization.

We now want to define the mean molecular weight per free electron $\mu_e$, a quantity which we shall need later. For a fully ionized gas each nucleus $i$ contributes $Z_i$ free electrons and we have

$$\mu_e = \left( \sum_i X_i Z_i / \mu_i \right)^{-1} \quad . \tag{13.8}$$

Since for all (not too rare) elements heavier than helium $\mu_i/Z_i \approx 2$ is a good approximation, we find

$$\mu_e = \left( X + \frac{1}{2} Y + \frac{1}{2}(1 - X - Y) \right)^{-1} = \frac{2}{1 + X} \quad , \tag{13.9}$$

where we have followed the custom of using $X := X_H$, $Y := X_{He}$ for the weight fractions of hydrogen and helium. Then $1 - X - Y$ is the mass fraction of the elements heavier than helium.

But the pressure in a star is not only given by that of the gas, because the photons in the stellar interior can contribute considerably to the pressure. Since the radiation is practically that of a black body (see § 5.1.1), its pressure $P_{rad}$ is given by

$$P_{rad} = \frac{1}{3} U = \frac{a}{3} T^4 \quad , \tag{13.10}$$

where $U$ is the energy density and $a$ is the radiation density constant $a = 7.56464 \times 10^{-15}$ erg cm$^{-3}$K$^{-4}$. Then the total pressure $P$ consists of the gas pressure $P_{gas}$ and radiation pressure $P_{rad}$:

$$P = P_{gas} + P_{rad} = \frac{\Re}{\mu} \varrho T + \frac{a}{3} T^4 \quad , \tag{13.11}$$

where on the right we have assumed that the gas is ideal. We now define a measure for the importance of the radiation pressure by

$$\beta := \frac{P_{gas}}{P} \quad , \quad 1 - \beta = \frac{P_{rad}}{P} \quad . \tag{13.12}$$

For $\beta = 1$ the radiation pressure is zero, while $\beta = 0$ means that the gas pressure is zero. The definition (13.12) can also be used if the gas is not ideal.

Two other relations which can be derived by differentiation of (13.12) are sometimes useful:

$$\left( \frac{\partial \beta}{\partial T} \right)_P = - \left[ \frac{\partial (1 - \beta)}{\partial T} \right]_P = - \frac{4}{T} (1 - \beta) \quad , \tag{13.13}$$

$$\left( \frac{\partial \beta}{\partial P} \right)_T = - \left[ \frac{\partial (1 - \beta)}{\partial P} \right]_T = \frac{1}{P} (1 - \beta) \quad . \tag{13.14}$$

## 13.2 Thermodynamic Quantities

From (13.11) we obtain

$$\varrho = \frac{\mu}{\Re} \frac{1}{T} \left( P - \frac{a}{3} T^4 \right) \quad , \tag{13.15}$$

and with the definitions (6.6) with (13.13,14) we find that

$$\alpha = \frac{1}{\beta} \quad , \quad \delta = \frac{4 - 3\beta}{\beta} \quad , \quad \varphi = 1 \quad . \tag{13.16}$$

Indeed, if the radiation pressure can be neglected ($\beta = 1$) we find $\alpha = \delta = 1$, as should be expected for an ideal monatomic gas.

If the gas components are monatomic, then the internal energy per unit mass is

$$u = \frac{3}{2} kT \frac{n}{\varrho} + \frac{aT^4}{\varrho} = \frac{3}{2} \frac{\Re}{\mu} T + \frac{aT^4}{\varrho} = \frac{\Re T}{\mu} \left[ \frac{3}{2} + \frac{3(1 - \beta)}{\beta} \right] \quad , \tag{13.17}$$

so that according to the definition (4.4) of $c_P$ we have

$$c_P = \left(\frac{\partial u}{\partial T}\right)_P + P\left(\frac{\partial v}{\partial T}\right)_P = \left(\frac{\partial u}{\partial T}\right)_P - \frac{P}{\varrho^2}\left(\frac{\partial \varrho}{\partial T}\right)_P . \tag{13.18}$$

Using (13.17), after some algebraic manipulations involving (13.13), we obtain

$$\left(\frac{\partial u}{\partial T}\right)_P = \frac{\Re}{\mu}\left[\frac{3}{2} + \frac{3(4+\beta)(1-\beta)}{\beta^2}\right] . \tag{13.19}$$

From the definition of $\delta$ with (13.16–18) we write

$$c_P = \frac{\Re}{\mu}\left[\frac{3}{2} + \frac{3(4+\beta)(1-\beta)}{\beta^2} + \frac{4-3\beta}{\beta^2}\right] , \tag{13.20}$$

and then the relation (4.21) may be applied in order to determine the adiabatic gradient $\nabla_{\mathrm{ad}}$ for the ideal gas plus radiation:

$$\nabla_{\mathrm{ad}} = \frac{\Re\delta}{\beta\mu c_P} = \frac{1 + \frac{(1-\beta)(4+\beta)}{\beta^2}}{\frac{5}{2} + \frac{4(1-\beta)(4+\beta)}{\beta^2}} . \tag{13.21}$$

For $\beta \to 1$, (13.20,21) give the well-known values for the ideal monatomic gas: $c_P = 5\Re/(2\mu)$ and $\nabla_{\mathrm{ad}} = 2/5$, while for $\beta \to 0$ one has $\nabla_{\mathrm{ad}} \to 1/4$ and $c_P$ becomes infinite.

Sometimes the derivative

$$\frac{1}{\gamma_{\mathrm{ad}}} := \left(\frac{d\ln\varrho}{d\ln P}\right)_{\mathrm{ad}} \tag{13.22}$$

is required. If in the definition

$$\frac{d\varrho}{\varrho} = \alpha\frac{dP}{P} - \delta\frac{dT}{T} \tag{13.23}$$

of $\alpha$ and $\delta$ the adiabatic condition $PdT/(TdP) = \nabla_{\mathrm{ad}}$ is introduced, one finds

$$\gamma_{\mathrm{ad}} = \frac{1}{\alpha - \delta\nabla_{\mathrm{ad}}} . \tag{13.24}$$

In the case of an ideal gas with radiation pressure we have to introduce the expressions (13.16), while for the limit $\beta = 1$ we find

$$\gamma_{\mathrm{ad}} = \frac{1}{1 - \nabla_{\mathrm{ad}}} . \tag{13.25}$$

For a monatomic gas without radiation pressure ($\beta = 1$) one has $\nabla_{\mathrm{ad}} = 0.4$ and therefore $\gamma_{\mathrm{ad}} = 5/3$, whereas in the limit $\beta \to 0$ – after $\alpha$, $\delta$, and $\nabla_{\mathrm{ad}}$ are inserted from (13.16,21) – we find for a gas dominated by radiation pressure that

$$\gamma_{\mathrm{ad}} \to \frac{4}{3} , \qquad \nabla_{\mathrm{ad}} \to \frac{1}{4} . \tag{13.26}$$

Instead of $\gamma_{ad}$, $\nabla_{ad}$, one often uses the "adiabatic exponents" introduced by Chandrasekhar, which are defined by

$$\Gamma_1 := \left(\frac{d\ln P}{d\ln \varrho}\right)_{ad} = \gamma_{ad} \quad , \tag{13.27}$$

$$\frac{\Gamma_2}{\Gamma_2 - 1} := \left(\frac{d\ln P}{d\ln T}\right)_{ad} = \frac{1}{\nabla_{ad}} \quad , \tag{13.28}$$

$$\Gamma_3 := \left(\frac{d\ln T}{d\ln \varrho}\right)_{ad} + 1 \quad , \tag{13.29}$$

and obey the relation

$$\frac{\Gamma_1}{\Gamma_3 - 1} = \frac{\Gamma_2}{\Gamma_2 - 1} \quad . \tag{13.30}$$

# § 14 Ionization

In § 13 we assumed complete ionization of all atoms. This is a good approximation in the very deep interior, where $T$ and $P$ are sufficiently large, but the degree of ionization certainly becomes smaller if one approaches the stellar surface, where $T$ and $P$ are small. In the atmosphere of the sun, for instance, hydrogen and helium atoms are neutral. When a gas is partially ionized the mean molecular weight and thermodynamic properties such as $c_P$ depend on the degree of ionization. It is the aim of this section to show how this can be calculated and how it influences the properties of the stellar gas.

## 14.1 The Boltzmann and Saha Formulae

We consider the atoms of a chemical element in a certain state of ionization, contained in a unit volume of gas in thermodynamical equilibrium. They are distributed over many states of excitation, which we denote by subscript $s$, and these different states can be degenerate such that the state of number $s$ consists in reality of $g_s$ substates. The number $g_s$ is the *statistical weight*. Consider in particular the atoms of a certain element in state $s$ and in the ground state $s = 0$, separated by the energy difference $\psi_s$, and the transition between both, say, by emission and absorption of photons. In equilibrium the rate of such upward transitions is equal to that of downward transitions. This gives as the ratio between the numbers of atoms in the two states

$$\frac{n_s}{n_0} = \frac{g_s}{g_0}\, e^{-\psi_s/kT} \quad . \tag{14.1}$$

Equation (14.1) is the well-known *Boltzmann formula*, which governs the distribution of particles over states of different energy.

Instead of referring to the atoms in the ground state, we want to compare the atoms of state $s$ with the number $n$ of *all* atoms of that element:

$$n = \sum_s n_s \quad . \tag{14.2}$$

From (14.1), multiplication by $g_0$ and summation over all states leads to

$$g_0\frac{n}{n_0} = g_0 \sum_{s=0}^{\infty} \frac{n_s}{n_0} = g_0 + g_1\, e^{-\psi_1/kT} + g_2\, e^{-\psi_2/kT} + \ldots := u_\mathrm{p} \quad , \tag{14.3}$$

where $u_\mathrm{p} = u_\mathrm{p}(T)$ is the so-called *partition function*. From (14.1,3) we obtain the Boltzmann formula in the form:

107

$$\frac{n_s}{n} = \frac{g_s}{u_p} e^{-\psi_s/kT} \quad . \tag{14.4}$$

We can also use the Boltzmann formula to determine the degree of ionization, but there are differences between excitation and ionization that require attention. Excitation concerns ions and bound electrons distributed over *discrete* states only. In the case of ionization the upper state consists of two separate particles, the ion and the electron; and the free electron has a *continuous* manifold of states. After ionization, say by absorption, the electron "thrown out" can have an arbitrary amount of kinetic energy and recombination can occur with electrons of arbitrary kinetic energy.

We say an atom is in the $r$th state of ionization if it has already lost $r$ electrons. The energy necessary to take away the next electron from the ground state is $\chi_r$. After ionization this electron is in general not at rest, but has a momentum relative to the atom of absolute value $p_e$. Then $p_e^2/(2m_e)$ is its kinetic energy; therefore relative to its original bound state the free electron has the energy $\chi_r + p_e^2/(2m_e)$, while the state of ionization of the atom is now $r + 1$.

Let us consider as the lower state an $r$-times ionized ion in the ground state. The upper state may be that of the $(r + 1)$ times ionized ion plus the free electron with momentum in the interval $[p_e, p_e + dp_e]$. The number densities of ions in these two states are $n_r$ and $dn_{r+1}$. The statistical weight of the upper state is the product of $g_{r+1}$ of the ion and of $dg(p_e)$, the statistical weight of the free electron. Transitions upwards and downwards occur between the two states with equal rates. In the case of thermodynamic equilibrium the Boltzmann formula (14.1) applies and gives

$$\frac{dn_{r+1}}{n_r} = \frac{g_{r+1} dg(p_e)}{g_r} \exp\left(-\frac{\chi_r + p_e^2/(2m_e)}{kT}\right) \quad . \tag{14.5}$$

What is the statistical weight $dg(p_e)$ of the electron in the momentum interval $[p_e, p_e + dp_e]$? The Pauli principle of quantum mechanics tells us that in phase space a cell of volume $dq_1 dq_2 dq_3 dp_1 dp_2 dp_3 = dV d^3p$ can contain up to $2dV d^3p/h^3$ electrons, namely up to two electrons per quantum cell of volume $h^3$. Here the $q$'s and the $p$'s are the space and momentum variables of the (6-dimensional) phase space, while $dV$ and $d^3p$ are the (3-dimensional) "volumes" and $h$ is the *Planck constant* ($h = 6.62620 \times 10^{-27}$ erg s). Then

$$dg(p_e) = \frac{2 \, dV \, d^3 p_e}{h^3} \quad . \tag{14.6}$$

If the electron density in (3-dimensional) space is $n_e$ then per electron the volume $dV = 1/n_e$ is available, while the volume in (3-dimensional) momentum space containing all points belonging to the interval $[p_e, p_e + dp_e]$ is $d^3 p_e = 4\pi p_e^2 dp_e$, since all these points are on a spherical shell of radius $p_e$ and thickness $dp_e$. We then have

$$dg(p_e) = \frac{8\pi p_e^2 dp_e}{n_e h^3} \tag{14.7}$$

and (14.5) yields

$$\frac{dn_{r+1}}{n_r} = \frac{g_{r+1}}{g_r} \frac{8\pi p_e^2 dp_e}{n_e h^3} \exp\left(-\frac{\chi_r + p_e^2/(2m_e)}{kT}\right) \quad . \tag{14.8}$$

*All* upper states (ions of degree $r+1$ in the ground state and free electrons of all momenta) are then obtained by integration over $p_e$:

$$\frac{n_{r+1}}{n_r} = \frac{g_{r+1}}{g_r} \frac{8\pi}{n_e h^3} e^{-\chi_r/kT} \int_0^\infty p_e^2 \exp\left(-\frac{p_e^2}{2m_e kT}\right) dp_e \quad . \tag{14.9}$$

Since for $a > 0$

$$\int_0^\infty x^2 e^{-a^2 x^2} dx = \frac{\sqrt{\pi}}{4a^3} \quad , \tag{14.10}$$

we obtain

$$\frac{n_{r+1}}{n_r} n_e = \frac{g_{r+1}}{g_r} f_r(T) \quad , \quad \text{with} \quad f_r(T) = 2\frac{(2\pi m_e kT)^{3/2}}{h^3} e^{-\chi_r/kT} \quad . \tag{14.11}$$

This is the *Saha equation* (named after the physicist Meghnad Saha) though it is still not yet in its final form, since we have considered only the ground states. Therefore in order to be more precise, we now use the quantities $n_{r+1,0}$, $n_{r,0}$, $g_{r+1,0}$, $g_{r,0}$, where the second subscript indicates the ground state for which these quantities are defined. By $n_{r+1}$, $n_r$, $g_{r+1}$, $g_r$, we from now on mean number densities of ions and statistical weights for *all* states of excitation. A particular state of excitation is indicated by a second subscript such that $n_{i,k}$ is the number density of atoms in the stage $i$ of ionization and in state $k$ of excitation, and $g_{i,k}$ is the corresponding statistical weight. The Saha equation (14.11) is then written more precisely as

$$\frac{n_{r+1,0}}{n_{r,0}} n_e = \frac{g_{r+1,0}}{g_{r,0}} f_r(T) \quad . \tag{14.12}$$

The number density of ions in the ionization state $r$ (in *all* states of excitation) is

$$n_r = \sum_s n_{r,s} \quad , \tag{14.13}$$

which corresponds to (14.2), and we now write the Boltzmann formula (14.1) for ions of state $r$ as

$$\frac{n_{r,s}}{n_{r,0}} = \frac{g_{r,s}}{g_{r,0}} e^{-\psi_{r,s}/kT} \quad , \tag{14.14}$$

where $\psi_{r,s}$ is the excitation energy of state $s$; then (14.13) can be written in the form

$$\frac{g_{r,0}}{n_{r,0}} n_r = g_{r,0} \sum_s \frac{n_{r,s}}{n_{r,0}}$$

$$= g_{r,0} + g_{r,1} e^{-\psi_{r,1}/kT} + g_{r,2} e^{-\psi_{r,2}/kT} + \ldots := u_r \quad , \tag{14.15}$$

where $u_r = u_r(T)$ is the partition function for the ion in state $r$. With the help of $n_r g_{r,0} = n_{r,0} u_r$, which follows from (14.15), the Saha equation can be written for all stages of excitation as

$$\frac{n_{r+1}}{n_r} n_e = \frac{u_{r+1}}{u_r} f_r(T) \quad , \tag{14.16}$$

where $f_r(T)$ is given in (14.11). With $P_e = n_e kT$ one has

$$\frac{n_{r+1}}{n_r} P_e = \frac{u_{r+1}}{u_r} 2\frac{(2\pi m_e)^{3/2}}{h^3} (kT)^{5/2} e^{-\chi_r/kT} \quad . \tag{14.17}$$

## 14.2 Ionization of Hydrogen

In order to see the consequences of the Saha equation we shall apply it to a pure hydrogen gas. We define the *degree of ionization* $x$ by

$$x = \frac{n_1}{n_0 + n_1} \quad , \tag{14.18}$$

i.e. $n_1/n_0 = x/(1 - x)$. If the gas is neutral, then $x = 0$; if it is completely ionized, $x = 1$. Also the left-hand side of (14.17) can be replaced by $xP_e/(1 - x)$, and if $n = n_0 + n_1$ is the total number of hydrogen atoms, then we can relate the partial pressure of the electrons to the total gas pressure:

$$P_e = n_e kT = (n + n_e)kT \frac{n_e}{n + n_e} = P_{gas} \frac{n_e}{n + n_e} \quad . \tag{14.19}$$

For each ionized atom there is just one electron ($n_e = n_1$); therefore

$$P_e = \frac{x}{1 + x} P_{gas} \tag{14.20}$$

and (14.17) can be written in the form

$$\frac{x^2}{1 - x^2} = K_H \quad , \quad \text{with} \quad K_H = \frac{u_1}{u_0} \frac{2}{P_{gas}} \frac{(2\pi m_e)^{3/2}}{h^3} (kT)^{5/2} e^{-\chi_H/kT} \quad . \tag{14.21}$$

Here $\chi_H = 13.6$ eV is the ionization energy of hydrogen. Now with (14.21) we have come up with a quadratic equation for the degree of ionization that can be solved if $T$ and $P_{gas}$ are given. If radiation pressure is important, it is sufficient to give $T$ and the total pressure $P$, and then $P_{gas}$ can be obtained from (13.11).

In order to compute the degree of ionization, the partition function has to be known. For this we need the statistical weights of the different states of excitation, which are given by quantum mechanics. Since the higher states contribute little to the partition function, we may approximate it by the weight of the ground state, $u_0 \approx g_{0,0} = 2$, while for ionized hydrogen $u_1 = 1$ (see, for instance, ALLEN, 1973, pp. 34, 35).

We now give some numerical examples. In the solar photosphere we have in cgs units $P_{gas} = 6.83 \times 10^4$, $T = 5636$K and we obtain $x = 10^{-4}$, while in a deeper layer with $P_{gas} = 1.56 \times 10^{12}$, $T = 7.15 \times 10^5$ K, hydrogen is almost completely ionized: $x = 0.993$.

Since in (14.21) $K_H$ increases with $T$ and decreases with $P_{gas}$, and since the left-hand side increases with $x$, one can see that the degree of ionization increases with temperature and decreases with the gas pressure. This can be easily understood: with increasing temperature the collisions become more violent, the photons more energetic, and the processes of "kicking off" the electrons from the atoms more frequent. If, on the other hand, the temperature is kept constant but the pressure increases, then the probability grows that the ion meets an electron and recombines.

In § 13 we have defined the mean molecular weight $\mu$ for a mixture of gases and have seen that it is different for ionized and non-ionized gases. Therefore mean molecular weights depend on the degree of ionization.

In order to determine $\mu$ for the hydrogen gas having the degree of ionization $x$, we define the number $E$ of free electrons per atom (neutral or ionized), which is here simply

$$E = \frac{n_e}{n} = x \quad . \tag{14.22}$$

Remember that $\mu m_u$, $\mu_0 m_u$ and $\mu_e m_u$ are defined as the average particle masses per free particle, per nucleus, and per free electron respectively. This means that the density can be written as

$$\varrho = (n + n_e)\mu m_u = n\mu_0 m_u = n_e \mu_e m_u \quad . \tag{14.23}$$

Using (14.22) and $n = n_0 + n_1$, we solve (14.23) for the mean molecular weight and find

$$\mu = \frac{\varrho}{m_u n} \frac{1}{1 + E} = \frac{\mu_0}{1 + E} = \mu_e \frac{E}{1 + E} \quad , \tag{14.24}$$

where we have neither replaced $\mu_0$ by its value 1 for hydrogen nor $E$ by $x$, since (14.24) also holds for a mixture of gases.

## 14.3 Thermodynamical Quantities for a Pure Hydrogen Gas

Many thermodynamic properties depend on the degree of ionization. We here indicate roughly how the formulae can be derived for the relatively simple case of the pure hydrogen gas. This is not because of its importance, but rather because the treatment is quite analogous to that in the much more involved case of mixtures. The gas is supposed to be ideal, since partial ionization usually occurs only in the stellar envelope, where effects of degeneracy can be neglected.

In §6.1 we defined the quantity $\delta = -(\partial \ln \varrho / \partial \ln T)_P$. In the case of pure hydrogen obeying the ideal gas equation we have $\delta = 1$ for $x = 0$ and $x = 1$, since $\mu$ is constant in both cases. (Remember that we wished to incorporate in $\alpha$ and $\delta$ the changes of $\mu$ due to partial ionization, while $\varphi$ should be reserved for changes of $\mu$ due to changing chemical composition.) For partial ionization, $x$ varies with $T$ and therefore $\delta$ is given by a complicated expression. From the ideal gas equation $\varrho \sim \mu P / T$ and (14.24) with $\mu_0 = \text{constant}$ we find

$$\delta = 1 + \frac{1}{1 + E} \left( \frac{\partial E}{\partial \ln T} \right)_P \quad , \tag{14.25}$$

which also holds for a mixture of gases. For pure hydrogen $E = x$ and we need the derivative of $x$, which can be obtained by differentiation of the Saha equation (14.21). This gives

$$\delta = 1 + \frac{1}{2} x \, (1 - x) \left( \frac{5}{2} + \frac{\chi_H}{kT} \right) \quad . \tag{14.26}$$

While the mean molecular weight as given by (14.24) depends only on the degree of ionization, $\delta$ depends also on $T$, and if in addition radiation pressure is taken into account, one has to add terms proportional to $(1 - \beta)/\beta$ to the right-hand sides of (14.25, 26).

The definition (4.4) of $c_P$ together with $P = \Re \varrho T / \mu$ gives

$$c_P = \left( \frac{\partial u}{\partial T} \right)_P + \frac{\Re}{\mu} \, \delta \quad . \tag{14.27}$$

So we need the internal energy per mass unit

$$u = \frac{3}{2} \frac{\Re}{\mu_0} \, (1 + E) \, T + u_{\text{ion}} \quad , \tag{14.28}$$

where the first term gives the kinetic energy of ions and electrons, and the second term $u_{\text{ion}}$ means the energy that has been used for ionization and that again becomes available if the ions recombine. Again (14.27, 28) also hold for mixtures. For pure hydrogen, $E = x$ and $u_{\text{ion}} = x \chi_H / (\mu_0 m_u) = x \chi_H / m_u$, and after lengthy manipulations one gets

$$c_P \frac{\mu_0}{\Re} = \frac{5}{2} \, (1 + x) + \frac{\Phi_H^2}{G(x)} \quad , \tag{14.29}$$

with the abbreviations

$$\Phi_H := \frac{5}{2} + \frac{\chi_H}{kT} \quad \text{and} \quad G(x) := \frac{1}{x(1 - x)} + \frac{1}{x(1 + x)} = \frac{2}{x(1 - x^2)} \quad . \tag{14.30}$$

If radiation plays a role, it appears not only in the equation for the pressure, but also in the internal energy. The result for $c_P$ is that in (14.29) the factor $5/2$ has to be replaced by $5/2 + 4(1 - \beta)(4 + \beta)/\beta^2$.

We can now easily derive an expression for $\nabla_{\text{ad}}$:

$$\nabla_{\text{ad}} = \frac{P \delta}{T \varrho \, c_P} = \frac{2 + x(1 - x) \Phi_H}{5 + x(1 - x) \Phi_H^2} \quad . \tag{14.31}$$

## 14.4 Hydrogen–Helium Mixtures

As a next step in the general problem we consider a gas of hydrogen and helium with weight fractions $X$, $Y$ respectively. This is important for stellar envelopes and shows the difficulties which arise if mixtures are treated. We now have six types of particles: neutral and ionized hydrogen; neutral, ionized, and double ionized helium; and electrons. There are three types of ionization energy: $\chi_H^0$ for hydrogen and

$\chi^0_{He}$, $\chi^1_{He}$ for neutral and single ionized helium ($\chi^0_H = 13.598$eV, $\chi^0_{He} = 24.587$eV, $\chi^1_{He} = 54.416$eV). Each ionized hydrogen atom contributes the energy $\chi^0_H$ to the internal energy, each helium atom in the first stage of ionization the energy $\chi^0_{He}$, and each helium atom completely stripped of its two electrons the energy $\chi^0_{He} + \chi^1_{He}$. By $x^0_H$, $x^1_H$, $x^0_{He}$, $x^1_{He}$, $x^2_{He}$ we define degrees of ionization, i.e. $x^r_i$ gives the number of atoms of type $i$ in ionization state $r$ ($= r$ electrons lost) divided by the total number of atoms of type $i$ (irrespective of their state of ionization):

$$x^0_H = \frac{n^0_H}{n_H} \quad , \quad x^1_H = \frac{n^1_H}{n_H} \quad , \quad x^0_{He} = \frac{n^0_{He}}{n_{He}} \quad ,$$

$$x^1_{He} = \frac{n^1_{He}}{n_{He}} \quad , \quad x^2_{He} = \frac{n^2_{He}}{n_{He}} \quad , \tag{14.32}$$

with $n_H = n^0_H + n^1_H$ and $n_{He} = n^0_{He} + n^1_{He} + n^2_{He}$, where the $n^r_i$ are number densities of ions of type $i$ in ionization state $r$. Note that the degrees of ionization $x^0_H$ and $x^1_H$ correspond to $1 - x$ and $x$ in § 14.2.

The contribution of the ionization energy to the internal energy per unit mass [cf. (14.28)] is

$$u_{ion} = \frac{1}{m_u} \left\{ X \, x^1_H \chi^0_H + \frac{1}{4} \, Y \left[ x^1_{He} \chi^0_{He} + x^2_{He} \left( \chi^0_{He} + \chi^1_{He} \right) \right] \right\} \quad , \tag{14.33}$$

since $X/m_u$, $Y/(4m_u)$ are the numbers of hydrogen and helium atoms (neutral and ionized) per unit mass. Correspondingly we have for the number $E$ of electrons per atom (irrespective of ionization state and chemical type)

$$E = \left[ X x^1_H + \frac{1}{4} \, Y \left( x^1_{He} + 2x^2_{He} \right) \right] \mu_0 \quad . \tag{14.34}$$

We now have three Saha equations:

$$\frac{x^1_H}{x^0_H} \frac{E}{E+1} = K^0_H \quad , \quad \frac{x^1_{He}}{x^0_{He}} \frac{E}{E+1} = K^0_{He} \quad , \quad \frac{x^2_{He}}{x^1_{He}} \frac{E}{E+1} = K^1_{He} \quad , \tag{14.35}$$

with

$$K^r_i = \frac{u_{r+1}}{u_r} \frac{2}{P_{gas}} \frac{(2\pi m_e)^{3/2}(kT)^{5/2}}{h^3} e^{-\chi^r_i/kT} \tag{14.36}$$

for $i = $ H, He, and by definition

$$x^0_H + x^1_H = 1 \quad , \quad x^0_{He} + x^1_{He} + x^2_{He} = 1 \quad . \tag{14.37}$$

We now consider $X$, $Y$, $P_{gas}$, and $T$ to be given. Then (14.34,35,37) are six equations for the six unknown quantities $x^0_H$, $x^1_H$, $x^0_{He}$, $x^1_{He}$, $x^2_{He}$, $E$. The equations (14.35) are coupled to each other via $E$, which, for instance, means that the degree of ionization of hydrogen also depends on the degree of ionization of helium. But this is to be expected, since a hydrogen ion can also recombine with free electrons that originally

113

came from helium, since it has no prejudices concerning the origin of a captured electron.

The coupling of the three Saha equations (14.35) makes an analytical treatment impossible: the degrees of ionization have to be obtained numerically. In general this is done by an iteration procedure, starting with a trial value of $E$, which is then gradually improved. In many cases, however, the situation is much simpler.

The ionization energies $\chi_H^0$, $\chi_{He}^1$, $\chi_{He}^2$ differ from each other to such an extent that the zones of partial ionization of the different particles are almost separated. Therefore one has to solve at most two of equations (14.35) simultaneously.

In Fig. 14.1 we give the degrees of ionization and $\nabla_{ad}$ for the outer layers of the sun. One can see that the regions of partial ionization of H and He are quite separate. The second helium ionization does not start until the hydrogen is almost completely ionized. Each of the three ionization layers produces a lowering of $\nabla_{ad}$ where influences of hydrogen and first helium ionization overlap.

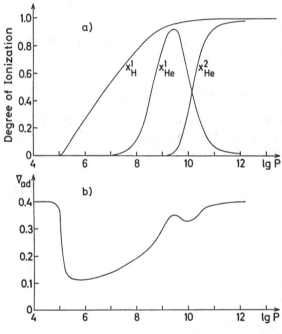

Fig. 14.1 a, b. Ionization in the outer layers of the sun. (a) Degrees of ionization of hydrogen and helium. (b) The influence of ionization on $\nabla_{ad}$

## 14.5 The General Case

If $X_i$ is the weight fraction of the chemical element $i$ with charge number $Z_i$ and molecular weight $\mu_i$, and if $x_i^r$ are the degrees of ionization (the numbers of atoms of type $i$ in ionization state $r$ in units of the total number of atoms of type $i$), then

$$E = \sum_i \nu_i \sum_{r=0}^{Z_i} x_i^r = \sum_i \frac{\mu_0}{\mu_i} X_i \sum_{r=0}^{Z_i} x_i^r r \quad , \tag{14.38}$$

where $\nu_i = n_i/n = X_i \mu_0 / \mu_i$ is the relative number of particles of type $i$. Equation

(14.34) is a special case of (14.38). Then the degrees of ionization are obtained from the set of Saha equations

$$\frac{x_i^{r+1}}{x_i^r} \frac{E}{E+1} = K_i^r \quad , \quad i = 1, 2, \ldots, \quad r = 0, 1, \ldots Z_i \quad , \tag{14.39}$$

where the $K_i^r$ are given by (14.36). In addition we have the relations

$$\sum_{r=0}^{Z_i} x_i^r = 1 \quad , \quad i = 1, 2, \ldots \quad . \tag{14.40}$$

For a given type $i$ of atoms, equations (14.39) in which $E$ is replaced by (14.38) represent $Z_i$ equations for the $Z_i+1$ degrees $r$ of ionization, and together with (14.40) one therefore has the same number of equations as of variables. The equations can be solved iteratively; thus the degrees of ionization can be used to determine the mean molecular weight according to $\mu = \mu_0/(1+E)$. The kinetic part of the internal energy [cf. (14.28)] is

$$u_{\text{kin}} = \frac{3}{2} \frac{\Re}{\mu} T = \frac{3}{2} \frac{\Re}{\mu_0} (1 + E)T \quad , \tag{14.41}$$

while the ionization energy per mass unit is

$$u_{\text{ion}} = \sum_i \frac{X_i}{\mu_i m_u} \sum_{r=0}^{Z_i} x_i^r \sum_{s=0}^{r-1} \chi_i^s \quad , \tag{14.42}$$

which is the general form of (14.33).

For the determination of $\delta$ and $c_P$ according to (14.25, 27) we need derivatives of the degrees of ionization: $(\partial x_i^r/\partial \ln T)_P$. They can be computed numerically by evaluating the $x_i^r$ for neighbouring arguments, though one has to be careful if the radiation pressure is not negligible. The derivatives of the $x_i^r$ are needed for constant *total* pressure $P$, whereas the argument for evaluating the degrees of ionization is the *gas* pressure. One therefore has to choose the neighbouring arguments $P_{\text{gas}}$ and $T$ such that $P = P_{\text{gas}} + P_{\text{rad}} = $ constant. The general theory of ionization and, in particular, the influence on the thermodynamic functions for arbitrary mixtures are given in BAKER, KIPPENHAHN (1962, Appendix A).

## 14.6 Limitation of the Saha Formula

In the derivation of the Saha formula we have assumed thermodynamic equilibrium. This is certainly fulfilled in the interior of stars, and the Saha formula is even a sufficient approximation for many atmospheres as long as one can assume so-called LTE (local thermodynamic equilibrium), which is the case when collisions dominate over radiative processes. One cannot apply it for non-LTE, as, for example, in the solar corona.

But even in the deep interior of a star, where thermodynamic equilibrium is certainly a very good approximation, the naïve application of the Saha formula gives wrong results. For instance let us apply it to the centre of the sun ($P_c \approx P_{gas} = 2.60 \times 10^{17}$ dyn/cm$^2$, $T_c = 1.60 \times 10^7$ K) and assume for simplicity pure hydrogen ($X = 1$); then (14.21) gives for the degree of ionization $x_H = 0.76$. This would mean that 24% of the hydrogen atoms are neutral. Indeed for sufficiently high temperatures the exponential in the Saha formula can be replaced by 1 and $x_H^1$ decreases inwards with $K_H$ if $\nabla \equiv d\ln T/d\ln P_{gas} < 2/5$, as can be seen from (14.21).

The solution of this paradox has to do with the decrease of the ionization energy with increasing density. Let us consider ions at a distance $d$ from each other: their electrostatic potentials have to be superimposed in order to obtain their total potential (Fig. 14.2). Obviously the higher quantum states of the ions are strongly disturbed, and the ionization energy is reduced for high density. This should be taken into account in the Saha formula, which would then give a higher degree of ionization. Furthermore, the neighbouring ions allow only a finite number of bound states. This has the consequence that in the partition function as given by (14.15) one has to sum over a finite number of excited states only.

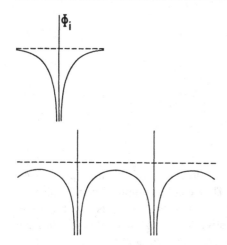

Fig. 14.2. Sketch of the electrostatic potential of an isolated ion (*above*) and the superposition of the potentials of neighbouring ions (*below*)

In order to estimate roughly at which density these effects become important, we consider a pure hydrogen gas. If the mean distance between two atoms is $d$, then there will be no bound states if the orbital radius $a$ of the electron is comparable with, or larger than, $d/2$. With

$$a = a_0\nu^2 \quad , \quad d \approx \left(\frac{3}{4\pi n_H}\right)^{1/3} \quad , \tag{14.43}$$

where $a_0 = 5.3 \times 10^{-9}$ cm is the Bohr radius, $\nu$ the quantum, number and $n_H$ the number density of the atoms, we obtain from the condition $a < d/2$ (which must be fulfilled for a bound state) that

$$\nu^2 < \left(\frac{3}{4\pi n_H}\right)^{1/3} \frac{1}{2a_0} \quad . \tag{14.44}$$

This allows a rough estimate of the principal quantum number of the highest bound state. In the centre of the sun, with $\varrho_c \approx 170$ g/cm$^3$, we have $n_H \approx \varrho_c/m_u \approx 10^{26}$ cm$^{-3}$ and therefore $\nu^2 < 0.13$, which means that even the ground state of hydrogen does not exist. Therefore all hydrogen atoms will be ionized.

For this so-called *pressure ionization* no good theory is at hand. The picture we have used above is a static one, since it does not take into account that the ions move relative to each other. It also ignores that at high densities electrons can tunnel from a bound state of one ion into a bound state of another ion in the neighbourhood.

For practical stellar-model calculations one often uses the Saha formula for the outer layers of the stars and then switches to complete ionization when the Saha formula gives degrees of ionization which decrease again towards deeper layers. This switching normally does not produce a noticeable discontinuity in the run of ionization, since the maximum often occurs close to complete ionization.

If we assume that pressure ionization can be neglected as long as $d > 10a_0$, then the Saha formula would be valid only for densities

$$\varrho = \mu_0 m_u n_{\text{ion}} < \frac{3\mu_0 m_u}{4\pi(10a_0)^3} = 2.66 \times 10^{-3}\mu_0 \text{ g cm}^{-3} \quad . \tag{14.45}$$

# §15 The Degenerate Electron Gas

## 15.1 Consequences of the Pauli Principle

We consider a gas of sufficiently high density in the volume $dV$ so that it is practically fully pressure ionized (§ 14.6). Here we shall deal with the free electrons, of number density $n_e$. If the velocity distribution of the electrons is given by Boltzmann statistics, then their mean kinetic energy is $3kT/2$. In momentum space $p_x$, $p_y$, $p_z$ each electron of a given volume $dV$ in local space is represented by a point and these points form a "cloud" which is spherically symmetric around the origin. If $p$ is the absolute value of the momentum ($p^2 = p_x^2 + p_y^2 + p_z^2$), then the number of electrons in the spherical shell $[p, p + dp]$ is, according to the Boltzmann distribution function,

$$f(p)dpdV = n_e \frac{4\pi p^2}{(2\pi m_e kT)^{3/2}} \exp\left(-\frac{p^2}{2m_e kT}\right) dp\, dV \quad . \tag{15.1}$$

Consider a reduction of $T$ with $n_e$ = constant. Then the maximum of the distribution function, which is at $p_{max} = (2m_e kT)^{1/2}$, tends to smaller values of $p$ and the maximum of $f(p)$ becomes higher, since $n_e$ is given by $\int_0^\infty f(p)dp$. This is indicated in Fig. 15.1 by the thin curves. But with this classical picture we can come into contradiction with quantum mechanics, since electrons are fermions, for which Pauli's exclusion principle holds: each quantum cell of the six-dimensional phase

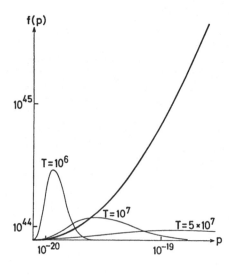

Fig. 15.1. For an electron gas with $n_e = 10^{28}$ cm$^{-3}$ (corresponding to a density of $\varrho = 1.66 \times 10^4$ g cm$^{-3}$ for $\mu_e = 1$), the Boltzmann distribution function $f(p)$ is shown by thin lines over the absolute value of the momentum $p$ (both in cgs units) for 3 different temperatures (in K). The heavy line shows the parabola that gives an upper bound to the distribution function owing to the Pauli principle. (Note that the coordinates are not logarithmic, but linear as in Figs. 15.2 and 15.5)

space ($x$, $y$, $z$, $p_x$, $p_y$, $p_z$) cannot contain more than two electrons (here $x$, $y$, $z$ are the space coordinates of the electrons with $dV = dx\,dy\,dz$). The volume of such a quantum cell is $dp_x\,dp_y\,dp_z\,dV = h^3$, where $h$ is Planck's constant. Therefore in the shell $[p, p + dp]$ of momentum space there are $4\pi p^2 dp\,dV/h^3$ quantum cells, which can contain not more than $8\pi p^2 dp\,dV/h^3$ electrons. Quantum mechanics therefore demands that

$$f(p)dp\,dV \leq 8\pi p^2 dp\,dV/h^3 \quad, \tag{15.2}$$

as indicated by the heavy parabola in Fig. 15.1, giving an upper bound for $f(p)$. One can immediately see that the Boltzmann distribution for $n_e$ = constant is in contradiction with quantum mechanics for sufficiently low temperatures. The same holds for $T$ = constant and sufficiently high density, since the Boltzmann distribution is proportional to $n_e$. We therefore have to include quantum-mechanical effects if the temperature of the gas is too low or if the electron density is too high, in order to avoid the distribution function exceeding its upper bound. One then says that the electrons become *degenerate*.

We first consider an electron gas of temperature zero, i.e. all the electrons have the lowest energy possible.

## 15.2 The Completely Degenerate Electron Gas

The state in which all electrons have the lowest energy without violating Pauli's principle is that in which all phase cells up to a certain momentum $p_F$ are occupied by two electrons, all other phase cells above $p_F$ being empty:

$$\begin{aligned} f(p) &= \frac{8\pi p^2}{h^3} \quad \text{for} \quad p \leq p_F \quad, \\ f(p) &= 0 \quad\quad \text{for} \quad p > p_F \quad. \end{aligned} \tag{15.3}$$

This distribution function is shown in Fig. 15.2, and the total number of electrons in the volume $dV$ is given by

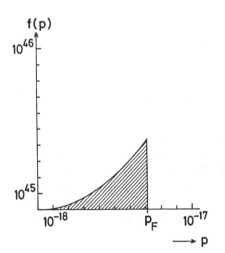

Fig. 15.2. The distribution function $f(p)$ against the momentum $p$ (both in cgs units) in the case of a completely degenerate electron gas with $T = 0$ K and $n_e = 10^{28}$ cm$^{-3}$ (cf. Fig. 15.1)

$$n_e dV = dV \int_0^{p_F} \frac{8\pi p^2 dp}{h^3} = \frac{8\pi}{3h^3} p_F^3 dV \quad . \tag{15.4}$$

If therefore the electron density is given, (15.4) gives the *Fermi momentum* $p_F \sim n_e^{1/3}$. Further, if the electrons are non-relativistic, then $E_F = p_F^2/2m_e \sim n_e^{2/3}$ is the *Fermi energy*, and, although the temperature of our electron gas is zero, the electrons have finite energies up to $E_F$. But there are no electrons of higher energy. If the electron density is sufficiently large, then according to (15.4) $p_F$ can become so high that the velocities of the fastest electrons may become comparable with $c$, the velocity of light. We therefore write the relations between velocity $v$, energy $E_{tot}$, and momentum $p$ of the electrons in the form given by special relativity (see, for instance, LANDAU, LIFSHITZ, vol. 2, 1961):

$$p = \frac{m_e v}{\sqrt{1 - v^2/c^2}} \quad , \tag{15.5}$$

$$E_{tot} = \frac{m_e c^2}{\sqrt{1 - v^2/c^2}} = m_e c^2 \sqrt{1 + \frac{p^2}{m_e^2 c^2}} \quad , \tag{15.6}$$

where $m_e$ is the *rest mass* of the electron. From (15.5,6) it follows that

$$\frac{1}{c} \frac{\partial E_{tot}}{\partial p} = \frac{p/(m_e c)}{[1 + p^2/(m_e^2 c^2)]^{1/2}} = \frac{v}{c} \quad . \tag{15.7}$$

In the following we have to distinguish between the *total energy* $E_{tot}$ as given by (15.6) and the *kinetic energy* $E$:

$$E = E_{tot} - m_e c^2 \quad . \tag{15.8}$$

For the equation of state we need the pressure, which by definition is the flux of momentum through a unit surface per second. We consider a surface element $d\sigma$ having a normal vector $n$, as indicated in Fig. 15.3. An arbitrary unit vector $s$, together with $n$, defines an angle $\vartheta$.

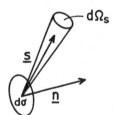

Fig. 15.3. A surface element $d\sigma$ with the normal vector $n$ and an arbitrary unit vector $s$ which is the axis of the solid angle $d\Omega_s$

Let us determine the number of electrons per second that go through $d\sigma$ into a small solid angle $d\Omega_s$ around the direction $s$. We restrict ourselves to electrons for which the absolute value of their momentum lies between $p$ and $p+dp$. At the location of the surface element there are $f(p)dp d\Omega_s/(4\pi)$ electrons per unit volume that have the right momentum (i.e. the right value of $p$ and the right direction). Therefore $f(p)dp d\Omega_s v(p) \cos \vartheta d\sigma/(4\pi)$ electrons per second move through the surface element

$d\sigma$ into the solid-angle element $d\Omega_s$. Here $v(p)$ is the velocity that according to (15.5) belongs to the momentum $p$. The factor $\cos\vartheta$ arises, since the electrons moving into the solid-angle element see only a projection of $d\sigma$. Each electron carries a momentum of absolute value $p$ and of direction $s$. The component in direction $n$ is therefore $p\cos\vartheta$. We obtain the total flux of momentum in direction $n$ by integration over all directions $s$ of a hemisphere and over all absolute values $p$; hence the pressure $P_e$ of the electrons is

$$P_e = \int_{2\pi} \int_0^\infty f(p)v(p)p\cos^2\vartheta \, dp d\Omega_s/(4\pi) = \frac{8\pi}{3h^3} \int_0^{p_F} p^3 v(p) dp \quad , \tag{15.9}$$

where we have replaced $f(p)$ by (15.3) and taken the value $4\pi/3$ for the integration of $\cos^2\vartheta$ over a hemisphere. It is obvious that the orientation of $d\sigma$ does not enter into the expression for $P_e$: the electron pressure is isotropic because $f$ is spherically symmetric in momentum space.

With (15.5) we obtain from (15.9) that

$$\begin{aligned}
P_e &= \frac{8\pi c}{3h^3} \int_0^{p_F} p^3 \frac{p/(m_e c)}{[1 + p^2/(m_e^2 c^2)]^{1/2}} dp \\
&= \frac{8\pi c^5 m_e^4}{3h^3} \int_0^x \frac{\xi^4 d\xi}{(1 + \xi^2)^{1/2}} \quad ,
\end{aligned} \tag{15.10}$$

where we have introduced new variables:

$$\xi = p/(m_e c) \quad , \qquad x = p_F/(m_e c) \quad . \tag{15.11}$$

The integral is

$$\int_0^x \frac{\xi^4 d\xi}{(1 + \xi^2)^{1/2}} = \frac{1}{8}\left[ x(2x^2 - 3)(1 + x^2)^{1/2} + 3\sinh^{-1} x \right] \quad ; \tag{15.12}$$

therefore

$$P_e = \frac{\pi m_e^4 c^5}{3h^3} f(x) \quad , \tag{15.13}$$

with

$$\begin{aligned}
f(x) &= x(2x^2 - 3)(x^2 + 1)^{1/2} + 3\sinh^{-1} x \equiv x(2x^2 - 3)(x^2 + 1)^{1/2} \\
&\quad + 3\ln\left[ x + (1 + x^2)^{1/2} \right] \quad .
\end{aligned} \tag{15.14}$$

We now write (15.4) in the form

$$n_e = \frac{\varrho}{\mu_e m_u} = \frac{8\pi m_e^3 c^3}{3h^3} x^3 \quad . \tag{15.15}$$

Equations (15.13–15) define the function $P_e(n_e)$, which is plotted in Fig. 15.4 for the fully degenerate electron gas. Before discussing this and deriving an equation of state $P_e = P_e(\varrho)$, we give an expression for the internal energy $U_e$ of the electron

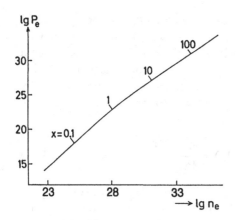

**Fig. 15.4.** The equation of state for the fully degenerate electron gas. On logarithmic scales the pressure $P_e$ (in dyn cm$^{-2}$) is plotted against the number density $n_e$ (in cm$^{-3}$). The relativity parameter $x = p_F/m_e c$ increases along the curve from the lower left to the upper right; values of $x$ are indicated above the curve

gas per volume:

$$U_e = \int_0^{p_F} f(p)E(p)dp = \frac{8\pi}{h^3} \int_0^{p_F} E(p)p^2 dp \quad , \tag{15.16}$$

where $E(p)$ has to be taken from (15.6,8). One obtains

$$U_e = \frac{\pi m_e^4 c^5}{3h^3} g(x) \quad , \tag{15.17}$$

with

$$g(x) = 8x^3 \left[ (x^2 + 1)^{1/2} - 1 \right] - f(x) \quad . \tag{15.18}$$

(For numerical values of the functions $f(x)$ and $g(x)$ see CHANDRASEKHAR, 1939, Table 23.)

## 15.3 Limiting Cases

The parameter $x$ as defined in (15.11) is a measure of the importance of relativistic effects for electrons with the highest momentum. With (15.5) we can write $x$ in the form

$$x = \frac{p_F}{m_e c} = \frac{v_F/c}{(1 - v_F^2/c^2)^{1/2}} \quad \text{or} \quad \frac{v_F^2}{c^2} = \frac{x^2}{1 + x^2} \quad , \tag{15.19}$$

where $v_F$ is the velocity of the electrons with $p = p_F$. If $x \ll 1$, then $v_F/c \ll 1$ and all electrons move much slower than the velocity of light (non-relativistic case). On the other hand if $x \gg 1$, then $v_F/c$ is very close to one: the bigger $x$, the more electrons with velocities near $v_F$ become relativistic, and for very high values of $x$ almost all electrons are relativistic.

The functions $f(x)$ and $g(x)$ as defined in (15.14,18) have the following asymptotic behaviour:

$$x \to 0 : f(x) \to \frac{8}{5}x^5 \quad , \quad g(x) \to \frac{12}{5}x^5 \quad . \tag{15.20}$$

$$x \to \infty : f(x) \to 2x^4 \quad , \quad g(x) \to 6x^4 \quad . \tag{15.21}$$

We first consider the case $x \ll 1$, where relativistic effects can be ignored, for which (15.13) yields

$$P_e = \frac{8\pi m_e^4 c^5}{15h^3} x^5 \quad , \tag{15.22}$$

and together with (15.15) we obtain the equation of state for a *completely degenerate non-relativistic* electron gas:

$$P_e = \frac{1}{20} \left(\frac{3}{\pi}\right)^{2/3} \frac{h^2}{m_e} n_e^{5/3} = \frac{1}{20} \left(\frac{3}{\pi}\right)^{2/3} \frac{h^2}{m_e m_u^{5/3}} \left(\frac{\varrho}{\mu_e}\right)^{5/3}$$

$$= 1.0036 \times 10^{13} \left(\frac{\varrho}{\mu_e}\right)^{5/3} \text{(cgs)} \tag{15.23}$$

where we have used $\varrho = n_e \mu_e m_u$. The internal energy $U_e$ of the electrons per unit volume and the electron pressure are related by

$$P_e = \frac{2}{3} U_e \quad , \tag{15.24}$$

which can be obtained from (15.17,20,22).

For the *extreme relativistic* case ($x \gg 1$) of a *completely degenerate* electron gas, one has according to (15.13,21)

$$P_e = \frac{2\pi m_e^4 c^5}{3h^3} x^4 \quad , \tag{15.25}$$

and therefore

$$P_e = \left(\frac{3}{\pi}\right)^{1/3} \frac{hc}{8} n_e^{4/3} = \left(\frac{3}{\pi}\right)^{1/3} \frac{hc}{8m_u^{4/3}} \left(\frac{\varrho}{\mu_e}\right)^{4/3}$$

$$= 1.2435 \times 10^{15} \left(\frac{\varrho}{\mu_e}\right)^{4/3} \text{(cgs)} \quad , \tag{15.26}$$

while (15.17, 21, 25) give

$$P_e = \frac{1}{3} U_e \quad . \tag{15.27}$$

## 15.4 Partial Degeneracy of the Electron Gas

For a finite temperature, not all electrons will be densely packed in momentum space in the cells of lowest possible momentum. Indeed, if the temperature is sufficiently high, we expect them to have a Boltzmann distribution. Further, there must be a smooth transition from the completely degenerate state (as discussed in § 15.2, 3) to the non-degenerate case.

The most probable occupation of the phase cells of the shell $[p, p + dp]$ in momentum space is determined by Fermi-Dirac statistics (see LANDAU, LIFSHITZ, vol. 5, 1959):

$$f(p)dpdV = \frac{8\pi p^2 dpdV}{h^3} \frac{1}{1 + e^{E/kT - \psi}} \tag{15.28}$$

(where the so-called degeneracy parameter $\psi$ will be discussed later). The first factor gives again the maximally allowed occupations for this shell, see (15.2). However, for $p \leq p_F$, there are fewer electrons in the shell than in the case of complete degeneracy: the second factor is smaller than one; it is a "filling factor", telling us what fraction of the cells is occupied. This factor depends on the temperature and the kinetic energy $E$ of a particle with momentum $p$ as defined in § 15.2.

With (15.28) $n_e$, $P_e$, and $U_e$ become

$$n_e = \frac{8\pi}{h^3} \int_0^\infty \frac{p^2 dp}{1 + e^{E/kT - \psi}} \quad , \tag{15.29}$$

$$P_e = \frac{8\pi}{3h^3} \int_0^\infty p^3 \, v(p) \frac{dp}{1 + e^{E/kT - \psi}} \quad , \tag{15.30}$$

$$U_e = \frac{8\pi}{h^3} \int_0^\infty \frac{E p^2 dp}{1 + e^{E/kT - \psi}} \quad . \tag{15.31}$$

We first deal only with the *non-relativistic case* for which $E = p^2/(2m_e)$ and the electron density $n_e$ is given by

$$n_e = \frac{8\pi}{h^3} \int_0^\infty \frac{p^2 dp}{1 + e^{p^2/2m_e kT - \psi}} = \frac{8\pi}{h^3}(2m_e kT)^{3/2} a(\psi) \quad , \tag{15.32}$$

with

$$a(\psi) = \int_0^\infty \frac{\eta^2}{1 + e^{(\eta^2 - \psi)}} d\eta \quad , \tag{15.33}$$

where we have used the variable $\eta = p/(2m_e kT)^{1/2}$.

We conclude from (15.32) that the *degeneracy parameter* $\psi$ is a function of $n_e/T^{3/2}$ only:

$$\psi = \psi\left(\frac{n_e}{T^{3/2}}\right) \quad . \tag{15.34}$$

We now discuss limiting cases for $\psi$, beginning with large negative values for $\psi$ (again non-relativistic). In this case $a(\psi)$ in (15.33) can be made arbitrarily small, and from (15.32) we infer that for a given electron density this is the case for high temperatures. We know that then $f(p)$ must become the Boltzmann distribution. Comparing (15.1) with (15.28) [where in the denominator the 1 can be neglected against $\exp(E/kT - \psi)$], we see that

$$e^\psi = \frac{h^3 n_e}{2(2\pi m_e kT)^{3/2}} \quad . \tag{15.35}$$

Here we have replaced $E/(kT)$ by its non-relativistic value $p^2/(2m_e kT)$. Indeed in this limit $\psi$ is a function of $n_e/T^{3/2}$, as concluded for the general case.

We now want to consider the case $\psi \to \infty$ (again non-relativistic) and introduce an energy $E_0$ by $\psi = E_0/(kT)$. We then have for large enough $\psi$

$$\frac{1}{1+e^{E/kT-\psi}} = \frac{1}{1+e^{\psi(E/E_0-1)}} \approx \begin{cases} 1 & \text{for} \quad E < E_0 \\ 0 & \text{for} \quad E > E_0 \end{cases} . \tag{15.36}$$

The transition of the numerical value of expression (15.36) from one to zero near $E_0$ becomes all the more steep, the larger the value of $\psi$. In the limiting case $\psi \to \infty$ it becomes a discontinuity, and comparison of (15.36) with (15.3) shows that $E_0$ is the Fermi energy $E_F = p_F^2/(2m_e)$. One immediately sees that $\psi \to \infty$ corresponds to the case of complete degeneracy, where the distribution function is given by (15.3).

We now deal with the (non-relativistic) case where the numerical value of $\psi$ is moderate. In (15.32) we replace the variable $p$ by $E$. With $m_e dE = p dp$ and $p = (2m_e E)^{1/2}$ we have

$$n_e = \frac{4\pi}{h^3}(2m_e)^{3/2} \int_0^\infty \frac{E^{1/2}dE}{1+e^{E/kT-\psi}} \quad , \tag{15.37}$$

and defining the so-called Fermi–Dirac integrals $F_\nu(\psi)$ by

$$F_\nu(\psi) := \int_0^\infty \frac{u^\nu}{e^{(u-\psi)}+1}du \quad , \tag{15.38}$$

we find that

$$n_e = \frac{\varrho}{\mu_e m_u} = \frac{4\pi}{h^3}(2m_e kT)^{3/2}F_{1/2}(\psi) \quad , \tag{15.39}$$

which again manifests the relation (15.34).

The distribution function for partial (non-relativistic) degeneracy as given by (15.28) is shown in Fig. 15.5 for $T = 1.9 \times 10^7$ K and $\psi = 10$ [$F_{1/2}(10) = 21.34$, see Table 15.1]. One can see that for small values of $p$ the function $f(p)$ is close to the Pauli parabola, but in contrast to the case $T = 0$ it is smooth near $p_F$. The higher the temperature the smoother the transition around $p_F$, until finally $f(p)$ resembles a Boltzmann distribution. The electron pressure $P_e$ is given in (15.30). Now (in the

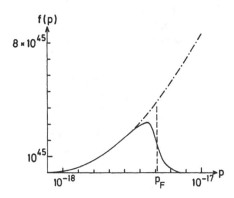

Fig. 15.5. The solid line gives the distribution function ($f(p)$ and $p$ in cgs) for a partially degenerate electron gas with $n_e = 10^{28}$ cm$^{-3}$ and $T = 1.9 \times 10^7$ K, which corresponds to a degeneracy parameter $\psi = 10$ (cf. the case of complete degeneracy of Fig. 15.2). The dot-dashed line shows the further increase of the parabola that defines an upper bound for the distribution function

**Table 15.1** Numerical values for Fermi–Dirac functions $F_{1/2}$, $F_{3/2}$ (after McDOUGALL, STONER, 1939) $F_2$, $F_3$ (after HILLEBRANDT, 1989)

| $\Psi$ | $\frac{2}{3}F_{3/2}(\Psi)$ | $F_{1/2}(\Psi)$ | $F_2(\Psi)$ | $F_3(\Psi)$ |
|---|---|---|---|---|
| −4.0 | 0.016179 | 0.016128 | 0.036551 | 0.109798 |
| −3.5 | 0.026620 | 0.026480 | 0.060174 | 0.180893 |
| −3.0 | 0.043741 | 0.043366 | 0.098972 | 0.297881 |
| −2.5 | 0.071720 | 0.070724 | 0.162540 | 0.490154 |
| −2.0 | 0.117200 | 0.114588 | 0.266290 | 0.805534 |
| −1.5 | 0.190515 | 0.183802 | 0.434606 | 1.321232 |
| −1.0 | 0.307232 | 0.290501 | 0.705194 | 2.160415 |
| −0.5 | 0.489773 | 0.449793 | 1.134471 | 3.516135 |
| 0.0 | 0.768536 | 0.678094 | 1.803249 | 5.683710 |
| 0.5 | 1.181862 | 0.990209 | 2.821225 | 9.100943 |
| 1.0 | 1.774455 | 1.396375 | 4.328723 | 14.393188 |
| 1.5 | 2.594650 | 1.900833 | 6.494957 | 22.418411 |
| 2.0 | 3.691502 | 2.502458 | 9.513530 | 34.307416 |
| 2.5 | 5.112536 | 3.196598 | 13.596760 | 51.496218 |
| 3.0 | 6.902476 | 3.976985 | 18.970286 | 75.749976 |
| 3.5 | 9.102801 | 4.837066 | 25.868717 | 109.179565 |
| 4.0 | 11.751801 | 5.770726 | 34.532481 | 154.252522 |
| 4.5 | 14.88489 | 6.77257 | 45.20569 | 213.80007 |
| 5.0 | 18.53496 | 7.83797 | 58.13474 | 291.02151 |
| 5.5 | 22.73279 | 8.96299 | 73.56744 | 389.48695 |
| 6.0 | 27.50733 | 10.14428 | 91.75247 | 513.13900 |
| 6.5 | 32.88598 | 11.37898 | 112.93904 | 666.29376 |
| 7.0 | 38.89481 | 12.66464 | 137.37668 | 853.64147 |
| 7.5 | 45.55875 | 13.99910 | 165.31509 | 1080.24689 |
| 8.0 | 52.90173 | 15.38048 | 197.00413 | 1351.54950 |
| 8.5 | 60.94678 | 16.80714 | 232.69369 | 1673.36371 |
| 9.0 | 69.71616 | 18.27756 | 272.63375 | 2051.87884 |
| 9.5 | 79.23141 | 19.79041 | 317.07428 | 2493.65928 |
| 10.0 | 89.51344 | 21.34447 | 366.26528 | 3005.64445 |
| 10.5 | 100.58256 | 22.93862 | 420.45675 | 3595.14883 |
| 11.0 | 112.45857 | 24.57184 | 479.89871 | 4269.86200 |
| 11.5 | 125.16076 | 26.24319 | 544.84118 | 5037.84863 |
| 12.0 | 138.70797 | 27.95178 | 615.53418 | 5907.54847 |
| 12.5 | 153.11861 | 29.69679 | 692.22772 | 6887.77637 |
| 13.0 | 168.41071 | 31.47746 | 775.17183 | 7987.72229 |
| 13.5 | 184.60190 | 33.29308 | 864.61653 | 9216.95127 |
| 14.0 | 201.70950 | 35.14297 | 960.81184 | 10585.40346 |
| 14.5 | 219.75048 | 37.02649 | 1064.00779 | 12103.39411 |
| 15.0 | 238.74150 | 38.94304 | 1174.45439 | 13781.61356 |
| 15.5 | 258.69893 | 40.89206 | 1292.40167 | 15631.12726 |
| 16.0 | 279.63888 | 42.87300 | 1418.09966 | 17663.37576 |
| 16.5 | 301.57717 | 44.88535 | 1551.79837 | 19890.17470 |
| 17.0 | 324.52939 | 46.92862 | 1693.74783 | 22323.71482 |
| 17.5 | 348.51087 | 49.00235 | 1844.19805 | 24976.56198 |
| 18.0 | 373.53674 | 51.10608 | 2003.39907 | 27861.65710 |
| 18.5 | 399.62188 | 53.23939 | 2171.60091 | 30992.31625 |
| 19.0 | 426.78099 | 55.40187 | 2349.05358 | 34382.23057 |
| 19.5 | 455.02855 | 57.59313 | 2536.00711 | 38045.46629 |
| 20.0 | 484.37885 | 59.81279 | 2732.71153 | 41996.46477 |

non-relativistic case), we have $p^3 v(p)dp = m_e^4 v^4 dv = m_e^3 v^3 dE = m_e^{3/2} 2^{3/2} E^{3/2} dE$
and

$$P_e = \frac{8\pi}{3h^3}(2m_e)^{3/2} \int_0^\infty \frac{E^{3/2}dE}{1+e^{E/kT-\psi}} \quad . \tag{15.40}$$

With $y = E/(kT)$ the integral becomes one of the type defined in (15.38):

$$P_e = \frac{8\pi}{3h^3}(2m_e kT)^{3/2} kT \, F_{3/2}(\psi) \quad . \tag{15.41}$$

For the internal energy $U_e$ per unit volume we have from (15.34) with the non-relativistic relation $p^2 = 2m_e E$

$$U_e = \frac{4\pi}{h^3}(2m_e kT)^{3/2} kT \, F_{3/2}(\psi) = \frac{3}{2}P_e \quad , \tag{15.42}$$

in agreement with (15.24)

Again, (15.39,41) define an equation of state for the electron gas. If $T$ and $n_e$ are given, then (15.39) gives $\psi$ (since $F_{1/2}(\psi)$ has a unique inverse function) and $P_e$ can be determined. Numerical values for some of the functions $F_\nu$ are given in Table 15.1. Much more detailed tables are given by McDOUGALL, STONER (1939).

Without proof we give an expansion of the integrals $F_\nu$ for large positive values of $\psi$, i.e. for strong degeneracy:

$$F_\nu(\psi) = \frac{\psi^{\nu+1}}{\nu+1}\Big\{1+2\Big[c_2(\nu+1)\nu\psi^{-2} \\ + c_4(\nu+1)\nu(\nu-1)(\nu-2)\psi^{-4} + \ldots\Big]\Big\} \quad , \tag{15.43}$$

with $c_2 = \pi^2/12$, $c_4 = 7\pi^4/720$. We therefore have for $\psi \gg 1$ that $F_{1/2}(\psi) \approx 2\psi^{3/2}/3$, $F_{3/2}(\psi) \approx 2\psi^{5/2}/5$. If we introduce these expressions into (15.39,41) and eliminate $\psi$, we come to the relation (15.23) for non-relativistic strong degeneracy.

On the other hand for $\psi \to -\infty$ (the electrons behave almost like an ideal gas) we can make the approximation

$$F_\nu(\psi) = \int_0^\infty \frac{y^\nu dy}{1+e^{(y-\psi)}} \approx e^\psi \int_0^\infty y^\nu e^{-y}dy \quad . \tag{15.44}$$

For $\nu = 1/2$ and $\nu = 3/2$ integration gives $F_{1/2}(\psi) \approx \sqrt{\pi}\, e^\psi/2$, $F_{3/2}(\psi) \approx 3\sqrt{\pi}\, e^\psi/4$. If we introduce these approximations into (15.39,41) and eliminate $\psi$, we find $P_e = n_e kT$, which is the equation of state for the ideal (non-degenerate) electron gas.

For the non-relativistic case we have derived the tools to deal with partial degeneracy. For the *extreme relativistic case* similar approximations are possible, since in the integrals (15.29,30) $p$ can be replaced by $E/c$, and $v$ by $c$. Then the same procedure which led to (15.39,41) now yields

$$n_e = 8\pi \left(\frac{kT}{hc}\right)^3 F_2(\psi) \quad , \tag{15.45}$$

127

$$P_e = \frac{8\pi}{3h^3c^3} (kT)^4 \, F_3(\psi) \quad , \tag{15.46}$$

where $F_2$ and $F_3$ are defined by (15.38). For strong degeneracy ($\psi \to \infty$) the first term of the expansion (15.43) is introduced into (15.45,46) and elimination of $\psi$ gives the already derived equation of state (15.26) for a completely degenerate, relativistic electron gas.

No analytical approach is known for the case of partial degeneracy if the electron gas is only moderately relativistic, because the relation between $E$ and $p$ cannot be approximated by a simpler expression and in the integrals (15.29,30) the full relation (15.6) has to be taken; hence the problem has to be treated numerically. The integrals can, for instance, be determined by using Laguerre polynomials as an approximation of the integrand (KIPPENHAHN, THOMAS, 1964).

# § 16 The Equation of State of Stellar Matter

In § 15 we dealt with degeneracy of arbitrary degree for the electron gas. We now discuss the combined effect of *all* components of stellar matter, starting with the ion gas.

## 16.1 The Ion Gas

In the non-degenerate case, electron pressure $P_e = n_e kT$ and ion pressure $P_{ion} = n_{ion} kT$ are of the same order of magnitude, they are even equal in the case of ionized hydrogen with $n_e = n_{ion}$. For sufficiently low temperature or sufficiently high density the ions can become degenerate, too. If they are Fermi particles such as protons, they will behave in phase space like the electrons, so that, for $P_{ion}$ and $n_{ion}$ relations, such as (15.29–31) hold if the mass of the ions $m_{ion}$ is used instead of $m_e$, and $\psi$ is now the degeneracy parameter for the ions. Again the transition between ideal-gas behaviour and degeneracy is roughly at $\psi = 0$. We write (15.39) in the form

$$\frac{n_j}{T^{3/2}} = \text{constant}\,(m_j)^{3/2} F_{1/2}(\psi) \quad , \tag{16.1}$$

where $n_j$ and $m_j$ refer to either electrons or ions. Suppose that the electron gas has a certain value of $\psi = \psi^*$ for $n_e = n_e^*$. An ion gas of the same temperature has the same degeneracy parameter $\psi = \psi^*$ for $n_{ion} = (m_{ion}/m_e)^{3/2} n_e^* \approx 8 \times 10^4 n_e^*$. Therefore the ions require much higher densities to become degenerate. For the interior of normal stars one can assume that even if the electrons are degenerate the ions still obey Boltzmann statistics; thus, because of the Pauli principle, the degenerate electrons have much higher momentum than the non-degenerate ions, and the electron pressure is much larger than the pressure of the ions: $P = P_{ion} + P_e \approx P_e$.

Even when the ion gas does not contribute noticeably to the pressure, it provides the main contribution to the mass density $\varrho$. This has already been taken into account by relating $n_e$ to $\varrho = n_e \mu_e m_u$, for example in (15.39). Furthermore, the ions can influence the thermodynamic properties of the plasma considerably.

One should be aware that, for certain types of stars, the treatment of the ions is not as simple as described here, since they can be subject to rather complicated interactions, for example those indicated in § 16.3,4.

## 16.2 The Equation of State

For normal stellar matter, the equation of state is then given by

$$P = P_{ion} + P_e + P_{rad} = \frac{\Re}{\mu_0}\varrho T + \frac{8\pi}{3h^3}\int_0^\infty p^3 v(p)\frac{dp}{e^{E/kT-\psi}+1} + \frac{a}{3}T^4 \quad , \quad (16.2)$$

$$\varrho = \frac{4\pi}{h^3}(2m_e)^{3/2}m_u\mu_e\int_0^\infty \frac{E^{1/2}dE}{e^{E/kT-\psi}+1} \quad , \quad (16.3)$$

where $v(p) = \partial E/\partial p$ according to (15.7) and where $E$ is given by (15.8). If the electron gas is highly degenerate, then also $P_{rad} \ll P_e$ and $P \approx P_e$.

For given $\varrho$ and $T$ and chemical composition ($\mu_0$), (16.3) can be used to determine $\psi$. Then $\varrho$, $\psi$, and $T$ determine $P$ via (16.2). The equation of state $P = P(\varrho, T)$ for all degrees of degeneracy, including relativistic effects, is therefore given here in implicit form.

An expression similar to (16.2) can be obtained for the internal energy $u$ per unit mass:

$$u = \frac{U_{ion} + U_e + U_{rad}}{\varrho} = \frac{3}{2}\frac{\Re}{\mu_0}T + \frac{8\pi}{h^3\varrho}\int_0^\infty \frac{p^2 E(p)dp}{e^{E/kT-\psi}+1} + \frac{aT^4}{\varrho} \quad , \quad (16.4)$$

where the $U$ are the energies per unit volume, and the first term on the right corresponds to the (ideal monatomic) ion gas.

Figure 16.1 shows the $\lg\varrho - \lg T$ plane for the ranges relevant for the interiors of most stars. In different regions different effects dominate the total pressure, e.g. in some places the electron degeneracy and in others the radiation pressure. We will derive rough borders between these different regimes.

Let us first consider the lines $\psi = $ constant for given $\mu_e$ in this diagram. In the non-relativistic regime, (15.39) shows that $\psi$ is constant for $T \sim \varrho^{2/3}$, i.e. on straight lines of slope 2/3 in the $\lg\varrho - \lg T$ plane. In the relativistic regime $\psi = $ constant for $T \sim \varrho^{1/3}$ according to (15.45), i.e. on straight lines with slope 1/3.

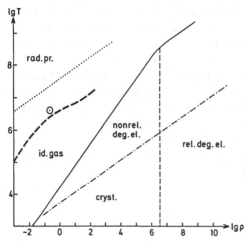

Fig. 16.1. Rough sketch of regions in the $\lg\varrho - \lg T$ plane ($\varrho$ in $g\,cm^{-3}$, $T$ in K), in which the equation of state is dominated by radiation pressure (above the dotted line given here by $P_{rad} = P_{gas}$ for $\mu = 0.5$), and by the degenerate electron gas (below the solid line given here by (16.6,8) for $\mu_e = 2$), which can be relativistic (right of the vertical broken line given by (16.7) for $\mu_e = 2$) or non-relativistic (left of the vertical broken line). The dot-dashed line indicates the melting temperature as given by (16.26) for $\mu_0 = 4$. By comparing with (14.45) one can see that the Saha formula is valid almost nowhere in the plotted domain. The heavy dashed curve on the left corresponds to a model of the present sun

We have already seen that the ideal-gas approximation $P_{gas} = \Re\varrho T/\mu$ becomes valid for large negative values of $\psi$. For large positive values of $\psi$ complete degeneracy is a good approximation for the electron gas, and $P \approx P_e$ for the non-relativistic case is given by (15.23). We can define the border between the two regimes by the condition that both approximations yield the same value for the pressure:

$$\frac{\Re}{\mu}\varrho T = \frac{1}{20}\left(\frac{3}{\pi}\right)^{2/3}\frac{h^2}{m_e}\left(\frac{\varrho}{\mu_e m_u}\right)^{5/3} . \qquad (16.5)$$

Equation (16.5) is equivalent to

$$\frac{T}{\varrho^{2/3}} = \frac{1}{20}\left(\frac{3}{\pi}\right)^{2/3}\frac{h^2}{m_e\Re m_u^{5/3}}\frac{\mu}{\mu_e^{5/3}} = 1.207 \times 10^5\frac{\mu}{\mu_e^{5/3}} , \qquad (16.6)$$

where the numerical constant is in cgs units. Equation (16.6) gives a straight line with slope 2/3 in Fig. 16.1 (lower left part of the solid line), which is obviously a line of $\psi$ = constant for given $\mu$, $\mu_e$. To the left of it the electrons behave almost like an ideal gas; to the right they are degenerate and dominate the pressure.

We now ask where relativistic effects become important. The transition between the non-relativistic and relativistic cases occurs around $x \approx 1$, where the relativity parameter $x$ is given by (15.11). Then (15.4) together with $\varrho = \mu_e m_u n_e$ gives

$$\varrho = \frac{8\pi m_u m_e^3 c^3}{3h^3}\mu_e = 9.74 \times 10^5\mu_e \text{ (cgs)} . \qquad (16.7)$$

In the plane of Fig. 16.1, (16.7) defines a vertical border line between relativistic (at larger $\varrho$) and non-relativistic degeneracy (at smaller $\varrho$). The same procedure which yielded (16.6) can be used with (15.26) in order to define the border between relativistic degeneracy and non-degeneracy:

$$\frac{T}{\varrho^{1/3}} = \left(\frac{3}{\pi}\right)^{1/3}\frac{hc}{8\Re}\frac{1}{m_u^{4/3}}\frac{\mu}{\mu_e^{4/3}} = 1.496 \times 10^7\frac{\mu}{\mu_e^{4/3}} , \qquad (16.8)$$

where the numerical constant is in cgs. The corresponding straight line of slope 1/3 is the upper-right part of the solid line in Fig. 16.1, again being a line of $\psi$ = constant for given $\mu$, $\mu_e$.

In a similar way we can determine a border between the regime of ideal gas pressure and that of dominating radiation pressure. From

$$\frac{\Re}{\mu}\varrho T = \frac{a}{3}T^4 \qquad (16.9)$$

we find

$$\frac{T}{\varrho^{1/3}} = \left(\frac{3\Re}{a\mu}\right)^{1/3} = \frac{3.2 \times 10^7}{\mu^{1/3}} , \qquad (16.10)$$

where the constant is in cgs. This line of slope 1/3 is dotted in Fig. 16.1.

In Fig. 16.1 it is indicated how $T$ grows with increasing density in the sun. As one can see, the interior regions of the sun avoid the area in the diagram where radiation pressure is important, as well as that of degeneracy. However, we will have to deal with other cases in which the equation of state is more complicated. This concerns highly evolved stars, but also unevolved stars of very low mass. (For a review see VAN HORN, 1986.)

## 16.3 Thermodynamic Quantities

With the implicit form (16.2,3) and with the expression (16.4) for the internal energy we are in principle able to determine $\delta$, $c_P$, and $\nabla_{ad}$. Since in general no analytic methods are known one can try to determine the thermodynamic quantities numerically. Here we just give them for some limiting cases for which analytic expressions can be derived. For the sake of simplicity we neglect the effects of radiation and we suppose the ions to be an ideal gas.

In the cases of complete degeneracy of a non-relativistic or an extremely relativistic electron gas, it is obvious from equations (15.23,26) that the quantities $\alpha$, $\delta$ as defined in (4.2,3) are $\alpha = 3/5$, $\delta = 0$ or $\alpha = 3/4$, $\delta = 0$ respectively.

We define the ratio $\eta$ of ion pressure to total pressure

$$\eta := \frac{P_{\text{ion}}}{P_{\text{ion}} + P_{\text{e}}} \quad . \tag{16.11}$$

For strong non-relativistic degeneracy (15.39,41), and (15.43) for $\psi \gg 1$, imply that

$$P_{\text{e}} \approx \frac{4}{15} B_1 (\psi k T)^{5/2} \quad , \quad B_1 = \frac{4\pi}{h^3} (2m_{\text{e}})^{3/2} \quad ,$$

$$\varrho \approx \frac{2}{3} \mu_{\text{e}} m_{\text{u}} B_1 (\psi k T)^{3/2} \quad , \tag{16.12}$$

which together with $P_{\text{ion}} = \Re \varrho T / \mu_0 = k \varrho T / (m_{\text{u}} \mu_0)$ and (16.11) result in

$$\eta \approx \frac{5}{2} \frac{\mu_{\text{e}}}{\mu_0} \frac{1}{\psi} \quad . \tag{16.13}$$

The larger $\psi$ (the stronger the degeneracy), the smaller $\eta$, and therefore the smaller the contribution of the ion gas to the total pressure.

The value of $\delta$ can be obtained from the relation

$$\delta = -\left(\frac{\partial \ln \varrho}{\partial \ln T}\right)_P = -\left(\frac{\partial \ln \varrho}{\partial \ln T}\right)_\psi + \frac{\left(\frac{\partial \ln \varrho}{\partial \ln \psi}\right)_T \left(\frac{\partial \ln P}{\partial \ln T}\right)_\psi}{\left(\frac{\partial \ln P}{\partial \ln \psi}\right)_T} \quad , \tag{16.14}$$

which follows from the total differentials of the functions $\varrho = \varrho(\psi, T)$, $P = P(\psi, T)$. For $P = P_{\text{e}}$ the partial derivatives can be taken from (16.12), and (16.14) gives $\delta = 0$. For a small but non-vanishing contribution $P_{\text{ion}}$ we write according to (16.11) the total pressure $P = P_{\text{e}}/(1-\eta) \approx (1+\eta)P_{\text{e}}$. If we then use the expressions (16.12,13), we obtain for the non-relativistic case

$$\delta \approx \frac{3}{5}\eta \approx \frac{3}{2}\frac{\mu_e}{\mu_0}\frac{1}{\psi} \quad . \tag{16.15}$$

For the extremely relativistic electron gas we find from (15.45,46), with the lowest terms of the expansion (15.43), that

$$P_e = \frac{B_2}{4}(\psi kT)^4 \quad , \quad B_2 = \frac{8\pi}{3c^3h^3} \quad ,$$
$$\varrho = \mu_e m_u B_2(\psi kT)^3 \tag{16.16}$$

and in the same way we obtained (16.13, 15) we now get

$$\eta \approx 4\frac{\mu_e}{\mu_0}\frac{1}{\psi} \quad , \quad \delta = \frac{3}{4}\eta = \frac{3\mu_e}{\mu_0}\frac{1}{\psi} \quad . \tag{16.17}$$

In order to derive $c_P$ we need the internal energy $u$. Let us again neglect the radiation field here; then $u$ contains a component $u_e$ of the (degenerate) electron gas and a component $u_{ion}$ of the (ideal) ion gas: $u = u_e + u_{ion}$. In the non-relativistic case, (15.42) gave $U_e = 3P_e/2$ for the internal energy $U_e$ per unit volume of the electron gas. A corresponding relation $U_{ion} = 3P_{ion}/2$ holds for the non-degenerate ions, and therefore

$$u = \frac{U}{\varrho} = \frac{3}{2}\frac{P_{ion} + P_e}{\varrho} = \frac{3}{2}\frac{P}{\varrho} \quad . \tag{16.18}$$

This gives the derivative

$$\left(\frac{\partial u}{\partial T}\right)_P = -\frac{3}{2}\frac{P}{\varrho T}\left(\frac{\partial \ln \varrho}{\partial \ln T}\right)_P = \frac{3}{2}\frac{P\delta}{\varrho T} \quad , \tag{16.19}$$

which is used in the definition (4.4) of $c_P$:

$$c_P = \left(\frac{\partial u}{\partial T}\right)_P - \frac{P}{\varrho^2}\left(\frac{\partial \varrho}{\partial T}\right)_P = \left(\frac{\partial u}{\partial T}\right)_P + \frac{P\delta}{\varrho T} = \frac{5}{2}\frac{P\delta}{\varrho T} \quad . \tag{16.20}$$

Then (4.21) gives $\nabla_{ad} = 2/5$, the same value we obtained for the ideal gas with $\beta = 0$ [see (13.21)]. Since we have derived it without making use of the degree of degeneracy, the numerical value 2/5 for $\nabla_{ad}$ is independent of $\psi$, but holds only for non-relativistic degeneracy.

In the extreme relativistic case, (15.27) shows that $U_e = 3P_e$, while again $U_{ion} = 3P_{ion}/2$ for the non-degenerate ions. The total energy density is then

$$u = u_e + u_{ion} = 3\frac{P_e}{\varrho} + \frac{3}{2}\frac{P_{ion}}{\varrho} = 3\frac{P}{\varrho} - \frac{3}{2}\frac{P_{ion}}{\varrho} = 3\frac{P}{\varrho} - \frac{3}{2}\frac{\Re}{\mu_0}T \quad , \tag{16.21}$$

the specific heat is

$$c_P = \left(\frac{\partial u}{\partial T}\right)_P - \frac{P}{\varrho^2}\left(\frac{\partial \varrho}{\partial T}\right)_P = -\frac{4P}{\varrho^2}\left(\frac{\partial \varrho}{\partial T}\right)_P - \frac{3}{2}\frac{\Re}{\mu_0} = \frac{4P}{\varrho T}\delta - \frac{3}{2}\frac{\Re}{\mu_0} \quad . \tag{16.22}$$

133

so that we can now determine $\nabla_{ad}$:

$$\nabla_{ad} = \frac{P\delta}{\varrho T c_P} = \frac{1}{4 - \frac{3}{2}\frac{\Re}{\mu_0}\frac{\varrho T}{P\delta}} . \qquad (16.23)$$

From (16.16,17) we find that

$$P \approx P_e = \frac{B_2}{4}(\psi kT)^4 \quad , \quad \varrho = B_2\mu_e m_u(\psi kT)^3 \quad , \quad \delta = 3\frac{\mu_e}{\mu_0}\frac{1}{\psi} \quad , \qquad (16.24)$$

and therefore $3\Re\varrho T/\mu_0 = 4P\delta$, which with (16.23) gives $\nabla_{ad} = 1/2$. This is the value for the fully degenerate, extreme relativistic case.

## 16.4 Crystallization

Up to now we have treated the ions as an ideal gas, which means we have neglected their interaction. However, this no longer suffices for high densities and particularly low temperatures, in which case the Coulomb interaction of the ions must be considered: instead of moving freely, the ions tend to form a rigid lattice, which minimizes their total energy. This occurs when the thermal energy $3kT/2$ becomes comparable with the Coulomb energy per ion of charge $-Ze$. If we define a volume $V_{ion}$ per ion by $n_{ion}V_{ion} = 1$ (where $n_{ion}$ is the number density of ions) and a mean separation $r_{ion}$ between the ions, we have $V_{ion} = 4\pi r_{ion}^3/3$. Then the ratio

$$\Gamma_C := \frac{(Ze)^2}{r_{ion}kT} = 2.7 \times 10^{-3}\frac{Z^2 n_{ion}^{1/3}}{T} \qquad (16.25)$$

is a measure for the importance of this effect, the numerical constant having units of cgs. $\Gamma_C \ll 1$ would mean that the electrostatic energy plays a minor role and the ions have a Boltzmann distribution, while $\Gamma_C \gg 1$ indicates that the kinetic energy of the ions is negligible and that they try to form a conglomerate that has a lower energy, i.e. they form a crystal.

More detailed considerations (see, for instance, SHAPIRO, TEUKOLSKY, 1983) indicate that $\Gamma_C \approx 100$ is a critical value for the transition between the two types of behaviour of the ion gas. With this value for $\Gamma_C$ and using the relation $\varrho = \mu_0 m_u n_{ion}$ we obtain the critical temperature $T_m$ (melting temperature)

$$T_m \approx \frac{Z^2 e^2}{\Gamma_c k}\left(\frac{4\pi\varrho}{3\mu_0 m_u}\right)^{1/3} = 2.3 \times 10^3 Z^2 \mu_0^{-1/3}\varrho^{1/3} \quad , \qquad (16.26)$$

where the numerical constant is in cgs units. The corresponding straight line is plotted (dot-dashed) in Fig. 16.1.

In the interior of evolved stars we have high densities, but the temperature is well above the melting temperature. The situation is different in cooling white dwarfs, where the temperature becomes smaller with time, while the density remains virtually unchanged. We will come back to this in §35, which deals with white dwarfs.

## 16.5 Neutronization

If in a plasma the electrons have sufficient energy, they can combine with the protons to form neutrons. If $m_n$ and $m_p$ are the masses of neutron and proton, then the electron must have the total energy $E_{tot} > E^* = c^2(m_n - m_p)$. At low densities the neutron will decay within 11 minutes back into a proton–electron pair, where the electron has the total energy $E^*$ and a kinetic energy $E^*_{kin} = E^* - m_e c^2$; however, the situation can be different if the gas is completely degenerate and the phase space is filled up to the (kinetic) Fermi energy $E_F$. If the Fermi energy $E_F$ exceeds $E^*_{kin}$, the electrons released do not have enough energy to find an empty cell in phase space and the neutrons cannot decay, i.e. the Fermi sea has stabilized the neutrons.

In order to estimate under which conditions this occurs we write the relation (15.6) between $E$ and $p$ in the form

$$p = \frac{1}{c} \left( E^2 - m_e^2 c^4 \right)^{1/2} . \tag{16.27}$$

If we put $E = E_{kin} + m_e c^2 = E_F + m_e c^2 = c^2(m_n - m_p) = 1.294 \times 10^6$ eV, we can determine the corresponding Fermi momentum $p_F$ from (16.27) and obtain $x = p_F/(m_e c) \approx 2.2$. Then according to (15.15) and taking $\varrho = \mu_e m_u n_e$ with $\mu_e = 2$, we find $\varrho \approx 2.4 \times 10^7$ g cm$^{-3}$. Therefore if a proton–electron gas is compressed to a density above this value, then the gas undergoes a transition into a neutron gas ("neutronization").

For stellar matter the situation is more complicated, since at sufficiently high densities the plasma contains heavier nuclei, and not just protons. The nuclei capture electrons (inverse $\beta$ decay) and become neutron-rich isotopes. This requires much higher electron energies than those just estimated, since the neutrons in the nucleus are degenerate and the new ones have to be raised above the Fermi energy. Correspondingly higher plasma densities are required to provide the electrons with the necessary energy. If the nuclei become too neutron rich they start to break up, releasing free neutrons. This "neutron drip" starts at $\varrho_{drip} \approx 4 \times 10^{11}$ g cm$^{-3}$.

Let us briefly consider the effect on the equation of state. Up to $\varrho_{drip}$ the total pressure $P \approx P_e$ is provided by relativistic electrons. With further increases of $\varrho$, the number density $n_e$ increases by less than an amount proportional to $\varrho$, owing to the capture of some electrons. Therefore the pressure rises by less than $\varrho^{4/3}$. Consequently $\gamma_{ad} \equiv (d \ln P/d \ln \varrho)_{ad}$ is reduced below 4/3, which can be seen in Fig. 16.2, where the slope of the curve $P = P(\varrho)$ is suddenly reduced for $\log \varrho \gtrsim$

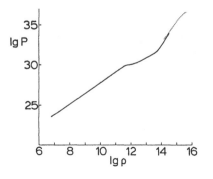

Fig. 16.2. The equation of state for very high densities. On logarithmic scales the pressure $P_e$ (in dyn cm$^{-2}$) is plotted against the density $\varrho$ (in g cm$^{-3}$). Solid line after HEINTZMANN et al. (1974); dotted line after ARPONEN (1972)

11.7. At still higher $\varrho$ the increasing number of free neutrons contribute gradually more to $P$.

With increasing $\varrho$ the neutrons become increasingly degenerate – as an ideal Fermi gas they would give the slope 5/3. But then interaction between neutrons becomes important, and the details of the equation of state are very uncertain, for example depending on rather badly known properties of the particles. For more details see § 35.2, § 36.1 and SHAPIRO, TEUKOLSKY (1983).

# § 17 Opacity

In this section we deal with the material function $\kappa(\varrho, T)$. While for the equation of state it was possible to use certain approximations (for instance, that of an ideal gas) without introducing too much error, this is almost impossible for the opacity. Although there are similar approximations (such as those for electron scattering or free–free transitions) they never hold for the whole star and are used only in simplifying approaches. Therefore nowadays, when solving the stellar-structure equations, one uses numerical opacity tables for different chemical mixtures, which give $\kappa(\varrho, T)$ in the full range of $\varrho$ and $T$.

In the following we describe the basic processes that contribute to the opacity and give approximate analytic formulae without deriving them from quantum mechanics. The reader who wants to learn more of the methods by which opacities are computed is referred to COX, GIULI (1968) and to the original papers quoted there.

## 17.1 Electron Scattering

If an electromagnetic wave passes an electron, the electric field makes the electron oscillate. The oscillating electron represents a classical dipole that radiates in other directions, i.e. the electron scatters part of the energy of the incoming waves. The weakening of the original radiation due to scattering is equivalent to that by absorption, and we can describe it by way of a cross-section at frequency $\nu$ per unit mass (which we called $\kappa_\nu$ in § 5.1). This can be calculated classically giving the result

$$\kappa_\nu = \frac{8\pi}{3} \frac{r_e^2}{\mu_e m_u} = 0.20(1 + X) \quad , \tag{17.1}$$

where $r_e$ is the classical electron radius, $X$ the mass fraction of hydrogen, and the constant is in $\text{cm}^2\text{g}^{-1}$. The term $\mu_e m_u$ arises because $\kappa_\nu$ is taken per unit mass; and $\mu_e$ is replaced by (13.9). Since $\kappa_\nu$ does not depend on the frequency, we immediately obtain the Rosseland mean for electron scattering:

$$\kappa_{\text{sc}} = 0.20(1 + X)\,\text{cm}^2\,\text{g}^{-1} \quad . \tag{17.2}$$

The "Thomson scattering" just described neglects the exchange of momentum between electron and radiation. If this becomes important, then $\kappa_\nu$ will be reduced compared to the value given in (17.1), though this effect plays a role only at temperatures sufficiently high for the scattered photons to be very energetic. In fact during the scattering process the electron must obtain such a large momentum that its velocity is comparable to $c$, say $v \gtrsim 0.1c$ for (17.2) to become a bad approximation. The momentum of the photon is $h\nu/c$, which after scattering is partly transferred to the electron, $m_e v \sim h\nu/c$. Therefore relativistic corrections ("Compton scattering") become important if the average energy of the photons is $h\nu \gtrsim 0.1 m_e c^2$. For $h\nu$ we take the frequency at which the Planck function has a maximum; then according to Wien's law this is at $h\nu = 4.965\, kT$ and the full Compton scattering cross-section has to be taken into account if $T > 0.1 m_e c^2/(4.965k)$, or roughly $T > 10^8$K. In fact even at $T = 10^8$K Compton scattering reduces the opacity by only 20% of that given by (17.2).

## 17.2 Absorption Due to Free–Free Transitions

If during its thermal motion a free electron passes an ion, the two charged particles form a system which can absorb and emit radiation. This mechanism is only effective as long as electron and ion are sufficiently close. Now, the mean thermal velocity of the electrons is $v \sim T^{1/2}$, and the time during which they form a system able to absorb or emit is proportional to $1/v \sim T^{-1/2}$; therefore, if in a mass element the numbers of electrons and ions are fixed, the number of systems temporarily able to absorb is proportional to $T^{-1/2}$.

The absorption properties of such a system have been derived classically by Kramers, who calculated that the absorption coefficient per system is proportional to $Z^2 \nu^{-3}$, where $Z$ is the charge number of the ion. We therefore expect the absorption coefficient $\kappa_\nu$ of a given mixture of (fully ionized) matter to be

$$\kappa_\nu \sim Z^2 \varrho T^{-1/2} \nu^{-3} \quad . \tag{17.3}$$

Here the factor $\varrho$ appears because for a given mass element the probability that two particles are accidentally close together is proportional to the density.

For the determination of the Rosseland mean $\kappa$ of this absorption coefficient we make use of a simple theorem which can be easily proved by carrying out the integration (5.19): a factor $\nu^\alpha$ contained in $\kappa_\nu$ gives a factor $T^\alpha$ in $\kappa$. With this and with (17.3) we find

$$\kappa_{\mathrm{ff}} \sim \varrho\, T^{-7/2} \quad . \tag{17.4}$$

All opacities of the form (17.4) are called *Kramers opacities* and give only a classical approximation. One normally multiplies the Kramers formula (17.4) by a correction factor $g$, the so-called *Gaunt factor*, in order to take care of the quantum-mechanical correction (see, for instance, COX, GIULI, 1968). In (17.4) we have still omitted the factor $Z^2$ which appears in (17.3). In general one has a mixture of different ions, and therefore one has to add the contributions of the different chemical species. The (weighted) sum over the values of $Z^2$ is taken into the constant of proportionality in (17.4), which then depends on the chemical composition. For a fully ionized mixture a good approximation is given by

$$\kappa_{\mathrm{ff}} = 3.8 \times 10^{22}(1+X)[(X+Y)+B]\varrho\, T^{-7/2} \quad , \tag{17.5}$$

with the numerical constant in cgs. The mass fractions of H and He are $X$ and $Y$ respectively. Here the factor $1 + X$ arises, since $\kappa_{ff}$ must be proportional to the electron density – which is proportional to $(1+X)\varrho$. The term $(X+Y)$ in the brackets can be understood in the following way: there are $X/m_u$ hydrogen ions and $Y/(4m_u)$ helium ions. The former have the charge number 1, the latter the charge number 2. But since $\kappa_\nu \sim Z^2$ [see (17.3)], when adding the contributions of H and He to the total absorption coefficient we obtain the factor $X/m_u + 4Y/(4m_u) = (X + Y)m_u$. Correspondingly the term $B$ gives the contribution of the heavier elements:

$$B = \sum_i \frac{X_i Z_i^2}{A_i} \quad , \tag{17.6}$$

where the summation extends over all elements higher than helium and $A_i$ is the atomic mass number.

## 17.3 Bound–Free Transitions

We first consider a (neutral) hydrogen atom in its ground state, with an ionization energy of $\chi_0$, i.e. a photon of energy $h\nu > \chi_0$ can ionize the atom. Energy conservation then demands that

$$h\nu = \chi_0 + \frac{1}{2}m_e v^2 \quad , \tag{17.7}$$

where $v$ is the velocity of the electron released (relative to the ion, which is assumed to be at rest before and after ionization).

If we define an absorption coefficient $a_\nu$ per ion ($a_\nu = \kappa_\nu \varrho / n_{ion}$), we expect $a_\nu = 0$ for $\nu < \chi_0/h$ and $a_\nu > 0$ for $\nu \geq \chi_0/h$. Classical considerations similar to those which lead to the Kramers dependence (17.3) of $\kappa_\nu$ for free–free transitions give $a_\nu \sim \nu^{-3}$ for $\nu \geq \chi_0/h$. Quantum-mechanical corrections can again be taken into account by a Gaunt factor (see, for instance, COX, GIULI, 1968). The absorption coefficient of the hydrogen atom in its ground state has a frequency dependence as given in Fig. 17.1a. But if we have neutral hydrogen atoms in different stages

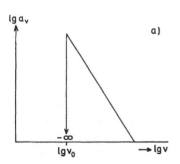
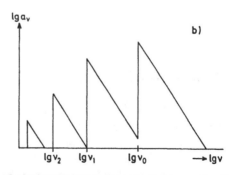

Fig. 17.1. (a) The absorption coefficient $a_\nu$ of a hydrogen atom in the ground state as a function of the frequency $\nu$. (b) The absorption coefficient of a mixture of hydrogen atoms in different stages of excitation

of excitation, the situation is different: an atom in the first excited stage has an absorption coefficient $a_\nu = 0$ for $h\nu < \chi_1$, where $\chi_1$ is the energy necessary to ionize a hydrogen atom from the first excited state, while $a_\nu \sim \nu^{-3}$ for $h\nu \geq \chi_1$. The absorption coefficient $\kappa_\nu$ for a mixture of hydrogen atoms in different states of excitation is a superposition of the $a_\nu$ for different stages of excitation. The resulting $\kappa_\nu$ is a saw-tooth function, as indicated in Fig. 17.1b. In order to obtain $\kappa_\nu$ for a certain value of the temperature $T$, one has to determine the relative numbers of atoms in the different stages of excitation by the Boltzmann formula; then their absorption coefficients $a_\nu$, weighted with their relative abundances, are to be summed. To obtain the Rosseland mean one has to carry out the integration (5.19).

If there are ions of different chemical species with different degrees of ionization, one has to sum the functions $a_\nu$ for all species in all stages of excitation and all degrees of ionization before carrying out the Rosseland integration. An important source of opacity are bound–free transitions of neutral hydrogen atoms, in which case the opacity must be proportional to the number of neutral hydrogen atoms and $\kappa$ can be written in the form

$$\kappa_{bf} = X(1 - x)\, \tilde{\kappa}(T) \quad . \tag{17.8}$$

Here $\tilde{\kappa}(T)$ is obtained by Rosseland integration over (weighted) sums of functions $a_\nu$ for the different stages of excitation, while $x$ is the degree of ionization as defined in § 14.2. The function $\tilde{\kappa}(T)$ is plotted in Fig. 17.2.

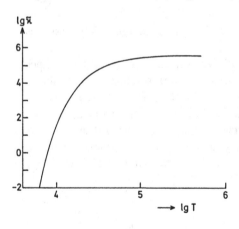

Fig. 17.2. The function $\tilde{\kappa}(T)$ of (17.8), where $\tilde{\kappa}$ is in cm$^2$ g$^{-1}$ and $T$ in K

## 17.4 Bound–Bound Transitions

For absorption by an electron bound to an ion, more than just the bound–free transitions discussed in § 17.3 contribute to the opacity. If, after absorption of a photon from a directed beam, the electron does not leave the atom but jumps to a higher bound state, the energy will later on be re-emitted in an arbitrary direction, so that the intensity of the directed beam is weakened. This mechanism is effective only at certain frequencies, and one would expect that absorption in a few lines gives only

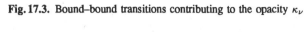

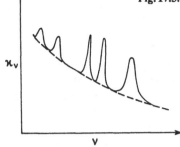

a small contribution to the overall opacity; however, the absorption lines in stars are strongly broadened by collisions, and as one can see in Fig. 17.3 they can occupy considerable regions of the spectrum. Bound–bound absorption can become a major contribution to the (Rosseland mean) opacity if $T < 10^6$K. It can then increase the total opacity by a factor 2, while for higher temperatures (say $T \approx 10^7$K) the contribution of bound–bound transitions to the total opacity is much smaller (10%).

## 17.5 The Negative Hydrogen Ion

Hydrogen can become a source of opacity in another way, by forming negative ions: a neutral hydrogen atom is polarized by a nearby charge and can then attract and bind another electron. This is possible since there exists a bound state for a second electron in the field of a proton, though this second electron is only loosely bound – the absorption of photons with $h\nu > 0.75$ eV is sufficient for its release. This energy is very small compared to the 13.6 eV ionization energy for neutral hydrogen and allows photons with $\lambda < 1655$ nm (infrared) to be absorbed, giving rise to a bound–free transition. The photon energy goes into the ionization energy and kinetic energy of the free electron in the same way as indicated in (17.7). The number of negative hydrogen ions in thermodynamic equilibrium is given by the Saha formula (14.17), where the ionization potential $\chi_r$ is the binding energy of the second electron. Replacing the partition functions by the statistical weights, we have $u_{-1} = 1$ for the negative ion and $u_0 = 2$ for neutral hydrogen; hence the Saha equation gives

$$\frac{n_0}{n_{-1}} P_{\mathrm{e}} = 4 \, \frac{(2\pi m_{\mathrm{e}})^{3/2}(kT)^{5/2}}{h^3} \, \mathrm{e}^{-\chi/kT} \quad , \tag{17.9}$$

with $\chi = 0.75$ eV. If we use $n_0 = (1-x)\varrho X/m_{\mathrm{u}}$, where $x$ is the degree of ionization of hydrogen as defined in (14.18) and $X$ the weight fraction of hydrogen, we find

$$n_{-1} = \frac{1}{4} \, \frac{h^3}{(2\pi m_{\mathrm{e}})^{3/2}(kT)^{5/2}m_{\mathrm{u}}} \, P_{\mathrm{e}}(1-x)X\varrho \, \mathrm{e}^{\chi/kT} \quad . \tag{17.10}$$

Now, for an absorption coefficient $a_\nu$ per $\mathrm{H}^-$ ion, it follows that $\kappa_\nu = a_\nu n_{-1}/\varrho$, which implies that the Rosseland mean is described by

$$\kappa_{H^-} = \frac{1}{4} \frac{h^3}{(2\pi m_e)^{3/2}(kT)^{5/2}m_u} P_e(1-x)X\, a(T)\, e^{\chi/kT} \quad , \qquad (17.11)$$

where $a = a(T)$ is obtained from $a_\nu$ by Rosseland integration (5.19). The opacity $\kappa_{H^-}$ is proportional to $n_{-1}$, which in turn is proportional to $n_0 n_e$ (or $n_0 P_e$), since the $H^-$ ions are formed from *neutral* hydrogen atoms and free electrons.

For a completely neutral, pure hydrogen gas there would be no free electrons and therefore no $H^-$ ions. If now the temperature is increased and the hydrogen becomes slightly ionized, giving $n_e \sim X$, the free electrons can combine with neutral hydrogen atoms. One therefore would expect an increase of $\kappa$ as long as $1 - x$ is not too small.

The situation is different in the case of a more realistic mixture of stellar material. Heavier elements have lower ionization potentials (a few eV) and provide electrons even at relatively low temperatures; hence, although there is only a small mass fraction of heavier elements, they determine the electron density at low temperatures where hydrogen is neutral. When the elements heavier than helium are singly ionized (say from 3000 K to 5000 K) one has

$$n_e = \varrho\left[x\,X + (1 - X - Y)/A\right]/m_u \quad , \qquad (17.12)$$

where $\varrho(1 - X - Y)/(Am_u)$ is the number density of atoms of higher elements ("metals") of mean mass number $A$. Even if the metals constitute only a small percentage in weight (and number), they still determine the opacity as long as $1 - X - Y > xXA$ (which becomes very small for low temperatures where $x$ is small). The metal content can therefore be of great influence on $\kappa$ for the surface layers and thus the outer boundary conditions of stars.

## 17.6 Conduction

Electrons, like all particles, can transport heat by conduction. Their contribution to the total energy transport can normally be neglected compared to that of photons, since the conductivity is proportional to the mean free path $\ell$, and in normal (non-degenerate) stellar material $\ell_{photon} \gg \ell_{particle}$.

However, conduction by electrons becomes important in the dense degenerate regions in the very interior of evolved stars, as well as in white dwarfs. The reason is that in the case of degeneracy all quantum cells in phase space below $p_F$ are filled up, and electrons, when approaching ions and other electrons, have difficulty exchanging their momentum. This is equivalent to saying that "encounters" are rare or that the mean free path is large. In § 5.2 we saw that the contribution to conduction can be formally taken into account in the equation of radiative transport by defining a "conductive opacity" $\kappa_{cd}$, as in (5.24). If $\kappa_{rad}$ is the Rosseland mean of the (radiative) opacity, then conduction reduces the "total" opacity $\kappa$, as can be seen from (5.25):

$$\frac{1}{\kappa} = \frac{1}{\kappa_{rad}} + \frac{1}{\kappa_{cd}} \quad . \qquad (17.13)$$

The thermal conductivity of the electron component of a gas is mainly determined by collisions between electrons and ions, but electron–electron collisions can also be important. Analytic formulae can be found in COX, GIULI (1968), while tables of the thermal conductivity due to electrons in stellar material have been computed by HUBBARD, LAMPE (1969). They list the conductivities of a pure hydrogen gas, a mixture of pure helium and pure carbon, a solar composition, and a mixture typical for the core of an evolved star.

Figure 17.4 shows the dependence of the conductive opacity on density for a given temperature. For extremely strong degeneracy, $\kappa_{cd}$ is proportional to $\varrho^{-2}T^2$.

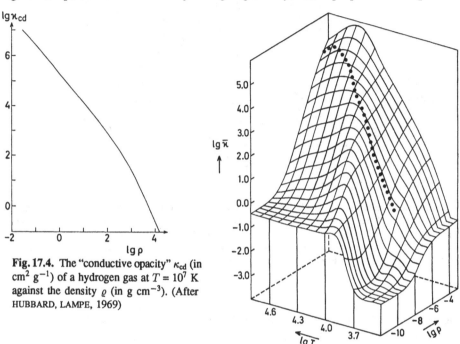

**Fig. 17.4.** The "conductive opacity" $\kappa_{cd}$ (in cm$^2$ g$^{-1}$) of a hydrogen gas at $T = 10^7$ K against the density $\varrho$ (in g cm$^{-3}$). (After HUBBARD, LAMPE, 1969)

**Fig. 17.5.** The Rosseland mean of the opacity $\kappa$ (in cm$^2$ g$^{-1}$) as a function of $\varrho$ (in g cm$^{-3}$) and $T$ (in K) for a mixture with a hydrogen and helium content $X = 0.739$, $Y = 0.240$, respectively, according to calculations using the Los Alamos code for the outer layers of stars. The dotted line indicates a solar model, starting at the right end with the photosphere. The dominant absorption mechanisms at different parts of the model are discussed in the text. The continuation towards deeper regions is shown in Fig. 17.6

## 17.7 Opacity Tables

Several authors have published extensive tables of opacities for different chemical mixtures over a wide range of temperatures and densities. In Figs. 17.5,6 we give a graphical representation of opacities obtained with the Los Alamos opacity code. Indeed one sees that, over the whole range of arguments, $\kappa(\varrho, T)$ is a rather complicated function. In order to give a feeling for the parts of the plotted surface that are relevant to stars, we discuss a model of the present sun computed for a chemical mixture of $X = 0.690$, $Y = 0.289$, and $Z = 1 - X - Y = 0.021$. This model is

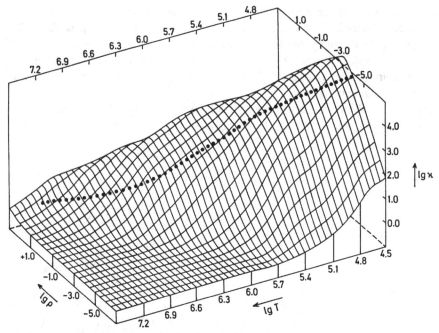

**Fig. 17.6.** Continuation of the display of opacity $\kappa$ of Fig. 17.5 to larger $\varrho$ and $T$, i.e. for the deeper regions of stars. (Note that the axes have different orientations in each illustration.) The dotted line continues to represent a solar model (for details see text). Electron scattering provides the flat region at the lower left. The plotted opacity surface drops away behind the visible part (beyond the ridge of the mountain, so to speak) owing to the reduction of effective opacity by conduction

plotted in Figs. 17.5,6 (heavily dotted line), and although the opacities there are for a somewhat different mixture, the main features are still visible.

The model starts with the photospheric values $\lg T = 3.76$, $\lg \varrho = -6.73$ (in cgs). The corresponding point lies on the right end of the dotted line in Fig. 17.5. On moving deeper into the sun the opacity sharply increases owing to the onset of hydrogen ionization, which provides the electrons for $H^-$ formation as described in § 17.5, and the opacity rises by several powers of 10 until it reaches a maximum value. This occurs when an appreciable amount of hydrogen becomes ionized and is not available for $H^-$ formation, because the factor $1-x$ in (17.11) reduces the opacity. In the regions below, bound–free transitions become the leading opacity source and still further inwards free–free transitions take over. There a simple power law seems to be a good approximation, as indicated in (17.4). Note that in the logarithmic representation the opacity surface for a power law is just a plane. Equation (17.4) therefore corresponds to a tangential plane which osculates the opacity surface. The line for the interior remains in the domain of free–free transitions. The region of dominant electron scattering is the horizontal plateau on the left of Fig. 17.6 at the foot of the "kappa mountain". In this figure the region where electron conduction reduces the (total) opacity cannot be seen, since this part of the surface is on the other side of the mountain.

In order to find the value of $\kappa(\varrho, T, X_i)$ for a given point with $\varrho_0, T_0, X_{i0}$ in a star, one has to interpolate in different opacity tables (for different compositions $X_i$) for the arguments $\varrho_0, T_0$ and then between these tables for $X_{i0}$. Such interpolations can be quite problematic.

The calculation of opacity tables requires very involved numerical computations, including approximations and procedures that can introduce appreciable uncertainties. Tables are published in many places. The classical Los Alamos opacities are found in COX, STEWART (1965, 1970) and in the review article by MEYER-HOFMEISTER (1982); see also CARSON (1976). For opacities including the effects of molecules, see ALEXANDER, JOHNSON, RYPMA (1983). For the Los Alamos opacities that are especially computed for a solar mixture, the reader is referred to HÜBNER, FRIEDLÄNDER (1978).

# §18 Nuclear Energy Production

We shall limit ourselves here to a very rough summary of the most important features of nuclear reactions in stars. This will suffice completely for the consideration of the main band of stellar structures, while the study of particular aspects of nuclear astrophysics anyway requires the consultation of specialized literature (see CLAYTON, 1968). For example, we will only deal with energy production of equilibrium nuclear burning, i.e. we will neglect the effects occurring when the time-scale of a rapidly changing star becomes comparable to that of an important nuclear reaction. On the other hand, we will also briefly touch on such topics as electron screening or neutrino production, about which a certain minimum of information seems to be indispensible for general discussions.

We begin with a few historical comments. That thermonuclear reactions can provide the energy source for the stars was first shown by R. Atkinson and F. Houtermans in 1929, after G. Gamow discovered the tunnel effect. Later, two important discoveries were published almost simultaneously in 1938: H. Bethe and Ch. Critchfield described the *pp* chain and C.F. von Weizsäcker and H. Bethe independently found the CNO cycle. The reactions of helium burning were then described in 1952 by E.E. Salpeter. Finally, a classic paper summarized the state of the art in 1957, "Synthesis of the Elements in Stars" (BURBIDGE, BURBIDGE, FOWLER, HOYLE, 1957).

## 18.1 Basic Considerations

Most observed stars (including the sun) live on so-called thermonuclear fusion. In such nuclear reactions, induced by the thermal motion, several lighter nuclei fuse to form a heavier one. Before this process, the involved nuclei $j$ have a total mass ($\sum M_j$) different from that of the product nucleus ($M_y$). The difference is called the *mass defect*

$$\Delta M = \sum_j M_j - M_y \quad . \tag{18.1}$$

It is converted into energy according to Einstein's formula

$$E = \Delta M c^2 \tag{18.2}$$

and is available (at least partly) for the star's energy balance. An example is the series of reactions called "hydrogen burning", where four hydrogen nuclei $^1$H with a total mass $4 \times 1.0081 m_u$ (atomic mass units, physical scale) are transformed into one

$^4$He nucleus of $4.0039m_u$. Obviously $2.85 \times 10^{-2}m_u$ per produced $^4$He nucleus have "disappeared", which is roughly 0.7% of the original masses and which corresponds, to an energy of about 26.5 MeV according to (18.2). As usual in nuclear physics, as the unit of energy we take the electron volt eV (1 eV = $1.6020 \times 10^{-12}$ erg) with the following equivalences:

$$1 \text{ keV} \cong 1.1605 \times 10^7 \text{ K} \quad,$$
$$931.1 \text{ MeV} \cong 1 \ m_u \quad. \tag{18.3}$$

The sun's luminosity corresponds to a mass loss rate of $L_\odot/c^2 = 4.25 \times 10^{12} \text{ g s}^{-1}$, which appears to be a lot, especially if it is read as "more than 4 million metric tons per second". If a total of 1 $M_\odot$ of hydrogen were converted into $^4$He, then the disappearing 0.7% of this mass would be $1.4 \times 10^{31}$ g, which could balance the sun's present mass loss by radiation for about $3 \times 10^{18}$ s $\approx 10^{11}$ years.

The deficiency of mass is just another aspect of the fact that the involved nuclei have different binding energies $E_B$. This is the energy required to separate the nucleons (protons and neutrons in the nucleus) against their mutual attraction by the strong, but short-range nuclear forces. Or else $E_B$ is the energy gained if they are brought together from infinity (which starts here at any distance large compared with, say, $10^{-12}$ cm).

Consider a nucleus of mass $M_{nuc}$ and atomic mass number $A$ (the integer "atomic weight"): it may contain $Z$ protons of mass $m_p$ and $(A-Z)$ neutrons of mass $m_n$. Its binding energy is then related to these masses by (18.2):

$$E_B = \left[(A-Z)m_n + Zm_p - M_{nuc}\right] c^2 \quad. \tag{18.4}$$

When comparing different nuclei, it is more instructive to consider the *average binding energy per nucleon*,

$$f = \frac{E_B}{A} \quad, \tag{18.5}$$

which is also called the *binding fraction*. With the exception of hydrogen, typical values are around 8 MeV, with relatively small differences for nuclei of very different $A$. This shows that the short-range nuclear forces due to a nucleon mainly affect the nucleons in its immediate neighbourhood only, such that with increasing $A$ a saturation occurs rather than an increase of $f$ proportional to $A$. An idealized plot of $f$ against $A$ is shown in Fig. 18.1. (The real curve zigzags around this smoothed curve as a consequence of the shell structure of the nucleus and pair effects.)

With increasing $A$, $f(A)$ rises steeply from hydrogen, then flattens out and reaches a maximum of 8.5 MeV at $A = 56$ ($^{56}$Fe), after which it drops slowly towards large $A$. The increase for $A < 56$ is a surface effect: particles at the surface of the nucleus experience less attraction by nuclear forces than those in the interior, which are completely surrounded by other particles. And in a densely packed nucleus, the surface area increases with radius slower than the volume (i.e. the number $A$) such that the fraction of surface particles drops. With increasing $A$, the number $Z$ of protons also increases. (The addition of neutrons only would require higher energy states, because the Pauli principle excludes more than two identical neutrons, and

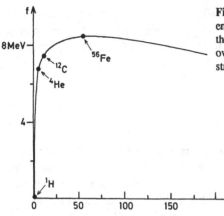

**Fig. 18.1.** A smoothed run of the fractional binding energy per nucleon, $f = E_B/A$, for stable nuclei, over the atomic mass number $A$. The curve is smoothed over the wiggles which are due to the nuclear shell structure and pair effects.

the nuclei would be unstable.) The positively charged protons experience a repulsive force which is far-reaching and therefore does not show the saturation of the nuclear forces. This increasing repulsion by the Coulomb forces brings the curve in Fig. 18.1 down again for $A > 56$.

Around the maximum, at $^{56}$Fe, we have the most tightly bound nuclei. In other words, the nucleus of $^{56}$Fe has the smallest mass per nucleon, so that any nuclear reaction bringing the nucleus closer to this maximum will be exothermic, i.e. will release energy. There are two ways of doing this:

(i)  either by fission of heavy nuclei, which happens, e.g., in radioactivity,

(ii) or by fusion of light nuclei, which is the prime energy source of stars (and possibly ours too in the future).

Clearly, both reach an end when one tries to extend them over the maximum of $f$, which is therefore a natural finishing point for the stellar nuclear engine. So if a star initially consisted of pure hydrogen, it could gain a maximum of about 8.5 MeV per nucleon by fusion to $^{56}$Fe; but 6.6 MeV of these are already used up when $^4$He is built up in the first step.

In order to obtain a fusion of charged particles, they have to be brought so close to each other that the strong, but very short-ranged, nuclear forces dominate over the weaker, but far-reaching, Coulomb forces. The counteraction of these two forces leads to a sharp potential jump at the interaction radius (Fig. 18.2),

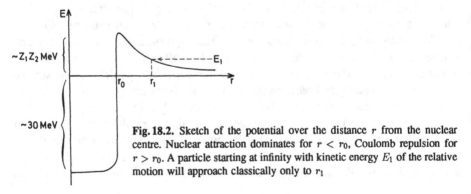

**Fig. 18.2.** Sketch of the potential over the distance $r$ from the nuclear centre. Nuclear attraction dominates for $r < r_0$, Coulomb repulsion for $r > r_0$. A particle starting at infinity with kinetic energy $E_1$ of the relative motion will approach classically only to $r_1$

$$r_0 \approx A^{1/3} 1.44 \times 10^{-13} \text{cm} \qquad (18.6)$$

(the "nuclear radius"). For distances less than $r_0$, the nuclear attraction dominates and provides a potential drop of roughly 30 MeV, while "outside" $r_0$ the repulsive Coulomb forces for particles with charges $Z_1$ and $Z_2$ yield

$$E_{\text{Coul}} = \frac{Z_1 Z_2 e^2}{r} . \qquad (18.7)$$

The height of the *Coulomb barrier* $E_{\text{Coul}}(r_0)$ is typically of the order:

$$E_{\text{Coul}}(r_0) \approx Z_1 Z_2 \text{ MeV} . \qquad (18.8)$$

If, in the stationary reference frame of the nucleus, a particle at "infinity" has kinetic energy $E_1$, it can come classically only to a distance $r_1$ given by $E_1 = E_{\text{Coul}}(r_1)$ from (18.7), as indicated in Fig. 18.2. Now, the kinetic energy available to particles in stellar interiors is that of their thermal motion, and hence the reactions triggered by this motion are called *thermonuclear*. Since in normal stars we observe a slow energy release rather than a nuclear explosion, we must certainly expect the *average* kinetic energy of the thermal motion, $E_{\text{th}}$, to be considerably smaller than $E_{\text{Coul}}(r_0)$. For the value $T \approx 10^7$ K estimated for the solar centre in §2.3, according to (18.3) $kT$ is only $10^3$ eV, i.e. $E_{\text{th}}$ is smaller than the Coulomb barrier (18.8) by a factor of roughly $10^3$. This is in fact so low that, with classical effects only, we can scarcely expect any reaction at all. In the high-energy tail of the Maxwell–Boltzmann distribution, the exponential factor drops here to $\exp(-1000) \approx 10^{-434}$, which leaves no chance for the "mere" $10^{57}$ nucleons in the whole sun (and even for the $\approx 10^{80}$ nucleons in the whole visible universe)!

The only possibility for thermonuclear reactions in stars comes from a quantum-mechanical effect found by G. Gamow: there is a small but finite probability of penetrating ("tunnelling") through the Coulomb barrier, even for particles with $E < E_{\text{Coul}}(r_0)$. This tunnelling probability varies as

$$P_0 = p_0 E^{-1/2} e^{-2\pi\eta} \quad ; \quad \eta = \left(\frac{m}{2}\right)^{1/2} \frac{Z_1 Z_2 e^2}{\hbar E^{1/2}} . \qquad (18.9)$$

Here $\hbar$ is $h/2\pi$, $m$ the reduced mass. The factor $p_0$ depends only on the properties of the two colliding nuclei. The exponent $2\pi\eta$ is here obtained as the only $E$-dependent term in an approximate evaluation of the integral over $\hbar^{-1}[2m(E_{\text{Coul}}-E)]^{1/2}$, which is extended from $r_0$ to the distance $r_c$ of closest approach (where $E = E_{\text{Coul}}$). For $Z_1 Z_2 = 1$ and $T = 10^7$ K, $P_0$ is of the order of $10^{-20}$ for particles with average kinetic energy $E$, and steeply increases with $E$ and decreases with $Z_1 Z_2$. Therefore, for temperatures as "low" as $10^7$ K, only the lightest nuclei (with smallest $Z_1 Z_2$) have a chance to react. For reactions of heavier particles, with larger $Z_1 Z_2$, the energy, i.e. the temperature, has to be correspondingly larger to provide a comparable penetration probability. This will result in well-separated phases of different nuclear "burning" during the star's evolution.

## 18.2 Nuclear Cross-sections

Consider a reaction of the nucleus $X$ with the particle $a$ by which the nucleus $Y$ and the particle $b$ are formed:

$$a + X \rightarrow Y + b \ , \tag{18.10}$$

represented by the notation $X(a, b)Y$. The reaction probability depends on nuclear details, some of which can be illustrated with the following simplified description. After penetration of the Coulomb barrier, an excited *compound nucleus* $C^*$ may form containing both original particles. (The level of excitation is dependent on the kinetic energy and binding energy brought along by the newly added particle.) $C^*$ may decay after a short time, which will still be long enough for the added nucleons to "forget" – owing to interactions within the compound nucleus – their history, a process for which only $\sim 10^{-21}$ s are necessary. The decay then depends only on the energy. $C^*$ can generally decay via one of several "channels" of different probability: $C^* \rightarrow X + a, \rightarrow Y_1 + b_1, \rightarrow Y_2 + b_2, \ldots, \rightarrow C + \gamma$. The first of these would be the reproduction of the original particles, while the last indicates a decay with $\gamma$-ray emission; the others are particle decays where the $b_1, b_2, \ldots$ may be, e.g., neutrons, protons, $\alpha$ particles. Compared to these, a decay with electron emission has negligible probability ($\beta$ decay times being of order 1s or larger). Outgoing particles will obtain a certain amount of kinetic energy, which (just as the energy of emitted $\gamma$ rays) will be shared with the surroundings, though an exception here are the neutrinos, which leave the star without interaction (§ 18.6). The possibility that a given energy level of $C^*$ can decay via a certain channel requires fulfillment of the conservation laws (energy, momentum, angular momentum, nuclear symmetries).

It is very important to know the energy levels of the compound nucleus $C^*$, which can be of different types. Let $E_{min}$ be the minimum energy required to remove a nucleon from the ground state to infinity with zero velocity (to the level $E = 0$ in Fig. 18.3). This corresponds to the atom ionization energy discussed in § 14. Levels below $E_{min}$ can obviously only decay by electromagnetic transitions with the emission of $\gamma$ rays, which are relatively improbable, and hence their lifetime $\tau$ is

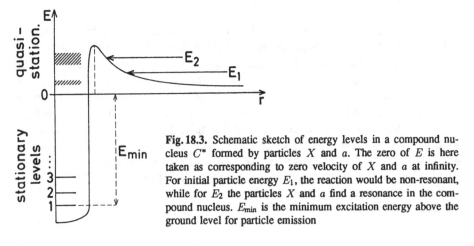

Fig. 18.3. Schematic sketch of energy levels in a compound nucleus $C^*$ formed by particles $X$ and $a$. The zero of $E$ is here taken as corresponding to zero velocity of $X$ and $a$ at infinity. For initial particle energy $E_1$, the reaction would be non-resonant, while for $E_2$ the particles $X$ and $a$ find a resonance in the compound nucleus. $E_{min}$ is the minimum excitation energy above the ground level for particle emission

150

large; these are "stationary" levels of small width $\Gamma$, since

$$\Gamma = \frac{\hbar}{\tau} \quad , \tag{18.11}$$

as follows from the Heisenberg uncertainty relation. These levels correspond to the discrete, bound atomic states.

The compound nucleus will not, however, immediately expel a particle if its energy is somewhat above $E_{min}$, since the sharp potential rise holds it back, at least for some time. Eventually it can leave the potential well by the tunnelling effect (which was, in fact, predicted by Gamow for explaining such outward escapes of particles from radioactive nuclei). So there can be "quasi-stationary" levels above $E_{min}$ that have an appreciably shorter lifetime $\tau$ (and are correspondingly wider) than those below $E_{min}$, since they can also decay via the much more probable particle emission. This probability will clearly increase strongly with increasing energy, which results in corresponding decreases of $\tau$ and increases of $\Gamma$, see (18.11). Above a certain energy $E_{max}$ the width $\Gamma$ will become larger than the distance between neighbouring levels, and their complete overlap yields a continuum of energy states, instead of separated, discrete levels.

The possible existence of quasi-stationary levels above $E_{min}$ requires particular attention. Consider an attempt to produce the compound nucleus $C^*$ by particles $X + a$ with gradually increasing energy $E$ of their relative motion at large distances. The reaction probability will simply increase with the penetration probability (18.9), if $E$ is in a region either without quasi-stationary levels, or between two of them. If, however, $E$ coincides with such a level, the colliding particles find a "resonance" and can form the compound nucleus much more easily. At such resonance energies $E_{res}$, the probability for a reaction (and hence the cross-section $\sigma$) is abnormally enhanced, as sketched in Fig. 18.4, with resonant peaks rising to several powers of ten above "normal". The energy dependence of the cross-section therefore has a factor which has the typical resonance form:

$$\xi(E) = \text{constant} \, \frac{1}{(E - E_{res})^2 + (\Gamma/2)^2} \quad . \tag{18.12}$$

At a resonance, the cross-section $\sigma$ for the reaction of particles $X$ and $a$ can nearly reach its maximum value (geometrical cross-section), given by quantum mechanics as $\pi \, \lambdabar^2$, where $\lambdabar$ is the de Broglie wavelength associated with a particle of relative momentum $p$,

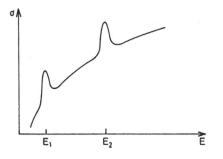

Fig. 18.4. Sketch of the reaction cross-section $\sigma$ over the energy $E$ of the relative motion of the reacting particles, with resonances at $E_1$ and $E_2$

$$\lambda = \frac{\hbar}{p} = \frac{\hbar}{(2mE)^{1/2}} \quad . \tag{18.13}$$

Here the non-relativistic relation betwen $p$ and $E$ is used, and $m$ is the reduced mass of the two particles. The meaning of $\pi\lambda^2$ is clear, because according to quantum mechanics the particles moving with momentum $p$ "see" each other not as a precise point, but smeared out over a length $\lambda$. The dependence of $\sigma$ on $E$ can now be seen from the relation

$$\sigma(E) \sim \pi\lambda^2 P_0(E)\xi(E) \quad , \tag{18.14}$$

where $\lambda$ is given by (18.13). For $E$ values well below the Coulomb barrier, $P_0$ can be taken from (18.9) with a pre-factor $p_0 = E_{\rm Coul}^{1/2}(r_0)\exp[(32mZ_1Z_2e^2r_0/\hbar^2)^{1/2}]$. In the range of a single resonance, $\xi(E)$ is given by (18.12), while far away from any resonances, $\xi \to 1$. In any case, with or without resonances, $\sigma$ is proportional to $\lambda^2 P_0$, which depends on $E$ as shown by (18.9,13). Therefore one usually writes

$$\sigma(E) = SE^{-1}e^{-2\pi\eta} \quad , \tag{18.15}$$

where all remaining effects are contained within the here-defined *"astrophysical cross-section factor"* $S$. This factor contains all intrinsic nuclear properties of the reaction under consideration, and can, in principle, be calculated, although one rather relies on measurements.

The difficulty with laboratory measurements of $S(E)$ – if they are possible at all – is that, because of the small cross-sections, they are feasible only at rather high energies, say above 0.1 MeV, but this is still roughly a factor 10 larger than those energies which are relevant for astrophysical applications. Therefore one has to extrapolate the measured $S(E)$ downward over a rather long range of $E$. This can be done quite reliably for non-resonant reactions, in which case $S$ is nearly constant or a very slowly varying function of $E$ [an advantage of extrapolating $S(E)$ rather than $\sigma(E)$]. The real problems arise from (suspected or unsuspected) resonances in the range over which the extrapolation is to be extended. Then the results can be quite uncertain.

## 18.3 Thermonuclear Reaction Rates

Let us denote the types of reacting particles, $X$ and $a$, by indices $j$ and $k$ respectively. Suppose there is one particle of type $j$ moving with a velocity $v$ relative to all particles of type $k$. Its cross-section $\sigma$ for reactions with the $k$ sweeps over a volume $\sigma v$ per second. The number of reactions per second will then be $n_k\sigma v$, if there are $n_k$ particles of type $k$ per unit volume. For $n_j$ particles per unit volume the total number of reactions per units of volume and time is

$$\tilde{r}_{jk} = n_j\, n_k\, \sigma\, v \quad . \tag{18.16}$$

This product may also be interpreted by saying that $n_jn_k$ is the number of pairs of possible reaction partners, and $\sigma v$ gives the reaction probability per pair and second. This indicates what we have to do in the case of reactions between identical particles ($j = k$). Then the number of pairs that are possible reaction partners is $n_j(n_j - 1)/2 \approx n_j^2/2$ for large particle numbers. This has to replace the product $n_j n_k$ in (18.16) so that we can generally write

$$\tilde{r}_{jk} = \frac{1}{1+\delta_{jk}}\, n_j n_k \sigma v \quad , \quad \delta_{jk} = \begin{cases} 0, & j \neq k \\ 1, & j = k \end{cases} \quad . \tag{18.17}$$

Now we have to allow for the fact that particles $j$ and $k$ do not attack each other with uniform velocities like well-organized squadrons, which is important since $\sigma$ depends strongly on $v$. Excluding extreme densities (as, e.g., in neutron stars) we can assume that both types have a Maxwell–Boltzmann distribution of their velocities. It is then well known that also their *relative velocity* $v$ is Maxwellian. If the corresponding energy is

$$E = \frac{1}{2}\, m\, v^2 \tag{18.18}$$

with the reduced mass $m = m_j m_k/(m_j + m_k)$, the fraction of all pairs contained in the interval $[E, E + dE]$ is given by

$$f(E)dE = \frac{2}{\sqrt{\pi}}\, \frac{E^{1/2}}{(kT)^{3/2}}\, e^{-E/kT} dE \quad . \tag{18.19}$$

This fraction of all pairs has a uniform velocity and contributes the amount $dr_{jk} = \tilde{r}_{jk}\, f(E)dE$ to the total rate. The total reaction rate per units of volume and time is then given by the integral $\int dr_{jk}$ over all energies

$$r_{jk} = \frac{1}{1+\delta_{jk}}\, n_j n_k \langle \sigma v \rangle \quad , \tag{18.20}$$

where the averaged probability is

$$\langle \sigma v \rangle = \int_0^\infty \sigma(E) v f(E) dE \quad . \tag{18.21}$$

Let us replace the particle numbers per unit volume $n_i$ by the mass fraction $X_i$ with

$$X_i \varrho = n_i m_i \quad , \tag{18.22}$$

cf. (8.2). If the energy $Q$ is released per reaction, then (18.20) gives the energy generation rate per units of mass and time:

$$\varepsilon_{jk} = \frac{1}{1+\delta_{jk}}\, \frac{Q}{m_j m_k}\, \varrho\, X_j X_k \langle \sigma v \rangle \quad . \tag{18.23}$$

Using (18.9,15,18,19) in (18.21), the average cross-section $\langle \sigma v \rangle$ can be written as

$$\langle \sigma v \rangle = \frac{2^{3/2}}{(m\pi)^{1/2}} \frac{1}{(kT)^{3/2}} \int\limits_0^\infty S(E)\, e^{-E/kT - \bar{\eta}/E^{1/2}}\, dE \quad , \tag{18.24}$$

where

$$\bar{\eta} = 2\pi\eta E^{1/2} = \pi(2m)^{1/2}\frac{Z_j Z_k e^2}{\hbar} \quad . \tag{18.25}$$

A further evaluation of $\langle \sigma v \rangle$ requires a specification of $S(E)$. We shall limit ourselves to the simplest, but for astrophysical applications very important, case of *non-resonant reactions*. Then we can set $S(E) \approx S_0 = $ constant, and take it out of the integral (18.24), since only a small interval of $E$ will turn out to contribute appreciably. The remaining integral may be written as

$$J = \int\limits_0^\infty e^{f(E)}dE \quad , \quad \text{with} \quad f(E) = -\frac{E}{kT} - \frac{\bar{\eta}}{E^{1/2}} \quad . \tag{18.26}$$

The integrand is the product of two exponential functions, one of which drops steeply with increasing $E$, while the other rises. The integrand will therefore have appreciable values only around a well-defined maximum (see Fig. 18.5), the so-called Gamow peak. This maximum occurs at $E_0$, where the exponent has a minimum. From the condition $f' = 0$, where $f'$ is the derivative with respect to $E$, one finds

$$E_0 = \left(\frac{1}{2}\bar{\eta}kT\right)^{2/3} = \left[\left(\frac{m}{2}\right)^{1/2} \pi \frac{Z_j Z_k e^2 kT}{\hbar}\right]^{2/3} \quad . \tag{18.27}$$

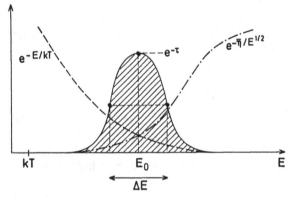

Fig. 18.5. The Gamow peak (solid curve) as the product of Maxwell distribution (*dashed*) and penetration factor (*dot-dashed*). The hatched area under the Gamow peak determines the reaction rate. All three curves are on different scales

It is usual to introduce now a quantity $\tau$ defined by

$$\tau = 3\frac{E_0}{kT} = 3\left[\pi\left(\frac{m}{2kT}\right)^{1/2}\frac{Z_j Z_k e^2}{\hbar}\right]^{2/3} \quad , \tag{18.28}$$

and to represent $f(E)$ near the maximum by the series expansion

154

$$f(E) = f_0 + f_0' \cdot (E - E_0) + \frac{1}{2} f_0'' \cdot (E - E_0)^2 + \dots$$

$$= -\tau - \frac{1}{4}\tau \left(\frac{E}{E_0} - 1\right)^2 + \dots \quad , \qquad (18.29)$$

from which we retain only these two terms (the linear term vanishes since $f_0' = 0$ at the maximum). Their substitution in (18.26) means to approximate the Gamow peak of the integrand by a Gaussian, as will become particularly clear when we transform $J$ to the new variable of integration $\xi = (E/E_0 - 1)\sqrt{\tau}/2$ :

$$J = \int_0^\infty \exp\left[-\tau - \frac{\tau}{4}\left(\frac{E}{E_0} - 1\right)^2\right] dE = \frac{2}{3} kT\tau^{1/2} e^{-\tau} \int_{-\sqrt{\tau}/2}^\infty e^{-\xi^2} d\xi . \quad (18.30)$$

The main contribution to $J$ comes from a range close to $E = E_0$, i.e. $\xi = 0$, so that no large errors are introduced when extending the range of integration to $-\infty$, the integral over the Gaussian becoming $\sqrt{\pi}$.

We then have

$$J \approx kT\frac{2}{3}\pi^{1/2}\tau^{1/2} e^{-\tau} \quad , \qquad (18.31)$$

and for non-resonant reactions (18.24) becomes

$$\langle\sigma v\rangle = \frac{4}{3}\left(\frac{2}{m}\right)^{1/2}\frac{1}{(kT)^{1/2}}S_0\tau^{1/2} e^{-\tau} \quad . \qquad (18.32)$$

From (18.28) one has $(kT)^{-1/2} \sim \tau^{3/2}$; hence the $kT$ can be substituted in (18.32), which then gives $\langle\sigma v\rangle \sim \tau^2 e^{-\tau}$.

The *properties of the Gamow peak* are so important that we should inspect some of them a bit further. In order to have convenient numerical values, we count the temperature in units of $10^7$ K (which is typical for many stellar centres) and denote this dimensionless temperature by $T_7 = T/10^7$ K, or generally

$$T_n := \frac{T}{10^n \mathrm{K}} \quad . \qquad (18.33)$$

We then have the following relations (some of which will be derived below):

$$W = Z_j^2 Z_k^2 A = Z_j^2 Z_k^2 \frac{A_j A_k}{A_j + A_k} \quad ,$$

$$\tau = 19.721 W^{1/3} T_7^{-1/3} \quad ,$$

$$E_0 = 5.665 \text{ keV} \cdot W^{1/3} T_7^{2/3} \quad ,$$

$$\frac{E_0}{kT} = \frac{\tau}{3} = 6.574 W^{1/3} T_7^{-1/3} \quad , \qquad (18.34)$$

$$\Delta E = 4.249 \text{ keV} \cdot W^{1/6} T_7^{5/6} \quad ,$$

$$\frac{\Delta E}{E_0} = 4(\ln 2)^{1/2}\tau^{-1/2} = 0.750 W^{-1/6} T_7^{1/6} \quad ,$$

$$\nu = \partial \ln\langle\sigma v\rangle / \partial \ln T = (\tau - 2)/3 = 6.574 W^{1/3} T_7^{-1/3} - 2/3 \quad .$$

The value of $W$ is determined by the reaction partners and is at least of order unity. Large $W$ discriminates against the reactions of heavy nuclei so much that only the lighter nuclei can react with appreciable rate. The Gamow peak occurs as a compromise in the counteraction between Maxwell distribution and penetration probability with a maximum at $E = E_0$, which is roughly 5 to 100 times the average thermal energy $kT$. This "effective stellar energy range" is, on the other hand, far below the $\gtrsim 100$ keV available to laboratory experiments. With increasing $T$, $E_0$ increases moderately, while the maximum height of the peak $H_0 = \mathrm{e}^{-\tau}$ increases very steeply owing to the decreasing $\tau$.

The width of the effective energy range is described by $\Delta E$, which is the full width of the Gamow peak at half maximum (see Fig. 18.5), i.e. between the points with height $0.5\,\mathrm{e}^{-\tau}$. Equating this to the integrand in the first form of (18.30), we obtain

$$\frac{\Delta E}{E_0} = 4\frac{(\ln 2)^{1/2}}{\tau^{1/2}} \ . \tag{18.35}$$

According to (18.34), this is always below unity and therefore one has a well-defined energy range in which the reactions occur effectively. With $\Delta E$ increasing with $T$ only slightly more than $E_0$, the relative form of the peak remains nearly constant.

The most striking feature of thermonuclear reactions is their strong sensitivity to the temperature. In order to demonstrate this, one represents the $T$ dependence of $\langle \sigma v \rangle$ (and thus of $r_{jk}$ and $\varepsilon_{jk}$) around some value $T = T_0$ by a power law such as

$$\langle \sigma v \rangle = \langle \sigma v \rangle_0 \left(\frac{T}{T_0}\right)^\nu \ , \quad \nu = \frac{\partial \ln\langle \sigma v \rangle}{\partial \ln T} \ . \tag{18.36}$$

From (18.28) we have $\tau \sim T^{-1/3}$, and then from (18.32) $\langle \sigma v \rangle \sim T^{-2/3}\,\mathrm{e}^{-\tau}$. Therefore

$$\ln\langle \sigma v \rangle = \text{constant} - \frac{2}{3}\ln T - \tau \ , \tag{18.37}$$

and

$$\frac{\partial \ln\langle \sigma v \rangle}{\partial \ln T} = -\frac{2}{3} - \frac{\partial \tau}{\partial \ln T} = -\frac{2}{3} - \tau\frac{\partial \ln \tau}{\partial \ln T} \ . \tag{18.38}$$

Since $\tau \sim T^{-1/3}$, we have $\partial \ln \tau / \partial \ln T = -1/3$, so that finally

$$\nu \equiv \frac{\partial \ln\langle \sigma v \rangle}{\partial \ln T} = \frac{\tau}{3} - \frac{2}{3} \ , \tag{18.39}$$

where for most reactions $\tau/3$ is much larger than $2/3$, and $\nu \approx \tau/3$. Then $\nu$ decreases with $T$ as $\nu \sim T^{-1/3}$. From (18.34) we see that even for reactions between the lightest nuclei $\nu \approx 5$, and it can easily attain values around (and even above) $\nu \approx 20$. With such values for the exponent (!) of $T$, the thermonuclear reaction rate is about the most strongly varying function treated in physics, and this temperature sensitivity has a clear influence on stellar models. Also, since small fluctuations of $T$ (which will certainly be present) must result in drastic changes in the energy production, we have to assume that there exists an effective stabilizing mechanism (a thermostat) in stars (§ 25.3.5).

We may easily see how the large $\nu$ values are related to the change of the Gamow peak with $T$: the value $\langle \sigma v \rangle$ is proportional to the integral $J$ in (18.30), and this is given by the area under the Gamow peak, which is roughly $J \approx \Delta E \cdot H_0$ ($H_0 = \mathrm{e}^{-\tau}$ is the height of the peak). According to (18.34), $\Delta E \sim T^{5/6}$, while $H_0$ increases strongly with $T$. In fact it is this height $H_0$ which provides the exponential $\mathrm{e}^{-\tau}$ in the expressions for $\langle \sigma v \rangle$ and is therefore responsible for the large values of $\nu$.

We should briefly mention a few corrections to the derived formulae for the reaction rates. The first concerns inaccuracies made by evaluating the integral in (18.24) with constant $S$ and with an integrand approximated by a Gaussian. This is usually corrected for by multiplying $\langle \sigma v \rangle$ with a factor

$$g_{jk} = 1 + \frac{5}{12\tau} + \frac{S'}{S} E_0 \left( 1 + \frac{105}{36\tau} \right) + \frac{1}{2} \frac{S''}{S} E_0^2 \left( 1 + \frac{267}{36\tau} \right) \quad , \qquad (18.40)$$

where $S$ and its derivatives with respect to $E$ have to be taken at $E = 0$ (COX, GIULI, 1968, Vol.I, p. 462).

Another correction factor, $f_{jk}$, allows for a partial shielding of the Coulomb potential of the nuclei, owing to the negative field of neighbouring electrons. This plays a role only at very high densities; it will be treated separately in § 18.4.

Concerning resonant reactions we shall only remark that the situation depends very much on the location of the resonance. For example, the integral in (18.24) can be dominated by a strong peak at the resonance energy. However, once $S(E)$ is given, (18.24) can in principle always be evaluated.

## 18.4 Electron Shielding

We have seen that the repulsive Coulomb forces of the nucleus play a decisive role in controlling the rate of thermonuclear reactions. Therefore any modification of its potential by influences from the outside can have an appreciable effect on these rates. An obvious effect to be considered comes from the surrounding free electrons. It is clear that beyond a certain distance an approaching particle will "feel" a neutral conglomerate of the target nucleus plus a surrounding electron cloud, rather than the isolated charge of the target nucleus.

The first step is to consider the polarization that the nucleus of charge $+Ze$ produces in its surrounding. The electrons of charge $-e$ are attracted and have a slightly larger density $n_e$ in the neighbourhood of the nucleus; the other ions are repelled and have a slightly decreased density $n_i$ in comparison with their average values $\bar{n}_e$ and $\bar{n}_i$ (without electric fields present). For non-degenerate gases the density of particles with charge $q$ is modified in the presence of an electrostatic potential $\phi$ according to

$$n = \bar{n} \, \mathrm{e}^{-q\phi/kT} \quad . \qquad (18.41)$$

In most normal cases one will find $|q\phi| \ll kT$ and can then approximate the exponential by $1 - q\phi/kT$. For ions and electrons, (18.41) now yields

$$n_i = \bar{n}_i \left( 1 - \frac{Z_i e\phi}{kT} \right) \quad , \quad n_e = \bar{n}_e \left( 1 + \frac{e\phi}{kT} \right) \quad , \tag{18.42}$$

which shows directly the decrease (ions), and increase (electrons) of the two densities.

Considering the $n_i$ for all types of ions present in the gas mixture, one can immediately write down the total charge density $\sigma$. For $\phi = 0$ one must have a neutral gas, with $\bar{\sigma} = 0$, i.e.

$$\bar{\sigma} = \sum_i (Z_i e)\bar{n}_i - e\bar{n}_e = 0 \quad , \tag{18.43}$$

whereas for non-vanishing $\phi$ we have

$$\begin{aligned} \sigma &= \sum_i (Z_i e)n_i - en_e \\ &= \sum_i -\frac{(Z_i e)^2 \phi}{kT}\bar{n}_i - \frac{e^2 \phi}{kT}\bar{n}_e \quad . \end{aligned} \tag{18.44}$$

Here we have already inserted (18.42) and made use of (18.43) to eliminate the $\phi$-independent terms. The second expression (18.44) suggests that we combine the two terms and write

$$\sigma = -\chi \frac{e^2 \phi}{kT} n \quad , \tag{18.45}$$

where we have introduced the total particle density $n = n_e + \sum_i n_i$, and the average value $\chi$:

$$\chi := \frac{1}{n} \left( \sum_i Z_i^2 \bar{n}_i + \bar{n}_e \right) \quad . \tag{18.46}$$

If one wishes to use the mass fraction $X_i = A_i \bar{n}_i / n\mu$ ($\mu$ = mean molecular weight per free particle, see § 13.1) instead of the particle numbers, the expression follows simply as

$$\chi = \mu\zeta = \mu \sum_i \frac{Z_i(Z_i + 1)}{A_i} X_i \quad . \tag{18.47}$$

The charge density $\sigma$ and the electrostatic potential $\phi$ are also connected by the Poisson equation

$$\nabla^2 \phi = -4\pi\sigma \quad . \tag{18.48}$$

If we assume spherical symmetry for the charge distribution surrounding the nucleus under consideration, the Laplace operator $\nabla^2$ then reduces to its well-known radial part. Introducing $\sigma$ from (18.45) on the right-hand side of (18.48), the Poisson equation becomes

$$\frac{r_{\mathrm{D}}^2}{r}\frac{d^2(r\phi)}{dr^2} = \phi \quad , \tag{18.49}$$

where we have scaled the distance $r$ by the so-called Debye–Hückel length

$$r_{\mathrm{D}} = \left(\frac{kT}{4\pi\chi e^2 n}\right)^{1/2} \quad . \tag{18.50}$$

One readily verifies that (18.49) is solved by

$$\phi = \frac{Ze}{r}e^{-r/r_{\mathrm{D}}} \quad , \tag{18.51}$$

and this shows that $\phi$ tends to the normal (unshielded) potential $Ze/r$ of a point charge $Ze$ for small distances, $r \to 0$, while we have an essential reduction of this "normal" potential at distances $r \gtrsim r_{\mathrm{D}}$. In a certain sense we can call $r_{\mathrm{D}}$ the "radius" of the electron cloud that envelopes the nucleus and shields part of its potential for an outside viewer.

The values of $\zeta$ in (18.47) are of order unity. For $T = 10^7$ K and $\varrho$ between 1 and $10^2$ g cm$^{-3}$, $r_{\mathrm{D}}$ has typical values of $10^{-8}\ldots 10^{-9}$ cm. In order to judge the influence of the shielding on nuclear reactions between nuclei of types 1 and 2, we should compare $r_{\mathrm{D}}$ with the closest distance $r_{c0}$ to which the particles can classically approach each other if their energy is that of the Gamow peak $E_0$ [given by (18.27)]. These particles will be the most effective ones for the energy production. According to (18.7) one has $r_{c0} = Z_1 Z_2 e^2 / E_0$, and convenient numerical expressions for $E_0$ are given in (18.34). We then find

$$\frac{r_{\mathrm{D}}}{r_{c0}} \approx 200\frac{E_0}{Z_1 Z_2}\left(\frac{T_7}{\zeta\varrho}\right)^{1/2} \quad , \tag{18.52}$$

where $E_0$ is in keV and $\varrho$ in g cm$^{-3}$. With rough values for the solar centre, $T_7 \approx 1$, $\varrho \approx 10^2$ g cm$^{-3}$, $\zeta \approx 1$, and for the most important hydrogen reactions, we have $Z_1 Z_2 = 1 \ldots 7$ and $E_0 \approx 5 \ldots 20$ keV; hence (18.52) gives $r_{\mathrm{D}}/r_{c0} \approx 50 \ldots 100$. For all such "normal" stars, $r_{\mathrm{D}} \gg r_{c0}$, which means that the incoming particle even classically (without the tunnelling effect) penetrates nearly the entire electron cloud and the shielding will have little effect at these critical distances.

The decrease of the Coulomb interaction energy $E_{\mathrm{Coul}}$ increases the probability $P_0$ for tunnelling through the Coulomb wall. The decisive exponent $\eta$ in $P_0$ [(18.9) and the following] is determined by the function $E_{\mathrm{Coul}} - E$. The energy $E_{\mathrm{Coul}}$ is now reduced according to (18.51) by the factor $\exp(-r/r_{\mathrm{D}})$, which is to a first approximation $1 - r/r_{\mathrm{D}}$ for $r/r_{\mathrm{D}} \ll 1$.

This gives

$$E_{\mathrm{Coul}} - E \equiv \frac{Z_1 Z_2 e^2}{r}e^{-r/r_{\mathrm{D}}} - E \approx \frac{Z_1 Z_2 e^2}{r} - \frac{Z_1 Z_2 e^2}{r_{\mathrm{D}}} - E \quad , \tag{18.53}$$

which shows that we will obtain the same result as without shielding, but with an enlarged energy

$$\tilde{E} = E + \frac{Z_1 Z_2 e^2}{r_D} = E + E_D \quad . \tag{18.54}$$

In order to see the influence on simple non-resonant reaction rates, consider the integrand in (18.21) and replace $\sigma(E)$ by $\sigma(\tilde{E})$. With (18.15,19) and $\tilde{\eta} = \eta(E/\tilde{E})^{1/2}$, we have the proportionality

$$\sigma(\tilde{E})vf(E) \sim \left(\tilde{E}^{-1}e^{-2\pi\tilde{\eta}}\right) E^{1/2} \left(E^{1/2} e^{-E/kT}\right)$$

$$\sim \left(1 - \frac{E_D}{\tilde{E}}\right) e^{E_D/kT - \tilde{E}/kT - 2\pi\tilde{\eta}} \quad . \tag{18.55}$$

We assume here that $E_D/kT \ll 1$, which is usually called the case of "weak screening". Considering the fact that only a small range of $E$ at values much larger than $kT$ contributes essentially to $\langle \sigma v \rangle$, we may as well neglect the factor $(1 - E_D/\tilde{E})$ in (18.55) and integrate over $\tilde{E}$ instead of $E$. The main change is then the additional constant exponent $E_D/kT$ such that $\langle \sigma v \rangle$ is multiplied by a "screening factor"

$$f = e^{E_D/kT} \quad , \tag{18.56}$$

which increases $\langle \sigma v \rangle$, since $E_D$ is positive. For weak screening we have numerically

$$\frac{E_D}{kT} = \frac{Z_1 Z_2 e^2}{r_D kT} = 5.92 \times 10^{-3} Z_1 Z_2 \left(\frac{\zeta\varrho}{T_7^3}\right)^{1/2} \quad , \tag{18.57}$$

with $\varrho$ in g cm$^{-3}$. For $\zeta \approx 1$, $\varrho = 1$ g cm$^{-3}$ and $T_7 = 1$, reactions with $Z_1 Z_2 \lesssim 16$ require correction factors $f$, which increase the rate by less than 10%.

Where very large densities are involved, however, one will leave the regime of weak screening. For $E_D/kT \gtrsim 1$, the treatment is much more complicated, and the limiting case of "strong screening" is described approximately by

$$\frac{E_D}{kT} \approx 0.0205 \left[(Z_1 + Z_2)^{5/3} - Z_1^{5/3} - Z_2^{5/3}\right] \frac{(\varrho/\mu_e)^{1/3}}{T_7} \quad , \tag{18.58}$$

with the molecular weight per free electron $\mu_e = \left(\sum X_i Z_i / A_i\right)^{-1}$, see (13.8), and $\varrho$ in g cm$^{-3}$.

Equations (18.57,58) show that the screening factor $f$ increases appreciably for increasing $\varrho$ and decreasing $T$. While $f$ was a minor correction factor to the rate for "normal" stars with weak screening, the situation changes completely in the high-density, low-temperature regime, where screening becomes the dominating factor in the reaction rate.

Consider the shielded reaction rate as represented by

$$f\langle \sigma v \rangle = f_0 \langle \sigma v \rangle_0 \left(\frac{\varrho}{\varrho_0}\right)^{\lambda} \left(\frac{T}{T_0}\right)^{\nu} \tag{18.59}$$

in the neighbourhood of $\varrho_0$, $T_0$. In a similar manner to the derivation of $\nu$ for the unshielded case in (18.36–39), we find now that

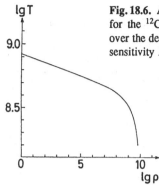

Fig. 18.6. A line of constant energy generation rate $\varepsilon$ (= $10^4$ erg $g^{-1}$ $s^{-1}$) for the $^{12}C+^{12}C$ burning in a diagram showing the temperature $T$ (in K) over the density $\varrho$ (in g cm$^{-3}$). The temperature sensitivity $\nu$ and the density sensitivity $\lambda$ are equal where the slope is $-1$

$$\nu = \frac{\tau}{2} - \frac{2}{3} - \frac{E_D}{kT} \quad ; \quad \lambda = 1 + \frac{1}{3}\frac{E_D}{kT} \ . \qquad (18.60)$$

For very high densities and moderate to low temperatures (say $\varrho > 10^6$ g cm$^{-3}$, $T < 10^7$ K), the temperature sensitivity $\nu$ decreases, while the density sensitivity $\lambda$ becomes larger. This can be seen from Fig. 18.6, where the line of constant $^{12}C$–$^{12}C$ burning turns steeply down for large $\varrho$. Finally, the reaction rates now depend mainly on the density (instead of the temperature) and one speaks of "*pycnonuclear reactions*". For $^{12}C$ burning in a pure $^{12}C$ plasma, (18.60) gives the transition $\lambda = \nu$ at $T_7 = 10$ for $\varrho = 1.60 \times 10^9$ g cm$^{-3}$.

Pycnonuclear reactions can play a role in very late phases of stellar evolution, where a burning may be triggered by a compression without temperature increase, and they can provide a certain amount of energy release even in cool stars, if only the density is high enough. Of course, other effects, such as the decrease of the mobility of the nuclei because of crystallization, must then also be considered.

## 18.5 The Major Nuclear Burnings

Although no chemical reactions are involved, one usually calls the thermonuclear fusion of a certain element the "burning" of this element. Owing to the properties of thermonuclear reaction rates, different burnings are well separated by appreciable temperature differences. A review of the cross-sections for all possible reactions shows that only very few reactions occur with non-negligible rates during a certain phase. The most important ones will be listed below. Their important properties, such as cross-section factors $S_0$, correction factors to (18.32), or energy release $Q$, can be found in the literature (for example, FOWLER, CAUGHLAN, ZIMMERMAN, 1967, 1975, 1983).

The $Q$ values usually contain all of the energy made available to the stellar matter by one such reaction. This includes the energies of the $\gamma$ rays that are either directly emitted or created by pair annihilation after $e^+$ emission. Excluded, however, is the energy carried away by neutrinos, since they normally do not interact with the stellar material.

A whole "network" of all simultaneously occurring reactions has to be calculated if one is interested in details such as the isotopic abundances produced by the reactions, or if the star changes on a time-scale comparable with that of one of the reactions. The total $\varepsilon$ is then obtained as a sum of (18.23) over all reactions, and one has to ensure the correct book-keeping of the changing abundances of all nuclei involved.

Often enough, a much simpler procedure suffices in which only the rate for the slowest of a chain of subsequent reactions is calculated, since it determines essentially the rate of the whole fusion process. An example of such a "bottleneck" is the $^{14}$N reaction in the CNO cycle (see below). Then (18.23) has to be used for this reaction, but with $Q$ equal to the sum of all energies released in the single reactions.

In this subsection, all formulae for $\varepsilon$ will be given in units of erg $g^{-1}s^{-1}$, $\varrho$ in g cm$^{-3}$, and $T$ in the dimensionless form $T_n = T/10^n$K. As usual we denote by $X_j$ the mass fraction of nuclei with mass number $A = j$.

### 18.5.1 Hydrogen Burning

The net result of hydrogen burning is the fusion of four $^1$H nuclei into one $^4$He nucleus. The difference in binding energy is 26.731 MeV, corresponding to a mass defect of about 0.71 per cent. This is roughly 10 times the energy liberated in any other fusion process, though not all of this energy is available to stellar matter. The fusion requires the transformation of two protons into neutrons, i.e. two $\beta^+$ decays, which must be accompanied by two neutrino emissions (conservation of lepton number). The neutrinos carry away 2 ... 30 per cent of the whole energy liberated, the amount depending strongly on the reaction in which they are emitted.

There are different chains of reactions by which a fusion process can be completed, and which in general will occur simultaneously in a star. The two main series of reactions are known as the proton–proton chain and the CNO cycle.

The *proton–proton chain* (*pp* chain) is named after its first reaction, between two protons forming a deuterium nucleus $^2$H, which then reacts with another proton to form $^3$He:

$$^1\text{H} + {}^1\text{H} \rightarrow {}^2\text{H} + e^+ + \nu \quad ,$$
$$^2\text{H} + {}^1\text{H} \rightarrow {}^3\text{He} + \gamma \quad . \tag{18.61}$$

The first of these reactions is unusual in comparison with most other fusion processes. In order to form $^2$H, the protons have to experience a $\beta^+$ decay at the time of their closest approach. This is a process governed by the weak interaction and is very unlikely. Therefore the first reaction has a very small cross-section.

The completion of a $^4$He nucleus can proceed via one of three alternative branches (*pp1*, *pp2*, *pp3*) all of which start with $^3$He. The first alternative requires two $^3$He nuclei, i.e. the reactions in (18.61) have first to be completed twice. The other alternatives require that $^4$He already exists (either produced in this burning, or primordial). The branching between *pp2* and *pp3* exists, since $^7$Be can react either with $e^-$ or with $^1$H. All possibilities can be seen from the following scheme:

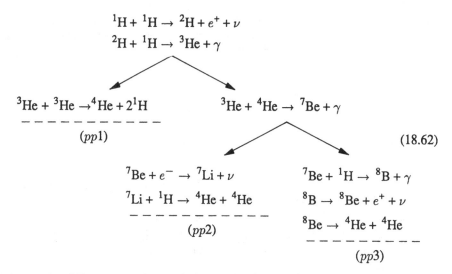

$$^1H + {}^1H \rightarrow {}^2H + e^+ + \nu$$
$$^2H + {}^1H \rightarrow {}^3He + \gamma$$

$$^3He + {}^3He \rightarrow {}^4He + 2{}^1H \qquad\qquad {}^3He + {}^4He \rightarrow {}^7Be + \gamma$$

$(pp1)$

$$^7Be + e^- \rightarrow {}^7Li + \nu \qquad\qquad {}^7Be + {}^1H \rightarrow {}^8B + \gamma$$
$$^7Li + {}^1H \rightarrow {}^4He + {}^4He \qquad\qquad {}^8B \rightarrow {}^8Be + e^+ + \nu$$

$(pp2)$ 

$$^8Be \rightarrow {}^4He + {}^4He$$

$(pp3)$

$$(18.62)$$

Owing to the different energies carried away by the neutrinos, the energies released to the stellar matter differ for the three chains. They are $Q = 26.20(pp1)$, $25.67(pp2)$, $19.20(pp3)$, in MeV per produced $^4He$ nucleus. For each quantity $Q$ released, the first two reactions of (18.61) have to be performed twice in the $pp1$ branch, but only once in the other branches.

The relative frequency of the different branches depends on the chemical composition, the temperature, and the density. The $^3He-{}^4He$ reaction has a 14% larger reduced mass, a 4.6% larger $\tau$, and thus a slightly larger temperature sensitivity $\nu$ than the $^3He-{}^3He$ reaction, cf. (18.34,39). With increasing $T$, $pp2$ and $pp3$ will therefore dominate more and more over $pp1$ (say above $T_7 \approx 1$) if $^4He$ is present with appreciable amounts. And with increasing $T$, the relative importance will gradually shift from the electron capture ($pp2$) to the proton capture ($pp3$) of $^7Be$.

The energy generation in the $pp$ chain should be calculated at small $T$ (say below $T_6 \approx 8$) by calculating all single reactions and their influence on the nuclei involved. For larger $T$, there will be an equilibrium abundance established for these nuclei (equal rates of consumption and production) and one can simply take the whole $\varepsilon_{pp}$ as proportional to that of the $pp1$ branch, which in turn may be calculated from the rate of the first reaction $^1H + {}^1H$:

$$\varepsilon_{pp} = 2.38 \times 10^6 \psi f_{11} g_{11} \varrho X_1^2 T_6^{-2/3} e^{-33.80/T_6^{1/3}} \quad,$$
$$g_{11} = \left(1 + 0.0123 T_6^{1/3} + 0.0109 T_6^{2/3} + 0.0009 T_6\right) \quad, \qquad (18.63)$$

where $\varepsilon_{pp}$ and $\varrho$ are in cgs and $f_{11}$ is the shielding factor for this reaction. The factor $\psi$ corrects for the additional energy generation in the branches $pp2$ and $pp3$ if there is appreciable $^4He$ present (see Fig. 18.7). For gradually increasing $T$, $\psi$ starts with the value 1 and can then increase to values close to 2 (at $T_7 \approx 2$), at which point $pp2$ takes over, since then *each* $^1H-{}^1H$ reaction gives one $^4He$ (compared to *every second* such reaction in the branch $pp1$). After this maximum, $\psi$ decreases again to about 1.5 where $pp3$ has taken over owing to its $Q$ being much smaller than those of the other branches.

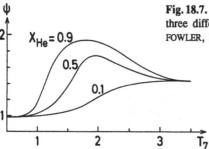

The temperature sensitivity of the *pp* chain is the smallest of all fusions. At $T_6 = 5$, we have $\nu \approx 6$, which decreases to 3.5 at $T_6 \approx 20$.

The *CNO cycle* is the other main series of reactions in hydrogen burning. It requires the presence of some isotopes of C, N, or O, which are reproduced in a manner similar to catalysts in chemical reactions. The sequence of reactions can be represented as follows:

$$
\begin{aligned}
{}^{12}\mathrm{C} + {}^{1}\mathrm{H} &\rightarrow {}^{13}\mathrm{N} + \gamma \\
{}^{13}\mathrm{N} &\rightarrow {}^{13}\mathrm{C} + e^{+} + \nu \\
{}^{13}\mathrm{C} + {}^{1}\mathrm{H} &\rightarrow {}^{14}\mathrm{N} + \gamma \\
{}^{14}\mathrm{N} + {}^{1}\mathrm{H} &\rightarrow {}^{15}\mathrm{O} + \gamma \\
{}^{15}\mathrm{O} &\rightarrow {}^{15}\mathrm{N} + e^{+} + \nu \\
{}^{15}\mathrm{N} + {}^{1}\mathrm{H} &\rightarrow {}^{12}\mathrm{C} + {}^{4}\mathrm{He} \\
\\
&\longrightarrow {}^{16}\mathrm{O} + \gamma \\
{}^{16}\mathrm{O} + {}^{1}\mathrm{H} &\rightarrow {}^{17}\mathrm{F} + \gamma \\
{}^{17}\mathrm{F} &\rightarrow {}^{17}\mathrm{O} + e^{+} + \nu \\
{}^{17}\mathrm{O} + {}^{1}\mathrm{H} &\rightarrow {}^{14}\mathrm{N} + {}^{4}\mathrm{He}
\end{aligned}
\tag{18.64}
$$

The main cycle (upper 6 lines of this scheme) is completed after the initially consumed $^{12}\mathrm{C}$ is reproduced by $^{15}\mathrm{N} + {}^{1}\mathrm{H}$. This reaction shows a branching via $^{16}\mathrm{O}$ into a secondary cycle (connected with the main cycle by dashed arrows), which is, however, roughly $10^4$ times less probable. Its main effect is that the $^{16}\mathrm{O}$ nuclei originally present in the stellar matter can also take part in the cycle, since they are finally transformed into $^{14}\mathrm{N}$ by the last three reactions of (18.64). The decay times for the $\beta^+$ decays are of the order of $10^2 \ldots 10^3$ s. As usual, a network of all simultaneous reactions has to be calculated for lower temperatures or rapid changes.

Most stars change slowly enough that, for sufficiently high temperature (say $T_7 \gtrsim 1.5$), the nuclei involved in the cycle reach their equilibrium abundance (i.e. the rate of production equals that of consumption). Then it suffices to calculate explicitly only the slowest reaction, which is $^{14}\mathrm{N} + {}^{1}\mathrm{H}$ and which essentially controls the time for completing the cycle. $\varepsilon_{CNO}$ will then be given by the rate of this reaction and by the energy gain of the whole cycle, which is 24.97 MeV. This slowest reaction acts like a bottleneck where the nuclei involved are congested in their "flow" through the cycle. Nearly all of the initially present C, N, and O nuclei will therefore be found as $^{14}$N, waiting to be transformed to $^{15}$O. The energy generation rate can be written as

$$\varepsilon_{CNO} = 8.67 \times 10^{27} g_{14,1} X_{CNO} X_1 \varrho T_6^{-2/3} e^{-152.28/T_6^{1/3}} \quad ,$$
$$g_{14,1} = \left(1 + 0.0027 T_6^{1/3} - 0.00778 T_6^{2/3} - 0.000149 T_6\right) \quad , \tag{18.65}$$

where $\varepsilon_{CNO}$ and $\varrho$ are in cgs. $X_{CNO}$ is the sum of $X_C$, $X_N$, and $X_O$. The temperature sensitivity $\nu$ is much higher here than in the $pp$ chain. For $T_6 = 10 \ldots 50$, we find $\nu \approx 23 \ldots 13$. This has the consequence that the $pp$ chain dominates at low temperatures ($T_6 < 15$), while it can be neglected against $\varepsilon_{CNO}$ for higher temperatures (see Fig. 18.8). Hydrogen burning normally occurs in the range $T_6 \approx 8 \ldots 50$, since at larger $T$ the hydrogen is very rapidly exhausted.

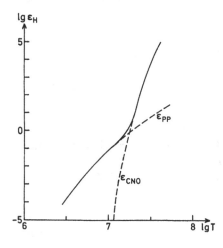

Fig. 18.8. Total energy generation rate $\varepsilon_H$ (in erg g$^{-1}$ s$^{-1}$) for hydrogen burning (solid line) over the temperature $T$ (in K), for $\varrho = 1$ g cm$^{-3}$, $X_1 = 1$, $X_{CNO} = 0.01$. The contributions of the $pp$ chain and the CNO cycle are dashed

### 18.5.2 Helium Burning

The reactions of helium burning consist of the gradual fusion of several $^4$He into $^{12}$C, $^{16}$O, .... This requires temperatures of $T_8 \gtrsim 1$, i.e. appreciably higher than those for hydrogen burning, because of the higher Coulomb barriers.

The first and key reaction is the formation of $^{12}$C from three $^4$He nuclei, which is called the *triple alpha reaction* (or $3\alpha$ reaction). A closer look shows that it is performed in two steps, since a triple encounter is too improbable:

$$^4He + {}^4He \rightleftarrows {}^8Be \quad ,$$
$$^8Be + {}^4He \rightarrow {}^{12}C + \gamma \quad . \tag{18.66}$$

In the first step, two $\alpha$ particles temporarily form a $^8$Be nucleus. Its ground state is nearly 100 keV higher in energy and therefore decays back into the two $\alpha$'s after a few times $10^{-16}$ s. This seems to be a very short time at a first glance, but it is roughly $10^5$ times larger than the duration of a normal scattering encounter. The probability for another reaction occurring during this time is correspondingly enhanced. In fact the lifetime of $^8$Be is sufficient to build up an average concentration of these nuclei of about $10^{-9}$ in the stellar matter. The high densities then ensure a sufficient rate of further $\alpha$ captures that form $^{12}$C nuclei [the second step in (18.66)]. Both these reactions are complicated owing to the involvement of resonances. The energy release per $^{12}$C nucleus formed is 7.275 MeV. This gives an energy release *per unit mass* that is 10.3 times smaller than in the case of the CNO cycle (where one-third fewer nucleons are processed): $E_{3\alpha} \approx 5.9 \times 10^{17}$ erg g$^{-1}$. The resulting energy generation rate is

$$\varepsilon_{3\alpha} = 5.09 \times 10^{11} f_{3\alpha} \varrho^2 X_4^3 T_8^{-3} e^{-44.027/T_8} \qquad (18.67)$$

($\varepsilon$ and $\varrho$ in cgs), with the screening factor $f_{3\alpha}$. This reaction has an enormous temperature sensitivity. For $T_8 = 1 \ldots 2$, (18.39) gives $\nu \approx 40 \ldots 19$!

Once a sufficient $^{12}$C abundance has been built up by the $3\alpha$ reaction, further $\alpha$ captures can occur simultaneously with (18.66) such that the nuclei $^{16}$O, $^{20}$Ne, ... are successively formed:

$$^{12}\text{C} + {}^4\text{He} \rightarrow {}^{16}\text{O} + \gamma \quad ,$$
$$^{16}\text{O} + {}^4\text{He} \rightarrow {}^{20}\text{Ne} + \gamma \quad ,$$
$$\ldots \qquad (18.68)$$

In a typical stellar-interior environment, reactions going beyond $^{20}$Ne are rare.

The energy release per $^{12}$C$(\alpha, \gamma)^{16}$O reaction is 7.162 MeV, corresponding to $E_{12,\alpha} = 4.320 \times 10^{17}$ erg g$^{-1}$ of produced $^{16}$O. (The whole formation of $^{16}$O from the initial four $\alpha$ particles has then yielded $8.71 \times 10^{17}$ erg g$^{-1}$.) This is a rather complicated reaction. For moderate temperatures (up to a few $10^8$ K), one may use the following simple approximation:

$$\varepsilon_{12,\alpha} = 1.3 \times 10^{27} f_{12,4} X_{12} X_4 \varrho T_8^{-2} \left( \frac{1 + 0.134 T_8^{2/3}}{1 + 0.01 T_8^{2/3}} \right)^2 e^{-69.20/T_8^{1/3}} \quad , \quad (18.69)$$

where $\varepsilon$ and $\varrho$ are in cgs.

In each reaction $^{16}$O $(\alpha, \gamma)$ $^{20}$Ne, an energy of 4.73 MeV is released. The rate is approximately

$$\varepsilon_{16,\alpha} \approx X_{16} X_4 \varrho f_{16,4} [1.82 \times 10^{27} T_8^{-2/3} e^{-85.65/T_8^{1/3}}$$
$$+ 9.22 \times 10^{19} T_8^{-3/2} e^{-103.59/T_8}] \quad , \qquad (18.70)$$

where $\varepsilon$ and $\varrho$ are in cgs; this rate is very uncertain.

Summarizing, we can say that during helium burning reactions (18.66) and (18.68) occur simultaneously, and the total energy generation rate is given by $\varepsilon_{\text{He}} = \varepsilon_{3\alpha} + \varepsilon_{12,\alpha} + \varepsilon_{16,\alpha}$. If the initial $^4$He is transformed into equal amounts of $^{12}$C and $^{16}$O, then the energy yield is $7.28 \times 10^{17}$ erg g$^{-1}$.

### 18.5.3 Carbon Burning etc.

For a mixture consisting mainly of $^{12}$C and $^{16}$O (as would be found in the central part of a star after helium burning), *carbon burning* will set in if the temperature or the density rises sufficiently. The typical range of temperature for this burning is $T_8 \approx 5 \dots 10$.

Here (and in the following types of burning) the situation is already so difficult that one often has to rely on rough approximations and guesses. The first complication is that the original $^{12}$C+$^{12}$C reaction produces an excited $^{24}$Mg nucleus, which can decay via many different channels (the last column gives $Q/1\mathrm{MeV}$):

$$
\begin{aligned}
^{12}\mathrm{C} + {}^{12}\mathrm{C} \rightarrow {}^{24}\mathrm{Mg} + \gamma \quad &, \quad 13.931 \\
\rightarrow {}^{23}\mathrm{Mg} + n \quad &, \quad -2.605 \\
\rightarrow {}^{23}\mathrm{Na} + p \quad &, \quad 2.238 \\
\rightarrow {}^{20}\mathrm{Ne} + \alpha \quad &, \quad 4.616 \\
\rightarrow {}^{16}\mathrm{O} + 2\alpha \quad &, \quad -0.114
\end{aligned}
\tag{18.71}
$$

The relative frequency of the channels is very different, and depends also on the temperature. The $\gamma$ decay (leaving $^{24}$Mg) is rather improbable, and the same is true for the two endothermic decays ($^{23}$Mg + $n$ and $^{16}$O + $2\alpha$). The most probable reactions are those which yield $^{23}$Na+$p$ and $^{20}$Ne+$\alpha$. These are believed to occur at about equal rates for temperatures that are not too high (say $T_9 < 3$).

The next problem is that the produced $p$ and $\alpha$ find themselves at temperatures extremely high for hydrogen and helium burning and will immediately react with some of the particles in the mixture (from $^{12}$C up to $^{24}$Mg). They may even start whole reaction chains, such as $^{12}\mathrm{C}(p,\gamma)^{13}\mathrm{N}(e^+\nu)^{13}\mathrm{C}(\alpha,n)^{16}\mathrm{O}$, where the neutron could immediately react further. All these details would have to be evaluated quantitatively in order to find the average energy gain and the final products. For a rough guess one may assume that on average $Q \approx 13$ MeV are released per $^{12}$C–$^{12}$C reaction (including all follow-up reactions). Then,

$$
\begin{aligned}
\varepsilon_{CC} \approx 5.49 \times 10^{43} f_{CC} \varrho X_{12}^2 T_9^{-3/2} T_{9a}^{5/6} \exp\left[-84.165/T_{9a}^{1/3}\right] \\
\times \left[\exp(-0.01T_{9a}^4) + 5.56 \times 10^{-3} \exp(1.685T_{9a}^{2/3})\right]^{-1} \quad ,
\end{aligned}
\tag{18.72}
$$

with $\varepsilon$ and $\varrho$ in cgs and with $T_{9a} = T_9/(1 + 0.067T_9)$. The screening factor $f_{CC}$ can become important (see Fig. 18.6), since this burning can start in very dense matter. The end products may be mainly $^{16}$O, $^{20}$Ne, $^{24}$Mg, $^{28}$Si.

For *oxygen burning*, $^{16}$O+$^{16}$O, the Coulomb barrier is already so high that the necessary temperature is $T_9 \gtrsim 1$. As in the case of carbon burning, the reaction can proceed via several channels:

$$
\begin{aligned}
^{16}\mathrm{O} + {}^{16}\mathrm{O} \rightarrow {}^{32}\mathrm{S} + \gamma \quad &, \quad 16.541 \\
\rightarrow {}^{31}\mathrm{P} + p \quad &, \quad 7.677 \\
\rightarrow {}^{31}\mathrm{S} + n \quad &, \quad 1.453 \\
\rightarrow {}^{28}\mathrm{Si} + \alpha \quad &, \quad 9.593 \\
\rightarrow {}^{24}\mathrm{Mg} + 2\alpha \quad &, \quad -0.393
\end{aligned}
\tag{18.73}
$$

Most frequent is the $p$ decay, followed by the $\alpha$ decays. Again, all released $p$, $n$, and $\alpha$ are captured immediately, giving rise to a multitude of secondary reactions. Among the end products one will find a large amount of $^{28}$Si. For an average energy $Q \approx 16$ MeV released per $^{16}$O+$^{16}$O reaction, the energy generation rate is roughly

$$\varepsilon_{OO} \approx 1.09 \times 10^{54} f_{OO} \varrho X_{16}^2 T_9^{-3/2} T_{9a}^{5/6} \exp(-135.93/T_{9a}^{1/3})$$

$$\times \left[ \exp(-0.032 T_{9a}^4) + 3.89 \times 10^{-4} \exp(2.659 T_{9a}^{-2/3}) \right]^{-1} \quad , \quad (18.74)$$

with $\varepsilon$ and $\varrho$ in cgs, with the screening factor $f_{OO}$, and $T_{9a} = T_9/(1 + 0.067 T_9)$.

For $T_9 > 1$, one also has to consider the possibility of *photodisintegration* of nuclei that are not too strongly bound. Here the radiation field contains a significant number of photons with energies in the MeV range, which can be absorbed by a nucleus, breaking it up, for example, by $\alpha$ decay. This is a complete analogue of photoionization of atoms, and, in equilibrium, a formula equivalent to the Saha formula [see (14.11)] holds for the number densities $n_i$ and $n_j$ of the final particles (after disintegration), relative to the number $n_{ij}$ of the original (compound) particles:

$$\frac{n_i n_j}{n_{ij}} \sim T^{3/2} e^{-Q/kT} \quad , \quad (18.75)$$

where $Q$ is the difference in binding energies between the original nucleus and its fragments. ($Q$ corresponds to the ionization energy $\chi$; however, it is about $10^2 \ldots 10^3$ times larger because of the much stronger nuclear forces.) The proportionality factor contains essentially the partition functions of the three types of particles. Equilibrium is usually not reached, and the details are very complicated and may differ from case to case, which is also true for the amount of energy released or lost.

The photodisintegration itself is, of course, endothermic. But the ejected particles ($X_j$) will be immediately recaptured. The capture can lead back to the original nucleus $X_{ij}$, i.e. the reaction would be $X_{ij} \rightleftarrows X_i + X_j$, or it can lead to quite different, even heavier, nuclei $X_{jk}$ that are more strongly bound than the original one $X_j + X_k \rightarrow X_{jk}$. The latter case would be exothermic and can outweigh the endothermic photodisintegration in the total energy balance.

An example is *neon disintegration*, which in stellar evolution occurs even before oxygen burning:

$$^{20}\text{Ne} + \gamma \rightarrow {}^{16}\text{O} + \alpha \quad , \quad Q = -4.73 \text{ MeV} \quad . \quad (18.76)$$

It dominates over the inverse reaction (known from helium burning) at $T_9 > 1.5$. The ejected $\alpha$ particle reacts mainly with other $^{20}$Ne nuclei, yielding $^{24}$Mg+$\gamma$. The net result will then be the conversion of Ne into O and Mg:

$$2\,^{20}\text{Ne} + \gamma \rightarrow {}^{16}\text{O} + {}^{24}\text{Mg} + \gamma \quad , \quad Q = +4.583 \text{ MeV} \quad . \quad (18.77)$$

Another example is the photodisintegration of $^{28}$Si, which may be the dominant reaction at the end of oxygen burning. Near $T_9 \approx 3$, $^{28}$Si can be decomposed by the photons and eject $n$, $p$, or $\alpha$. There follows a large number of reactions in which the thereby created nuclei (e.g. Al, Mg, Ne) will also be subject to photodisintegration, leading to the existence of an appreciable amount of free $n$, $p$, and $\alpha$ particles. These react with the remaining $^{28}$Si, thus building up gradually heavier nuclei, until $^{56}$Fe is reached. Since $^{56}$Fe is so strongly bound, it may survive this melting pot as the only (or dominant) species. So, forgetting all intermediate stages, we would ultimately have the conversion of two $^{28}$Si into $^{56}$Fe, which can be called *silicon burning*.

For $T_9 \gtrsim 5$, photodisintegration breaks up even the $^{56}$Fe nuclei into $\alpha$ particles and thus reverses the effect of all prior burnings. Such processes can occur during supernova explosions (see § 34).

## 18.6 Neutrinos

Neutrinos require special consideration because their cross-section $\sigma_\nu$ for interaction with matter is so extremely small. For scattering of neutrinos with energy $E_\nu$, one has roughly $\sigma_\nu \approx \left( E_\nu / m_e c^2 \right)^2 10^{-44}$ cm$^2$. Neutrinos in the MeV range then have $\sigma_\nu \approx 10^{-44}$ cm$^2$, which is a factor $10^{-18}$ smaller than the cross-section for typical photon–matter interactions. The corresponding mean free path in matter of density $\varrho = n \mu m_u$ and molecular weight $\mu (\approx 1)$ is about

$$\ell_\nu = \frac{1}{n \sigma_\nu} = \frac{\mu m_u}{\varrho \sigma_\nu} \approx \frac{2 \times 10^{20} \text{cm}}{\varrho} \quad , \tag{18.78}$$

with $\varrho$ in cgs. For "normal" stellar matter with $\varrho \approx 1$ g cm$^{-3}$, (18.78) would give a mean free path of the neutrinos of $\ell_\nu \approx 100$ parsec, and even for $\varrho = 10^6$ g cm$^{-3}$ one has $\ell_\nu \approx 3000 R_\odot$.

Therefore it is safe to say that neutrinos, once created somewhere in the central region, leave a normal star without interactions carrying away their energy. This neutrino energy has then to be excluded from all other forms of energies (e.g. that released by nuclear reactions), which are subject to some diffusive transport of energy according to the temperature gradient.

The situation can be completely different, however, during a collapse in the final evolutionary stage. The density can reach nuclear values, and for $\varrho = 10^{14}$ g cm$^{-3}$, (18.78) gives only $\ell_\nu \approx 20$ km. Considering the fact that neutrinos can then be rather energetic (which increases $\sigma_\nu$ appreciably) one sees that many of them will be reabsorbed within the star. Then it is necessary to consider a transport equation for neutrino energy, and to evaluate the amount of momentum the interacting neutrinos deliver to the overlying layers (see § 34.3.3).

Only electron neutrinos play a role in stellar interiors, and these can be created in quite different processes inside a star. We first mention those processes involving nuclear reactions, which have already been mentioned (§ 18.5) in connection with certain nuclear burnings. In this special case one usually allows for the neutrino energy loss by a corresponding reduction of the released energy. [This means that in (9.3) $\varepsilon_n$ is reduced and no separate $\varepsilon_\nu$ term is needed.]

The net balance of *hydrogen burning* is the transformation of 4 protons into a $^4$He nucleus. The conservation of charge requires two $\beta^+$ decays, each of which is accompanied by a neutrino emission in order to conserve the lepton number. In the reaction chains (18.62) and (18.64) we have the following $\nu$ reactions ($Q_\nu$ = average neutrino energy):

$$
\begin{array}{llll}
^1\text{H} + {}^1\text{H} \rightarrow {}^2\text{H} + e^+ + \nu & (pp1, 2, 3) & Q_\nu = 0.263\,\text{MeV} & \\
^7\text{Be} + e^- \rightarrow {}^7\text{Li} + \nu & (pp2) & 0.80\;\;\text{MeV} & \\
^8\text{B} \rightarrow {}^8\text{Be} + e^+ + \nu & (pp3) & 7.2\;\;\;\;\text{MeV} & (18.79) \\
^{13}\text{N} \rightarrow {}^{13}\text{C} + e^+ + \nu & (\text{CNO}) & 0.71\;\;\text{MeV} & \\
^{15}\text{O} \rightarrow {}^{15}\text{N} + e^+ + \nu & (\text{CNO}) & 1.0\;\;\;\;\text{MeV} &
\end{array}
$$

With an average energy yield of 25 MeV $\approx 4 \times 10^{-5}$ erg per cycle, the generation of one solar luminosity ($L_\odot \approx 4 \times 10^{33}$ erg s$^{-1}$) by hydrogen burning implies a production of about $2 \times 10^{38}$ neutrinos per second. Those neutrinos coming directly from the central region of the sun yield a flux of roughly $10^{11}$ neutrinos per cm$^2$ each second at the earth. For attempts to measure the *solar neutrinos* from the reactions of the first and the third line of (18.79) see § 29.2.

There are also neutrino-producing nuclear reactions that are not connected with nuclear burnings. For example, at extreme densities degenerate electrons can be pushed up to energies large enough for *electron capture* by protons in nuclei of charge $Z$ and atomic weight $A$: $e^- + (Z, A) \rightarrow (Z - 1, A) + \nu$.

Another interesting example is the so-called *Urca process*. For a suitable nucleus $(Z, A)$, an electron capture occurs which is followed by a $\beta$ decay:

$$
\begin{array}{ll}
(Z, A) + e^- \rightarrow (Z - 1, A) + \nu & , \\
(Z - 1, A) \rightarrow (Z, A) + e^- + \bar{\nu} & .
\end{array}
\qquad (18.80)
$$

The original particles are restored and two neutrinos are emitted. There are obvious restrictions on the nuclei $(Z, A)$ suitable for this process: they must have an isobaric nucleus $(Z - 1, A)$ of slightly higher energy that is unstable to $\beta$ decay. A possible example would be $^{35}\text{Cl}\,(e^-, \nu){}^{35}\text{S}$ (endothermic with $Q = -0.17$ MeV), followed by the decay $^{35}\text{S}\,(e^- \bar{\nu})\,{}^{35}\text{Cl}$, the energy for the first reaction being supplied by the captured electron. In this way, thermal energy of the stellar matter is converted into neutrino energy and lost from the star, while the composition remains unchanged. (*Urca* is the name of a Rio de Janeiro casino, where Gamow and Schönberg found that, as the only recognizable net effect, similar losses, little by little, occur with visitors' money.) Details depend very much on the stellar material. If appropriate nuclei for this are present, the energy loss will increase with $\varrho$ and $T$.

The following processes occur *without nuclear reaction*. These purely leptonic processes were predicted as a consequence of the generalized Fermi theory of weak interaction, which allows a direct electron–neutrino coupling, such that a neutrino pair can be emitted if an electron changes its momentum. It is clear that such processes may be reduced by degeneracy if the electrons do not find enough free cells in phase space.

The following processes of this type can be important for stellar interiors. Figure 18.9 shows the regions of the $\varrho - T$ plane where this is the case.

*Pair annihilation neutrinos*: $e^- + e^+ \rightarrow \nu + \bar{\nu}$. In very hot environments ($T_9 > 1$), there are enough energetic photons to create large numbers of $(e^- e^+)$ pairs. These will soon be annihilated, usually giving two photons, and a certain equilibrium abun-

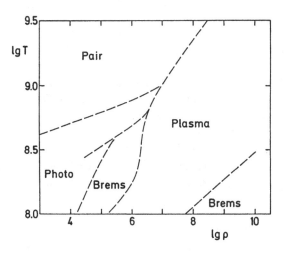

Fig. 18.9. Regions in which different types of neutrino loss dominate. $\varrho$ is in g cm$^{-3}$, $T$ in K. (After BARKAT, 1975)

dance of $e^+$ will be reached. In this continuous back and forth exchange, however, there is a small one-way leakage, since roughly once in $10^{19}$ times, the annihilation results in a pair $(\nu\bar{\nu})$ instead of the usual photons. This can lead to appreciable energy loss only in a very hot, not too dense plasma. $\varepsilon_\nu$ is a complicated function, but is always proportional to $\varrho^{-1}$. We quote only simple asymptotic expressions ($\varepsilon$ and $\varrho$ in cgs) for non-degeneracy:

$$\varepsilon_\nu^{(\text{pair})} = \begin{cases} \frac{4.9\times10^8}{\varrho} T_9^3\, e^{-11.86 T_9} & , \quad T_9 < 1 \\ \frac{4.45\times10^{15}}{\varrho} T_9^9 & , \qquad T_9 > 3 \end{cases} \tag{18.81}$$

*Photoneutrinos*: $\gamma + e^- \rightarrow e^- + \nu + \bar{\nu}$. This is the analogue of normal Compton scattering, in which a photon is scattered by an electron. In a very few cases it may happen that, after scattering, the photon is replaced by a neutrino–antineutrino pair. The rates of energy loss for this process are rather different for different limiting cases (depending on the degrees of degeneracy and the importance of relativistic effects). A rough interpolation formula (PETROSIAN, BEAUDET, SALPETER, 1967) is

$$\varepsilon_\nu^{(\text{phot})} = \varepsilon_1 + \varepsilon_2(\mu_e + \bar{\varrho})^{-1} \;,$$
$$\varepsilon_1 = 1.103 \times 10^{13} \varrho^{-1} T_9^9\, e^{-5.93/T_9} \;,$$
$$\varepsilon_2 = 0.976 \times 10^8 T_9^8 (1 + 4.2 T_9)^{-1} \;, \tag{18.82}$$
$$\bar{\varrho} = 6.446 \times 10^{-6} \varrho\, T_9^{-1} (1 + 4.2 T_9)^{-1} \;,$$

where the $\varepsilon$ and $\varrho$ are in cgs.

*Plasmaneutrinos*: $\gamma_{\text{plasm}} \rightarrow \nu + \bar{\nu}$. A so-called plasmon decays here to a neutrino–antineutrino pair. The plasma frequency $\omega_0$ is given by

$$\omega_0^2 \frac{m_e}{4\pi e^2 n_e} = \begin{cases} 1 & , \quad \text{non-degenerate} \\ \left[1 + \left(\frac{\hbar}{m_e c}\right)^2 (3\pi^2 n_e)^{2/3}\right]^{-1/2} & , \quad \text{degenerate} \end{cases} \tag{18.83}$$

This is important for an electromagnetic wave of frequency $\omega$ moving through the plasma, since its dispersion relation is

$$\omega^2 = K^2 c^2 + \omega_0^2 \quad , \tag{18.84}$$

where $K$ is the wave number. Here the wave is coupled to the collective motions of the electrons, and a propagating wave can occur only for $\omega > \omega_0$. Multiplication of (18.84) by $\hbar^2$ gives the square of the energy $E$ of a quantum, which therefore behaves as if it were a relativistic particle with a rest mass corresponding to the energy $\hbar\omega_0$. Such a quantum is called a *plasmon*. For the energy rate one has to add the rates of transversal and longitudinal plasmons: $\varepsilon_\nu^{(\text{plasm})} = \varepsilon_\nu^{(t)} + \varepsilon_\nu^{(l)}$. With the abbreviations $\gamma = \hbar\omega_0/kT$ and $\lambda = kT/m_e c^2$, one has the approximations for two limits of $\gamma$:

$$\varepsilon_\nu^{(\text{plasm})} = 3.356 \times 10^{19} \varrho^{-1} \lambda^6 (1 + 0.0158\gamma^2) T_9^3 \quad , \gamma \ll 1 \quad ,$$
$$\varepsilon_\nu^{(\text{plasm})} = 5.252 \times 10^{20} \varrho^{-1} \lambda^{7.5} T_9^{1.5} e^{-\gamma} \quad , \gamma \gg 1 \quad , \tag{18.85}$$

with $\varepsilon$ and $\varrho$ in cgs. The exponential decrease for large $\gamma$ (i.e. for increasing $\omega_0 \sim \varrho^{1/2}$ at constant $T$) comes from the fact that very few plasmons can be excited if $kT$ drops below $\hbar\omega_0$.

*Bremsstrahlung neutrinos.* Inelastic scattering (deceleration) of an electron in the Coulomb field of a nucleus will usually lead to emission of a "Bremsstrahlung" photon (free–free emission). This photon can be replaced by a neutrino–antineutrino pair. The rate of energy loss for very large $\varrho$ is

$$\varepsilon_\nu^{(\text{brems})} \approx 0.76 \frac{Z^2}{A} T_8^6 \quad , \tag{18.86}$$

(in cgs) where $Z$ and $A$ are the charge and mass number of the nuclei. For smaller densities $\varepsilon_\nu$ is smaller than this expression, the correction being roughly a factor 10 at $\varrho \approx 10^4$ g cm$^{-3}$. This process can dominate, in particular, at low temperature and very high density. The rate $\varepsilon_\nu^{(\text{brems})}$ does not decrease with increasing degeneracy (as other processes do), since the lack of free cells in phase space is compensated by an increasing cross-section for neutrino emission.

*Synchrotron neutrinos.* These can only occur in the presence of strong magnetic fields. The normal synchrotron photon emitted by an electron moving in this field is again replaced by a neutrino–antineutrino pair.

# IV Simple Stellar Models

# § 19 Polytropic Gaseous Spheres

## 19.1 Polytropic Relations

As we have seen in § 9.1 the temperature does not appear explicitly in the two mechanical equations (9.1, 2). Under certain circumstances this provides the possibility of separating them from the "thermo-energetic part" of the equations. For the following it is convenient to introduce once again the gravitational potential $\Phi$, as it was defined in § 1.3. We here treat stars in hydrostatic equilibrium, which requires [see (1.11),(2.3)]

$$\frac{dP}{dr} = -\frac{d\Phi}{dr}\varrho \quad , \tag{19.1}$$

together with Poisson's equation (1.10)

$$\frac{1}{r^2}\frac{d}{dr}\left(r^2\frac{d\Phi}{dr}\right) = 4\pi G\varrho \quad . \tag{19.2}$$

We have replaced the partial derivatives by ordinary ones since only time-independent solutions shall be considered.

In general the temperature appears in the system (19.1, 2) if the density is replaced by an equation of state of the form $\varrho = \varrho(P,T)$. However, we have already encountered examples for simpler cases. If $\varrho$ does not depend on $T$, i.e. $\varrho = \varrho(P)$ only, then this relation can be introduced into (19.1,2), which become a system of two equations for $P$ and $\Phi$ and can be solved without the other structure equations. An example is the completely degenerate gas of non-relativistic electrons for which $\varrho \sim P^{3/5}$ [see (15.23)].

We shall deal here with similar cases and assume that there exists a simple relation between $P$ and $\varrho$ of the form

$$P = K\varrho^\gamma \equiv K\varrho^{1+\frac{1}{n}} \quad , \tag{19.3}$$

where $K$, $\gamma$, and $n$ are constant. A relation of the form (19.3) is called a *polytropic relation*. $K$ is the *polytropic constant* and $\gamma$ the *polytropic exponent* (which we have to distinguish from the adiabatic exponent $\gamma_{ad}$). One often uses, instead of $\gamma$, the *polytropic index* $n$, which is defined by

$$n = \frac{1}{\gamma - 1} \quad . \tag{19.4}$$

Obviously for a completely degenerate gas the equation of state in its limiting cases has the polytropic form (19.3). In the non-relativistic limit (15.23) we have $\gamma = 5/3$, $n = 3/2$, while for the relativistic limit (15.26) holds, so that $\gamma = 4/3$, $n = 3$. For such cases, where the equation of state has a polytropic form, the polytropic constant $K$ is fixed and can be calculated from natural constants.

But there are also examples for a relation of the form (19.3) where $K$ is a free parameter which is constant within a particular star but can have different values from one star to another.

Let us consider an isothermal ideal gas of temperature $T = T_0$ and mean molecular weight $\mu$. Its equation of state $\varrho = \mu P/(\Re T)$ can be written in the form (19.3), with $K = \Re T_0/\mu$, $\gamma = 1$, and $n = \infty$. Here $K$ is not fixed but depends on $T_0$ and $\mu$, and if we then use (19.3) in the stellar-structure equations, we are free to give $K$ any (positive) value for a certain star.

In a star that is completely convective the temperature gradient (except for that in a region near the surface, which we shall ignore) is given, to a very good approximation, by $\nabla = (d \ln T/d \ln P)_{ad} = \nabla_{ad}$ (see §7.3). If radiation pressure can be ignored and the gas is completely ionized, we have $\nabla_{ad} = 2/5$ according to (13.21). This means that throughout the star $T \sim P^{2/5}$, and for an ideal gas with $\mu = $ constant, $T \sim P/\varrho$, and therefore $P \sim \varrho^{5/3}$. This again is a polytropic relation of the form (19.3) with $\gamma = 5/3$, $n = 3/2$. But now $K$ is not fixed by natural constants; it is a free parameter in the sense that it can vary from star to star.

The homogeneous gaseous sphere can also be considered a special case of the polytropic relation (19.3). Let us write (19.3) in the form

$$\varrho = K_1 P^{1/\gamma} \quad ; \tag{19.5}$$

then $\gamma = \infty$ (or $n = 0$) gives $\varrho = K_1 = $ constant.

These examples have shown that we can have two reasons for a polytropic relation in a star. (1) The equation of state is of the simple form $P = K\varrho^\gamma$, with a fixed value of $K$. (2) The equation of state contains $T$ (as for an ideal gas), but there is an additional relation between $T$ and $P$ (like the adiabatic condition) that together with the equation of state yields a polytropic relation; then $K$ is a free parameter.

On the other hand, if we assume a polytropic relation for an ideal gas, this is equivalent to adopting a certain relation $T = T(P)$. This means that one fixes the temperature stratification instead of determining it by the thermo-energetic equations of stellar structure. For example, a polytrope with $n = 3$ does not necessarily have to consist of relativistic degenerate gases, but can also consist of an ideal gas and have $\nabla = 1/(n+1) = 0.25$.

## 19.2 Polytropic Stellar Models

With the polytropic relation (19.3) (independent of whether $K$ is a free parameter or a constant with a fixed value), (19.1) can be written

$$\frac{d\Phi}{dr} = -\gamma K \varrho^{\gamma-2} \frac{d\varrho}{dr} \quad . \tag{19.6}$$

If $\gamma \neq 1$ (the case $\gamma = 1$, $n = \infty$, corresponding to the isothermal model, will be treated in § 19.8), (19.6) can be integrated:

$$\varrho = \left( \frac{-\Phi}{(n+1)K} \right)^n \quad , \tag{19.7}$$

where we have made use of (19.4) and chosen the integration constant to give $\Phi = 0$ at the surface ($\varrho = 0$). Note that in the interior of our model $\Phi < 0$, giving there $\varrho > 0$. If we introduce (19.7) into the right-hand side of the Poisson equation (19.2), we obtain an ordinary differential equation for $\Phi$:

$$\frac{d^2\Phi}{dr^2} + \frac{2}{r}\frac{d\Phi}{dr} = 4\pi G \left( \frac{-\Phi}{(n+1)K} \right)^n \quad . \tag{19.8}$$

We now define dimensionless variables $z$, $w$ by

$$z = Ar \quad , \quad A^2 = \frac{4\pi G}{(n+1)^n K^n} (-\Phi_c)^{n-1} = \frac{4\pi G}{(n+1)K} \varrho_c^{\frac{n-1}{n}} \quad , \tag{19.9}$$

$$w = \frac{\Phi}{\Phi_c} = \left( \frac{\varrho}{\varrho_c} \right)^{1/n} \quad ,$$

where the subscript c refers to the centre and where the relation between $\varrho$ and $\Phi$ is taken from (19.7). At the centre ($r = 0$) we have $z = 0$, $\Phi = \Phi_c$, $\varrho = \varrho_c$ and therefore $w = 1$. Then (19.8) can be written

$$\frac{d^2w}{dz^2} + \frac{2}{z}\frac{dw}{dz} + w^n = 0 \quad ,$$

$$\frac{1}{z^2}\frac{d}{dz}\left( z^2\frac{dw}{dz} \right) + w^n = 0 \quad . \tag{19.10}$$

This is the famous *Lane–Emden equation* (named after J.H. Lane and R. Emden). We are only interested in solutions that are finite at the centre, $z = 0$. Equation (19.10) shows that we then have to require $dw/dz \equiv w' = 0$. Let us assume we have a solution $w(z)$ of (19.10) that fulfils the central boundary conditions $w(0) = 1$ and $w'(0) = 0$; then according to (19.9) the radial distribution of the density is given by

$$\varrho(r) = \varrho_c w^n \quad , \quad \varrho_c = \left[ \frac{-\Phi_c}{(n+1)K} \right]^n \quad . \tag{19.11}$$

For the pressure we obtain from (19.3,4) that $P(r) = P_c w^{n+1}$, where $P_c = K\varrho_c^\gamma$.

Before trying to construct stellar polytropic models we shall discuss some of the mathematical properties of the solutions $w(z)$ of (19.10).

## 19.3 Properties of the Solutions

The Lane–Emden equation has a regular singularity at $z = 0$. In order to understand the behaviour of the solutions there, we expand into a power series:

$$w(z) = 1 + a_1 z + a_2 z^2 + a_3 z^3 + \ldots \quad , \tag{19.12}$$

with $a_1 = w'(0)$, $2a_2 = w''(0), \ldots$. Since the gravitational acceleration $|g| = d\Phi/dr \sim dw/dz$ must vanish in the centre, we have $a_1 = 0$. Inserting (19.12) into the Emden equation (19.10), by comparing coefficients one finds

$$w(z) = 1 - \frac{1}{6}z^2 + \frac{n}{120}z^4 + \ldots \quad , \tag{19.13}$$

where again we have excluded the isothermal sphere $n = \infty$. Equation (19.13) shows that $w(z)$ has a maximum at $z = 0$.

Only for three values of $n$ can the solutions be given by analytic expressions. The first case is

$$n = 0: \quad w(z) = 1 - \frac{1}{6}z^2 \quad , \tag{19.14}$$

and we have already mentioned that this corresponds to the homogeneous gas sphere. Indeed $\varrho = \varrho_c w^n$ gives constant density for $n = 0$. The two other cases are

$$n = 1: \quad w(z) = \frac{\sin z}{z} \quad , \tag{19.15}$$

$$n = 5: \quad w(z) = \frac{1}{(1 + z^2/3)^{1/2}} \quad . \tag{19.16}$$

The surface of the polytrope of index $n$ is defined by the value $z = z_n$, for which $\varrho = 0$ and thus $w = 0$. While for $n = 0$ and $n = 1$ the surface is obviously reached for a finite value of $z_n$, the case $n = 5$ yields a model of infinite radius. It can be shown that for $n < 5$ the radius of polytropic models is finite; for $n \geq 5$ they have infinite radius. This also holds for the limiting case $n = \infty$ (cf. § 19.8).

Apart from the three cases where analytic solutions are known, the Emden equation (19.10) has to be solved numerically, beginning with the expansion (19.13) for the neighbourhood of the centre. Here the solution starts with zero tangent and $w = 1$ and decreases outwards. This can be seen from (19.13) and is illustrated in Fig. 19.1.

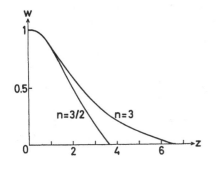

Fig. 19.1. If $n < 5$ the solution of the Lane–Emden equation (19.10) of index $n$ starting with $w(0) = 1$ becomes zero at a finite value of $z = z_n$. Here the solutions for $n = 3/2$ and $n = 3$ are plotted

For a given value of $n < 5$ the integration comes to a point $z$ where $w(z)$ vanishes, i.e. $\varrho = 0$. This value of $z$, which corresponds to the surface of the polytrope, will be called $z_n$. From (19.14–16) one finds $z_0 = \sqrt{6}$, $z_1 = \pi$, $z_5 = \infty$. It is a general property of the solutions that $z_n$ grows monotonically with the polytropic index $n$. Table 19.1 gives some values of $z_n$ and the values of certain functions at $z = z_n$ which will later turn out to be useful for the construction of models.

Table 19.1 Numerical values for polytropic models with index $n$ (after CHANDRASEKHAR, 1939)

| $n$ | $z_n$ | $\left(-z^2 \dfrac{dw}{dz}\right)_{z=z_n}$ | $\varrho_c/\bar{\varrho}$ |
|-----|-------|-------------------------------------------|---------------------------|
| 0   | 2.4494   | 4.8988  | 1.0000   |
| 1   | 3.14159  | 3.14159 | 3.28987  |
| 1.5 | 3.65375  | 2.71406 | 5.99071  |
| 2   | 4.35287  | 2.41105 | 11.40254 |
| 3   | 6.89685  | 2.01824 | 54.1825  |
| 4   | 14.97155 | 1.79723 | 622.408  |
| 4.5 | 31.8365  | 1.73780 | 6189.47  |
| 5   | $\infty$ | 1.73205 | $\infty$ |

So far, we have discussed only solutions that are regular at the centre. But solutions with a singularity at $z = 0$ can also be important if one uses them for stellar regions outside the centre. Let us for instance consider a star that is convective in its outer layer, while in the inner part the energy may be transported by radiation. If the convective envelope is adiabatic, with $\nabla = \nabla_{ad} = 2/5$, it is polytropic and therefore $\varrho \sim w^{3/2}$ and $P \sim w^{5/2}$. But it is unimportant whether this solution is finite at the centre, since anyway the equations do not hold in the radiative interior. On the other hand, one may have to fit a polytropic central core to an envelope with different properties. In this case the polytropic solution has to be regular at the centre, but its behaviour for $w = \varrho = 0$ is unimportant, since it is used only up to the core surface where $\varrho$ and $P$ are non-vanishing. In the following we mainly deal with *complete polytropes*, which have a polytropic relation of the form (19.3) from surface to centre.

## 19.4 Application to Stars

We now construct polytropic models for a given index $n < 5$ and for given values of $M$ and $R$. This will turn out to be possible as long as $K$ is not fixed by the equation of state. We first derive some more relations for polytropes.

From (9.1) and (19.11) it follows that

$$m(r) = \int_0^r 4\pi \varrho r^2 dr = 4\pi \varrho_c \int_0^r w^n r^2 dr = 4\pi \varrho_c \frac{r^3}{z^3} \int_0^z w^n z^2 dz \quad , \qquad (19.17)$$

where we have made use of relations (19.9) and of the fact that $r^3/z^3$ is constant and can be brought in front of the integral. According to the Lane–Emden equation

(19.10) the integrand $w^n z^2$ on the right is a derivative and can immediately be integrated, so that the integral becomes $-z^2 dw/dz$. We obtain

$$m(r) = 4\pi \varrho_c r^3 \left( -\frac{1}{z} \frac{dw}{dz} \right) \quad , \tag{19.18}$$

where the simultaneously appearing $z$ and $r$ are related to each other by $r/z = 1/A = R/z_n$. For the special case of the surface we have

$$M = 4\pi \varrho_c R^3 \left( -\frac{1}{z} \frac{dw}{dz} \right)_{z=z_n} \quad . \tag{19.19}$$

The quantity in brackets can be derived from Table 19.1 for several values of $n$. If we introduce the mean density $\bar{\varrho} := 3M/(4\pi R^3)$, we find

$$\frac{\bar{\varrho}}{\varrho_c} = \left( -\frac{3}{z} \frac{dw}{dz} \right)_{z=z_n} \quad . \tag{19.20}$$

The right-hand side of this equation depends only on $n$: for $n = 0$ it is $1$ – as one can see from (19.11). The higher $n$, the smaller $\bar{\varrho}/\varrho_c$, which means the higher the density concentration, as can be seen in Table 19.1.

We now have all the means at hand to construct the whole polytropic stellar model for given values of $n$, $M$, and $R$ for the case that $K$ is not fixed by the equation of state.

If $n$ is given, a numerical solution of the Lane–Emden equation (19.10) yields the functions $w(z)$, $w'(z)$ and the values of $z_n$ and of $-z_n/(3dw/dz)_n$. If we now use $M$ and $R$ to determine the mean density $\bar{\varrho}$, (19.20) gives $\varrho_c$. On the other hand, we know the constant $A = z/r = z_n/R$ by which we adjust the dimensionless $z$ scale to the $r$ scale. We therefore know the density distribution in the model $\varrho(r) = \varrho_c w^n(z)$ from (19.11). With $\varrho_c$ and the constant $A$ we can determine $K$ from (19.9) and obtain the pressure distribution $P(r) = K \varrho^{(n+1)/n} = K \varrho_c^{(n+1)/n} w^{n+1}$. The local mass $m$ then follows from (19.18) and the (known) relation between the $z$ scale and the $r$ scale. The whole mechanical structure is now determined. It has to be emphasized that this method of constructing models for given values of $n$, $M$, and $R$ is only applicable if $K$ is a free parameter, otherwise the problem would be overdetermined. (The case that $K$ has fixed value will be discussed in § 19.6.)

As an example we try to construct a polytropic model of index 3 for the sun ($M = 1.989 \times 10^{33}$ g, $R = 6.96 \times 10^{10}$ cm). For $n = 3$ Table 19.1 gives $z_3 = 6.897$, $\varrho_c/\bar{\varrho} = 54.18$. The mean density becomes $\bar{\varrho} = 1.41$ g cm$^{-3}$; consequently the central density $\varrho_c = 76.39$ g cm$^{-3}$ and, further, $A = z_3/R = 9.91 \times 10^{-11}$. From (19.9) we find $K = 3.85 \times 10^{14}$ and consequently $P_c = 1.24 \times 10^{17}$ dyn/cm$^2$. For the ideal gas equation with $\mu = 0.62$ corresponding to $X \approx 0.7$, $Y \approx 0.3$ we find for the temperature $T_c = 1.2 \times 10^7$ K. A proper numerical solution of the full set of stellar-structure equations for a chemically homogeneous model of $1M_\odot$ gives $T_c = 1.4 \times 10^7$K. We see that a polytropic estimate with $n = 3$ comes considerably closer to the honestly computed value than our crude estimate in § 2.3.

## 19.5 Radiation Pressure and the Polytrope $n = 3$

We consider here only the case that $K$ is a free parameter. In the example at the end of the previous section we approximated the sun by a polytrope of $n = 3$. This is formally equivalent to the assumption of an ideal gas ($P \sim \varrho T$) together with a constant temperature gradient $\nabla = 1/4 \, (T \sim P^{1/4})$. We will now show that this polytropic relation with $n = 3$ can also be obtained by a certain assumption on the radiation pressure. For an ideal gas with radiation pressure

$$P = \frac{\Re}{\mu} \varrho T + \frac{a}{3} T^4 = \frac{\Re}{\mu \beta} \varrho T \tag{19.21}$$

we assume that the ratio $\beta = P_{\text{gas}}/P$ is constant throughout the star. Now

$$1 - \beta = \frac{P_{\text{rad}}}{P} = \frac{aT^4}{3P} \tag{19.22}$$

shows that $\beta = $ constant means a relation of the form $T^4 \sim P$, which we introduce into (19.21). This gives

$$P = \left( \frac{3 \Re^4}{a \mu^4} \right)^{1/3} \left( \frac{1 - \beta}{\beta^4} \right)^{1/3} \varrho^{4/3} \quad , \tag{19.23}$$

which indeed is a polytropic relation with $n = 3$ for constant $\beta$. Here the polytropic constant $K$ is again a free parameter, since we can choose $\beta$ in the interval $0, 1$.

In § 19.10 we shall apply this to very massive stars. They are fully convective ($\nabla = \nabla_{\text{ad}}$) and dominated by radiation pressure.

Relation (19.23) goes back to A.S. Eddington, who obtained it for his famous "standard model". He found that the full set of stellar-structure equations (including the thermo-energetic equations) could be solved very simply by the assumption $\kappa l/m = $ constant throughout the star. One then obtains $\beta = $ constant and therefore the polytropic relation (19.23).

## 19.6 Polytropic Stellar Models with Fixed $K$

As a typical example we have already mentioned the non-relativistic degenerate electron gas for which the equation of state (15.23) is polytropic with $n = 3/2$ and polytropic constant

$$K = \frac{1}{20} \left( \frac{3}{\pi} \right)^{2/3} \frac{h^2}{m_e} \frac{1}{(\mu_e m_u)^{5/3}} \quad . \tag{19.24}$$

We consider the chemical composition to be given ($\mu_e$ fixed). Then in this expression there is no room for the choice of a free parameter as in (19.23). Although $n = 3/2$ is a particularly interesting case, we shall derive our relation for general values of the polytropic index with $n < 5$.

Let us see how to construct a model with index $n$ for a given value of $\varrho_c$. The functions $w(z)$ and $w'(z)$ can be considered known from an integration of the Emden equation. Then $\varrho = \varrho_c w^n$ is known as a function of $z$. According to (19.9) the relation between $r$ and $z$ is

$$\left(\frac{r}{z}\right)^2 = \frac{1}{4\pi G}(n+1)K\varrho_c^{\frac{1-n}{n}} \quad . \tag{19.25}$$

This can be used to derive the density also as a function of $r$, where the radius of the model is $R = z_n/A$ and the value $z_n$ is obtained from the integration. The constant $A$ depends on $\varrho_c$, as shown by (19.25), and

$$R \sim \varrho_c^{\frac{1-n}{2n}} \quad . \tag{19.26}$$

As long as $n > 1$, the radius $R$ becomes smaller with increasing central density $\varrho_c$, becoming zero for infinite $\varrho_c$. On the other hand, the mass $M$ of the model varies with $\varrho_c$ according to (19.19) as $M \sim \varrho_c R^3$, or

$$M = C_1\,\varrho_c^{\frac{3-n}{2n}} \quad ; \quad C_1 = 4\pi\left(-\frac{w'}{z}\right)_{z_n} z_n^3 \left(\frac{n+1}{4\pi G}\right)^{3/2} K^{3/2} \quad . \tag{19.27}$$

Elimination of $\varrho_c$ from (19.26) and (19.27) shows that there is a mass–radius relation of the form

$$R \sim M^{\frac{1-n}{3-n}} \quad . \tag{19.28}$$

We see that for given $K$ and $n$ there is a one-dimensional manifold of models only, the parameter being *either $M$ or $R$* (or $\varrho_c$), whereas there was a two-dimensional manifold (*$M$ and $R$* as parameters) when $K$ was a free parameter.

Consider again the case of the non-relativistic degenerate electron gas, which is not too bad an approximation for white dwarfs of small mass. With $n = 3/2$, (19.28) gives $R \sim M^{-1/3}$ and the surprising result that the larger the mass the smaller the radius. (This is made plausible by simple considerations in § 35.1.) The model will shrink with increasing mass and should finally end as a point mass for infinite $M$. But long before this, our assumed equation of state will not be valid any more, since from (19.27) we see that $\varrho_c$ is proportional to $\sim M^2$. For ever increasing densities the electrons will become relativistic (see § 16.2) and the equation of state (15.23) has to be replaced by (15.26). This means a transition from a polytrope $n = 3/2$ to one with $n = 3$ (and a different, but also given, polytropic constant $K$). In this case we shall encounter a new problem, hinted at by the exponent in (19.28).

## 19.7 Chandrasekhar's Limiting Mass

In § 19.6 we have seen that a polytropic model in which the pressure is provided by a non-relativistic degenerate electron gas reaches higher and higher central and mean densities with growing total mass $M$. But with increasing density the electrons become gradually more relativistic. This starts in the central region where the density is highest, the outer parts remaining non-relativistic. Although we know that the transition between equations of state (15.23,26) does not occur abruptly, but smoothly via the more general equation of state (15.13), one can imagine that an idealized stellar model consisting of degenerate matter can be constructed by fitting two regions smoothly together: a (relativistic) polytropic core with $n = 3$ surrounded by a (non-relativistic) polytropic envelope with $n = 3/2$. Indeed Chandrasekhar constructed his first white-dwarf model in this way.

Let us consider how this idealized model changes with growing mass $M$. At small $M$ the whole model is still non-relativistic. The relativistic core will occur for $\varrho_c \gtrsim 10^6$ g cm$^{-3}$ (Fig. 16.1) and gradually encompass larger parts of the model as $\varrho_c$ increases. One would therefore expect the model finally to approach the state where all its mass (except a small surface region) is relativistic, so that a polytrope of index $n = 3$ would describe the whole model properly; however, there is a difficulty. As one can see from (19.27) the mass does not vary with central density in the case of a polytrope of index $n = 3$ if $K$ is fixed. In this case, (19.27) gives $M = C_1$:

$$M = 4\pi \left( -\frac{w'}{z} \right)_{z_3} z_3^3 \left( \frac{K}{\pi G} \right)^{3/2} . \tag{19.29}$$

This is the only possible mass for relativistic degenerate polytropes and is called the *Chandrasekhar mass*, which after insertion of the proper numerical values yields

$$M_{Ch} = \frac{5.836}{\mu_e^2} M_\odot . \tag{19.30}$$

We therefore can expect that our series of models constructed by fitting an $n = 3/2$ envelope to an $n = 3$ core finds its end at a critical total mass $M = M_{Ch}$ as given by (19.30). Or in other words our models of increasing central density tend to a finite mass and approach zero radius for $\varrho_c \to \infty$. Of course, this final state is physically unrealistic, since the equation of state is changed by different effects at very high density (see § 16, § 35, § 36).

Although we have discussed the problem only from the standpoint of polytropic models, the result for $M_{Ch}$ remains numerically the same if one uses Chandrasekhar's more general equation of state (15.13), (compare the treatment in § 35.1). The reason is that for extremely high density (15.13) approaches the polytropic relation (19.3) with $\gamma = 4/3$ or $n = 3$.

It is surprising that the limiting mass not only is finite, but that it is so small that many stars exceed it. But their equation of state is not dominated by degenerate electrons and therefore Chandrasekhar's limiting mass (19.30) has no meaning for them. White dwarfs seem to be formed of material where all the hydrogen is transformed into helium, carbon, or oxygen, such that we expect $\mu_e = 2$ and therefore $M_{Ch} = 1.46 M_\odot$. Indeed no white dwarf has been found which exceeds this mass.

In the above considerations we have approached the relativistic degenerate polytrope by way of a sequence with $\varrho_c \to \infty$ (and consequently $R \to 0$). However, this polytrope is a particular case: we have already mentioned that according to (19.27) $M$ and $\varrho_c$ are then no longer coupled. In other words, for $M = M_{Ch}$ the central density can be arbitrary (and therefore also the radius $R$), i.e. there is a whole series of relativistic degenerate polytropes (having $\varrho_c$ or $R$ as parameter) that all have the same mass $M_{Ch}$. This is a case of neutral equilibrium (see § 25.3.2).

## 19.8 Isothermal Spheres of an Ideal Gas

We now deal with the case $\gamma = 1$ or $n = \infty$, which we omitted in § 19.2. Here $K = \Re T/\mu$ is a free parameter. If $\gamma = 1$, integration of (19.6) gives

$$-\frac{\Phi}{K} = \ln \varrho - \ln \varrho_c \quad , \tag{19.31}$$

where we have now chosen the constant of integration in such a way that the gravitational potential is zero at the centre. With

$$\varrho = \varrho_c \, e^{-\Phi/K} \tag{19.32}$$

and with the Poisson equation (19.2) we find

$$\frac{d^2\Phi}{dr^2} + \frac{2}{r}\frac{d\Phi}{dr} = 4\pi G\varrho_c \, e^{-\Phi/K} \quad . \tag{19.33}$$

We now introduce dimensionless variables $z$, $w$ by

$$z = Ar \quad , \quad A^2 = \frac{4\pi G\varrho_c}{K} \quad , \quad \Phi = Kw \tag{19.34}$$

and obtain the "isothermal" Lane–Emden equation

$$\frac{d^2w}{dz^2} + \frac{2}{z}\frac{dw}{dz} = e^{-w} \quad , \tag{19.35}$$

which now has to be integrated with the central conditions

$$w(0) = 0 \quad , \quad \left(\frac{dw}{dz}\right)_{z=0} = 0 \quad . \tag{19.36}$$

Again, a power series expansion can be derived and has to be used to describe the behaviour near the centre. The solution is given in Fig. 19.2.

As already mentioned, the isothermal sphere consisting of an ideal gas has an infinite radius, like all polytropes of $n \geq 5$. It also has an infinite mass. Certainly

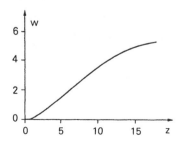

Fig. 19.2. The solution of the Lane–Emden equation (19.35) for the case of an isothermal ideal gas ($n = \infty$)

there can be no such stars, but polytropes with $n = \infty$ can be used in order to construct models with non-degenerate isothermal cores. Such models play a role in connection with the so-called Schönberg–Chandrasekhar limit (see § 30.5).

## 19.9 Gravitational and Total Energy for Polytropes

We now give a general expression for the gravitational energy $E_g$ of polytropes. We first show that quite generally

$$E_g = \frac{1}{2} \int_0^M \Phi \, dm - \frac{1}{2} \frac{GM^2}{R} \quad . \tag{19.37}$$

From the definition (3.3) of $E_g$ we find

$$E_g = -G \int_0^M \frac{m}{r} dm = -\frac{1}{2} \frac{GM^2}{R} - \frac{1}{2} G \int_0^R \frac{m^2}{r^2} dr \quad , \tag{19.38}$$

where the last expression has been obtained by partial integration and where we have used the fact that $m/r$ vanishes at the centre. But on the other hand

$$\frac{d\Phi}{dr} = \frac{Gm}{r^2} \tag{19.39}$$

and therefore

$$\begin{aligned} E_g &= -\frac{1}{2} \frac{GM^2}{R} - \frac{1}{2} \int_0^R \frac{d\Phi}{dr} m \, dr \\ &= -\frac{1}{2} \frac{GM^2}{R} + \frac{1}{2} \int_0^M \Phi \, dm \quad , \end{aligned} \tag{19.40}$$

where again we have integrated partially and used the fact that $m\Phi$ vanishes at the centre ($m = 0$) and at the surface [$\Phi = 0$, according to our choice of the integration constant in connection with (19.7)], so we have indeed recovered (19.37). For a polytrope we can use (19.3,7) and write

$$\Phi = -\frac{K\gamma}{\gamma - 1} \varrho^{\gamma - 1} = -\frac{\gamma}{\gamma - 1} \frac{P}{\varrho} \tag{19.41}$$

and therefore, with (19.37),

$$E_g = -\frac{1}{2} \frac{GM^2}{R} - \frac{1}{2} \frac{\gamma}{\gamma - 1} \int_0^M \frac{P}{\varrho} dm \quad . \tag{19.42}$$

According to (3.2,3) the last term on the right can be expressed by $E_g$. If we replace $\gamma$ by $n$, then

$$E_g = -\frac{1}{2} \frac{GM^2}{R} + \frac{1}{6}(n + 1) E_g \tag{19.43}$$

and therefore

$$E_g = -\frac{3}{5-n}\frac{GM^2}{R} \quad . \tag{19.44}$$

We now derive a similar expression for the internal energy $E_i$. In (3.8) we defined a quantity $\zeta$ by

$$\zeta := 3P/(\varrho u) \tag{19.45}$$

($u$ = internal energy per mass unit).

We saw that for an ideal gas

$$\zeta = 3(\gamma_{ad} - 1) \quad . \tag{19.46}$$

This relation also holds for a more general equation of state as long as $\zeta$ is constant. In order to show this, we take the total differentials from (19.45) and obtain

$$\zeta\, du = 3\frac{dP}{\varrho} - 3\frac{P}{\varrho^2}d\varrho \quad . \tag{19.47}$$

We now assume that the differentials describe adiabatic changes. The first law of thermodynamics gives

$$du = \frac{P}{\varrho^2}d\varrho \quad . \tag{19.48}$$

Then with

$$\gamma_{ad} = \frac{\varrho}{P}\frac{dP}{d\varrho} \quad , \tag{19.49}$$

(19.47) yields

$$\zeta = 3\frac{\varrho}{P}\frac{dP}{d\varrho} - 3 = 3(\gamma_{ad} - 1) \quad . \tag{19.50}$$

For an ideal gas with $\gamma_{ad} = 5/3$ one has $\zeta = 2$, while for an ideal gas with $\gamma_{ad} = 4/3, \zeta = 1$. In the case of a gas dominated by radiation pressure ($P = aT^4/3$ and $u = aT^4$) one finds $\zeta = 1$. Assuming $\zeta$ to be constant throughout the star and using (19.44) we find with (3.9)

$$E_i = -\frac{1}{\zeta}E_g = \frac{3}{\zeta(5-n)}\frac{GM^2}{R} \quad . \tag{19.51}$$

The total energy then becomes

$$W = E_i + E_g = \frac{3}{5-n}\left(\frac{1}{\zeta}-1\right)\frac{GM^2}{R} \quad . \tag{19.52}$$

We can conclude from (19.52) that the total energy for a polytrope of finite radius vanishes when $\zeta = 1$ and in particular for the above cases of an ideal gas with $\gamma_{ad} = 4/3$ and a radiation-dominated gas.

## 19.10 Supermassive Stars

Let us consider an ideal gas with radiation pressure and assume that $\beta = P_{\mathrm{gas}}/P = $ constant throughout the star. We have seen in (19.23) that this yields a polytrope with $n = 3$.

Relation (19.23) defines the polytropic constant $K$:

$$K = \left(\frac{3\mathfrak{R}^4}{a\mu^4}\right)^{1/3} \left(\frac{1 - \beta}{\beta^4}\right)^{1/3} \quad . \tag{19.53}$$

On the other hand, from (19.9) for $n = 3$ we have

$$K = \pi G \varrho_c^{2/3} \frac{R^2}{z_3^2} \quad , \tag{19.54}$$

where we have used $A = z_3/R$. The numerical value of $z_3$ is 6.897 (Table 19.1). With (19.20) $\varrho_c$ can be expressed by $M$ and $R$:

$$\varrho_c = 54.18\bar{\varrho} = 54.18\frac{3M}{4\pi R^3} = c_1\frac{M}{R^3} \quad , \tag{19.55}$$

where we have taken the numerical value from Table 19.1. From (19.53) we eliminate $K$ with (19.54) and then $\varrho_c$ with (19.55) and obtain "Eddington's quartic equation":

$$\frac{1 - \beta}{\mu^4 \beta^4} = \frac{a}{3\mathfrak{R}^4} \frac{(\pi G)^3 c_1^2}{z_3^6} M^2 = 3.02 \times 10^{-3} \left(\frac{M}{M_\odot}\right)^2 \quad . \tag{19.56}$$

In the interval $0 \le \beta \le 1$ the left-hand side is a monotonically decreasing function of $\beta$, which therefore becomes smaller with growing $M$; this means that radiation pressure becomes the more important the larger the stellar mass.

For a pure hydrogen star of $10^6\, M_\odot$ and $\mu = 0.5$, (19.56) gives $(1 - \beta)/\beta^4 = 1.9 \times 10^8$, or $\beta \approx 0.0086$.

Supermassive stars are therefore dominated by radiation pressure. One consequence is that $\nabla_{\mathrm{ad}}$ is appreciably reduced [$\nabla_{\mathrm{ad}} \to 1/4$, for $\beta \to 0$; see (13.21)] and the star becomes convective with $\nabla = \nabla_{\mathrm{ad}}$. This can also be seen from an extrapolation of the main-sequence models towards large $M$ (§ 22.3). The adiabatic structure requires constant specific entropy $s$. For a gas dominated by radiation pressure (the density being determined by the gas, the pressure by the photons) the energy $u$ per mass unit and the pressure are given by

$$u = \frac{aT^4}{\varrho} \quad , \quad P = \frac{a}{3}T^4 \quad . \tag{19.57}$$

Then with the first law of thermodynamics we have

$$\begin{aligned}
ds &= \frac{dq}{T} = \frac{1}{T}\left(du - \frac{P}{\varrho^2}d\varrho\right) \\
&= \frac{4aT^2}{\varrho}dT - \frac{4aT^3}{3\varrho^2}\,d\varrho = d\left(\frac{4aT^3}{3\varrho}\right)
\end{aligned} \tag{19.58}$$

and

$$s = \frac{4aT^3}{3\varrho} \quad . \tag{19.59}$$

Constant specific entropy means $\varrho \sim T^3$, which together with the pressure equation $P \sim T^4$ immediately gives $P \sim \varrho^{4/3}$. Indeed supermassive stars are polytropic with $n = 3$ as we assumed initially.

The supermassive star polytropes have a free $K$, which means that $M$ can be chosen arbitrarily (in contrast to the relativistic degenerate polytrope of the same index, where $K$ and $M$ were fixed). For each mass, $(1 - \beta)/(\mu\beta)^4$ can be obtained from (19.56), and then (19.53) gives the corresponding value of $K$. But if the mass is given, there still exists an infinite number of models for different $R$. This is possible in spite of the fact that $K$ is already determined by $M$: since according to (19.55) $\varrho_c \sim \bar{\varrho} \sim M/R^3$, (19.54) shows $K$ to be independent of $R$. This is typical for the polytropic index $n = 3$.

Equation (19.59) shows that for an adiabatic change ($ds = 0$) of a given mass element $\varrho \sim T^3$, and therefore with (19.57) $P \sim \varrho^{4/3}$ or $\gamma_{ad} = 4/3$. Then $\zeta = 1$ and (19.52) gives the total energy of the model $W = 0$. The supermassive configuration is in neutral equilibrium. No energy is needed to compress or expand it. In § 25 we will find that $\gamma_{ad} = 4/3$ corresponds to the case of marginal dynamical stability. There a simple interpretation is given for this peculiar behaviour.

### 19.11 A Collapsing Polytrope

Up to now we have only treated polytropic gaseous spheres in hydrostatic equilibrium. One can also find solutions for polytropes of $n = 3$ for which the inertia term, neglected in (19.1), is important (GOLDREICH, WEBER, 1980). Then (19.1) has to be replaced by

$$\frac{\partial v_r}{\partial t} + v_r \frac{\partial v_r}{\partial r} + \frac{1}{\varrho} \frac{\partial P}{\partial r} + \frac{\partial \Phi}{\partial r} = 0 \quad , \tag{19.60}$$

with $v_r = \partial r/\partial t$.

Let us consider a relativistic degenerate polytrope with $n = 3$, or $\gamma = \gamma_{ad} = 4/3$. In a manner similar to that of § 19.2 we define a dimensionless length-scale $z$ by

$$r = a(t)z \quad , \quad v_r = \dot{a}z \tag{19.61}$$

such that $z$ is time independent, the whole time dependence of $r$ being contained in $a(t)$. [Note that $a$ corresponds to $1/A$ in (19.9)]. The form (19.61) describes a homologous change (compare with § 20.3). If we introduce a velocity potential $\psi$ by $v_r = \partial\psi/\partial r$, we can write

$$av_r = a\dot{a}z = a\frac{\partial\psi}{\partial r} = \frac{\partial\psi}{\partial z} \quad , \quad \psi = \frac{1}{2}a\dot{a}z^2 \quad , \tag{19.62}$$

where we have fixed the constant of integration in the velocity potential by $\psi = 0$ at $z = 0$. Note that the time derivative of $\psi$ in the comoving frame is

$$\frac{d\psi}{dt} = \frac{\partial\psi}{\partial t} + v_r \frac{\partial\psi}{\partial r} = \frac{\partial\psi}{\partial t} + (\dot{a}z)^2 \quad . \tag{19.63}$$

With the new variables, Poisson's equation (19.2) can be written

$$\frac{1}{z^2} \frac{\partial}{\partial z} \left( z^2 \frac{\partial\psi}{\partial z} \right) = 4\pi G\varrho a^2 \quad , \tag{19.64}$$

while the continuity equation (1.4) becomes with (19.62)

$$\frac{1}{\varrho} \frac{d\varrho}{dt} + \frac{1}{z^2 a^2} \frac{\partial}{\partial z} \left( z^2 \frac{\partial\psi}{\partial z} \right) \equiv \frac{1}{\varrho} \frac{d\varrho}{dt} + 3\frac{\dot{a}}{a} = 0 \quad . \tag{19.65}$$

This means that $\varrho \sim a^{-3}$ (in the comoving frame), a result that is obvious from (19.61). As in (19.9) we define $w(z)$ by $\varrho = \varrho_c w^3(z)$. This $w(z)$ will turn out to be related to the Emden function of index 3, as we shall see later. Note that $\varrho_c$ is a function of time. In order to stay as close as possible to the formalism of hydrostatic equilibrium, we fix $a = r/z$ [rather as we did with $1/A$ in (19.9)] by

$$\frac{1}{a^2} = \frac{\pi G}{K} \varrho_c^{2/3} \tag{19.66}$$

such that

$$\varrho = \varrho_c w^3(z) = \left( \frac{K}{\pi G} \right)^{3/2} \frac{1}{a^3} w^3(z) \quad . \tag{19.67}$$

We now come to the equation of motion and define

$$h := \int \frac{dP}{\varrho} = 4K\varrho^{1/3} \quad , \tag{19.68}$$

where we have made use of (19.3) for $\gamma = 4/3$. Inserting $\psi$ and $h$ from (19.62,68) into the equation of motion (19.60) gives

$$\frac{\partial^2\psi}{\partial r \partial t} + \frac{1}{2} \frac{\partial}{\partial r} \left( \frac{\partial\psi}{\partial r} \right)^2 + \frac{\partial\Phi}{\partial r} + \frac{\partial h}{\partial r} = 0 \quad , \tag{19.69}$$

which can be integrated with respect to $r$. If we set the integration constant to zero, replace $\partial\psi/\partial r$ by $\dot{a}z$, and consider (19.63), we find that

$$\frac{d\psi}{dt} = \frac{1}{2} \dot{a}^2 z^2 - \Phi - h \tag{19.70}$$

and therefore with (19.62)

$$\frac{1}{2} a\ddot{a}z^2 = -\Phi - h \quad . \tag{19.71}$$

From (19.67,68) follows

$$h = 4K\varrho^{1/3} = 4\frac{K^{3/2}}{(\pi G)^{1/2}} \frac{1}{a} w(z) \quad . \tag{19.72}$$

We try a similar dependence of $\Phi$ on $t$ and write

$$\Phi = 4\frac{K^{3/2}}{(\pi G)^{1/2}}\frac{1}{a}g(z) \quad , \tag{19.73}$$

which defines the dimensionless function $g(z)$. If we insert (19.72,73) into (19.71) we find

$$\frac{1}{2}a^2\ddot{a} = -\frac{4K^{3/2}}{(\pi G)^{1/2}}(g+w)\frac{1}{z^2} \quad . \tag{19.74}$$

Since the left-hand side is a function of $t$ only and the right-hand side is a function of $z$ only, both sides must be constant; therefore

$$\frac{3}{4}\frac{(\pi G)^{1/2}}{K^{3/2}}a^2\ddot{a} = -\lambda \quad , \tag{19.75}$$

$$6\frac{g+w}{z^2} = \lambda \tag{19.76}$$

($\lambda$ = constant). The first of these equations can be integrated twice. After multiplication with $\dot{a}/a^2$, the first integration gives

$$\dot{a}^2 = \frac{8}{3}\lambda\left(\frac{K^3}{\pi G}\right)^{1/2}\frac{1}{a} \quad , \tag{19.77}$$

where the constant of integration is set equal to zero (assuming a zero velocity when the sphere is expanded to infinity). Multiplication of (19.77) with $a$ gives

$$a^{1/2}\dot{a} \equiv \frac{2}{3}\frac{d}{dt}(a^{3/2}) = \pm\left[\frac{8\lambda}{3}\left(\frac{K^3}{\pi G}\right)^{1/2}\right]^{1/2} \tag{19.78}$$

(the signs representing exploding or collapsing models respectively). This can immediately be integrated, yielding for a collapse ($\dot{a} < 0$) that starts at $a_0$ for $t = 0$

$$a^{3/2}(t) = a_0^{3/2} - \frac{3}{2}\left[\frac{8\lambda}{3}\left(\frac{K^3}{\pi G}\right)^{1/2}\right]^{1/2}t \quad . \tag{19.79}$$

This expression gives the time dependence of the scaling factor $a(t)$ and therefore by way of (19.67) of the density as a function of time.

We now investigate the spatial dependence of our solution. In particular the function $w(z)$ in (19.67) has to be determined. For this purpose we write Poisson's equation (19.2) in the dimensionless variable $z$

$$\frac{1}{z^2}\frac{\partial}{\partial z}\left(z^2\frac{\partial\Phi}{\partial z}\right) = 4\pi G\varrho a^2 \quad . \tag{19.80}$$

If we here replace $\Phi$ by (19.73), $g(z)$ by (19.76), and $\varrho$ by (19.67), we find

$$\frac{1}{z^2} \frac{d}{dz} \left( z^2 \frac{dw}{dz} \right) + w^3 = \lambda \quad . \tag{19.81}$$

For $\lambda = 0$ this is the classical Emden equation. Solutions for $\lambda \neq 0$ deviate from hydrostatic equilibrium, the value of $\lambda$ being a measure for this deviation. From numerical integrations it follows that physically relevant solutions $w(z)$ are obtained only for very small values of $\lambda$, namely for $\lambda < \lambda_m = 0.0065$. Otherwise the solution $w(z)$ and therefore $\varrho(r)$ do not become zero at a finite radius; they rather increase again to infinity after a minimum has been reached (see Fig. 19.3). This figure shows also that for $\lambda < \lambda_m$ the solutions deviate appreciably from the "classical" one ($\lambda = 0$) only in the outer layers, where $\lambda \ll w^3$ no longer applies.

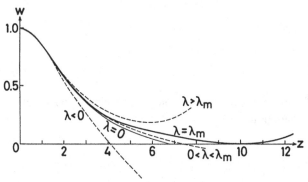

**Fig. 19.3.** Solutions of (19.81) for different values of $\lambda$. In the range $0 < \lambda \leq \lambda_m$ they describe homologously collapsing polytropes of index 3. The solution for $\lambda = \lambda_m$ reaches the abscissa with slope zero. The broken lines indicate the behaviour of the solutions for different values of $\lambda$

The time-dependent solution discussed here has to be understood in the following way. Let us consider a polytrope with $n = 3$ in equilibrium; then the equilibrium is independent of radius. We have already seen that the total energy is $W = 0$, independent of the radius, see (19.52). Therefore the polytrope $n = 3$ is indifferent to radial changes. If we now assume that suddenly the pressure is slightly reduced, say, because the constant $K$ is slightly diminished, then the gaseous sphere begins to contract. This contraction can be described by the two equations (19.75,76). The solution of the first gives the behaviour in time (19.79), while the second is used to derive the modification of the Lane–Emden equation due to the inertia terms. The parameter $\lambda$ is a measure of the deviation from hydrostatic equilibrium, caused by the assumed reduction of $K$.

The solutions for collapsing polytropes have been discussed by GOLDREICH, WEBER (1980) with respect to collapsing stellar cores causing supernova outbursts (§ 34).

# § 20 Homology Relations

In physical problems it often happens that from one solution others can be obtained by simple transformations. When comparing different stellar models that are calculated under similar assumptions (concerning parameters or material functions), one therefore expects to find similarities in the solutions. It would be very helpful if we could find simple analytic expressions that transform one solution into another. It would then only be necessary to produce *one* numerical solution in order to find new ones by a transformation. There is indeed often a kind of "similarity" between different solutions, which is called *homology*, though the conditions for this are so severe that real stars will scarcely match them. There are a few cases, however, for which homology relations offer a rough, but helpful, indication for interpreting or predicting the numerical solutions. We indicate this in two examples, the main-sequence models and the homologous contraction. Except for this classical homology there is another type of homology, which applies to certain red giants (see § 32.2.)

## 20.1 Definitions and Basic Relations

When comparing different models (say of masses $M$ and $M'$, and radii $R$ and $R'$), one considers in particular *homologous points* at which the relative radii are equal: $r/R = r'/R'$. We now speak of *homologous stars* if their homologous mass shells ($m/M = m'/M'$) are situated at homologous points. To be more precise, let us consider all radii as functions of the *relative* mass values $\xi$, which are the same for homologous masses:

$$\xi := m/M = m'/M' \quad . \tag{20.1}$$

We can then write the homology condition as

$$\frac{r(\xi)}{r'(\xi)} = \frac{R}{R'} \tag{20.2}$$

for all $\xi$. In homologous stars the ratio of the radii $r/r'$ for homologous mass shells is constant throughout the stars. Going from one homologous star to another, all homologous mass shells are compressed (or expanded) by the same factor $R/R'$. (Note that therefore any two polytropic models of the same index $n$ are homologous to each other.)

Since both models have to fulfil the stellar-structure equations, the transition has, of course, consequences for all other variables. We derive these by comparing two homologous stars of masses $M$ and $M'$, and of two different compositions that are supposed to be homogeneous and represented by the mean molecular weights $\mu$ and $\mu'$. The ratio of these basic parameters will be called

$$x = M/M' \quad ; \quad y = \mu/\mu' \quad . \tag{20.3}$$

The variables in the two models are always considered functions of the relative mass variable $\xi$ and may be called $r$, $P$, $T$, $l$ (for $M$, $\mu$), and $r'$, $P'$, $T'$, $l'$ (for $M'$, $\mu'$) respectively. We try the following "ansatz": for homologous mass values the variables are supposed to have the ratios

$$\frac{r}{r'} = z = \frac{R}{R'} \quad ; \quad \frac{P}{P'} = p = \frac{P_c}{P_c'} \quad ; \quad \frac{T}{T'} = t = \frac{T_c}{T_c'} \quad ; \quad \frac{l}{l'} = s = \frac{L}{L'} \quad , \tag{20.4}$$

where $z$, $p$, $t$, $s$ *have the same values for all* $\xi$ and where the subscript c indicates central values.

We start with homologous main-sequence models. Since they evolve within the long nuclear time-scale, one can neglect the inertia term in (9.2) as well as the time derivatives in the energy equation (9.3). Let us assume that in these two stars in complete equilibrium (hydrostatic and thermal) the energy transport is radiative. The basic equations to be fulfilled are then (9.1,4,16,17) together with (9.6). We write them for the first star in terms of the relative mass variable $\xi$ as

$$\frac{dr}{d\xi} = c_1 \frac{M}{r^2 \varrho} \quad , \qquad c_1 = \frac{1}{4\pi} \quad ,$$

$$\frac{dP}{d\xi} = c_2 \frac{\xi M^2}{r^4} \quad , \qquad c_2 = -\frac{G}{4\pi} \quad ,$$

$$\frac{dl}{d\xi} = \varepsilon M \quad , \tag{20.5}$$

$$\frac{dT}{d\xi} = c_4 \frac{\kappa l M}{r^4 T^3} \quad , \qquad c_4 = -\frac{3}{64\pi^2 ac} \quad .$$

Since no time derivatives appear, the differentiations with respect to $\xi$ are written as ordinary derivatives. In these equations we transform the variables $r$, $P$, $T$, $l$ into $r'$, $P'$, $T'$, $l'$ by use of (20.4). Noting that the $z$, $p$, $t$, $s$ are independent of $\xi$, and that $\xi$ contains the total mass as scaling factor, which has to be transformed by (20.3), one immediately finds the transformed equations:

$$\frac{dr'}{d\xi} = c_1 \frac{M'}{r'^2 \varrho'} \left[ \frac{x}{z^3 d} \right] \quad ,$$

$$\frac{dP'}{d\xi} = c_2 \frac{\xi M'^2}{r'^4} \left[ \frac{x^2}{z^4 p} \right] \quad ,$$

$$\frac{dl'}{d\xi} = \varepsilon' M' \left[ \frac{\varepsilon x}{s} \right] \quad , \tag{20.6}$$

$$\frac{dT'}{d\xi} = c_4 \frac{\kappa' l' M'}{r'^4 T'^3} \left[ \frac{k s x}{z^4 t^4} \right] \quad .$$

$c_1, \ldots, c_4$ are the same constants as before, and we have used the abbreviations

$$\frac{\varrho}{\varrho'} = d \quad ; \quad \frac{\varepsilon}{\varepsilon'} = e \quad ; \quad \frac{\kappa}{\kappa'} = k \tag{20.7}$$

for the ratios of the material functions at homologous points.

Since the variables $r'$, $P'$, $T'$, $l'$ must fulfil the same basic equations as the $r$, $P$, $T$, $l$, a comparison of (20.6) with (20.5) shows immediately that the four factors in brackets in (20.6) must be equal to one:

$$\frac{x}{z^3 d} = 1 \quad , \quad \frac{x^2}{z^4 p} = 1 \quad , \quad \frac{ex}{s} = 1 \quad , \quad \frac{ksx}{z^4 t^4} = 1 \quad . \tag{20.8}$$

In order to find solutions, we represent the material functions by power laws:

$$\varrho \sim P^\alpha T^{-\delta} \mu^\varphi \quad , \quad \varepsilon \sim \varrho^\lambda T^\nu \quad , \quad \kappa \sim P^a T^b \quad , \tag{20.9}$$

which from (20.7) with (20.4) give

$$d = p^\alpha t^{-\delta} y^\varphi \quad , \quad e = p^{\lambda\alpha} t^{\nu - \lambda\delta} y^{\lambda\varphi} \quad , \quad k = p^a t^b \quad . \tag{20.10}$$

These can be introduced into the equations (20.8), which are then four conditions for the powers of $z$, $p$, $t$, and $s$. We will try to represent them in terms of $x$ and $y$, which, according to (20.3), describe the change of the basic parameters $M$ and $\mu$:

$$z = x^{z_1} y^{z_2} \quad ; \quad p = x^{p_1} y^{p_2} \quad ; \quad t = x^{t_1} y^{t_2} \quad ; \quad s = x^{s_1} y^{s_2} \quad . \tag{20.11}$$

Introducing these and (20.10) into (20.8), we obtain four conditions which contain only products of powers of $x$ and $y$. In each condition, the exponents of $x$ and of $y$ must sum up to zero, since the right-hand sides of (20.8) are independent of $x$ and $y$. This yields 8 linear equations for the exponents $z_1, \ldots, s_2$, which are written in matrix form as:

$$\begin{pmatrix} -3 & -\alpha & \delta & 0 \\ -4 & -1 & 0 & 0 \\ 0 & \lambda\alpha & (\nu - \lambda\delta) & -1 \\ -4 & a & (b-4) & 1 \end{pmatrix} \begin{pmatrix} z_1 \\ p_1 \\ t_1 \\ s_1 \end{pmatrix} = \begin{pmatrix} -1 \\ -2 \\ -1 \\ -1 \end{pmatrix} \tag{20.12}$$

and

$$\begin{pmatrix} -3 & -\alpha & \delta & 0 \\ -4 & -1 & 0 & 0 \\ 0 & \lambda\alpha & (\nu - \lambda\delta) & -1 \\ -4 & a & (b-4) & 1 \end{pmatrix} \begin{pmatrix} z_2 \\ p_2 \\ t_2 \\ s_2 \end{pmatrix} = \begin{pmatrix} \varphi \\ 0 \\ -\lambda\varphi \\ 0 \end{pmatrix} \quad . \tag{20.13}$$

The solutions are

$$z_1 = \frac{1}{2}(1 + A) \quad , \quad p_1 = -2A \quad ,$$

$$t_1 = \frac{1}{2\delta}[1 + (3 - 4\alpha)A] \quad , \tag{20.14}$$

$$s_1 = 1 + \frac{4 - b}{2\delta} + \left[2 + 2a + \frac{3 - 4\alpha}{2\delta}(4 - b)\right] A \quad ,$$

and

$$z_2 = \varphi B \quad , \quad p_2 = -4\varphi B \quad , \quad t_2 = \frac{\varphi}{\delta}[1 + (3 - 4\alpha)B] \quad ,$$

$$s_2 = \frac{\varphi}{\delta}(4 - b) + \varphi\left[4 + 4a + \frac{3 - 4\alpha}{\delta}(4 - b)\right]B \quad ,$$

(20.15)

$$A = \left[\frac{4\delta(1 + a + \lambda\alpha)}{\nu + b - 4 - \lambda\delta} + 4\alpha - 3\right]^{-1} \quad , \quad B = A\left(1 - \frac{\lambda\delta}{\nu + b - 4}\right)^{-1} \quad . \quad (20.16)$$

Without further specification of the material functions, we obtain two useful relations from the first and second of equations (20.8). They can be rewritten as

$$\frac{\varrho}{\varrho'} = \frac{M/M'}{(R/R')^3} \quad , \quad \frac{P}{P'} = \frac{(M/M')^2}{(R/R')^4} \quad . \tag{20.17}$$

Therefore, for all homologous points, the density changes simply as the mean density for the whole star, while $P$ varies like $M^2 R^{-4}$.

## 20.2 Applications to Simple Material Functions

### 20.2.1 The Case $\delta = 0$

A special situation arises for the case that the density is independent of $T$, i.e. $\delta = 0$ in (20.9). The equation of state then is polytropic, the polytropic index being $n = \alpha/(1 - \alpha)$, and we must recover the typical properties of polytropic stars (see § 19.3). This can, in fact, be easily verified. To start with, the first two equations of system (20.12) (which represent the mechanical part) can be solved independently of the rest (the thermo-energetic part). For $\delta = 0$ we find from (20.14) and (20.16) that $A = (4\alpha - 3)^{-1}$ and $z_1 = (2\alpha - 1)/(4\alpha - 3)$. The first of equations (20.11) gives for homologous stars of equal composition ($y = 1$) the mass-radius relation

$$R \sim M^{z_1} \quad . \tag{20.18}$$

For a non-relativistic degenerate electron gas one has $\alpha = 3/5$, which gives the exponent $z_1 = -1/3$ as already obtained in § 19.6.

### 20.2.2 The Case $\alpha = \delta = \varphi = 1$, $a = b = 0$

Further discussion of the above homology solutions will concentrate on the simplest case, an ideal gas ($\alpha = \delta = \varphi = 1$) with constant opacity ($a = b = 0$), [cf. (20.9)]. This extremely rough approximation to reality suffices for outlining some general properties of main-sequence stars. (The assumption of homology introduces a much severer limitation on the results.)

From (20.14–16), one finds

$$z_1 = \frac{\nu + \lambda - 2}{\nu + 3\lambda} \quad , \qquad z_2 = \frac{\nu - 4}{\nu + 3\lambda} \quad ,$$
$$p_1 = 2 - 4z_1 \quad , \qquad p_2 = -4z_2 \quad ,$$
$$t_1 = 1 - z_1 \quad , \qquad t_2 = 1 - z_2 \quad ,$$
$$s_1 = 3 \quad , \qquad s_2 = 4 \quad . \tag{20.19}$$

The first surprising result concerns the exponents of the luminosity, $s_1$ and $s_2$. In this simple case the square brackets in the equations for $s_1$ and $s_2$ in (20.14) and (20.15) vanish, and $s_1$ and $s_2$ become simple constant numbers. In particular, *they are independent* of $\nu$ and $\lambda$, i.e. *of the special mode of energy generation*. In fact the energy equation [giving the third of equations (20.12)] has no influence on the luminosity, which is determined by hydrostatic equilibrium, the equations of state, and radiative energy *transfer* only. The model has to adjust so that the energy sources ($\varepsilon$) provide this luminosity. Introducing the exponents into (20.11), we have from (20.4) that

$$\frac{L}{L'} = \left(\frac{M}{M'}\right)^3 \left(\frac{\mu}{\mu'}\right)^4 \quad . \tag{20.20}$$

There thus exists a mass–luminosity relation that gives a steeply increasing $L$ with increasing $M$. And $L$ varies even more strongly with the molecular weight $\mu$. (The precise values of the exponents vary for other values of $a$ and $b$ roughly in a range from 3 to 6, but the principle result remains.)

All other exponents depend on $\nu$ and $\lambda$. $z_1$ and $z_2$ describe the variation of the radius:

$$\frac{R}{R'} = \left(\frac{M}{M'}\right)^{z_1} \left(\frac{\mu}{\mu'}\right)^{z_2} \quad . \tag{20.21}$$

The exponent $z_1$ of the $M$–$R$ relation is positive for all relevant combinations of $\lambda$ and $\nu$ but smaller than one, i.e. $R$ increases slightly with $M$. Values for typical parameters of hydrogen burning ($\lambda = 1$) via the $pp$ chain ($\nu = 4 \ldots 5$) and the CNO cycle ($\nu \approx 15 \ldots 18$) are given in Table 20.1. Over this very large range of $\nu$, $z_1$ varies relatively little, roughly from 0.4 to 0.8.

The $M$–$R$ relation together with the $M$–$L$ relation immediately give the locus of these stars in the Hertzsprung–Russell (HR) diagram, where $\lg L$ is plotted over $-\lg T_{\text{eff}}$ (see Fig. 20.1).

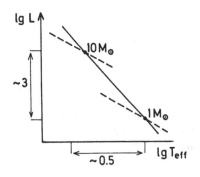

Fig. 20.1. Sketch of the Hertzsprung–Russell diagram with the locus of homologous main-sequence stars (*solid line*) of different masses for a certain constant value of $\nu$. The dashed lines indicate lines of $R = $ constant

**Table 20.1** Exponents in equations (20.11) for various temperature sensitivities $\nu$ of the nuclear reactions, and for $\alpha = \delta = \varphi = 1$, $a = b = 0$, $\lambda = 1$, calculated from (20.19). The exponents describe the dependence of $R$, $P$, $T$, $L$ on $M$ and $\mu$ ($R \sim M^{z_1}\mu^{z_2}$; $P \sim M^{p_1}\mu^{p_2}$; $T \sim M^{t_1}\mu^{t_2}$; $L \sim M^{s_1}\mu^{s_2}$)

| $\nu$ : | 4 | 5 | 15 | 18 |
|---|---|---|---|---|
| $z_1$ | 0.43 | 0.5 | 0.78 | 0.81 |
| $z_2$ | 0 | 0.13 | 0.61 | 0.67 |
| $p_1$ | 0.29 | 0 | −1.11 | −1.24 |
| $p_2$ | 0 | −0.5 | −2.44 | −2.67 |
| $t_1$ | 0.57 | 0.5 | 0.22 | 0.19 |
| $t_2$ | 1.0 | 0.88 | 0.39 | 0.33 |
| $s_1$ | 3 | 3 | 3 | 3 |
| $s_2$ | 4 | 4 | 4 | 4 |

From (20.20) and (20.21) we have $R \sim L^{z_1/3}$. Introducing this into the definition of the effective temperature

$$\sigma T_{\text{eff}}^4 = \frac{L}{4\pi R^2} \quad , \tag{20.22}$$

we obtain the locus as given by

$$\lg L = \frac{12}{3 - 2z_1} \lg T_{\text{eff}} + \text{constant} \quad . \tag{20.23}$$

For an average value $z_1 = 0.6$, the slope is 6.67.

Let us consider how a star of fixed $M$ moves in the HR diagram if $\mu$ changes. From (20.20,21) we have $L \sim \mu^4$, $R \sim \mu^{z_2}$, which with (20.22) gives $T_{\text{eff}}^8 \sim L^{2-z_2} \approx L^{1.5}$ for $z_2 \approx 0.5$. This defines in the HR diagram a straight line of smaller slope ($\approx 5.3$) than that of the main sequence. This line for $M = $ constant and $\mu$ increasing goes to the upper left with a slope between that of the main sequence and that of the lines $R = $ constant.

The expression for $t_1$ in (20.19) means that

$$T \sim M/R \quad , \tag{20.24}$$

which simply reflects the virial theorem (thermal energy $\sim$ potential energy). Of special interest are the central values of temperature and density, $T_c$ and $\varrho_c$, for which one has

$$T_c \sim M^{1-z_1} \quad , \quad \varrho_c \sim M^{1-3z_1} \quad . \tag{20.25}$$

The values in Table 20.1 show that for increasing $M$, $T_c$ increases relatively slowly, while $\varrho_c$ decreases. This trend is especially pronounced for CNO burning, where $T_c$ scarcely changes at all, typical variations being $T_c \sim M^{0.2}$ and $\varrho_c \sim M^{-1.4}$

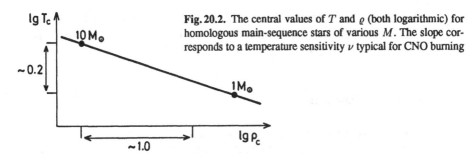

Fig. 20.2. The central values of $T$ and $\varrho$ (both logarithmic) for homologous main-sequence stars of various $M$. The slope corresponds to a temperature sensitivity $\nu$ typical for CNO burning

(see Fig. 20.2). The predictions of the homology relations are at least qualitatively recovered in the numerical solutions for main-sequence stars (§ 22).

### 20.2.3 The Role of the Equation of State

The procedure by which the homology solutions were obtained shows that their existence rests entirely on the fact that the right-hand sides of (20.5) contain only products of the variables, but no sums. This property is destroyed if the material functions, instead of being products of powers of $P$ and $T$, contain additive terms as is in general the case with the equation of state. The simplest example is the addition of radiation pressure to an ideal gas such that $P = \Re\varrho T/\mu + aT^4/3$. No strict homology relations are then possible. But one can try to make rough approximations.

One usually writes the corresponding equation of state as

$$\varrho \sim (\mu\beta)\frac{P}{T} \quad , \quad \beta = \frac{P_{\text{gas}}}{P} = \frac{1 - P_{\text{rad}}}{P} \quad . \tag{20.26}$$

The situation would be simple and homology relations would hold if $\beta$ were constant throughout the model. Then a variation of $\beta$ obviously has the same effect as that of $\mu$ and we would find $R \sim \beta^{z_2}$, $P \sim \beta^{p_2}$, $T \sim \beta^{t_2}$, $L \sim \beta^{s_2}$. In reality $\beta$ is determined by $P$ and $T$. For simultaneous variations of $M$ and $\beta$, therefore

$$1 - \beta = \frac{P_{\text{rad}}}{P} \sim \frac{T^4}{P} \sim \frac{M^{4t_1}}{M^{p_1}} \frac{\beta^{4t_2}}{\beta^{p_2}} \quad , \tag{20.27}$$

which, if we simply use (20.19), gives

$$\frac{1 - \beta}{\beta^4} \sim M^2 \quad . \tag{20.28}$$

Now, $\beta$ is generally *not* constant inside a star [except for the polytrope $n = 3$ as treated in § 19.5; compare with the identical relation (19.56)], but we can consider (20.28) as a relation between $M$ and some kind of mean value of $\beta$. One then sees that $\beta$ decreases strongly with $M$, i.e. the contribution of the radiation pressure to $P$ increases with mass. Quite similarly we can write

$$L \sim M^{s_1}\beta^{s_2} \quad . \tag{20.29}$$

197

Since $\beta$ decreases with increasing $M$, (20.29) can be written as $L \sim M^{s_1-c}$ ($c > 0$ for $s_2 > 0$) and the $M$–$L$ relation becomes less steep. For $\beta \ll 1$ (large $P_{rad}$), relation (20.28) gives $\beta \sim M^{-1/2}$ such that $L \sim M^{s_1-s_2/2} = M$.


## 20.3 Homologous Contraction

Now we briefly consider the homologous contraction. This may apply to a chemically homogeneous star of given mass in hydrostatic equilibrium, if its radius is not fixed by an $M$–$R$ relation but changes in time. Let us assume that consecutive models are homologous to each other. An example in which this assumption is fulfilled is the contraction of a polytrope that does not change its polytropic index $n$. The solution of the Lane–Emden equation for given $n$ yields the mass value $m$ as a unique function of $z$ only, where $z$ is Emden's dimensionless radius variable, i.e. $z \sim r/R$ (see §19.2). Therefore the mass elements remain at homologous points, since their values of $z$ do not change in time.

Homologous mass shells ($\xi$ = constant) are here simply those which have the same value of $m$, since the normalizing factor $M$ remains constant. The radius of any such shell is supposed to change by a rate $\dot{r} = \partial r/\partial t$. In two neighbouring models, separated by a time interval $\Delta t$, we have the values $r$ and $r'$ connected by $r' = r + \dot{r}\Delta t$. This gives

$$\frac{r'}{r} = 1 + \frac{\dot{r}}{r}\Delta t \quad . \tag{20.30}$$

For a homologous contraction, we must require that $r'/r = R'/R = $ constant throughout the star. Then also

$$\frac{\dot{r}}{r} = \frac{\dot{R}}{R} \tag{20.31}$$

must be constant, or

$$\frac{\partial}{\partial m}\left(\frac{\partial \ln r}{\partial t}\right) = 0 \quad . \tag{20.32}$$

The relative rate of change of the other variables can then be easily expressed in terms of $\dot{r}/r$. From (20.32) we find by exchange of the two derivatives, and by using (9.1),

$$\frac{\partial}{\partial t}\left(\frac{1}{r}\frac{\partial r}{\partial m}\right) = \frac{\partial}{\partial t}\left(\frac{1}{4\pi r^3 \varrho}\right) = \frac{1}{4\pi r^3 \varrho}\left(-3\frac{\dot{r}}{r} - \frac{\dot{\varrho}}{\varrho}\right) = 0 \quad , \tag{20.33}$$

which gives

$$\frac{\dot{\varrho}}{\varrho} = -3\frac{\dot{r}}{r} \quad . \tag{20.34}$$

The pressure at a layer of mass value $m$ is given by an integration of the hydrostatic equation as

$$P = \int_m^M \frac{Gm}{4\pi r^4}\,dm \quad . \tag{20.35}$$

Differentiating this with respect to time and observing that $\dot{r}/r$ is constant throughout the model, we have

$$\dot{P} = \int_m^M \frac{\partial}{\partial t} \left( \frac{1}{r^4} \right) \frac{Gm}{4\pi} \, dm = -4\frac{\dot{r}}{r} \int_m^M \frac{Gm}{4\pi r^4} \, dm \quad . \tag{20.36}$$

Equations (20.35) and (20.36) yield

$$\frac{\dot{P}}{P} = -4\frac{\dot{r}}{r} \quad . \tag{20.37}$$

If we have an equation of state with $\varrho \sim P^\alpha T^{-\delta}$, then $\dot{\varrho}/\varrho = \alpha \dot{P}/P - \delta \dot{T}/T$. Solving this for $\dot{T}/T$ and replacing $\dot{\varrho}$ and $\dot{P}$ by (20.34,37), we have

$$\frac{\dot{T}}{T} = -\frac{4\alpha - 3}{\delta} \frac{\dot{r}}{r} \quad . \tag{20.38}$$

The energy generation due to contraction is according to (4.27)

$$\varepsilon_g = c_P T \left( \nabla_{\text{ad}} \frac{\dot{P}}{P} - \frac{\dot{T}}{T} \right) \quad . \tag{20.39}$$

We introduce (20.37,38,31), thus obtaining

$$\varepsilon_g = c_P T \left( -4\nabla_{\text{ad}} + \frac{4\alpha - 3}{\delta} \right) \frac{\dot{R}}{R} \quad . \tag{20.40}$$

For an ideal monatomic gas ($\nabla_{\text{ad}} = 2/5$, $\alpha = \delta = 1$) this becomes

$$\varepsilon_g = -\frac{3}{5} c_P T \frac{\dot{R}}{R} \quad . \tag{20.41}$$

Therefore $\varepsilon_g > 0$ for contraction ($\dot{R} < 0$). We also see that $|\varepsilon_g| \sim |\dot{R}/R|$; and since $\varepsilon_g$ is proportional to $T$, it represents an energy source that is only rather moderately concentrated towards the centre.

As already mentioned, homology considerations are important for rough interpretations of numerical results, but their strict applicability is very limited. This is ultimately because homology requires a very well concerted action of all mass elements. It can hold approximately only for homogeneous stars. In § 32.2 we will encounter another type of homology which considers only certain parts inside a star, and which applies to some very inhomogeneous stellar configurations.

# §21 Simple Models in the $U$–$V$ Plane

There are stars in which the nuclear energy generation proceeding close to the centre creates such a high energy flux that the whole central region is convective. These stars can be described by models with a convective core and a radiative envelope. In later stages of stellar evolution the nuclear fuel in the central region of the star is exhausted and nuclear burning takes place only at the surface of a burned-out core. Under certain circumstances these models with shell burning can be described by a core that is isothermal, since no energy has to be transported there, and that is surrounded by a radiative envelope. In both cases a core solution of one type has to be fitted to an envelope solution of another type. In the following we shall deal with a classical fitting procedure which in the past was often used to construct models for such stars (see SCHWARZSCHILD, 1958; WRUBEL, 1958) and which gives valuable insight into some of their general properties. Moreover, procedures like this can be helpful in certain special cases where the usual, iterative numerical methods are not practicable.

## 21.1 The $U$–$V$ Plane

We define two dimensionless quantities using (1.2) and (2.4):

$$U := \frac{d\ln m}{d\ln r} = \frac{4\pi r^3 \varrho}{m} \quad , \quad V := -\frac{d\ln P}{d\ln r} = \frac{\varrho}{P}\frac{Gm}{r} \quad . \tag{21.1}$$

A solution which is regular in the stellar centre has the central values $U = 3$, $V = 0$, as can easily be seen: a small sphere around the centre has the mass $m = 4\pi r^3 \varrho_c/3$, so that there $U \to 3$ and $V \sim r^2 \to 0$. Near the surface the numerical value of $U$ becomes very small (as $\varrho$ does), as well as $P/\varrho$ ($\sim T$ for the ideal gas or $\sim \varrho^{\gamma-1}$ for polytropes). Therefore $V$ becomes very large.

Compare two homologous models. Then $U$ as well as $V$ have the same value in homologous mass shells. Indeed with $r/r' = R/R'$, $m/m' = M/M'$, and (20.17) it follows that

$$U = \frac{4\pi r^3 \varrho}{m} = \frac{4\pi r'^3 \varrho'}{m'} = U' \quad \text{and correspondingly} \quad V = V' \quad . \tag{21.2}$$

$U$ and $V$ are therefore also called *homology invariants*.

We now determine the quantities $U$ and $V$ for polytropes. From (19.11,18), we find

$$U = -w^n \left( \frac{1}{z} \frac{dw}{dz} \right)^{-1} . \tag{21.3}$$

With the expansion (19.12) one can see that indeed $U \rightarrow 3$ for $z \rightarrow 0$, independent of the value of $n$. We furthermore find – with $\varrho = \varrho_c w^n$, $P = P_c(\varrho/\varrho_c)^{1+1/n} = P_c w^{n+1}$, and (19.18) – from (21.1) that

$$V = \frac{4\pi G \varrho_c^2 r^2}{P_c} \left( -\frac{1}{z} \frac{dw}{dz} \right) \frac{1}{w} , \tag{21.4}$$

and with (19.3,9)

$$V = -(n+1) \frac{z}{w} \frac{dw}{dz} , \tag{21.5}$$

which indeed vanishes at the centre and becomes large near the surface where $w \rightarrow 0$. Note that the functions $U(z)$ and $V(z)$ depend only on $n$: they are independent of any other parameter of the model. This is the property which makes a discussion of the $U$–$V$ plane worthwhile. The function $V = V(U)$ for $n = 3/2$ is plotted in Fig. 21.1.

The above polytropic relations hold for finite $n$ only. The isothermal polytrope for an ideal gas ($n = \infty$) again is an exceptional case. Instead of (21.3,5) one finds from (21.1) and the relations of § 19.8

$$U = e^{-w} \left( \frac{1}{z} \frac{dw}{dz} \right)^{-1} , \quad V = z \frac{dw}{dz} , \tag{21.6}$$

where $w$ now is the solution of (19.35). This case is shown in Fig. 21.2: although the corresponding polytropic model has an infinite radius, its image curve in the $U$–$V$

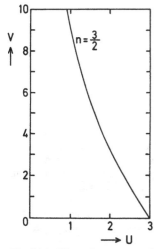

Fig. 21.1. The polytrope $n = 3/2$ in the $U$–$V$ plane. The stellar centre is in the lower-right corner ($U = 3$, $V = 0$)

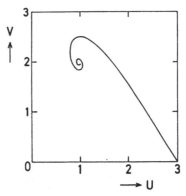

Fig. 21.2. The isothermal sphere for an ideal gas in the $U$–$V$ plane. The centre ($r = 0$) is in the lower-right corner ($U = 3$, $V = 0$), while for the surface ($r \rightarrow R = \infty$) the curve spirals into the point $U = 1$, $V = 2$

plane spirals into the point $U = 1$, $V = 2$, which represents the surface ($z = \infty$). The spiral of the isothermal gaseous sphere unwinds and reaches higher and higher values of $V$ if degeneracy becomes important. In the limit case of complete non-relativistic degeneracy the image curve approaches that of the polytrope $n = 3/2$ of Fig. 21.1.

The $U$–$V$ plane has often been used to construct simple stellar models by fitting core and envelope solutions. Clearly this is most profitable when the core is polytropic with given index $n$ and, therefore all possible cores are represented by a single, known curve in the plane. This is the case for stars with convective cores (polytropic with $n = 3/2$) or with non-degenerate isothermal cores ($n = \infty$).

The fitting requires continuity of $r$, $P$, $T$, $l$ at the interface. If $\mu$ is continuous, then also $\varrho$ – and according to (21.1) – $U$ and $V$ have to be continuous at the fitting point: core and envelope curves intersect (compare Figs. 21.3, 4). If $\mu$ is discontinuous at the interface having there the values $\mu_1$, $\mu_2$, then the continuity of $P$ and $T$ for an ideal gas requires $\varrho_1/\varrho_2 = \mu_1/\mu_2$, and (21.1) shows that

$$\frac{U_1}{U_2} = \frac{V_1}{V_2} = \frac{\varrho_1}{\varrho_2} = \frac{\mu_1}{\mu_2} \quad , \tag{21.7}$$

where subscripts 1 and 2 refer to core and envelope solutions at the interface respectively. This means that the points $(U_1, V_1)$ and $(U_2, V_2)$ lie on a straight line through the origin.

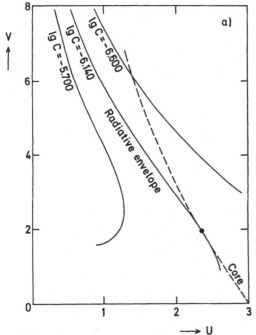
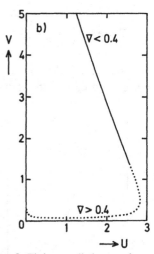

Fig. 21.3. a, b Fitting a radiative-envelope solution with a convective core in the $U$–$V$ plane. (a) Three envelope solutions with different values of the parameter $C$ come from the upper left downwards (*solid lines*). One of them fits to the convective-core solution (*dashed line*), which is given by the polytrope of $n = 3/2$ and starts in the centre at $U = 3$, $V = 0$. At the fitting point, both curves have the same gradient $\nabla = \nabla_{ad} = 0.4$ and the same tangent. (b) A radiative-envelope solution in the $U$–$V$ plane. The solution is shown by a solid line as far as $\nabla < 0.4$, and by a dotted line where $\nabla > 0.4$ such that the assumption of radiative transport breaks down. (After SCHWARZSCHILD, 1958)

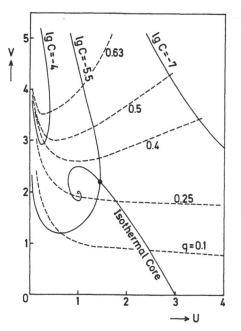

**Fig. 21.4.** Three envelope solutions with different parameters $C$ and the curve of the non-degenerate isothermal core in the $U$–$V$ plane. The dashed lines combine those points of the envelope solutions where $q = m/M$ reaches certain values. Since, in the case of a homogeneous model, envelope and core solution must be fitted continuously in the $U$–$V$ plane, one can see that no complete models are possible for isothermal cores with more than about $0.38M$. (This limit is even lower if the core has a higher molecular weight than the envelope.) A possible fit for $q \approx 0.3$ between the envelope curve for $\lg C = -5.5$ and the isothermal-core curve is indicated by a heavy dot

## 21.2 Radiative Envelope Solutions

We first consider solutions for the envelope where $\varepsilon = 0$ and therefore $l = $ constant $= L$. The gas is supposed to be ideal and the opacity is approximated by a power law

$$\kappa = \kappa_0 \varrho^a T^{-b} \quad , \tag{21.8}$$

where $\kappa_0 = $ constant. (Note that here a representation in $\varrho$ and $T$ is used which gives a different exponent $b$ than a representation in $P$ and $T$.)

We want to obtain many different solutions from a given one by simple scaling. For this aim we replace $P$, $T$, $m$, $r$ by the dimensionless Schwarzschild variables $y$, $t$, $q$, $x$ (SCHWARZSCHILD, 1946):

$$P = \frac{GM^2}{4\pi R^4} y \quad , \quad T = \frac{\mu}{\Re} \frac{GM}{R} t \quad , \quad m = qM \quad , \quad r = xR \quad . \tag{21.9}$$

The equation of state gives the density as

$$\varrho = \frac{M}{4\pi R^3} \frac{y}{t} \quad . \tag{21.10}$$

One can easily see that then the homology variables become $U = x^3 y/(qt)$ and $V = q/(tx)$. The stellar-structure equations (9.1, 16) give

$$\frac{dx}{dq} = \frac{t}{x^2 y} \quad , \quad \frac{dy}{dq} = -\frac{q}{x^4} \quad , \tag{21.11}$$

203

while the equation for energy transport (9.4) with expression (9.6) gives

$$\frac{dt}{dq} = -C \frac{y^a}{t^{a+b+3} x^4} \quad , \tag{21.12}$$

with

$$C = \frac{3\kappa_0}{4ac(4\pi)^{a+2}} \left(\frac{\Re}{\mu G}\right)^{b+4} L R^{b-3a} M^{a-b-3} \quad . \tag{21.13}$$

At the surface $q = 1$, and the solutions have to fulfil the boundary conditions

$$y = 0 \quad , \quad x = 1 \quad , \quad y/t = 0 \quad , \tag{21.14}$$

the last of which guarantees that according to (21.10) the density vanishes there.

The singularity of the system (21.11,12) at the surface can be overcome by an approximation. If one puts $q$ = constant = 1 for the whole near-surface region, one finds from (21.11,12) that

$$\frac{dy}{dt} = \frac{1}{C} \frac{t^{a+b+3}}{y^a} \quad , \quad \frac{dt}{dx} = -\frac{a+1}{a+b+4} \frac{1}{x^2} \quad . \tag{21.15}$$

The first equation has been integrated (the integration constant being chosen in such a way that $y = t = 0$ at the surface). This is used for eliminating $y$ from (21.11,12), which then give the second equation (21.15).

The two ordinary differential equations (21.15) are integrated by separation of the variables. The solutions can be used near the surface down to a safe distance from the singularity. From there on the normal equations (21.11,12) can be numerically integrated inwards.

Obviously one obtains a one-parameter set of solutions, the parameter being $C$. Three such envelope solutions in the $U$–$V$ plane are shown in Fig. 21.3a. All of them come from the upper left and miss the central boundary condition ($U = 3$, $V = 0$), since they have a singularity there. This does not matter, since anyway we have to fit them to a core solution (compare also with § 11.1). From (21.11,12) it results that

$$\nabla \equiv \frac{d \ln T}{d \ln P} = \frac{y}{t} \frac{dt}{dy} = C \frac{y^{a+1}}{t^{a+b+4} q} \quad , \tag{21.16}$$

from which one can see that owing to the factor $q^{-1}$ the value of $\nabla$ tends to infinity near the centre. In fact, $\nabla$ is small near the surface and increases inwards until it reaches the critical value $\nabla_{ad}$ (see Fig. 21.3b). Further inwards the Schwarzschild criterion (6.13) requires convection and the radiative envelope solutions are no longer valid.

## 21.3 Fitting of a Convective Core

In order to obtain a model with a convective core inside a radiative envelope we have to fit the solutions of §21.2 with a polytropic solution of $n = 3/2$ starting at the centre ($U = 3$, $V = 0$). The fit has to be done at the point where the envelope solution reaches $\nabla = \nabla_{ad}$. Joining all these points on the different envelope solutions (different $C$) gives a line $\nabla = \nabla_{ad}$ in the $U$–$V$ plane, which intersects the core polytrope at the fitting point $U^*$, $V^*$. The envelope solution through this point has the value $C = C^*$. Because of the condition that the gradient $\nabla$ is also continuous there, the solutions for core and envelope are tangential to each other, as can be seen in Fig. 21.3a. At the fitting point the variables of the envelope solution may be $q^*$, $y^*$, $x^*$, $t^*$, while the core polytrope has the variables $z^*$, $w^*$.

Let us assume a certain value for the mean molecular weight $\mu$ in the envelope. The fit has fixed $C = C^*$, which according to (21.13) gives a relation between $L$, $R$, and $M$. But $L$ is determined by the energy generation in the core, for which we assume a rate of

$$\varepsilon = \varepsilon_0 \varrho T^\nu \quad . \tag{21.17}$$

In the convective core we can connect the Emden variable $z$ with $r$ by $r = zr^*/z^*$, where $r^* = x^*R$ from the outer solution. Then $r^* dl/dr = z^* dl/dz$, and with $\varrho = \varrho_c w^{3/2}$, $T = T_c w$, we have the energy equation with $\lambda = l/L$

$$\frac{d\lambda}{dz} = Bz^2 w^{\nu+3} \quad , \quad B = \frac{4\pi\varepsilon_0}{L} \left(\frac{x^*R}{z^*}\right)^3 \varrho_c^2 T_c^\nu \quad . \tag{21.18}$$

Continuity of $\varrho$ and $T$ in core and envelope solutions requires

$$\varrho^* = \varrho_c w^{*3/2} = \frac{M}{4\pi R^3} \frac{y^*}{t^*} \quad , \tag{21.19}$$

$$T^* = T_c w^* = \frac{\mu}{\Re} \frac{GM}{R} t^* \quad . \tag{21.20}$$

With these two equations we can express $\varrho_c$, $T_c$ as functions of $w^*$, $y^*$, $t^*$ (all known from the integrations) and of $M$ and $R$. The expressions inserted into (21.18) give

$$B = B_0\varepsilon_0 \left(\frac{\mu G}{\Re}\right)^\nu \frac{M^{\nu+2}}{LR^{\nu+3}} \quad , \tag{21.21}$$

where $B_0$ is known from the numerical integrations to the fitting point. Since $L$ is to be generated in the core, $\lambda = l/L = 1$ at the fitting point. Therefore integration of (21.18) gives

$$1 = \int_0^{z^*} \frac{d\lambda}{dz} dz = B \int_0^{z^*} z^2 w^{\nu+3} dz \quad . \tag{21.22}$$

This fixes the value $B = B^*$, since $z^*$ is known, and the integral follows from a simple quadrature.

The fitting procedure now has yielded two numerical values $C^*$, $B^*$. Therefore for a given value of $M$ one obtains $L$ and $R$ from (21.13,21). Of course, one has to check afterwards that (21.17) only gives negligible contributions to $L$ in the envelope solution (where $l$ = constant was assumed).

Models of this type were first constructed by COWLING (1935). They have the advantage that $l$ appears in the structure equations only for the envelope where it is constant (= $L$).

## 21.4 Fitting of an Isothermal Core

In stellar evolution we shall have to discuss models with an isothermal helium core surrounded by a hydrogen-rich envelope. The luminosity is generated in a thin shell at the interface. This will be idealized by assuming a discontinuity of $l$ (from 0 to $L$) at the interface.

Let us discuss here a model in which $\mu$ is continuous at the interface so that the image curve in the $U$–$V$ plane is continuous at the fit.

In Fig. 21.4 we have plotted envelope solutions together with the isothermal-core solution for an ideal gas. Along each envelope curve the value of $q$ decreases inwards. We have also plotted some lines $q$ = constant. As one can see from the figure there are no fits possible with $q > q_{max} \approx 0.38$, i.e. when more than 38% of the total mass lies within the isothermal core. For given $q < q_{max}$ a fit is possible. An example for a fit at $q \approx 0.3$ is shown in Fig. 21.4. One can show that such a fit determines a model completely for given $M$. Physically more realistic is a model in which $\mu$ is higher in the core than in the envelope, which we idealize by a jump of $\mu$ at the interface. Then the curve in the $U$–$V$ plane is discontinuous, fulfilling the conditions (21.7) at the interface ($\mu_1 > \mu_2$). If one tries to fit core and envelope with this condition, and say $\mu_1/\mu_2 = 1.333/0.62$, one finds that $q_{max}$ is considerably smaller: no fits are possible at $q > q_{max} \approx 0.1$. This gives the Schönberg–Chandrasekhar limit for isothermal cores consisting of an ideal gas (see § 30.5) enclosed by the stellar envelope.

# § 22 The Main Sequence

We consider here a sequence of chemically homogeneous models in complete (mechanical and thermal) equilibrium with central hydrogen burning. All of them are composed of the same hydrogen-rich mixture, while the stellar mass $M$ varies from model to model along the sequence.

These models can represent very young stars which have just formed from the interstellar medium, and in which the foregoing contraction (see § 28) has raised the central temperature so far that hydrogen burning has started. This provides a long-lasting energy source, and consequently the stars change only on the very long nuclear time-scale $\tau_n$. Within the much shorter Kelvin–Helmholtz time-scale (see § 3.3) the stars will "forget" the details of their thermal history long before the nuclear reactions have noticeably modified the composition. This is why one can reasonably treat them as homogeneous models in thermal equilibrium. The now-beginning evolution, in which hydrogen is slowly consumed in the stellar core, has such a long duration that most visible stars are presently found in this phase. Our homogeneous models define its very beginning and their sequence is therefore more precisely called the *zero-age main sequence* (ZAMS), since one usually counts the age of a star from this point on.

## 22.1 Surface Values

Homogeneous, hydrogen-burning equilibrium models can be very easily calculated and are available for many different chemical compositions. We limit ourselves to discussing a set of calculations with $X_H = 0.685$, $X_{He} = 0.294$, such that all heavier elements amount only to the fraction $Z = 0.021$ of the mass.

Figure 22.1 shows the Hertzsprung–Russell diagram for the models in the wide range of stellar masses from 0.1 $M_\odot$ to more than 20 $M_\odot$. $L$ and $T_{eff}$ increase with increasing $M$, thus forming the ZAMS, which coincides more or less with the lower border of the observed main-sequence band.

The important mass–radius and mass–luminosity relations for these models are shown in Fig. 22.2 and 22.3 by the solid lines. As predicted already by the simple homology relations for main-sequence models [see (20.20,21)] $R$ increases slowly, and $L$ increases strongly with increasing $M$. For an interpolation over a certain range of $M$ we may again write

$$R \sim M^\xi \quad , \quad L \sim M^\eta \quad . \tag{22.1}$$

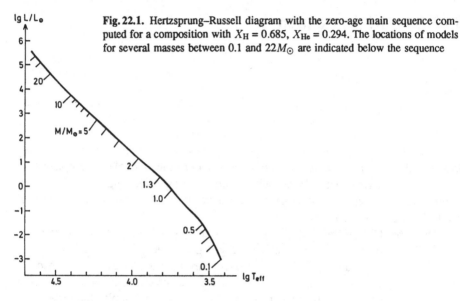

lg L/L⊙

**Fig. 22.1.** Hertzsprung–Russell diagram with the zero-age main sequence computed for a composition with $X_H = 0.685$, $X_{He} = 0.294$. The locations of models for several masses between 0.1 and $22 M_\odot$ are indicated below the sequence

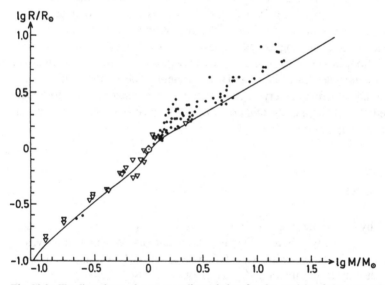

**Fig. 22.2.** The line shows the mass–radius relation for the models of the zero-age main sequence plotted in Fig. 22.1. For comparison, the best measurements (as selected by POPPER, 1980) of main-sequence components of detached (*dots*) and visual (*triangles*) binary systems are indicated

From the slopes of the curve in Fig. 22.2 we find roughly $\xi = 0.57$ and 0.8 in the upper and lower mass ranges respectively. In the range of small values of $M$, there is a pronounced maximum of the slope around $M = 1\ M_\odot$, indicating a remarkable deviation from homologous behaviour in this range. With decreasing effective temperature these models have outer convective zones of strongly increasing extension (cf. § 10.3.2, § 10.3.3, and Fig. 22.7). This tends to decrease $R$, in addition to other effects.

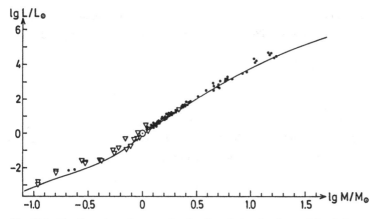

**Fig. 22.3.** The line gives the mass–luminosity relation for the models of the main sequence shown in Fig. 22.1. Measurements of binary systems are plotted for comparison (the symbols have the same meaning as in Fig. 22.2)

Also the slope of the $M$–$L$ relation in Fig. 22.3 varies with $M$. Over the whole mass range plotted, the average of $\eta$ is about 3.2. For $M = 1 \ldots 10\ M_\odot$ the average exponent is 3.88, while in the larger range $M = 1 \ldots 40 M_\odot$ it is 3.35. The decreasing slope towards larger $M$ is an effect of the increasing radiation pressure (see below).

Let us consider the way in which the variation of the exponents $\xi$ and $\eta$ influences the slope of the main sequence in the Hertzsprung–Russell diagram. Eliminating $M$ from the two relations (22.1), we find immediately that

$$R \sim L^{\xi/\eta} \quad . \tag{22.2}$$

We introduce this into the relation $L \sim R^2 T_{\text{eff}}^4$ and obtain for the main sequence in the Hertzsprung–Russell diagram the proportionality

$$L \sim T_{\text{eff}}^\zeta \quad , \quad \zeta = \frac{4}{1 - 2\xi/\eta} \quad . \tag{22.3}$$

We have seen that for large stellar masses, $\eta$ decreases and $\xi$ remains about constant with further increasing $M$. Equation (22.3) then gives an increase of $\zeta$, which means that the main sequence must become gradually steeper towards high luminosities.

We should mention that these two relations belong to the rare instances for which a reasonable quantitative test of the theory is possible. Even here one is rather restricted, since it is extremely difficult to obtain sufficiently precise measurements of $R$, $L$, and $M$. From this point of view, the $M - R$ relation should be the more reliable one. In Figs. 22.2 and 22.3 a selection of the best observed main-sequence double stars are plotted (POPPER, 1980). When comparing the scattering in the two diagrams one should note that Fig. 22.3 has an appreciably more compressed ordinate. The theoretical curves map out roughly the lower border of the measured values. They would be shifted slightly upwards, for example, by the assumption of a smaller hydrogen content. However, we have compared zero age main sequence stars with real stars here. In view of the uncertainties and difficulties involved in theory as well as in observation, one can scarcely expect a better fit, particularly when considering the enormous range of values involved (a factor 160 in $M$, nearly 8 powers of 10 in $L$).

## 22.2 Interior Solutions

The behaviour of the interior may be illustrated by characteristic variables as functions of $m/M$. They are plotted in Fig. 22.4 for two stellar masses in order to demonstrate typical dependencies of the solutions on $M$.

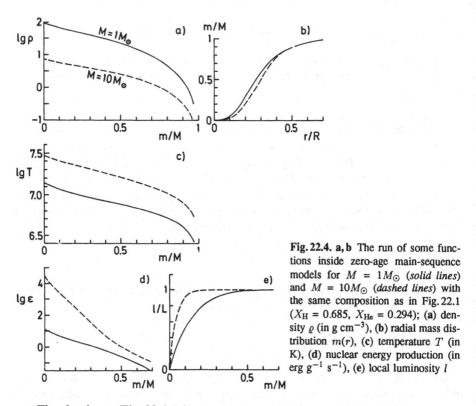

Fig. 22.4. a, b The run of some functions inside zero-age main-sequence models for $M = 1M_\odot$ (*solid lines*) and $M = 10M_\odot$ (*dashed lines*) with the same composition as in Fig. 22.1 ($X_H = 0.685$, $X_{He} = 0.294$); (a) density $\varrho$ (in g cm$^{-3}$), (b) radial mass distribution $m(r)$, (c) temperature $T$ (in K), (d) nuclear energy production (in erg g$^{-1}$ s$^{-1}$), (e) local luminosity $l$

The density $\varrho$ (Fig. 22.4a) increases appreciably towards the centre where we have $\varrho_c \approx 10^2$ g cm$^{-3}$ for 1 $M_\odot$, i.e. roughly a factor $10^9$ larger than in the photosphere. For $10M_\odot$, the central density is smaller by more than a factor 10. The inward increase of $\varrho$ indicates a very strong concentration of the mass elements towards the centre, illustrated in Fig. 22.4b. For 1 $M_\odot$, the inner 30% of the radius (i.e. only 3% of the total volume) contains 60% of the mass; and in the outer 50% of $R$ (i.e. 88% of the volume) only about 10% of $M$ can be found.

The temperature (Fig. 22.4c) also increases towards the centre. For 1 $M_\odot$, the central value of nearly $1.40 \times 10^7$ K is a factor 2500 larger than the photospheric value. Values of $T > 3 \times 10^6$ K extend to $m \approx 0.95\, M$, so that the average $T$ value (averaged over the mass elements) is roughly $7.7 \times 10^6$ K. In a $10M_\odot$ star, $T$ has slightly more than twice the values of corresponding mass elements for 1 $M_\odot$.

The behaviour of $T$ is necessarily reflected by that of the rate of energy generation due to hydrogen burning (Fig. 22.4d). The dependence of $\varepsilon$ on $T$ (cf. § 18.5.1), together with the $T$ gradient, yields a strong decrease of $\varepsilon$ from the centre outwards. In the 1 $M_\odot$ star, $\varepsilon$ has dropped by a factor $10^2$ from the centre to $m = 0.6 \, M$, and still further outward it is quite negligible. This is particularly well seen in Fig. 22.4e: 90% of $L$ is generated in the inner 30% of $M$; and $l$ reaches about 99% of $L$ at $m/M = 0.5$. In the central part of the $10M_\odot$ star, where $T_c = 3 \times 10^7$ K, the dominant energy source is the CNO cycle (instead of the $pp$ chain in $1M_\odot$). The much larger $T$ dependence of $\varepsilon$ gives an even more pronounced concentration of $\varepsilon$ towards the centre (Fig. 22.4d). In the innermost 30% of $M$, $\varepsilon$ drops by about a factor $10^3$ (as compared to a factor 10 in the same interval of $1M_\odot$). This corresponds to an $\varepsilon$ with an exponent of $T$ roughly 3 times larger. Further outwards, where $T$ is low enough for the $pp$ chain to dominate, the slope of $\varepsilon$ becomes the same in both stars. In the $10M_\odot$ star, 90% of the total luminosity is generated within the innermost 10% of the mass (Fig. 22.4e).

We have seen that in spite of all similarities there are characteristic differences between the interior solutions for different values of $M$. Some of these can be found in the plot of the central values of temperature and density (Fig. 22.5). This diagram exhibits at least qualitatively another prediction of the homology considerations in § 20: with increasing $M$ there is a slight increase of $T_c$ together with a substantial decrease of $\varrho_c$. Between $M = 2 \, M_\odot$ and 50 $M_\odot$ the differences are $\varDelta \lg T_c = +0.26$ and $\varDelta \lg \varrho_c = -1.44$. But the striking change of the curve around and below 1 $M_\odot$ reveals clearly enough the deviations from homology. These are connected partly with the changes of the central values, partly with those at the surface (especially $T_{eff}$ and the depth of the outer convection zone). The extension of convective regions,

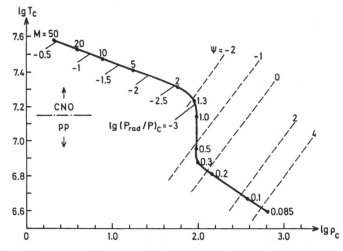

**Fig. 22.5.** The heavy solid line gives the central temperature $T_c$ (in K) over the central density $\varrho_c$ (in g cm$^{-3}$) for the same zero-age main-sequence models as in Fig. 22.1. The dots give the positions of some models with masses between $M = 0.085$ and $M = 50$ (in solar masses). The labels below the curve indicate the fractional contribution of the radiation pressure $P_{rad}$ to the total pressure in the centre. The dot-dashed line at the left gives roughly the border between dominating CNO-cycle and dominating $pp$-chain reactions. The dashed lines give the constant degeneracy parameter $\psi$ of the electron gas

for example, should certainly influence the centre, since they have a less pronounced mass concentration than radiative regions. Note that both flat parts of the $T_c$–$\varrho_c$ curve in Fig. 22.5 belong to models in which the central part is convective (cf. Fig. 22.7).

In the upper range of masses degeneracy is negligible, while it becomes increasingly important towards smaller $M$ owing to the increasing density. Below $0.5M_\odot$, say, other deviations from the ideal gas approximation also become important in the equation of state, e.g. electrostatic interaction between the ions.

On the other hand, the radiation pressure $P_{rad}$ must increase towards larger $M$ owing to the increasing $T$, since $P_{rad} \sim T^4$. At $M = 1M_\odot$, radiation contributes only the negligible fraction of a few $10^{-4}$ to the total central pressure. This fraction becomes about 1% at $4M_\odot$, while in the centre of the $50M_\odot$ star, $P_{rad}$ contributes no less than 1/3 to the total pressure (see Fig. 22.5).

Another effect of the growing $T_c$, which also occurs around $1M_\odot$, is the transition from the $pp$ chain to the CNO cycle as the dominant energy source (compare also Fig. 18.8). For models in the transition region from $M = 1M_\odot$ to $3M_\odot$, Fig. 22.6 shows the contribution of $\varepsilon_{CNO}$ to the local energy generation rate as a function of $l/L$. The integral over such a curve gives the fraction of $L$ due to burning in the CNO cycle. This amounts only to a few percent for $M = 1M_\odot$. In the $1.5M_\odot$ star, the CNO cycle already contributes 73% at the centre, and nearly one half of the total luminosity. It clearly dominates the whole energy generation for $1.7M_\odot$ and more massive stars.

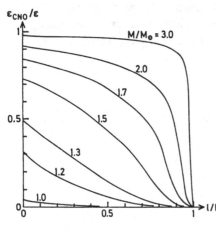

Fig. 22.6. For seven zero-age main-sequence models of the same composition as in Fig. 22.1, the fraction that the CNO cycle contributes to the total energy generation rate at different places inside the model (characterized by the corresponding local luminosity $l$ at the abscissa) is shown

## 22.3 Convective Regions

Knowledge of the extension of convective regions is very important in view of their influence on the ensuing chemical evolution. A rough overview can be obtained from Fig. 22.7, where $m/M$ and $\lg M$ are ordinate and abscissa. For any given stellar mass $M$ along a line parallel to the ordinate it is indicated what conditions we would encounter when drilling a radial borehole from the surface to the centre. In particular, one can see whether the corresponding mass elements are convective

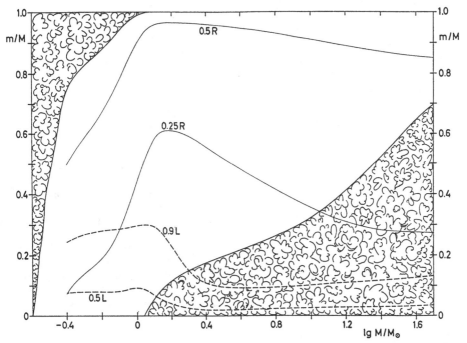

**Fig. 22.7.** The mass values $m$ from centre to surface are plotted against the stellar mass $M$ for the same zero-age main-sequence models as in Fig. 22.1. "Cloudy" areas indicate the extension of convective zones inside the models. Two solid lines give the $m$ values at which $r$ is 1/4 and 1/2 of the total radius $R$. The dashed lines show the mass elements inside which 50% and 90% of the total luminosity $L$ are produced

or radiative. Aside from the stars of smallest mass ($M < 0.25\ M_\odot$), we can roughly distinguish between two types of model:

convective core + radiative envelope (upper main sequence);

radiative core + convective envelope (lower main sequence).

The transition from one type to the other again occurs near $M = 1M_\odot$.

The distinction between convective and radiative regions is made here by using the Schwarzschild criterion (see §6.1), which predicts convection if the radiative gradient of temperature $\nabla_{\text{rad}}$ exceeds the adiabatic gradient $\nabla_{\text{ad}}$. (The gradient $\nabla_\mu$ of the molecular weight appearing in the Ledoux criterion is zero in these homogeneous models. Possible effects of overshooting will be discussed in §30.) The variation of these gradients (together with that of the actual gradient $\nabla$) throughout the star is plotted in Fig. 22.8 for $M = 1M_\odot$ and $10M_\odot$. For the abscissa, $\lg T$ is chosen, since this conveniently stretches the scale in the complicated outer layers.

Let us start with the simpler situation concerning the convective core. When comparing Fig. 22.8a and b, we see that the convective core in the more massive models is caused by a steep increase of $\nabla_{\text{rad}}$ towards the centre. The reason for this is that the dominating CNO cycle, with its extreme temperature sensitivity, concentrates the energy production very much towards the centre (cf. the curve $l/L = 0.5$ in Fig. 22.7, and Fig. 22.4e). Therefore we find in these stars very high

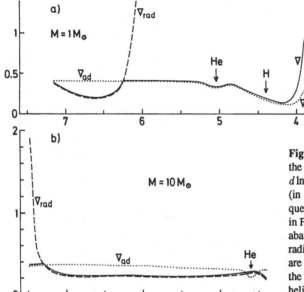

Fig. 22.8 (a, b). The solid lines show the actual temperature gradient $\nabla = d\ln T/d\ln P$ over the temperature $T$ (in K) inside two zero-age main-sequence models (same composition as in Fig. 22.1). The corresponding adiabatic gradients $\nabla_{ad}$ (*dotted lines*) and radiative gradients $\nabla_{rad}$ (*dashed lines*) are also plotted, and the location of the ionization zones of hydrogen and helium are indicated (*arrows*)

fluxes of energy $(l/4\pi r^2)$ at small $r$, which produce large values of $\nabla_{rad}$. Figure 22.7 shows a remarkable increase in the extent of the convective core for increasing $M$. The core covers as much as 70% of the stellar mass in a star of $50M_\odot$, an increase caused by the increasing radiation pressure (cf. § 22.2, and Fig. 22.5), which depresses the value of $\nabla_{ad}$ well below its standard value of 0.4 for an ideal monatomic gas [see (13.21)]. In the centre of the $50M_\odot$ model, roughly 1/3 of $P$ is radiation pressure, and $\nabla_{ad} \approx 0.27$. Figure 22.8b shows that the depression of $\nabla_{ad}$ in the central region shifts the intersection with $\nabla_{rad}$ (i.e. the border of the convective core) outwards to smaller $T$. When we increase $M$ to much larger values still, the top of the convective core will finally approach the surface such that we should obtain fully convective stars. We then approach models of the so-called supermassive stars (see § 19.10).

In less massive stars, the $pp$ chain with its smaller temperature sensitivity dominates. This distributes the energy production over a much larger area, so that the flux and $\nabla_{rad}$ are much smaller in the central region, which thus remains radiative.

Outer convective envelopes can generally be expected to occur in stars of low effective temperature, as the discussion of the boundary conditions in § 10.3.2 has already shown. When studying the different gradients in the outer layers of cool stars (Fig. 22.8a), one finds a variety of complicated details. The variation of $\nabla_{ad}$ clearly shows depressions in those regions where the most abundant elements, hydrogen ($T \gtrsim 10^4$ K) and helium ($T \approx 10^5$ K), are partially ionized (see § 14). The most striking feature is that $\nabla_{rad}$ reaches enormous values (more than $10^5$). This is due to the large opacity $\kappa$, which here increases by several powers of 10 (cf. § 17). Therefore the Schwarzschild criterion indicates convective instability: the models have an outer convective zone. In the largest part of it, the density is so high that convection is very effective and the actual gradient $\nabla$ is close to $\nabla_{ad}$. Convective transport becomes ineffective only in the outermost, superadiabatic part, where $\nabla$ is clearly above $\nabla_{ad}$. Scarcely anything of all these features appears in the hot envelope of the $10 M_\odot$ star (Fig. 22.8b). $\nabla_{rad}$ remains nearly at the same level; even the photosphere is too hot for hydrogen to be neutral, and only the small dip from the second He ionization is seen immediately below the photosphere. This causes such a shallow zone with convective instability that it is doubtful whether convective motions can set in at all.

The outer convection zone gradually penetrates deeper into the star with decreasing $T_{eff}$. Its lower border finally reaches the centre at $M \approx 0.25 M_\odot$ (left end of Fig. 22.7), such that the main-sequence stars of even smaller masses are fully convective.

## 22.4 Extreme Values of $M$

Only a few calculations are available for main-sequence stars of very large and very small $M$. In the latter range, the results suffer particularly from the fact that the input physics is not reliable. This concerns the notorious problem of the treatment of convection, as well as the opacity values for mixtures containing many molecules. Both these effects are important in very cool envelopes. Complications for the interior structure are equally severe. They arise, e.g., from the difficult treatment of particle interaction in the low-temperature high-density regime and influence the equation of state and the electron screening of nuclear reactions.

Quite another problem concerns the relevance of the calculated equilibrium models for real, evolving stars. At the low central temperatures in models of extremely small masses, for example, the time for reaching equilibrium burning can become exceedingly long. A preceding phase in which the original $^3$He is burned may be at least equally important, but this $^3$He content is very uncertain. And below about $M = 0.1 M_\odot$, even the original contraction leads so far into electron degeneracy that hydrogen burning is no longer ignited (refer to §28). In this sense one may speak of the "lower end of the main-sequence" at this mass value. Disregarding this evolutionary argument, however, one can ask whether solutions for main-sequence models (homogeneous, hydrogen burning, complete equilibrium) exist down to arbitrary small values of $M$. In terms of linear series (§12.2,3) we ask how far the branch of thermally stable main-sequence models extends. It turns out that it ends in a turning point at $M \approx 0.08 M_\odot$. This termination of the stable main-sequence branch will be discussed in §23.

In the direction towards large $M$, on the other hand, the sequence of equilibrium models can principally be continued up to the "supermassive" stars (see §19.10). Long before they are reached, however, an instability occurs which sets in at about $M \approx 90 M_\odot$ (depending on the composition). It is a vibrational instability caused by the so-called $\varepsilon$ mechanism (see §39.5) and supported by the large amount of radiation pressure. Such stars, instead of sitting quietly at their proper place on the main sequence, will start to oscillate with growing amplitude. This may go so far as to throw off matter from the surface.

# §23 Other Main Sequences

The simplicity and the importance of the results obtained for the main sequence suggest the extension of this concept to stars of quite different composition. We can then describe a main sequence as any sequence of homogeneous models with various masses $M$ in complete equilibrium, consisting (mainly) of a certain element which burns in the central region. In this sense, *the* (normal) main sequence as treated before is a special case and is more precisely called the *hydrogen main sequence* (H-MS). In a further step of generalization, we will even drop the assumption of chemical homogeneity, thus arriving at the so-called generalized main sequences (§23.3,4). Of course, compared with the H-MS, the other sequences are far less important for real, observed stars. But their properties yield valuable information for understanding certain types of evolved stars, for example.

## 23.1 The Helium Main Sequence

The helium main sequence (He-MS) contains chemically homogeneous equilibrium models that consist almost completely of He (with the usual few per cent of heavier elements) and have central helium burning. In principle one could imagine them to be the descendants of perfectly mixed hydrogen-burning stars (however, perfect mixing during evolution is very improbable). Or they may represent the remnants of originally more massive stars that have developed a central helium core and then lost their hydrogen-rich envelope.

In the Hertzsprung–Russell diagram (Fig. 23.1) the He-MS is situated far to the left of the (normal) H-MS at fairly high luminosities. If we compare the same stellar mass $M$ on each sequence, we see that the helium stars have smaller radii and much higher luminosities. The remarkable difference in $L$ for given $M$ is particularly well illustrated by the $M-L$ relations in Fig. 23.2. The main cause is certainly the difference in the mean molecular weight $\mu$, which is 0.624 for the mixture used for the stars on the H-MS, and 1.343 for the helium stars. If everything else were the same and the models were homologous, then we would expect from (20.20) for stars with the same $M$ a difference in luminosity given by $\Delta \lg L = 4 \Delta \lg \mu = 1.33$. This is in fact very nearly the shift between the two $M-L$ relations in Fig. 23.2 at $M = 10 M_\odot$, while for $M = 1 M_\odot$ we even have $\Delta \lg L \approx 2.5$.

The interior structure resembles roughly that of models on the upper H-MS. The extreme temperature sensitivity of helium burning concentrates the energy production into a small central sphere where the large energy flux produces a convective core. This contains about $0.27M$ in the $1 M_\odot$ star, and nearly $0.7M$ for

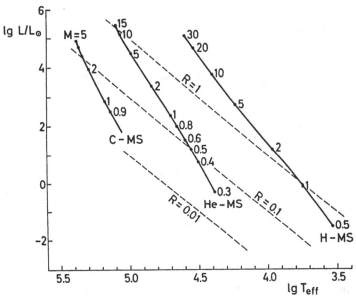

**Fig. 23.1.** In the Hertzsprung–Russell diagram the solid lines show the normal hydrogen main sequence (H-MS; $X_H = 0.685$, $X_{He} = 0.294$), the helium main sequence (He-MS; $X_H = 0$, $X_{He} = 0.979$) and the carbon main sequence (C-MS; $X_H = X_{He} = 0$, $X_C = X_O = 0.497$). The labels along the sequences give stellar masses $M$ (in units of $M_\odot$). Three lines of constant stellar radius ($R$ in units of $R_\odot$) are plotted (*dashed*)

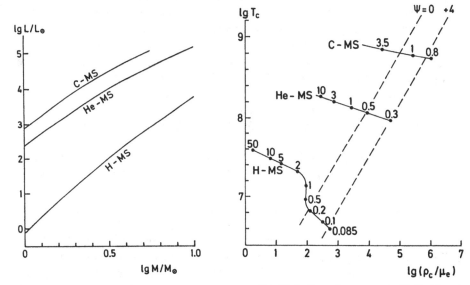

**Fig. 23.2.** Mass–luminosity relations for the models of the hydrogen, helium, and carbon main sequences of Fig. 23.1

**Fig. 23.3.** Central temperature $T_c$ (in K) and central density $\varrho_c/\mu_e$ ($\varrho_c$ in g cm$^{-3}$, $\mu_e$ = molecular weight per electron) of the models on the hydrogen, helium and carbon main sequences of Fig. 23.1. The labels along the lines give the stellar mass $M$ (in $M_\odot$). The dashed lines indicate constant degeneracy parameters $\psi$ of the electron gas

$10 M_\odot$. The increase of the convective core is again a consequence of the increasing radiation pressure: it contributes 1.5% to the total pressure in the centre of the $1 M_\odot$ star, 18% for $5 M_\odot$, and 32% for $10 M_\odot$, which is very much more than for the corresponding stars on the H-MS ($6 \times 10^{-4}$, 0.018, and 0.063 respectively). The difference is due to the fact that helium burning requires temperatures roughly 6 times higher, as can be seen in Fig. 23.3, which shows the central values of $T$ and $\varrho$. The high radiation pressure provides relatively large amplitudes of pulsation in the central region. This again produces a vibrational instability due to the $\varepsilon$ mechanism, the onset of which occurs around $M = 15 M_\odot$, depending somewhat on the content of heavier elements.

Another property of the helium stars to be seen in Fig. 23.3 is their much larger central density: for $M = 0.3 M_\odot$, $\varrho_c$ reaches $10^5$ g cm$^{-3}$, and, in spite of the larger $T$, the electron gas has about the same degree of degeneracy as at the lower end of the H-MS. [In order to plot a unique degeneracy parameter $\psi$ (see §15) for compositions with different molecular weight per electron $\mu_e$, the abscissa of Fig. 23.3 gives $\lg(\varrho_c/\mu_e)$. The He-MS and the C-MS (see below) have $\mu_e = 2$, while $\mu_e = 1.19$ for the plotted H-MS.] The increasing degeneracy causes the sequence of stable helium-burning stars to terminate at about $M \approx 0.3 M_\odot$ (however, cf. §23.3).

## 23.2 The Carbon Main Sequence

The next major step in the nuclear history of a star is carbon burning. Thus we now consider a carbon main sequence (C-MS) consisting of homogeneous models in complete equilibrium that have central carbon burning. Except for the usual admixture of a few per cent of heavy elements, the composition can be either pure $^{12}$C, or a mixture of $^{12}$C and $^{16}$O in equal amounts, which represents roughly the end products of stellar helium burning. (For both assumptions the basic results, in particular the luminosities, are not too different, since the molecular weights are nearly the same.) The models of the C-MS are not so much used for describing homogeneous carbon stars, but rather for the purpose of surveying carbon-burning cores in highly evolved stars.

In the Hertzsprung–Russell diagram (Fig. 23.1) the C-MS is at $T_{\text{eff}} > 10^5$ K even to the left of the He-MS. For equal masses, models on the C-MS have remarkably smaller $R$ and larger $L$. The $M$–$L$ relation for carbon stars is $\Delta \lg L \approx 0.5$ above that for helium stars (Fig. 23.2) because of the larger mean molecular weight ($\Delta \lg \mu \approx 0.11$).

The interior solutions of carbon stars have similar properties to those of the helium stars, for example large convective cores and an appreciable amount of radiation pressure. In a model of $M = 3.5 M_\odot$, the convective core encompasses about 45% of the total mass, and the radiation pressure contributes more than 20% to the central pressure. Figure 23.3 shows that, according to the requirements of carbon burning, the central temperatures are between 5 and $8 \times 10^8$ K. But the central density is even more increased compared to helium stars. Therefore appreciable degeneracy of the electron gas is already found in carbon stars around $1 M_\odot$. And the sequence of stars with a stable carbon burning terminates at masses in the range

$M \approx 0.9\ldots0.8M_\odot$. The exact value of this limiting mass depends somewhat on the assumptions in the physical parameters. A well-known uncertainty comes, for example, from neutrino losses, which can become noticeable in these very hot and dense stars (§ 18.6). Large neutrino losses have the tendency to increase the lower limit of $M$ for stable carbon burning. Figure 23.3 shows that in all three main sequences the limiting mass occurs at roughly the same degree of degeneracy of the electron gas ($\psi \approx 4.5$). The C-MS and the He-MS have a much simpler structure than the H-MS, which is affected by the complications occurring near $1M_\odot$, namely the transition from convective to radiative cores and the growth of outer convection zones with decreasing $T_{\mathrm{eff}}$.

### 23.3 Main Sequences as Linear Series of Stellar Models

As described in § 12.2.3, we speak of a linear series if we have a continuous sequence of models (generated by the continuous variation of a parameter) where a model can be obtained from a neighbouring one by linearized equations. This no longer holds at so-called critical points (which in the simplest and most frequently occuring case are turning points). They are of special interest, since at critical points several branches of the linear series can merge, local uniqueness is violated, and the thermal stability properties change (the eigenvalue being $\sigma = 0$, see § 12.4). We concentrate on discussing carbon and helium star sequences. For these, the physical properties are sufficiently simple that even models beyond the turning points have been calculated (HANSEN, SPANGENBERG, 1971; PACZYŃSKI, KOSLOWSKI, 1972; MARISKA, HANSEN, 1972), on which the following is based.

It is obvious that the models on a main sequence represent a linear series along which the stellar mass $M$ varies as the parameter. This is illustrated in Fig. 23.4, where each model is represented by its radius $R$. The sequences have two critical points (turning points) where the parameter $M$ reaches an extremum. One of them (labelled 1) corresponds to the previously described "lower end of the main sequence". Here the stable main-sequence branch merges with a new branch of ther-

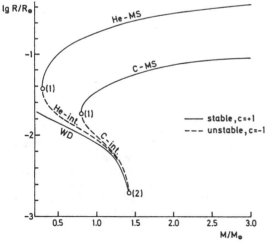

**Fig. 23.4.** Mass–radius relations for models of the helium and carbon main sequences of Fig. 23.1. Open circles indicate turning points (see text)

mally unstable models (dashed). This intermediate (sometimes called "high-density") branch extends to a second turning point (labelled 2) near 1.4 $M_\odot$, which is about the Chandrasekhar limiting mass for degenerate configurations with $\mu_e = 2$ (see § 19.7). Then follows a third, stable branch, which is called the white-dwarf branch, on which $R$ increases again for decreasing $M$. In contrast to evolving (cooling) white dwarfs, the models of this branch are cold configurations in thermal equilibrium. (Note that for all models along a sequence the composition is assumed to be the same, either helium or carbon, thus neglecting any effects such as inverse $\beta$ decay that can change the composition near the Chandrasekhar limit, see § 35.2.)

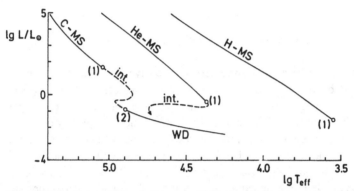

Fig. 23.5. Hertzsprung–Russell diagram with linear series of homogeneous hydrogen-rich models, helium models, and carbon models. Thermally stable branches are solid; thermally unstable branches are dashed. Circles indicate turning points. Turning point (1) separates the main sequences (*above*) from the intermediate branch, which leads to the white dwarf branch starting at turning point (2)

The models on these three branches are plotted in the Hertzsprung–Russell diagram in Fig. 23.5. They extend down to very low values of $L$. The interior of the models on the intermediate and white-dwarf branches is governed by degeneracy of the electron gas. Figure 23.6 shows that the central density reaches very high values. On the white-dwarf branch, the central temperature drops drastically with decreasing $M$. The minute energy output of these stars is supplied by pycnonuclear burning, which depends relatively weakly on $T$, but strongly on $\varrho$ (cf. § 18.4 and Fig. 18.6). The temperature distribution in the stars adjusts to allow the produced energy to be transported by conduction. Since this is very effective and $L$ is very small, the stars are nearly isothermal at very low temperature.

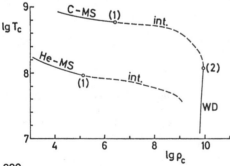

Fig. 23.6. Central temperature ($T_c$ in K) against central density ($\varrho_c$ in g cm$^{-3}$) for sequences of homogeneous helium and carbon models. At turning points (1) the stable main-sequence branch merges with the thermally unstable intermediate branch; at turning point (2) the stable white-dwarf branch starts. Thermal instability is indicated by dashes

As seen from Fig. 23.4, the existence of the three branches obviously means that, for given parameter $M$, we find in Fig. 23.2 either one stable solution (for $M$ above turning point 2 and below turning point 1), or two stable and one unstable solution (between turning points 1 and 2). There may be even more solutions in certain ranges of $M$, for example with negative $T$ gradients due to strong neutrino losses (see PACZYŃSKI, KOSLOWSKI, 1972). They then form additional branches not presented here, which always have to occur in pairs with opposite stability properties (for instance as additional closed curves).

Note that the normal main-sequence branch shows no critical point for large values of $M$, although we have seen that such models also become unstable. This, however, is the onset of a vibrational instability that is not connected with a zero eigenvalue, but with a complex conjugate pair of modes becoming unstable; and without a zero eigenvalue there is no critical point.

Little is known about high-density stars with hydrogen-rich mixtures, since for them the physical properties are much more complicated than for carbon or helium stars. But the principal structure of the linear series for hydrogen-rich stars should be similar to the other cases. An intermediate branch might in principle extend to larger values of $M$ (near $4M_\odot$), since the limiting mass corresponding to the turning point 2 is proportional to $\mu_e^{-2}$ (see § 19.7).

## 23.4 Generalized Main Sequences

The logical next step in extending the concept of main sequences is to drop the condition of chemical homogeneity. This is suggested by the chemical evolution we encounter in all stars: the conversion of hydrogen to helium by nuclear reactions (which are concentrated towards the centre) produces a central helium core, while the outer envelope retains its original hydrogen-rich mixture. If the temperatures are high enough, helium burning will occur around the centre, and hydrogen burning continues in a so-called shell source, i.e. a concentric shell starting at the bottom of the hydrogen-rich envelope. Based on this picture, different types of significant sequences may be defined. We will limit ourselves in the following to the simplest case, which nevertheless finds useful applications.

For these *generalized main sequences* (GMS), we consider models in complete equilibrium, with a chemical profile as shown in Fig. 23.7: a central helium core of mass $M_{He}$, i.e. of the mass fraction $q_0 = M_{He}/M$, is surrounded by an envelope of mass $(1 - q_0)M$ with the usual hydrogen-rich mixture of unevolved stars. At the interface of the two regions, the hydrogen content $X_H$ changes discontinuously ("step profile"), while the hydrogen content in the envelope as well as the small admixture of heavier elements in both regions is assumed to be fixed at some reasonable values. The energy is supplied by central helium burning and (possibly) by an additional hydrogen burning in a shell source at $q_0$.

Each of these models is characterized by two parameters, the stellar mass $M$, and the relative core mass $q_0$. We then obtain a generalized main sequence by keeping $q_0$ constant, and varying $M$ as a parameter. For each value of $q_0$ there is one GMS. In the evolution the value of $q_0$ is not constant: $q_0$ can slowly increase because of

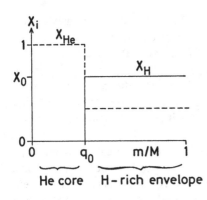

**Fig. 23.7.** Chemical composition inside the models on the generalized main sequences. The mass concentrations of hydrogen $X_H$ (*solid line*) and helium $X_{He}$ (*dashed line*) are plotted over the mass variable $m/M$ from centre to surface. $X_0$ is the hydrogen content in the envelope. The relative core mass is $M_{He}/M = q_0$

the shell source burning, and it can increase by mass loss from the surface. We will therefore consider GMS of various values of $q_0$.

The upper limit is obviously $q_0 = 1$, implying that the "core" encompasses the whole star, which is then a homogeneous helium star. The GMS for $q_0 = 1$ is therefore identical with the well-known He-MS discussed in § 23.1.

For values of $q_0$ slightly below 1, the GMS are shifted appreciably to the right in the Hertzsprung–Russell diagram (Fig. 23.8). They have already passed the H-MS for $q_0 \approx 0.9 \ldots 0.85$, depending on the value of $M$. In other words, the addition of a relatively small hydrogen-rich layer on top of a helium star will remarkably increase its radius and decrease $T_{eff}$.

This behaviour changes completely if $q_0$ drops below a certain value, which is about $0.8 \ldots 0.7$, depending on $M$. Figure 23.8 shows that the GMS are then

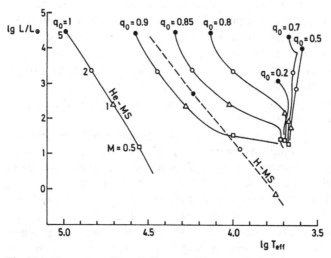

**Fig. 23.8.** Hertzsprung–Russell diagram with generalized main sequences for models with helium cores of relative mass $q_0$ and hydrogen-rich envelopes of relative mass $1 - q_0$ (cf. Fig. 23.7). The sequences plotted here cover only the range from $q_0 = 1$ (helium main sequence) to $q_0 = 0.2$. For comparison, the limiting case of the hydrogen main sequence ($q_0 = 0$, dashed) is shown. Models with a stellar mass $M = 5$ (in $M_\odot$) are indicated by solid dots, $M = 2$ by open circles, $M = 1$ by triangles, and $M = 0.5$ by squares. (After GIANNONE, KOHL, WEIGERT, 1968)

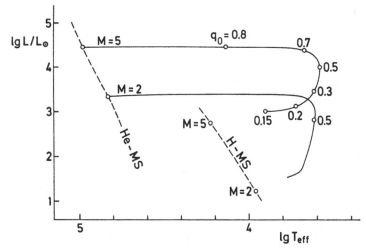

Fig. 23.9. The solid lines connect models of the same stellar mass $M$ (in $M_\odot$) on the different generalized main sequences of Fig. 23.8. Labels along the lines give the $q_0$ values of the generalized main sequences. (After LAUTERBORN, REFSDAL, WEIGERT, 1971)

compressed towards a limiting line far to the right-hand side of the Hertzsprung–Russell diagram. This will turn out to be the Hayashi–line, a limit for all stars in hydrostatic equilibrium (§ 24). The closest approach to it is found roughly for the GMS with $q_0 = 0.5$. For even smaller $q_0$, the GMS move slowly back to the left in the Hertzsprung–Russell diagram. We conclude that the upper part of this diagram can be covered at least once by these GMS, i.e. by very simple equilibrium models depending on two parameters ($M$, $q_0$) only.

Let us compare models with the same $M$ on different GMS. If we connect their points in Fig. 23.8, we obtain curves such as those plotted in Fig. 23.9 for two values of $M$. This shows that the luminosity remains roughly constant in the range $q_0 = 1 \ldots 0.7$. This is caused by two opposite effects nearly cancelling each other: when we decrease $q_0$ at $M = $ constant, $M_{He}$ decreases, which reduces the luminosity of the core, $L_{He}$, approximately as given by the $M$–$L$ relation for the He-MS (Fig. 23.2, if here we take $M_{He}$ for $M$). At the same rate, the mass of the envelope $M(1 - q_0)$ increases, which gives an increasing energy production $L_H$ of the hydrogen shell source, such that the total luminosity $L = L_{He} + L_H$ can remain almost constant. The situation changes when $q_0$ drops below, say, 0.7. The "helium luminosity" $L_{He}$ then decreases so strongly that it is compensated no longer by the increase of $L_H$, which eventually dominates $L$ completely.

The GMS offer the possibility of defining a variety of interesting linear series. For simplicity we postpone the complicated discussion of all existing solutions (with helium burning or degenerate cores, stable or unstable) until their relevance has become clear. We have therefore limited the discussion here to the branch of stable solutions with helium-burning cores, which have similar properties to homogeneous helium stars of the same mass on the He-MS. We can expect that these models of the GMS extend down to $M_{He} = q_0 M \approx 0.3 M_\odot$.

# § 24 The Hayashi Line

We have seen that convection can occur in quite different regions of a star. In this section we consider the limiting case of *fully convective stars*, i.e. stars which are convective in the whole interior from centre to photosphere, while only the atmosphere remains radiative.

The Hayashi Line (HL) is defined as *the locus in the Hertzsprung–Russell diagram of fully convective stars of given parameters* (mass $M$ and chemical composition). Note that for each set of the parameters, such as mass or chemical composition, there is a separate Hayashi line. These lines are located far to the right in the Hertzsprung–Russell diagram, typically at $T_{eff} \approx 3000 \ldots 5000$ K, and they are very steep, in large parts almost vertical.

From the foregoing definition one may not immediately realize the importance of this line. However, the HL also represents *a borderline between an "allowed" region* (on its left) *and a "forbidden" region* (on its right) in the Hertzsprung–Russell diagram for all stars with these parameters, provided that they are in hydrostatic equilibrium and have a fully adjusted convection. The latter means that, at any time, the convective elements have the properties (for instance the average velocity) required by the mixing-length theory. Changes in time of the large-scale quantities of the stars are supposed to be slow enough for the convection to have time to adjust to the new situation; otherwise one would have to use a theory of time-dependent convection. Since hydrostatic and convective adjustment are very rapid, stars could survive on the right-hand side of the HL only for a very short time.

In addition, parts of the early evolutionary tracks of certain stars may come close to, or even coincide with, the HL. It is certainly significant for the later evolution of stars, which is clearly reflected by observed features (e.g. the ascending branches of the Hertzsprung–Russell diagrams of globular clusters). One may even say that the importance of the HL is only surpassed by that of the main sequence. It is all the more surprising that its role was not recognized until the early 1960s when the work of C. HAYASHI (1961) appeared. The late recognition of the HL may partly be because its properties are derived from involved numerical calculations. In the following we will use extreme simplifications in order to make some basic characteristics of the HL plausible.

## 24.1 Luminosity of Fully Convective Models

Let us consider the different ways in which the luminosity is coupled to the pressure–temperature stratification of radiative and convective stars.

For regions with radiative transport of energy, we can write the "radiative luminosity" $l_{rad} = 4\pi r^2 F_{rad}$ according to (7.2) as

$$l_{rad} = k'_{rad} \nabla \quad , \tag{24.1}$$

with the usual notation $\nabla = d\ln T/d\ln P$ and the "radiative coefficient of conductivity"

$$k'_{rad} = \frac{16\pi acG}{3} \frac{T^4 m}{\kappa P} \quad . \tag{24.2}$$

If a stratification of $P$ and $T$ is given, then the luminosity $l_{rad}$ is obviously determined and can be easily calculated from (24.1).

For convective transport of energy by adiabatically rising elements we can write accordingly from (7.7) the convective luminosity as

$$l_{con} = k'_{con}(\nabla - \nabla_{ad})^{3/2} \tag{24.3}$$

with the coefficient

$$k'_{con} = \frac{\pi}{\sqrt{2}} \left(\frac{\ell_m}{H_P}\right)^2 r^2 c_P T(\varrho P\delta)^{1/2} \quad . \tag{24.4}$$

Here we have made use of the hydrostatic equation and the definition (6.8) of the pressure scale height. The mixing length $\ell_m$ was defined in §7.1.

In principle, we can again assume the luminosity to be determined using (24.3) for a given $P$–$T$ stratification. In practice, however, we would never be able to calculate $l_{con}$ from this equation for the stellar interior, since it would require the knowledge of the value of $\nabla$ with inaccessible accuracy. The point is that $l_{con}$ is not proportional to the gradient $\nabla$ itself, but rather to a power of the excess over the adiabatic gradient, $\nabla - \nabla_{ad}$, which may be as small as $10^{-7}$ for very effective convection (see §7.3). Therefore the convective conductivity $k'_{con}$ must be very high, since large luminosities $l_{con}$ are carried. This may be looked at in another way: by solving (24.3) for $\nabla$ and writing

$$\nabla = \nabla_{ad}(1 + \varphi) \quad , \tag{24.5}$$

we see that the luminosity influences the $T$ gradient only through the tiny correction $\varphi(\approx 10^{-7})$:

$$\varphi = \left[\frac{l_{con}}{\nabla_{ad}^{3/2} k'_{con}}\right]^{2/3} \quad . \tag{24.6}$$

Therefore one usually neglects this correction in the case of effective convection and takes simply

$$\nabla = \nabla_{ad} \quad , \tag{24.7}$$

which is equivalent to assuming an infinite conductivity $k'_{con}$. Then de facto the luminosity is decoupled from the $T$–$P$ structure.

In order to fix the luminosity of a fully convective star, we have to appeal to the only region where the gradient is sufficiently non-adiabatic. This is the radiative atmosphere and a layer immediately below where the convection is ineffective, i.e. strongly superadiabatic. We have seen that then the transport of energy is essentially radiative (in spite of violent convective motions), and we can again use (24.1). By this argumentation one arrives at the statement that the structure of the outermost layers determines the luminosity of a fully convective star (which reminds of the situation in white dwarfs models). This means, on the other hand, that such stars are very sensitive to all influences and uncertainties near their outer boundary.

Of course, if the energy production is prescribed, one would rather say that the outer layers have to adjust to this value of $L$ (for this point of view, see § 24.5).

### 24.2 A Simple Description of the Hayashi Line

In order to derive some typical properties of the HL analytically, we shall use an extremely crude model for fully convective stars. (Further refinements of the picture, though possible, would not be worth the large additional complications involved.)

We have seen that nearly all of the interior part of convective stars has an adiabatic stratification, such that $d \ln T / d \ln P = \nabla_{ad}$. We shall assume that this simple relation between $P$ and $T$ holds for the whole interior up to the photosphere, i.e. we neglect the superadiabaticity in the range immediately below the photosphere. We also neglect the depression of $\nabla_{ad}$ in those regions near the surface where H and He are partially ionized (see Figs. 10.2 and 14.1). We thus simply assume $\nabla_{ad}$ to be constant throughout the star's interior, say $\nabla_{ad} = 0.4$, which is the value for a fully ionized ideal gas. With these simplifications we certainly introduce errors in the $P$–$T$ stratification. However, they will be nearly the same for neighbouring models and we can hope to obtain at least the correct differential behaviour.

We then have for the whole interior the simple $P$–$T$ relation

$$P = CT^{1+n} \quad , \tag{24.8}$$

i.e. the star is polytropic with an index $n = 1/\nabla_{ad} - 1 = 3/2$ and we can use the earlier results for such stars (see § 19). The constant $C$ is related to the polytropic constant $K$ defined in (19.3). With $P = \Re \varrho T/\mu$, one finds $C = K^{-n}(\Re/\mu)^{1+n}$. $K$ and $C$ are constant only within one model, but vary from star to star, which means that we do not have a mass–radius relation. From (19.9,19) it follows that

$$K \sim \varrho_c^{1/3} A^{-2} \sim \varrho_c^{1/3} R^2 \sim M^{1/3} R \quad , \tag{24.9}$$

so that

$$C = C' R^{-3/2} M^{-1/2} \quad , \tag{24.10}$$

where the constant $C'$ is known for given $n$ and $\mu$.

Relation (24.8) is now assumed to hold as far as the photosphere, where the optical depth $\tau = 2/3$, $P = P_0$, $T = T_{eff}$, $r = R$, and $m = M$. Above this point we suppose to have a radiative atmosphere with a simple absorption law of the form

$$\kappa = \kappa_0 P^a T^b \quad . \tag{24.11}$$

Integration of the hydrostatic equation through the atmosphere yields the photospheric pressure [cf. (10.13), where $\bar{\kappa}$ is replaced by (24.11)] as

$$P_0 = \text{constant} \left( \frac{M}{R^2} T_{\text{eff}}^{-b} \right)^{\frac{1}{a+1}} \quad . \tag{24.12}$$

We now fit this to the interior solution by setting $P = P_0$, $T = T_{\text{eff}}$ in (24.8) and then eliminating $P_0$ with (24.12). For given values of $M$ and $\mu$ this yields a relation between $R$ and $T_{\text{eff}}$, or between $R$ and $L$, since $L \sim R^2 T_{\text{eff}}^4$. Thus any value of $R$ corresponds to a certain point in the Hertzsprung–Russell diagram. The interior solutions form a one-dimensional manifold, since the constant $C$ contains the free parameter $R$ for given $M$ [and given $\mu$, see (24.10)]. In the Hertzsprung–Russell diagram this is reflected by a one-dimensional manifold of points defining the Hayashi line.

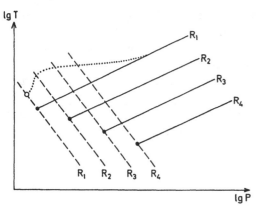

Fig. 24.1. Fit of a polytropic ($n = 3/2$) interior solution (*solid line*) with an atmospheric condition (*dashed line*) for different values of $R$ ($R_1 > R_2 > R_3 > R_4$). The photospheric points obtained by this fit are marked by dots. The dotted line illustrates schematically the effects of superadiabatic convection and depression of $\nabla_{\text{ad}}$ in an ionization zone for $R = R_1$

The fitting procedure is illustrated in Fig. 24.1. Each interior solution of the form (24.8) with $n = 3/2$ is represented in this diagram by a straight line

$$\lg T = 0.4 \lg P + 0.4 \left( \frac{3}{2} \lg R + \frac{1}{2} \lg M - \lg C' \right) \quad . \tag{24.13}$$

For fixed values of $M$ and $\mu$, each of these lines is characterized by a value of $R$. The atmospheric solutions (24.12) are another set of straight lines in Fig. 24.1:

$$(a + 1) \lg P_0 = \lg M - 2 \lg R - b \lg T_{\text{eff}} + \text{constant} \tag{24.14}$$

The intersection of a line of the first set with a line of the second set, both with the same value of $R$, fixes the corresponding value of $T_{\text{eff}}$ (and of $P_0$). From $R$ and $T_{\text{eff}}$ we have $L$, i.e. a point in the Hertzsprung–Russell diagram. We then obtain the Hayashi line by a continuous variation of $R$.

The formalism for this procedure, as described, yields immediately an equation for the Hayashi line in the Hertzsprung–Russell diagram:

$$\lg T_{\text{eff}} = A \lg L + B \lg M + \text{constant} \tag{24.15}$$

with the coefficients

$$A = \frac{0.75a - 0.25}{b + 5.5a + 1.5} \quad , \quad B = \frac{0.5a + 1.5}{b + 5.5a + 1.5} \quad . \tag{24.16}$$

We now need typical values for the exponents $a$ and $b$ in the atmospheric absorption law (24.11). An important property of fully convective stars can immediately be concluded from the discussion in § 10.3: such stars must have very low values of $T_{eff}$, i.e. *the Hayashi line must be far to the right in the Hertzsprung–Russell diagram.* For atmospheres this means that in most parts $T \lesssim 5 \times 10^3$ K, and $H^-$ absorption will provide the dominant contribution to $\kappa$. If hydrogen is essentially neutral, the free electrons necessary for the formation of $H^-$ ions are provided by the heavier elements (see § 17.5). A very rough interpolation gives $a \simeq 1$, $b \simeq 3$. With these values (24.16) yields the coefficients

$$A = 0.05 \quad , \quad B = 0.2 \quad . \tag{24.17}$$

According to (24.15), the slope of the Hayashi line in the Hertzsprung–Russell diagram is $\partial \lg L / \partial \lg T_{eff} = 1/A$. Since $A \ll 1$, we conclude that *the Hayashi line must be very steep.* The value of $B \equiv \partial \lg T_{eff} / \partial \lg M$ means that *the Hayashi line shifts slightly to the left in the Hertzsprung–Russell diagram for increasing $M$.* These qualitative predictions, although derived from very crude assumptions, are fully supported by the numerical results.

Let us consider once more the reason for the steepness of the HL. At the photosphere the pressures $P_{0i}$ of the interior solution (24.8, 24.10) and $P_{0a}$ of the atmospheric solution (24.12) vary for constant $M$ as

$$P_{0i} \sim \frac{T_{eff}^{2.5}}{R^{3/2}} \quad , \quad P_{0a} \sim \frac{T_{eff}^{-\frac{b}{a+1}}}{R^{\frac{2}{a+1}}} \quad . \tag{24.18}$$

First of all, we expect a very steep HL for small positive values of $a$. In fact, for $a = 1/3$, $P_{0i}$ and $P_{0a}$ have the same dependence on $R$; then $T_{eff}$ does not vary with $R$ (and $L$), and the line is vertical. If this is not quite fulfilled, the fit $P_{0i} = P_{0a}$ requires the smaller variations of $T_{eff}$ with varying $R$, the more different the two exponents of $T_{eff}$ in (24.18) are, i.e. the larger $b$.

The basic approximations made were to neglect the depression of $\nabla_{ad}$ in ionization zones and to ignore superadiabatic convection. The dotted line in Fig. 24.1 indicates how these effects change the $P - T$ structure relative to a simple polytrope. One sees that they tend to increase the effective temperature. The precise value of $T_{eff}$ obviously depends on the detailed structure of the outermost envelope. The extension and the depth of the ionization zones and the superadiabatic layers change systematically with $L$. This has the consequence that, in better approximations, the coefficient $A$ in (24.15) changes sign at $L \simeq L_\odot$. It is positive for smaller $L$, and negative for larger $L$, so that the HL is convex relative to the main sequence.

Another important conclusion is that the whole uncertainty which remained in the mixing-length theory of ineffective convection must occur as a corresponding uncertainty in the precise value of $T_{eff}$ for the HL.

Finally, we note that the chemical composition enters into the position of the HL in two ways. The interior is affected, since the polytropic constant $C$ depends on $\mu$ via $C'$ [see (24.10)], and the outer layers are particularly affected via the opacity $\kappa$.

## 24.3 The Neighbourhood of the Hayashi Line and the Forbidden Region

We now consider stars in hydrostatic equilibrium that are close to, but not exactly on, their HL. Certainly the stars cannot be fully convective with an adiabatic interior (otherwise they would be on the HL). Their interior is then no longer a simple polytrope. They do not even have to be chemically homogeneous, since they are not fully mixed by the turbulent motions. We must therefore expect that an analytical treatment will be much more complicated. We will nevertheless try to give some simple arguments which may help to make the numerical results plausible. In the following, we treat models with a fixed value of $M$ and the same chemical composition (at least in their outer layers).

An important indication can be obtained from the discussion of the envelope integrations in § 10.3. When integrating inwards into models with different $T_{\text{eff}}$ (but with the same parameters $M$ and $\mu$ and, say, the same $L$), we will reach a radiative region the earlier, the larger $T_{\text{eff}}$. In other words, in models left of the HL we will encounter a radiative region before reaching the centre. In these regions, the gradient $\nabla < \nabla_{\text{ad}}$. Let us consider some average $\bar{\nabla}$ obtained by averaging over the whole interior (where we again neglect the complications in the outermost parts of the envelope). On the HL we have $\bar{\nabla} = \nabla_{\text{ad}}$. In a model to the left of the HL the radiative part decreases the average value such that $\bar{\nabla} < \nabla_{\text{ad}}$. This suggests that we would have to allow $\bar{\nabla} > \nabla_{\text{ad}}$ in models to the right of the HL.

In order to prove this we treat models with a constant gradient $\nabla = \bar{\nabla}$ in the interior and vary $\bar{\nabla}$ slightly around $\nabla_{\text{ad}}$. We then have again polytropic stars with slightly different $n$ (around 3/2). The interior solution is written as

$$P = C_n T^{1+n} \quad , \tag{24.19}$$

where $\bar{\nabla} = (1 + n)^{-1}$ and, similarly to (24.10),

$$C_n = C'_n \mu^{-n-1} M^{1-n} R^{n-3} \quad . \tag{24.20}$$

From now on we measure $R$ and $M$ in solar units. Then

$$C'_n = \frac{\Re^{n+1}}{4\pi G^n}(n+1)^n \left[ -\left(\frac{dw}{dz}\right)_{z=z_n} \right]^{n-1} z_n^{n+1} R_{\odot}^{n-3} M_{\odot}^{1-n} \quad . \tag{24.21}$$

We extend relation (24.19) to the photosphere ($P = P_0$, $T = T_{\text{eff}}$), where we again eliminate $P_0$ by (24.12) and $R$ by the relation $R = c_2 L^{1/2} T_{\text{eff}}^{-2}$. This gives the locus in the Hertzsprung–Russell diagram. The factor of proportionality in (24.12) may be called $c_1$. Choosing for simplicity $a = 1$, $b = 3$ in the opacity law, we obtain

$$\lg T_{\text{eff}} = \alpha_1 \lg L + \alpha_2 \lg M + \alpha_3 \lg \mu + \alpha_4 \lg C'_n + \alpha_5 \lg c_1 + \alpha_6 \lg c_2 \, , \tag{24.22}$$

where the coefficients depend on $n$:

$$\alpha_1 = \frac{2-n}{13-2n} \quad , \quad \alpha_2 = \frac{2n-1}{13-2n} \quad , \quad \alpha_3 = \frac{2(1+n)}{13-2n} \quad ,$$

$$\alpha_4 = \frac{-2}{13-2n} \quad , \quad \alpha_5 = -\alpha_4 \quad , \quad \alpha_6 = 2\alpha_1 \quad . \tag{24.23}$$

The $\alpha_i$ do not vary too much with small deviations of $n$ from 3/2. This means, for example, since $\alpha_1$ determines the slope, lines of neighbouring values of $n$ are nearly parallel to the HL. Without loss of generality, we may consider particular models on and close to the HL with $L = M = \mu = 1$. The variation of $\lg T_{\text{eff}}$ with $n$ is then only due to the variation of the last three terms in (24.22). One finds that $\partial \lg T_{\text{eff}} / \partial n > 0$: the stars move to the right in the Hertzsprung–Russell diagram with decreasing $n$ (i.e. increasing $\bar{\nabla}$).

Thus we have to expect the following situation (see Fig. 24.2): left of the HL we have $\bar{\nabla} < \nabla_{\text{ad}}$ and some part of the model is radiative. On the HL, the model is fully convective with $\bar{\nabla} = \nabla_{\text{ad}}$. Models to the right of the HL should have $\bar{\nabla} > \nabla_{\text{ad}}$, which means that they should have a superadiabatic stratification in their very interior (aside from the outermost zone of ineffective convection).

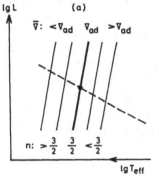
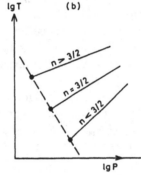

Fig. 24.2. (a) In the Hertzsprung–Russell diagram, the Hayashi line ($n = 3/2$, *heavy line*) is indicated, together with some neighbouring lines for interior polytropes with $n > 3/2$ and $< 3/2$. (b) The same as Fig. 24.1, but with three different polytropic interior solutions for the same value of $R$

The mixing-length theory has shown that a negligibly small excess of $\nabla$ over $\nabla_{\text{ad}}$ suffices in order to transport any reasonable luminosity in the deep interior of stars. Then, what happens with a star that by some arbitrary means (e.g. initial conditions) has been brought to a place to the right of the HL, such that some region in its deep interior has remarkably large values of $\nabla - \nabla_{\text{ad}} > 0$? The results are large convective velocities $v_{\text{conv}} \sim (\nabla - \nabla_{\text{ad}})^{1/2}$ and corresponding convective fluxes [cf. (24.3)]. These cool the interior and heat the upper layers rapidly until the gradient is lowered to $\nabla \approx \nabla_{\text{ad}}$ and the star has moved to the HL. This will happen within the short time-scale for the adjustment of convection.

Another possibility for a star being situated to the right of its HL is, of course, that it is not in hydrostatic equilibrium (which is assumed for the interior solution). But a deviation from this equilibrium will be removed in the time-scale for hydrostatic adjustment, which is even shorter.

Therefore the HL is in fact a borderline between an "allowed" region (left) and a "forbidden" region (right) for stars of given $M$ and composition that are in hydrostatic equilibrium and have a fully adjusted convection.

## 24.4 Numerical Results

There are many results available giving the position of Hayashi lines for stars of widely ranging mass and chemical composition, and for different assumptions in the convection theory. The latter concerns in particular the ratio of mixing length to pressure scale height used for calculating the superadiabatic envelope.

Figure 24.3 shows typical results of calculations for stellar masses in the range $M = 0.5 \ldots 10 \, M_\odot$. One sees that indeed the HLs plotted here are very steep, the exact slope depending mainly on $L$. The dependence on $M$ is roughly given by $\partial \lg T_{\rm eff} / \partial \lg M \approx 0.15$, i.e. we find the expected weak increase of $T_{\rm eff}$ with $M$ [cf. (24.22)]. The HLs are far away from the main-sequence in the upper part of the diagram, and approach it in the lower part. This fact will turn out to influence the evolutionary tracks of stars of different $M$. Recall that the main sequence stars were found to be fully convective for $M \lesssim 0.25 M_\odot$ (see § 22.3). This obviously means that the corresponding Hayashi lines cross the main sequence there.

As mentioned earlier the chemical composition enters in several ways. A very important factor certainly is the opacity in the atmosphere. For $T_{\rm eff} \lesssim 5000$ K the dominant absorption is due to $H^-$, and $\kappa$ then is proportional to the electron pressure, which in turn is proportional to the abundance of the easily ionized metals. It turns out that a *decrease* of their abundance (usually comprised in $X_{\rm rest}$) by a factor 10 shifts the HL by $\Delta \lg T_{\rm eff} \approx +0.05$ to the left in the Hertzsprung–Russell diagram. However, Fig. 24.4 shows that roughly the same shift can be obtained by the comparatively small increase of $\ell_{\rm m}/H_P$ from 1 to 1.5. The uncertainty of the convection theory, therefore, severely limits our knowledge of the HL.

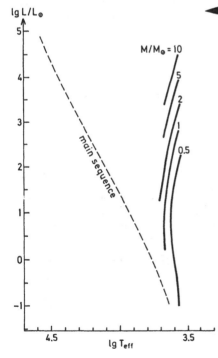

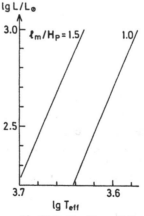

◄ **Fig. 24.3.** The position of Hayashi lines for $M = 0.5 \ldots 10 M_\odot$, for a composition with $X_{\rm H} = 0.739$, $X_{\rm He} = 0.24$, $X_{\rm rest} = 0.021$, and $\ell_{\rm m}/H_P = 2$ (after EZER, CAMERON, 1967). A main sequence is plotted (*dashed*) for comparison

**Fig. 24.4.** The Hayashi line for $M = 5 M_\odot$ with two different assumptions for the ratio of mixing length to pressure-scale height. (After HENYEY et al., 1965)

231

## 24.5 Limitations for Fully Convective Models

In order to describe the HL, we have considered models for which the convection was postulated to reach from centre to surface. This provided a polytropic interior structure with typical decoupling from the luminosity. We have not yet asked whether the physical situation will in fact allow the onset of convection throughout the star. This depends on the distribution of the energy sources.

According to the Schwarzschild criterion (6.13), a chemically homogeneous layer will be convective if

$$\nabla_{\text{rad}} \geq \nabla_{\text{ad}} \quad , \tag{24.24}$$

where the radiative gradient [see (5.28)] is

$$\nabla_{\text{rad}} \sim \frac{\kappa l P}{T^4 m} \quad . \tag{24.25}$$

If the energy sources were completely arbitrary, we could choose their distribution so that (24.24) is violated at some point and the model could not be *fully* convective. A trivial example would be a central core without any sources, with the result that there $l = 0$, i.e. $\nabla_{\text{rad}} = 0$. Then the core must be radiative. On the other hand, we have the best chance of finding convection throughout a star of given $L$ if the sources are highly concentrated towards the centre (in the extreme: a point source), which gives almost $l = L$ everywhere.

We consider a contracting polytrope (see § 20.3) without nuclear energy sources, which is of interest for early stellar evolution. According to (20.41) the energy generation rate is then only proportional to $T$, which means a rather weak central concentration. For the sake of simplicity we even go a step further and assume constant energy sources with

$$\frac{l}{m} = \frac{L}{M} = \text{constant} \quad . \tag{24.26}$$

We again use the opacity law (24.11) and the polytropic relation (24.8) with $n = 1.5$ (corresponding to $\nabla = \nabla_{\text{ad}} = 0.4$). Equation (24.25) then gives

$$\nabla_{\text{rad}} \sim \frac{L}{M} C^{1+a} T^{b-4+2.5(1+a)} \quad . \tag{24.27}$$

For a typical Kramers opacity with $a = 1$, $b = -4.5$ this becomes $\nabla_{\text{rad}} \sim T^{-3.5}$. Indeed, for all reasonable interior opacities, $\nabla_{\text{rad}}$ has a minimum at the centre and increases outwards. Therefore the centre is the first point in a fully convective star where $\nabla_{\text{rad}}$ drops below $\nabla_{\text{ad}}$ (and a radiative region starts to develop) if $L$ decreases below a minimum value $L_{\text{min}}$.

The constant $C$ depends on $M$ and $R$ as given by (24.10), and $T \sim T_{\text{c}} \sim M/R$ after (20.24). Introducing this into (24.27) we obtain

$$\nabla_{\text{rad}} \sim L M^{b-5+2(1+a)} R^{-b+4-4(1+a)} \quad . \tag{24.28}$$

Let us again set $a = 1$, $b = -4.5$, which gives

$$\nabla_{\text{rad}} \sim L M^{-5.5} R^{0.5} \quad .$$ (24.29)

For models on the HL, the effective temperatures vary only a very little and we simply take $R \sim L^{1/2}$. Then,

$$\nabla_{\text{rad}} \sim L^{1.25} M^{-5.5} \quad .$$ (24.30)

For any given value of $M$ the luminosity reaches $L_{\text{min}}$ if the central value of $\nabla_{\text{rad}}$ has dropped to 0.4. According to (24.30), $L_{\text{min}}$ depends on $M$ as

$$L_{\text{min}} \sim M^{4.4} \quad .$$ (24.31)

This minimum luminosity (down to which models of the specified type on the HL remain fully convective) decreases strongly with $M$. The decrease is in fact steeper than that given by the $M-L$ relation of the main sequence. This provides the possibility that the HL for very small $M$ can cross the main sequence without reaching $L_{\text{min}}$.

Note, however, that strictly speaking a "minimum luminosity" always refers to a fixed distribution of the energy sources.

# § 25 Stability Considerations

Even the most beautiful stellar model is not worth anything if one does not know whether it is stable or not. Stability is discussed again and again throughout this book. Here we review the different types of stability considerations necessary for stars. We intend to make the basic mechanisms and concepts plausible rather than present the full formalism; the reader will find this, for example, in the review article by LEDOUX (1958).

## 25.1 General Remarks

It is not easy to give a very general concept of stability that is applicable to all possible cases. Different definitions are discussed in LA SALLE, LEFSCHETZ (1961). We may use for example the following: let the solution of a system of (time-dependent) differential equations be a set of functions $y_1(t)$, $y_2(t)$,... which we comprise in the symbol $y(t)$. We define a "distance" between two such solutions $y^a(t)$, $y^b(t)$ by

$$\|y^a(t) - y^b(t)\| := \sum_i \left[ \left( y_i^a(t) - y_i^b(t) \right)^2 \right] \quad . \tag{25.1}$$

We then call the solution $y^a(t)$ stable at $t = t_0$ if for any $t_1 > t_0$ and for any small positive number $\varepsilon$ there exists a small positive number $\delta$ such that any other solution $y^b(t)$ having the distance $\|y^a(t_0) - y^b(t_0)\| < \delta$ at $t = t_0$ will keep a distance $\|y^a(t_1) - y^b(t_1)\| < \varepsilon$.

This definition in plain words says that a solution is stable at a given point $t_0$ if all solutions that at $t = t_0$ are in its neighbourhood remain neighbouring solutions. The problems we are interested in can be reduced to first-order systems in time. Therefore the above definition of neighbouring solutions also guarantees neighbouring derivatives.

One normally is familiar with stability problems in mechanics. We recall a few simple examples, the first being the freely rolling ball on a curved surface which is concave in the direction opposite to gravity (see Fig. 25.1a). One solution is that of equilibrium, where the ball rests in the lowest position. The initially neighbouring position is obtained by a small perturbation, say by a slight horizontal displacement. The ball will then move about the equilibrium position, but it will never increase its distance above its initial value: the equilibrium position is stable and friction would merely restore the ball to its equilibrium position. In the case of a convex surface (see Fig. 25.1b) the equilibrium is unstable, since after a small displacement the ball will move further and further from the equilibrium position. While these examples

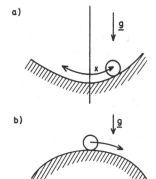

a)

b)

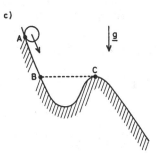

c)

**Fig. 25.1 a–c.** An example of stability in mechanics. A ball on a surface under the influence of gravity (a) in stable and (b) in unstable equilibrium. In (c) the motion starting at point $A$ is stable, but, starting with zero velocity at point $B$, the motion is unstable

deal with the stability of an equilibrium in which the solution is time-independent, our general definition also concerns time-dependent solutions. The motion of a ball rolling on the surface in Fig. 25.1c can be stable or unstable. The motion is stable if it starts with zero velocity at a point $A$ *above* $B$ (non-periodic motion), or *below* $B$ (periodic motion). But a motion starting exactly at $B$ with zero velocity and ending at rest at $C$ is unstable: a slight perturbation of the initial conditions can either produce a periodic motion (the ball never overcomes the summit $C$) or cause the ball to roll beyond $C$ and never come back.

When considering the influence of friction, one may naïvely expect that it stabilizes an otherwise unstable motion, since it uses up energy. But the following example will show that friction can also produce instability.

We again consider the ball in the spherical bowl (Fig. 25.1a). But now we assume that the bowl is rotating with an angular velocity $\omega$ around a vertical axis through the minimum. Without friction no angular momentum can be transferred to the ball which therefore does not know anything of rotation and behaves as in the non-rotating case: the lowest position is stable. If there is friction, however, and the ball is "kicked" out of its lowest (equilibrium) position, it will take up angular momentum from the rotating bowl. For sufficiently large $\omega$ the ball goes to a new equilibrium position outside the axis around which it rotates with $\omega$ and where the tangential components of centrifugal and gravity forces balance each other. The lowest position has obviously become unstable by the inclusion of friction.

## 25.2 Stability of the Piston Model

Closer to stars than the above mechanical examples is the piston model introduced in §2.7, since it also incorporates thermal effects. We consider the stability of an equilibrium solution with a certain constant height $h$. Will a solution originating from a small displacement of the piston remain in its neighbourhood? This stability problem has already been discussed in §6.6, where we made approximations appropriate for the illustration of the stability of convective blobs. We now improve the model by adding some complications typical of stars.

### 25.2.1 Dynamical Stability

In this case one assumes that there is no heat leakage, no nuclear energy generation, and no absorption, i.e. $\varepsilon = \kappa = \chi = 0$ in (5.39). Therefore the entropy of the gas remains constant during the displacement of the piston. In §6.6, we investigated the resulting (adiabatic) oscillations of the model around the equilibrium position, though with constant weight $G^*$ only. We now allow $G^*$ to vary with height $[G^* = G^*(h)]$ as we did in §3.2. This can be achieved, for instance, by putting the piston model into an inhomogeneous gravitational field. Then the equation of motion (2.34)

$$M^* \frac{d^2 h}{dt^2} = -G^* + PA \tag{25.2}$$

with the perturbations (6.30) gives after linearization, instead of (6.32),

$$M^* h_0 \omega^2 x + P_0 A p - G_h^* G_0^* x = 0 \quad . \tag{25.3}$$

Here $G_h^* := d \ln G^* / d \ln h \; (< 0)$, while $G_0^* = P_0 A = g_0 M^*$ is the equilibrium value of $G^*$ and $g_0$ is that of $g$. With the perturbed ideal gas equation (6.31) we find

$$\left[ \frac{\omega^2 h_0}{g_0} - G_h^* - 1 \right] x + \vartheta = 0 \quad . \tag{25.4}$$

This together with the adiabatic equation (6.36),

$$(\gamma_{\text{ad}} - 1)x + \vartheta = 0 \quad , \tag{25.5}$$

gives for the eigenvalues of adiabatic oscillations $\omega = +\omega_{\text{ad}}$ and $\omega = -\omega_{\text{ad}}$ with

$$\omega_{\text{ad}} = \left[ (\gamma_{\text{ad}} + G_h^*) \frac{g_0}{h_0} \right]^{1/2} \quad , \tag{25.6}$$

which replaces (6.37). Recall that the perturbation changes with time as $e^{i\omega t}$. We see that $\omega_{\text{ad}}$ is a real number only as long as $\gamma_{\text{ad}} > -G_h^*$. In this case the small perturbation is followed by a periodic oscillation which remains small for all times. It is therefore stable in the sense of our definition of stability at the beginning of this paragraph. But if $\gamma_{\text{ad}} < -G_h^*$, then $\omega_{\text{ad}}$ is imaginary and one of the eigenvalues $\omega$ gives an amplitude growing exponentially in time: the equilibrium solution is unstable. (We will see in §25.3.2 that for stars the analogue of $\gamma_{\text{ad}} > -G_h^*$ is $\gamma_{\text{ad}} > 4/3$).

### 25.2.2 Inclusion of Non-adiabatic Effects

We now drop the assumption of strict adiabaticity. Non-adiabatic changes were previously included in §5.4 (refer also to the last part of §6.6). The energy equation of the piston model (5.39) includes the non-adiabatic terms for nuclear generation $\varepsilon$, absorption $\kappa$, and heat leakage $\chi$. We consider $\varepsilon$ and $\kappa$ as functions of $P$ and $T$, while $\chi$ shall be constant. In the case of thermal equilibrium (vanishing time derivatives) we have

$$\varepsilon_0 m^* + \kappa_0 m^* F = \chi(T_0 - T_s) \quad , \tag{25.7}$$

where subscript 0 indicates the equilibrium and subscript s the surroundings. If we perturb this equilibrium according to (6.30), we find for the perturbations after linearization

$$i\omega(c_v m^* T_0 \vartheta + P_0 A h_0 x)$$
$$= \varepsilon_0 m^*(p\varepsilon_P + \vartheta\varepsilon_T) + \kappa_0 m^* F(p\kappa_P + \vartheta\kappa_T) - \chi T_0 \vartheta \quad , \tag{25.8}$$

where the derivatives

$$\varepsilon_P = \left(\frac{\partial \ln \varepsilon}{\partial \ln P}\right)_T \quad , \quad \varepsilon_T = \left(\frac{\partial \ln \varepsilon}{\partial \ln T}\right)_P \quad ,$$

$$\kappa_P = \left(\frac{\partial \ln \kappa}{\partial \ln P}\right)_T \quad , \quad \kappa_T = \left(\frac{\partial \ln \kappa}{\partial \ln T}\right)_P \tag{25.9}$$

are taken at the values $P_0$, $T_0$.

The equation of motion (25.2) yielded (25.4) for which we now assume constant weight of the piston ($G_h^* = 0$, giving dynamical stability):

$$\left(\frac{\omega^2 h_0}{g_0} - 1\right) x + \vartheta = 0 \quad . \tag{25.10}$$

Since $\varrho \sim h^{-1}$, the equation of state for an ideal gas gives

$$p = \vartheta - x \quad . \tag{25.11}$$

System (25.8,10,11) comprises three linear homogeneous algebraic equations for the perturbations $p$, $\vartheta$, $x$. To find a solution it is necessary that the determinant of the coefficients vanishes:

$$\frac{h_0}{g_0} i u_0 \omega^3 - \frac{h_0}{g_0}(e_P + e_T)\,\omega^2 - \frac{5}{3}u_0 i\omega + e_T = 0 \tag{25.12}$$

with

$$e_P = \varepsilon_0 \varepsilon_P + \kappa_0 F \kappa_P \quad , \quad e_T = \varepsilon_0 \varepsilon_T + \kappa_0 F \kappa_T - \frac{\chi T_0}{m^*} \quad , \quad u_0 = c_v T_0 \ , \tag{25.13}$$

where for the last relation we have assumed the gas to be ideal and monatomic. (Note that $P_0 A h_0 / m^* = P_0/\varrho_0 = 2u_0/3$.) Equation (25.12) becomes one with *real* coefficients if instead of $\omega$ we use the eigenvalue $\sigma := i\omega$,

$$\frac{h_0}{g_0} u_0 \sigma^3 - \frac{h_0}{g_0}(e_P + e_T)\,\sigma^2 + \frac{5}{3}u_0\sigma - e_T = 0 \quad . \tag{25.14}$$

This is a third order equation for the eigenvalue $\sigma$ (or $\omega$). While in the adiabatic case ($e_P = e_T = 0$) we obtained two solutions $\sigma = \pm\sigma_{ad} = \pm i\omega_{ad}$ (where $\omega_{ad}$ was real), we now have *three* eigenvalues. If the non-adiabatic terms $e_P$, $e_T$ are small, we can expect that two (conjugate complex) eigenvalues lie near the adiabatic ones:

$$\sigma = \sigma_r \pm i\omega_{ad} \quad , \quad \omega_{ad} = \left(\gamma_{ad}\frac{g_0}{h_0}\right)^{1/2} \quad , \tag{25.15}$$

237

where $\sigma_r$ is real and $|\sigma_r| \ll \omega_{ad}$. While in the adiabatic case the oscillation was strictly periodic, the real part $\sigma_r$ causes the amplitude of the oscillation to grow or decrease in time, depending on the sign of $\sigma_r$. Because of $|\sigma_r| \ll \omega_{ad}$ these changes take place over a time much longer than the oscillation period, actually on a scale corresponding to $\tau_{adj}$ in (5.41). This type of stability behaviour is called the vibrational stability (compare § 6.6). If the oscillation grows in time, the solution leaves the neighbourhood of equilibrium, which therefore is unstable.

We now turn to the third root of (25.12) or (25.14), which occurs necessarily with the dissipative terms $e_P$, $e_T$. Instead of solving the third-order equation (25.14), we will follow some heuristic arguments. The addition of non-adiabatic terms has changed the rapid oscillations only to the extent that their amplitude varies on long time-scales (of the order of $\sigma_r^{-1}$). We now look for the existence of a third solution changing with this long time-scale only. Then the inertia terms can be neglected and, consequently, the terms with $\sigma^3$ and $\sigma^2$ disappear in (25.14). The solution of (25.14) for this so-called secular stability problem is

$$\sigma = \sigma_{sec} = i\omega_{sec} = \frac{3}{5}\frac{e_T}{u_0} \quad . \tag{25.16}$$

For sufficiently small nonadiabaticity $e_T$, we can achieve $|\sigma_{sec}| \ll \omega_{ad}$, and neglecting the $\sigma^2$ and $\sigma^3$ terms in (25.14) was justified. If $\sigma_{sec} < 0$, any perturbation will decay within a kind of thermal adjustment time $\tau_{adj} \approx \sigma_{sec}^{-1}$ and the equilibrium is secularly stable. But if $\sigma_{sec} > 0$, then it will grow on that time-scale (independently of vibrational stability): The equilibrium is secularly unstable.

We have now found the three well-known types of stability behaviour: dynamical, vibrational, and secular stability. This classification is possible since $|\omega_{ad}| \gg |\omega_{sec}|$, which is equivalent to saying that $\tau_{hydr} \ll \tau_{adj}$. From one type of stability one cannot draw any conclusions about the behaviour of another type, e.g. a dynamically stable model can still be vibrationally or secularly unstable. If the model were dynamically unstable, the other instabilities would be of no interest since the model would move out of equilibrium long before any other instability can develop.

We will find more or less the same behaviour in stars where also $\tau_{hydr} \ll \tau_{adj} \approx \tau_{KH}$. However, there we cannot solve the eigenvalue problem analytically any more. This is the reason why we dwelt in such length on the stability of the piston model.

### 25.3 Stellar Stability

For the problem of *stellar* stability a very general definition, like that given at the beginning of § 25.1, has to be taken with care. For example, a star may be stable in one phase (e.g. on the main sequence) and later on become unstable (e.g. in the cepheid phase). At any stage of evolution the solution (the stellar model) is obtained for certain parameters, for instance a certain chemical composition or a certain distribution of entropy. It is reasonable to ask whether this solution is stable in the following sense: Does a small perturbation decay rapidly compared to the change of the parameters of the model (for example its chemical composition)? Then we would call the model stable. Therefore, the question of the cepheid stability is irrelevant for the stability of its main sequence progenitor since the chemical composition is different. The solution for a certain phase of evolution, in general, is obtained by solving approximate equations. For example, complete equilibrium is assumed in the case of the main sequence, while only the inertia terms are dropped for the evolution through the cepheid phase. If such approximate models approach an instability in the run of their evolution, the neglected time derivatives become important and have to be taken into account. In general, then, the solution obtained from better approximations tells us in which direction the evolution really goes.

The problem of stellar stability is closely connected to that of local uniqueness and to certain properties of linear series of models. This has already been discussed in § 12.4.

### 25.3.1 Perturbation Equations

We want to investigate the stability of a stellar model in complete equilibrium for given input parameters $M$ and chemical composition. Let the model be described by $r_0(m)$, $P_0(m)$, $T_0(m)$, $l_0(m)$ which solve the time independent stellar structure equations. We test its stability by investigating how a neighbouring (perturbed) solution evolves in time. We here restrict ourselves to spherically symmetric perturbations which depend on $m$ and $t$ in such a way that the perturbed variables become

$$r(m,t) = r_0(m)\left[1 + x(m)\,e^{i\omega t}\right] \quad ,$$

$$P(m,t) = P_0(m)\left[1 + p(m)\,e^{i\omega t}\right] \quad ,$$

$$T(m,t) = T_0(m)\left[1 + \vartheta(m)\,e^{i\omega t}\right] \quad ,$$

$$l(m,t) = l_0(m)\left[1 + \lambda(m)\,e^{i\omega t}\right] \quad , \tag{25.17}$$

where the absolute values of $x$, $p$, $\vartheta$ and $\lambda$ are $\ll 1$. These variables have to fulfill the time-dependent equations (9.1–4). As an example let us introduce (25.17) into the equation of motion (9.2). If we linearize with respect to $p$ and $x$, this becomes

$$P_0'\left(1 + p\,e^{i\omega t}\right) + P_0 p'\,e^{i\omega t}$$
$$= -\frac{Gm}{4\pi r_0^4}\left(1 - 4x\,e^{i\omega t}\right) + \frac{\omega^2}{4\pi r_0}x\,e^{i\omega t} \quad , \tag{25.18}$$

where primes indicate derivatives with respect to $m$. Since $P_0$, $r_0$ obey (9.16), we have $P_0' = -Gm/(4\pi r_0^4)$: The time independent terms in (25.18) cancel each other, the exponentials drop out and we are left with (25.19). By a similar procedure, we find for the case of a radiative layer and an equation of state of the form $\varrho \sim P^\alpha T^{-\delta}$ from (9.1,3,4) the equations (25.20–22):

$$p' = -\frac{P_0'}{P_0}\left[p + \left(4 + \frac{r_0^3}{Gm}\omega^2\right)x\right] \quad , \tag{25.19}$$

$$x' = -\frac{1}{4\pi r_0^3\varrho_0}(3x + \alpha p - \delta\vartheta) \quad . \tag{25.20}$$

$$\lambda' = -\frac{\varepsilon_0}{l_0}(\lambda - \varepsilon_P\, p - \varepsilon_T \vartheta) - i\omega \frac{P_0 \delta}{l_0 \varrho_0}\left(\frac{\vartheta}{\nabla_{ad}} - p\right) \quad , \tag{25.21}$$

$$\vartheta' = \frac{P_0'}{P_0}\nabla_{rad}\left[\kappa_P p + (\kappa_T - 4)\vartheta + \lambda - 4x\right] \quad . \tag{25.22}$$

Equations (25.19–22) are four linear homogeneous differential equations of first order for the variables $p$, $\vartheta$, $x$, $\lambda$ which have to obey certain boundary conditions corresponding to those of the unperturbed solutions. They have to be regular in the centre and to be fitted to an atmosphere. We will deal with the boundary conditions in §38 and §39, where they are shown to be equivalent to four linear homogeneous equations. Therefore, solutions exist only for certain *eigenvalues* of $\omega$, which have to be found numerically. There exists an infinite number of eigenvalues for which the system can be solved. For each eigenvalue $\omega^*$ one obtains a set of *eigenfunctions* $p^*(m)$, $\vartheta^*(m)$, $x^*(m)$, $\lambda^*(m)$.

The term with $\omega^2(\sim \ddot{r})$ in (25.19) comes from the inertial terms in the equation of motion, while in (25.21) the term with $i\omega(\sim \dot{P}, \dot{T})$ is due to the time derivatives in the energy equation. The two corresponding time-scales are $\tau_{hydr}$ and $\tau_{adj} = \tau_{KH}$. Since $\tau_{hydr} \ll \tau_{KH}$, we have a situation similar to that described for the piston model in §25.2. Correspondingly, in general, we can speak of dynamical, vibrational and secular stability.

There are, however, more complicated cases where this classification of stability behaviour is not possible. For example, the relevant thermal time-scale may not be that of the whole star, but a much shorter one for a small subregion. If the characteristic wavelength of a thermal perturbation is short enough, the corresponding adjustment time can become comparable or shorter than $\tau_{hydr}$ (of the whole star). Another example is the case of a dynamically stable model which evolves in such a way that it approaches marginal stability ($\omega_{ad} \to 0$). Then the oscillations become so slow that they certainly will not be adiabatic anymore: $1/\omega_{ad} \gg \tau_{KH}$ (although $\tau_{hydr} \ll \tau_{KH}$ still).

### 25.3.2 Dynamical Stability

Since in §38 we will treat this problem thoroughly, we merely present some general results here. Instead of solving all four equations (25.19–22), one can consider oscillations taking place on the time-scale $\tau_{hydr}$. Since $\tau_{hydr} \ll \tau_{adj}$, the temperature of the matter changes almost adiabatically. Instead of solving (25.21,22) one just replaces $\vartheta$ by $p\nabla_{ad}$ in (25.20). Therefore (25.19,20) present two equations for $p$ and $x$ with the eigenvalue $\omega^2$. As we will see in §38 the eigenvalue problem is self-adjoint. Then there exists an infinite series of eigenvalues $\omega_n^2$ which are real. ($\omega_n$ is either real or purely imaginary). Therefore, they either correspond to periodic oscillations ($\omega_n^2 > 0$) or exponentially decreasing/ increasing solutions ($\omega_n^2 < 0$). The same behaviour was found for the adiabatic case of the piston model. But now, with an infinite number of eigenvalues, stability demands that for *all* eigenvalues $\omega_n^2 > 0$, while even a *single* eigenvalue with $\omega_n^2 < 0$ is sufficient for instability.

How a star behaves after it is adiabatically compressed or expanded depends on the numerical value of $\gamma_{ad}$. This can be most easily seen in the case of homologous changes. Let us consider a concentric sphere $r = r(m)$ in a star of hydrostatic equilibrium.

The pressure there is equal to the weight of the layers above a unit area of the sphere, as shown by integrating the hydrostatic equation:

$$P = \int_m^M \frac{Gm}{4\pi r^4} dm \quad .$$ 

(25.23)

We now compress the star artificially and assume the compression to be adiabatic and homologous. In general, after this procedure the star will no longer be in hydrostatic equilibrium.

If a prime indicates values after the compression, then homology demands that the right-hand side of (25.23) varies like $(R'/R)^{-4}$ [cf. (20.37)] where $R$ is the stellar radius, while adiabaticity *and* homology demand that the left-hand side varies as

$$(\varrho'/\varrho)^{\gamma_{ad}} = (R'/R)^{-3\gamma_{ad}}$$ 

(25.24)

according to (20.17). Therefore, if $\gamma_{ad} > 4/3$, the pressure on the left-hand side of (25.23) increases stronger with the contraction than the weight on the right: The resulting force is directed outwards and the star will move back towards equilibrium: it is dynamically stable.

For $\gamma_{ad} < 4/3$ the weight increases stronger than the pressure and the star would collapse after the initial compression (dynamical instability). For $\gamma_{ad} = 4/3$ the compression leads again to hydrostatic equilibrium: One has neutral equilibrium. The condition $\gamma_{ad} > 4/3$ corresponds to the dynamical stability condition $\gamma_{ad} > -G_h^*$ for the piston model (§ 25.2.1).

In § 38 we will see that $\gamma_{ad} = 4/3$ is also a critical value for nonhomologous perturbations. If $\gamma_{ad}$ is not constant within a star, for instance because of ionization, then marginal stability occurs if a certain mean value of $\gamma_{ad}$ over the star reaches the critical value 4/3.

It should be noted that radiation pressure can bring $\gamma_{ad}$ near the critical value 4/3 (see § 13.2). This is the reason why supermassive stars are in indifferent equilibrium, i.e. they are marginally stable (see § 19.10).

The critical value 4/3 depends strongly on spherical symmetry and Newtonian gravitation. The 4 in the numerator comes from the fact that the weight of the envelope in Newtonian mechanics varies as $\sim r^{-2}$ and has to be distributed over the surface of our sphere, giving another $r^{-2}$. The 3 in the denominator comes from the $r^3$ in the formula for the volume of a sphere. Therefore, effects of general relativity change the critical value (see § 36.2) of $\gamma_{ad}$ and make the models less stable. Since we have assumed spherical symmetry in deriving the critical value of $\gamma_{ad}$, rotation changes it, too. It can decrease the critical value of $\gamma_{ad}$ and make the models more stable.

### 25.3.3 Non-adiabatic Effects

The inclusion of nonadiabatic effects in a dynamically stable model brings us to the question of its vibrational and secular stability. (A dynamical instability makes a perturbation grow so rapidly that another possible instability of vibrational or secular type is irrelevant because of their much longer time-scales.) Vibrational stability means an oscillation with nearly adiabatic frequency but with slowly decreasing (stability) or increasing amplitude (instability). Such oscillations describe the behaviour of pulsating stars and therefore are treated in detail in § 39.

Secular (or thermal) stability is governed by thermal relaxation processes. In general, these proceed on time-scales long compared to $\tau_{hydr}$ and, therefore, the inertia terms in the equation of motion can be dropped. This means that the term $\sim \omega^2$ in (25.19) can be omitted. Equations (25.19–22) together with proper boundary conditions can then be solved, yielding an infinite number of secular eigenvalues $\omega_{sec}$. Normally they are purely imaginary (as in the case of the piston model). This is what one expects from a thermal relaxation process, such as in the problem of diffusion of heat. It is therefore all the more surprising that in certain cases a few complex eigenvalues occur (AIZENMAN, PERDANG, 1971). The oscillatory behaviour here comes from heat flowing back and forth between different regions in the star. (Obviously this could not occur in the single layer of the piston model). If instead of $\omega$ we again use $\sigma := i\omega$, the system (25.19–22) has real coefficients. Therefore the eigenvalues $\sigma$, if complex, appear in conjugate complex pairs. Again, the sign of the real part of $\sigma$ (the imaginary part of $\omega$) distinguishes between secular stability or instability.

The most important application of the secular problem to stellar evolution concerns the question whether a nuclear burning is stable or not. Secular instability in degenerate regions leads to the flash phenomenon, while in thin (nondegenerate) shell sources it results in quasiperiodic thermal pulses.

In order to make the secular stability of a central burning plausible, we treat a simple model of the central region, assuming homologous changes of the rest of the star. Other secular instabilities which occur in burning shells or which are due to nonspherical perturbations will be discussed later (§ 32.6, § 33.2).

### 25.3.4 The Gravothermal Specific Heat

Let us consider a small sphere of radius $r_s$ and mass $m_s$ around the centre of a star in hydrostatic equilibrium. If the sphere is sufficiently small, then $P$ at $r_s$ and the mean density in the sphere are good approximations for the central values $P_c$, $\varrho_c$. Suppose that, as a reaction to the addition of a small amount of heat to the central sphere, the whole star is slightly expanding and let the expansion be homologous. Then any mass shell of radius $r$ after expansion has the radius $r + dr = r(1 + x)$, where $x$ is constant for all mass shells. If after the expansion the pressure in the sphere is $P_c + dP_c$, then, similarly to (20.34,37), the resulting changes of $\varrho_c$ and $P_c$ are:

$$\frac{d\varrho_c}{\varrho_c} = -3x \quad , \quad p_c := \frac{dP_c}{P_c} = -4x \quad . \tag{25.25}$$

We now write the equation of state in differential form,

$$\frac{d\varrho_c}{\varrho_c} = \alpha p_c - \delta \vartheta_c \quad , \tag{25.26}$$

242

$(\vartheta_c := dT_c/T_c)$ as in (6.5) but here with constant chemical composition. Elimination of $d\varrho_c/\varrho_c$ and of $x$ from (25.25,26) gives

$$p_c = \frac{4\delta}{4\alpha - 3} \vartheta_c \quad . \tag{25.27}$$

According to the first law of thermodynamics the heat $dq$ per mass unit added to the central sphere is

$$dq = du + Pdv = c_P \, T_c \, (\vartheta_c - \nabla_{ad} p_c) := c^* T_c \vartheta_c \quad , \tag{25.28}$$

where we have used (4.18,21) and where according to (25.27)

$$c^* = c_P \left( 1 - \nabla_{ad} \frac{4\delta}{4\alpha - 3} \right) \quad . \tag{25.29}$$

This quantity has the dimension of a specific heat per mass unit. Indeed, $dT = dq/c^*$ gives the temperature variation in the central sphere if the heat $dq$ is added. In thermodynamics we are used to defining specific heats with some mechanical boundary conditions, for example $c_P$ and $c_v$. For $c^*$ the mechanical condition is that the gas pressure is kept in equilibrium with the weight of all the layers with $r > r_s$. This $c^*$ is called the *gravothermal specific heat*.

For an ideal monatomic gas ($a = \delta = 1$, $\nabla_{ad} = 2/5$), as we have approximately in the central region of the sun, one finds from (25.29) that $c^* < 0$. This is fortunate, since if in the sun the nuclear energy generation is accidentally enhanced for a moment ($dq > 0$), then $dT < 0$, the region cools, thereby reducing the overproduction of energy immediately. Therefore the negative specific heat acts as a stabilizer. At first glance it seems as if the decrease of temperature after an injection of heat contradicts energy conservation. But one has also to take into account the $Pdv$ work done by the central sphere. Indeed, while the centre cools ($\vartheta_c < 0$) the whole star expands, since elimination of $p_c$ and $d\varrho_c/\varrho_c$ from (25.25,26) gives $x = -\delta\vartheta_c/(4\alpha - 3)$, which in the case $\alpha = \delta = 1$ yields $x > 0$. It turns out that, if heat is added to the central sphere, more energy is used up by the expansion and therefore some must be taken from the internal energy. This behaviour is essentially connected with the virial theorem (see § 3.1). A corresponding property can be found for the piston model by assuming a variable weight $G^*$ of the piston as in § 3.2.

For a nonrelativistic degenerate gas ($\delta \to 0$, $\alpha \to 3/5$) equation (25.29) gives $c^* > 0$: The addition of energy to the central sphere heats up the matter, which can lead to thermal runaway.

### 25.3.5 Secular Stability Behaviour of Nuclear Burning

Having derived a handy expression for $dq$, we shall now use it in the energy balance of the central sphere considered in § 25.3.4. Energy is released in the sphere by nuclear reactions and transported out of it by radiation (we assume here that the central region is not convective). In the steady state gains and losses compensate each other. Let $\varepsilon$ be the mean energy generation rate, and $l_s$ the energy per unit time which leaves the sphere; then $\varepsilon m_s - l_s = 0$. Now the equilibrium is supposed to be

243

perturbed on a time-scale $\tau$, such that $\tau$ is much larger than $\tau_{hydr}$, but short compared to the thermal adjustment time of the sphere. Then, while hydrostatic equilibrium is maintained, the thermal balance is perturbed.

For the perturbed state the energy balance is

$$m_s d\varepsilon - dl_s = m_s \frac{dq}{dt} \equiv m_s c^* \frac{dT_c}{dt} \quad . \tag{25.30}$$

Here, $dq$ is the heat gained per mass unit, which is expressed by $c^* dT_c$ according to (25.28).

If we now perturb the equation for radiative heat transfer (5.12),

$$l \sim \frac{T^3 r^4}{\kappa} \frac{dT}{dm} \quad , \tag{25.31}$$

we obtain for $l_s$

$$\frac{dl_s}{l_s} = 4\vartheta_c + 4x - \kappa_P \, p_c - \kappa_T \, \vartheta_c \quad . \tag{25.32}$$

For the perturbation of $dT/dm$ we have made use of the fact that for homology $\vartheta = dT/T = $ constant and therefore $d(dT/dm) = d(T\vartheta)/dm = \vartheta dT/dm$. From (25.25,27,32) it follows that

$$\frac{dl_s}{l_s} = \left[ 4 - \kappa_T - \frac{4\delta}{4\alpha - 3}(1 + \kappa_P) \right] \vartheta_c \quad . \tag{25.33}$$

This, introduced into (25.30), gives

$$\frac{m_s}{l_s} \frac{dq}{dt} = (m_s d\varepsilon - dl_s)/l_s = \varepsilon_T \, \vartheta_c + \varepsilon_P \, p_c - \frac{dl_s}{l_s}$$
$$= \left[ (\varepsilon_T + \kappa_T - 4) + \frac{4\delta}{4\alpha - 3} \left( \varepsilon_P + \kappa_P + 1 \right) \right] \vartheta_c \quad , \tag{25.34}$$

where we have made use of $l_s = \varepsilon m_s$ and of (25.27). Then with (25.30) we find

$$\frac{m_s c^* T_c}{l_s} \frac{d\vartheta_c}{dt} = \left[ (\varepsilon_T + \kappa_T - 4) + \frac{4\delta}{4\alpha - 3} \left( 1 + \varepsilon_P + \kappa_P \right) \right] \vartheta_c \quad . \tag{25.35}$$

The sign of the bracket tells us whether for $dT_c > 0$ the additional energy production exceeds the additional energy loss of the sphere ($[\ldots] > 0$). The sign of $c^*$ tells us whether in this case the sphere heats up ($c^* > 0$) or cools ($c^* < 0$). Normally $\varepsilon_T$ is the leading term in the bracket, so that indeed $[\ldots] > 0$. We first assume an ideal gas ($\alpha = \delta = 1$, $c^* < 0$) and obtain

$$\frac{m_s c^* T_c}{l_s} \frac{d\vartheta_c}{dt} = \left[ \varepsilon_T + \kappa_T + 4 \left( \varepsilon_P + \kappa_P \right) \right] \vartheta_c \quad . \tag{25.36}$$

Since $c^* < 0$, one finds from (25.36) that $(d\vartheta_c/dt)/\vartheta_c < 0$, meaning that the perturbation $dT_c$ decays and the equilibrium is stable if

$$\varepsilon_T + \kappa_T + 4(\varepsilon_P + \kappa_P) > 0 \quad . \tag{25.37}$$

This criterion is normally fulfilled. The only "dangerous" term is $\kappa_T$, which can be as low as to $-4.5$ for Kramers opacity. But then, even $\varepsilon_T = 5$ for the $pp$ chain suffices to fulfill (25.37), since the other terms are positive.

Any temperature increase $dT_c > 0$ would cause a large additional energy overproduction $\varepsilon_0 \varepsilon_T dT_c / T_c$. But since the gravothermal heat capacity $c^* < 0$, the sphere reacts with $dT_c < 0$, and this cooling brings energy production back to normal. We then can say that the burning in a sphere of ideal gas proceeds in a stable manner, the negative gravothermal specific heat acts like a thermostat. This, for example, is the case in the sun.

We go back to (25.35) for the general equation of state. Since normally $\varepsilon_T$ dominates the other terms in the square bracket (in some case $\varepsilon_T > 20$), we neglect them for simplicity. Then (25.35) can be written

$$\frac{d\vartheta_c}{dt} = \frac{l_s \varepsilon_T}{m_s T_c c^*} \vartheta_c := \frac{1}{D} \vartheta_c \quad . \tag{25.38}$$

Obviously $D < 0$ indicates stability, $D > 0$ instability. Since $\varepsilon_T > 0$ and, for an ideal gas, $c^* < 0$, the quantity $D$ is negative: The nuclear burning is stable.

For a nonrelativistic degenerate gas we have $\delta = 0$, $\alpha = 3/5$. Therefore, $c^* > 0$ and $D > 0$: Any nuclear burning with a sufficiently strong temperature dependence will then be unstable. This is the reason, for instance, why in the central regions of a white dwarf there can be no strong nuclear energy source [as first shown by MESTEL (1952)]; the star would be destroyed by thermal runaway, or at least heat up until it was not degenerate and then expand. Of course, then it would no longer be a white dwarf. The same instability is also responsible for the phenomenon of the so-called flash (compare §32.4) which occurs if a new nuclear burning starts in a degenerate region. Note that the appearance of $4\alpha - 3$ in the denominator in several equations, including (25.29) for $c^*$ and (25.35), does not become serious even if $\alpha \to 3/4$ for partial nonrelativistic degeneracy, since the singularity can be removed from the equation which one obtains if $c^*$ is inserted in (19.35) by multiplication with $4\alpha - 3$.

From (25.38) one can draw another conclusion. Let us assume that in the central region of a star there is no nuclear burning but that energy losses by neutrinos (§18.6) are important. The nuclear energy production in the star may take place in a concentric shell of finite radius. Part of this energy flows outwards, providing the star's luminosity, while part of it flows from the shell inwards towards the centre where it goes into neutrinos. The maximum temperature then is in the shell and not in the stellar centre. In §32.6 we shall see that this really can be the case in models of evolved stars. If we now again look at (25.38), we have to be aware that $l_s < 0$. If $\varepsilon_T > 0$, as it is for neutrino losses (see §18.6), all the above conclusions are contradicted because of the different sign of $l_s$: The equilibrium is stable if $c^* > 0$, that is for degeneracy, but unstable if $c^* < 0$, which is the case for an ideal gas.

All our discussions here were based on the assumption of homologous changes in the stellar model. Although stars clearly never change precisely in such a simple way, it turns out that the above conclusions describe qualitatively correctly the secular stability behaviour of stars. Deviations from homology only influence the factors [e.g. in the bracket in (25.36)], thus modifying the exact position of the border between secular stability and instability.

# V    Early Stellar Evolution

# § 26 The Onset of Star Formation

Observational evidence favours the picture that stars form out of interstellar matter. Indeed a homogeneous cloud of compressible gas can become gravitationally unstable and collapse. In this section we shall deal with gravitational instability and then discuss some of its consequences. But before we do so it may be worth comparing this instability with those discussed in § 25. For gravitational instability the inertia terms are important as well as heat exchange of the collapsing mass with its surroundings. But it is not a vibrational instability, since the classification scheme of § 25 holds only if the free-fall time is much shorter than the time-scale of thermal adjustment. As we will see later, just the opposite is the case here.

## 26.1 The Jeans Criterion

### 26.1.1 An Infinite Homogeneous Medium

We start with an infinite homogeneous gas at rest. Then density and temperature are constant everywhere. However, we must be aware that this state is not a well-defined equilibrium. For symmetry reasons the gravitational potential $\Phi$ must also be constant. But then Poisson's equation $\nabla^2 \Phi = 4\pi G \varrho$ demands $\varrho = 0$. Indeed the gravitational stability behaviour should be discussed starting from a better equilibrium state, as we will do later. Nevertheless we first assume a medium of constant non-vanishing density. If we here apply periodic perturbations of sufficiently small wavelength, the single perturbation will behave approximately like one with the same wavelength in an isothermal sphere in hydrostatic equilibrium (which is a well-defined initial state).

The gas has to obey the equation of motion of hydrodynamics

$$\frac{\partial v}{\partial t} + (v \cdot \nabla)v = -\frac{1}{\varrho}\nabla P - \nabla \Phi \tag{26.1}$$

(Euler equation), together with the continuity equation

$$\frac{\partial \varrho}{\partial t} + v\nabla \varrho + \varrho \nabla \cdot v = 0 \quad . \tag{26.2}$$

In addition we have Poisson's equation

$$\nabla^2 \Phi = 4\pi G \varrho \tag{26.3}$$

and the equation of state for an ideal gas

$$P = \frac{\mathfrak{R}}{\mu}\varrho T = v_s^2 \varrho \quad , \tag{26.4}$$

where $v_s$ is the (isothermal) speed of sound. For equilibrium we assume $\varrho = \varrho_0 =$ constant, $T = T_0 =$ constant, and $v_0 = 0$. $\Phi_0$ may be determined by $\nabla^2\Phi_0 = 4\pi G\varrho_0$ and by boundary conditions at infinity.

We now perturb the equilibrium

$$\varrho = \varrho_0 + \varrho_1 \quad , \quad P = P_0 + P_1 \quad , \quad \Phi = \Phi_0 + \Phi_1 \quad , \quad v = v_1 \quad , \tag{26.5}$$

where the functions with subscript 1 depend on space and time. In (26.5) we have already used that $v_0 = 0$. If we substitute (26.5) in (26.1,4), assuming that the perturbations are isothermal ($v_s$ is not perturbed), and if we ignore non-linear terms in the these quantities, we find

$$\frac{\partial v_1}{\partial t} = -\nabla\left(\Phi_1 + v_s^2\frac{\varrho_1}{\varrho_0}\right) \quad , \tag{26.6}$$

$$\frac{\partial \varrho_1}{\partial t} + \varrho_0\nabla \cdot v_1 = 0 \quad , \tag{26.7}$$

$$\nabla^2\Phi_1 = 4\pi G\varrho_1 \quad . \tag{26.8}$$

This is a linear homogeneous system of differential equations with constant co-efficients. We therefore can assume that solutions exist with the space and time dependence proportional to $\exp[i(kx + \omega t)]$ such that

$$\frac{\partial}{\partial x} = ik \quad , \quad \frac{\partial}{\partial y} = \frac{\partial}{\partial z} = 0 \quad , \quad \frac{\partial}{\partial t} = i\omega \quad . \tag{26.9}$$

With $v_{1x} = v_1$, $v_{1y} = v_{1z} = 0$ we find from (26.6–8) that

$$\omega v_1 + \frac{kv_s^2}{\varrho_0}\varrho_1 + k\Phi_1 = 0 \quad , \tag{26.10}$$

$$k\varrho_0 v_1 + \omega\varrho_1 = 0 \quad , \tag{26.11}$$

$$4\pi G\varrho_1 + k^2\Phi_1 = 0 \quad . \tag{26.12}$$

This homogeneous linear set of three equations for $v_1$, $\varrho_1$, $\Phi_1$ can only have non-trivial solutions if the determinant

$$\begin{vmatrix} \omega & \dfrac{kvs^2}{\varrho_0} & k \\ k\varrho_0 & \omega & 0 \\ 0 & 4\pi G & k^2 \end{vmatrix}$$

is zero. Assuming a non-vanishing wave number $k$ we obtain

$$\omega^2 = k^2 v_s^2 - 4\pi G\varrho_0 \quad . \tag{26.13}$$

For sufficiently large wave numbers the right-hand side is positive, i.e. $\omega$ is real. The perturbations vary periodically in time. Since the amplitude does not increase, the equilibrium is stable with respect to perturbations of such short wavelengths.

In the limit $k \to \infty$, (26.13) gives $\omega^2 = k^2 v_s^2$, which corresponds to isothermal sound waves. Indeed for very short waves gravity is not important, any compression is restored by the increased pressure and the perturbations travel with the speed of sound through space.

If $k^2 < 4\pi G \varrho_0 / v_s^2$, the eigenvalue $\omega$ is of the form $\pm i\xi$, where $\xi$ is real. Therefore there exist perturbations $\sim \exp(\pm \xi t)$ which grow exponentially with time, so that the equilibrium is unstable. If we define a characteristic wave number $k_J$ by

$$k_J^2 := \frac{4\pi G \varrho_0}{v_s^2} \quad , \tag{26.14}$$

or a characteristic wavelength

$$\lambda_J := \frac{2\pi}{k_J} \quad , \tag{26.15}$$

then perturbations with a wave number $k < k_J$ (or a wavelength $\lambda > \lambda_J$) are unstable, otherwise they are stable with respect to the perturbations applied here. The condition for instability $\lambda > \lambda_J$, where

$$\lambda_J = \left(\frac{\pi}{G\varrho_0}\right)^{1/2} v_s \tag{26.16}$$

is called the *Jeans criterion* after James Jeans, who derived it in 1902. Depending on the detailed geometrical properties of equilibrium and perturbation the factors on the right-hand side of (26.16) can differ.

For our special choice of perturbations the case of instability can be described as follows: after a slight compression of a set of plane-parallel slabs, gravity overcomes pressure and the slabs collapse to thin sheets. If we estimate $\omega$ for the collapsing sheets only from the gravitational term in (26.13) (which indeed is larger than the pressure term), we have $i\omega \approx (G\varrho_0)^{1/2}$, and the corresponding time-scale is $\tau \approx (G\varrho_0)^{-1/2}$, which corresponds to the free-fall time as defined in §2.4.

### 26.1.2 A Plane Parallel Layer in Hydrostatic Equilibrium

We have already mentioned the contradictions connected with the assumption of an infinite homogeneous gas as initial condition. One way out of this difficulty is to investigate the equilibrium of an isothermal plane-parallel layer stratified according to hydrostatic equilibrium in the $z$ direction. Perpendicular to the $z$ direction all functions are constant, the layer extending to infinity. This defines a one-dimensional problem: $\varrho_0$, $P_0$, $T_0$ depend only on one coordinate, say $z$. Poisson's equation then is

$$\frac{d^2\Phi_0}{dz^2} = 4\pi G \varrho_0 \quad , \tag{26.17}$$

while hydrostatic equilibrium, $dP_0/dz = -\varrho_0 d\Phi_0/dz$, can be written with (26.4) as

$$v_s^2 \frac{d \ln \varrho_0}{dz} = -\frac{d\Phi_0}{dz} \quad . \tag{26.18}$$

After differentiation of (26.18) one obtains from (26.17)

$$\frac{d^2 \ln \varrho_0}{dz^2} = -\frac{4\pi G}{v_s^2} \varrho_0 \quad . \tag{26.19}$$

With the boundary condition $\varrho_0 = 0$ for $z = \pm\infty$, (26.19) has the solution

$$\varrho_0(z) = \frac{\varrho_0(0)}{\cosh^2(z/H)} \quad , \tag{26.20}$$

with

$$H = \left(\frac{\Re T}{2\pi \mu G \varrho_0(0)}\right)^{1/2} = \frac{v_s}{[2\pi G \varrho_0(0)]^{1/2}} \quad , \tag{26.21}$$

which can be seen if (26.20,21) are inserted into (26.19). The (stratified) disc does not cause problems similar to those enountered in the case of the infinite homogeneous gas.

In order to investigate the stability of this disc one defines cartesian coordinates $x$, $y$ in the plane perpendicular to the $z$ axis and considers perturbations of the form $\varrho_1 \sim f(z) \exp[\mathrm{i}(kx + \omega t)]$. Since the perturbations do not depend on $y$ the layer collapses to a set of plane-parallel slabs in the case of instability. We shall not go into the details of the stability analysis, which has been described by SPITZER (1968). The result is that again there is a critical wave number

$$k_{\mathrm{J}} = \frac{1}{H} = \frac{[2\pi G \varrho_0(0)]^{1/2}}{v_s} \tag{26.22}$$

and that instability occurs for wave numbers $k < k_{\mathrm{J}}$, while perturbations with $k > k_{\mathrm{J}}$ remain finite. This is very similar to what we have obtained in the homogeneous case, as can be seen by comparing (26.22) and (26.14). The difference in the numerical factors is due to the different geometry.

The two cases discussed above have in common that for smaller wave numbers (larger wavelengths and therefore larger amounts of mass involved in the resulting collapse) the equilibrium is unstable, while for larger wave numbers it is stable. In hydrostatic equilibrium the force due to the pressure gradient and the gravitational force cancel each other. In general this balance is disturbed after a slight compression. If only a small amount of mass is compressed, the pressure increases more than the force due to gravity, and the gas is pushed back towards the equilibrium state. This is the case if a toy balloon is slightly compressed. Only the increase of pressure counts, since the gravity of the trapped gas is negligible. The same is true for the compressions which occur in sound waves where gravity plays no rôle. But if a sufficient amount of gas is compressed simultaneously, the increase of gravity overcomes that of pressure and makes the compressed gas contract even more.

## 26.2 Instability in the Spherical Case

In order to investigate the Jeans instability for interstellar gas in a configuration more realistic than the two examples of §26.1, we now consider an isothermal sphere of finite radius imbedded in a medium of pressure $P^* > 0$. The sphere is supposed to consist of an ideal gas. The structure of the sphere can be obtained from a solution of the Lane–Emden equation (19.35) for an isothermal polytrope. The solution is cut off at a certain radius where $P$ has dropped to the surface pressure $P = P^*$. The stratification outside the sphere is not relevant as long as it is spherically symmetric with respect to the centre, since then there is no gravitational influence of the outside on the inside. Its only influence will be via the surface pressure, which we assume to be constant during the perturbation.

The essential points of this problem can be easily seen if one discusses the virial theorem for the sphere, as described in §3.4. Since our sphere of mass $M$ and radius $R$ is isothermal, its internal energy is $E_i = c_v MT$. For the gravitational energy we write $E_g = -\Theta GM^2/R$, where $\Theta$ is a factor of order one. It can be obtained by numerical integration of the Lane–Emden equation. With these expressions and with $\zeta = 2$ (ideal monatomic gas) the virial theorem (3.21) can be solved for the surface pressure $P_0$ giving

$$P_0 = \frac{c_v MT}{2\pi R^3} - \frac{\Theta GM^2}{4\pi R^4} \ . \tag{26.23}$$

The first term on the right is due to the internal gas pressure, which tries to expand the sphere. It is proportional to the mean density. The second term is due to the self-gravity of the sphere, which tries to bring all matter to the centre.

We now discuss how $P_0$ varies with $R$ for fixed values of $M$, $T$, and $\Theta$. For small $R$ the value of $P_0$ is negative. It changes sign with increasing $R$, while it approaches zero from positive values for $R \rightarrow \infty$. $P_0$ has a (positive) maximum at $R = R_m$, a value which can be obtained by differentiation of (26.23). After replacing $c_v$ by $3\Re/(2\mu)$ we find that $dP_0/dR$ vanishes at

$$R_m = \frac{4\Theta}{9} \frac{G\mu M}{\Re T} \ . \tag{26.24}$$

Suppose the sphere to be in equilibrium with the surroundings: $P_0 = P^*$. For $R < R_m$, the surface pressure $P_0$ decreases with decreasing $R$. Therefore, after a slight compression, $P_0 < P^*$ and the sphere will be compressed even more; it is unstable. For $R > R_m$, the pressure $P_0$ increases during a slight compression and the sphere will expand back to equilibrium; it is stable. (These simple plausibility arguments are supported by the results of decent stability analysis.) We have obviously recovered the Jeans instability discussed in §26.1. This can be seen if in (26.24) $M$ is replaced by $4\pi R_m^3 \bar{\varrho}/3$, where $\bar{\varrho}$ is the mean density of the sphere. We then obtain

$$R_m^2 = \frac{27}{16\pi\Theta} \frac{\Re T}{G\mu\bar{\varrho}} \ . \tag{26.25}$$

Here $R_m$ is the critical radius of a gaseous mass of mean density $\bar{\varrho}$ and temperature

$T$ which is marginally stable. We compare it with the critical Jeans wavelength obtained in (26.16), which with $v_s^2 = \Re T/\mu$ becomes

$$\lambda_J^2 = \frac{\Re T \pi}{G \mu \varrho} \ . \tag{26.26}$$

Clearly $\lambda_J$ and $R_m$ are of the same order of magnitude.

Obviously for a given equilibrium state there exists a critical mass $M_J$, the so-called *Jeans mass*. Masses larger than $M_J$ are gravitationally unstable. If slightly compressed they fall together. According to (26.25)

$$M_J = \frac{4\pi}{3} \varrho R_m^3 = \frac{27}{16} \left(\frac{3}{\pi}\right)^{1/2} \left(\frac{\Re}{\Theta G}\right)^{3/2} \left(\frac{T}{\mu}\right)^{3/2} \left(\frac{1}{\varrho}\right)^{1/2} \ . \tag{26.27}$$

Depending on the treatment of the perturbation problem and its geometry, one finds slightly differing pre-factors in the expression for $M_J$, but they all give the same order of magnitude.

An often used expression derived from perturbation considerations is

$$M_J = \left(\frac{\pi \Re}{G \mu}\right)^{3/2} T^{3/2} \varrho^{-1/2}$$

$$= 1.2 \times 10^5 M_\odot \left(\frac{T}{100 \mathrm{K}}\right)^{3/2} \left(\frac{\varrho}{10^{-24} \mathrm{g\ cm}^{-3}}\right)^{-1/2} \mu^{-3/2} \ . \tag{26.28}$$

With $\varrho = 10^{-24}$ g cm$^{-3}$, $T = 100$ K and $\mu = 1$ (typical for the conditions in interstellar clouds of neutral hydrogen) we obtain $M_J \approx 10^5 M_\odot$. Only masses large compared to the stellar masses ($0.1 \ldots 100 M_\odot$) seem to be able to collapse because of the Jeans instability.

We have already shown, following (26.16), that the time-scale for the growth of the instability is $\tau \approx (G\varrho)^{-1/2}$, the free-fall time. This is of course also valid for the present spherical case. For a density of $\varrho \approx 10^{-24}$ g cm$^{-3}$, the collapse takes place on a time-scale of some $10^8$ years. During collapse, $\tau$ becomes shorter, since the density increases.

This time-scale $\tau$ is long compared to that for thermal adjustment $\tau_{\mathrm{adj}}$. Since the cloud is optically thin, $\tau_{\mathrm{adj}}$ is the internal energy per unit mass divided by the rate of energy losses owing to radiation. For typical neutral hydrogen clouds SPITZER (1968) and LOW, LYNDEN-BELL (1976) estimate a loss $\Lambda$ of the order 1 erg g$^{-1}$ s$^{-1}$. With $T = 100$ K we find $\tau_{\mathrm{adj}} \approx c_v T/\Lambda \approx 100$ years. Comparison with the free-fall time of some $10^8$ years shows that the collapse proceeds in thermal adjustment (which turns out to mean that it is almost isothermal). In §26.3 we will show where this breaks down.

## 26.3 Fragmentation

As shown above, only masses large compared to the stellar masses can become gravitationally unstable. So one may wonder how stars can actually form out of the interstellar medium. The explanation nowadays generally believed is that a cloud exceeding the Jeans mass and therefore collapsing undergoes *fragmentation*, i.e. while the cloud falls together, fragments of it become unstable and collapse faster than the cloud as a whole. If this is true, then out of the collapsing cloud of mass $> M_J$ smaller submasses can condense.

At first glance this seems to be a promising mechanism for producing collapsing objects with masses much smaller than $M_J$. Indeed, if the cloud collapses isothermally, then $M_J$ decreases as $\varrho^{-1/2}$. If, however, the gas were to change adiabatically, then for a monatomic ideal gas $\nabla_{ad} = (d\ln T/d\ln P)_{ad} = 2/5$ or $T \sim P^{2/5}$, and from $P \sim \varrho T$ the temperature would change as $T \sim \varrho^{2/3}$, and therefore $M_J \sim T^{3/2}\varrho^{-1/2} \sim \varrho^{1/2}$. So the Jeans mass would *grow* during an adiabatic collapse. But we have seen already in §26.2 that under interstellar conditions the thermal-adjustment time-scale is much shorter than the free-fall time, which is of the order $(G\varrho)^{-1/2}$, and this also holds when the density increases during collapse. One can therefore assume the collapse to be isothermal rather than adiabatic. Then the Jeans mass becomes smaller than the mass of the originally collapsing cloud. If it has dropped, say, to one half its original value, the cloud can split into two independently collapsing parts. This kind of fragmentation can go on as long as the collapse remains roughly isothermal. (Note that in principle it is not justified to apply the concept of the Jeans mass to an already collapsing medium, since it has been derived for an equilibrium state. But we may do it for order-of-magnitude estimates.)

What are the final products of this fragmentation process? Will the cloud finally fall apart into a swarm of cloudlets of planetary masses or even smaller? We cannot follow strictly the hydrodynamics and thermodynamics of this complicated process (which soon shows no symmetry at all). So we just estimate when the thermal adjustment time of the fragments becomes comparable with the free-fall time. Then the collapse can certainly not be isothermal any more and must approach an adiabatic one. But, as we have seen, then the Jeans mass no longer decreases with increasing $\varrho$. This means that subregions of the fragments do not fall together on their own and fragmentation stops.

For a detailed estimate one has to know the radiation processes that cool the gas during collapse. One can then find how long the gained work $-Pdv$ can be radiated away, as for instance has been evaluated by HOYLE (1953). Instead of this procedure we shall follow REES (1976), who gave an estimate of the mass limit of fragmentation without specifying the detailed radiation processes.

The characteristic time of the free-fall of a fragment is $(G\varrho)^{-1/2}$, and the total energy to be radiated away during collapse is of the order of the gravitational energy $E_g \approx GM^2/R$ (see §3.1), where $M$ and $R$ are the mass and radius of the fragment. Therefore the rate $A$ of energy to be radiated away in order to keep the fragment always at the same temperature is of the order

$$A \approx \frac{GM^2}{R}(G\varrho)^{1/2} = \left(\frac{3}{4\pi}\right)^{1/2}\frac{G^{3/2}M^{5/2}}{R^{5/2}} \quad . \tag{26.29}$$

But the fragment at temperature $T$ cannot radiate more than a black body of that temperature. (This implies approximate thermal equilibrium, which is not too bad an assumption for the final stage of fragmentation, where matter starts to become opaque.) Therefore the rate of radiation loss of the fragment is

$$B = 4\pi f \sigma T^4 R^2 \quad, \tag{26.30}$$

where $\sigma = 2\pi^5 k^4/(15c^2 h^3)$ is the Stefan–Boltzmann constant, while $f$ is a factor less than 1 taking into account that the fragment radiates less than the corresponding black body. For isothermal collapse it is necessary that $B \gg A$. The transition to adiabatic collapse will occur if $A \approx B$. From (26.29,30) we find that this is the case when

$$M^5 = \frac{64\pi^3}{3} \frac{\sigma^2 f^2 T^8 R^9}{G^3} \quad. \tag{26.31}$$

We assume that fragmentation has reached its limit when $M_J$ is equal to this $M$. We therefore replace $M$ in (26.31) by $M_J$, $R$ by

$$R = \left(\frac{3}{4\pi}\right)^{1/3} \frac{M_J^{1/3}}{\varrho^{1/3}} \quad, \tag{26.32}$$

and eliminate $\varrho$ with the help of (26.28). The Jeans mass at the end of fragmentation is then obtained as

$$M_J = \left(\frac{\pi^9}{9}\right)^{1/4} \frac{1}{(\sigma G^3)^{1/2}} \left(\frac{\Re}{\mu}\right)^{9/4} f^{-1/2} T^{1/4}$$

$$= 3.86 \times 10^{31}\mathrm{g}\ f^{-1/2} T^{1/4} = 0.02 M_\odot \frac{T^{1/4}}{f^{1/2}} \quad, \tag{26.33}$$

where $T$ is in K and where we have set $\mu = 1$.

Let us assume that the temperature $T$ of the smallest elements is 1000 K and, further, that appreciable deviations from isothermal collapse occur when the radiation losses have to exceed 10% of the maximal possible (black-body) radiation losses ($f = 0.1$). We then find from (26.33) that $M = 1/3 M_\odot$. This estimate would not be very different if we had assumed a different value for $T$ or for $f$ within reasonable ranges. The point is that *fragmentation terminates if the fragments are of the order of the solar mass*, not of the order of planetary masses or of a whole star cluster.

It should be noted that our result is not particularly dependent on the chemical composition. Therefore this estimate also holds for stars of the first generation, which are formed shortly after the Big Bang (so-called Population III stars), when heavy elements were far less abundant. Although this matter does not contain metals, star formation may have produced stars of a similar mass to those produced today.

# § 27 The Formation of Protostars

The Jeans criterion derived in the foregoing section follows from a first-order perturbation theory and gives conditions under which perturbations of an equilibrium stage will grow exponentially. But the linear theory does not give information, for instance, about the fully developed collapse, to say nothing about the final product. For this, one has to follow the perturbation into the non-linear regime. We first begin with some very simple cases, assuming always spherical symmetry for the collapsing cloud.

## 27.1 Free-Fall Collapse of a Homogeneous Sphere

If, according to the Jeans criterion, a gaseous mass has become unstable and the collapse has started, gravity increases relatively more than the pressure gradient. The collapse is more and more governed by gravity alone, which is easily seen from the following arguments. For spherical symmetry the gravitational acceleration is of the order $GM/R^2$, where $M$ and $R$ are the mass and radius of the cloud. On the other hand, an estimate of the acceleration due to the pressure gradient is

$$\left|\frac{1}{\varrho}\frac{\partial P}{\partial r}\right| \approx \frac{P}{\varrho R} \approx \frac{\mathfrak{R}}{\mu}\frac{T}{R} \quad . \tag{27.1}$$

The ratio of gravitational force to pressure gradient is therefore $\sim M/(RT)$, which during isothermal collapse increases as $1/R$. Consequently we here neglect the gas pressure.

The free collapse of a homogeneous sphere can be treated analytically. At a distance $r$ from the centre the gravitational acceleration is $Gm/r^2$, where $m$ is the mass within the sphere of radius $r$. If the pressure can be neglected, the sphere collapses in free fall, according to the equation of motion

$$\ddot{r} = -\frac{Gm}{r^2} \quad , \tag{27.2}$$

where the dots indicate the time derivatives of the radius $r(m,t)$. We now replace $m$ by $4\pi\varrho_0 r_0^3/3$, where the subscript zero indicates the values at the beginning of the collapse, by assumption $\varrho_0 =$ constant. Multiplication of (27.2) by $\dot{r}$ and integration gives

$$\frac{1}{2}\dot{r}^2 = \frac{4\pi r_0^3}{3r}G\varrho_0 + \text{constant} \quad . \tag{27.3}$$

Choosing the integration constant so that $\dot{r} = 0$ at the beginning, when $r = r_0$, we get

$$\frac{\dot{r}}{r_0} = \pm \left[ \frac{8\pi G}{3} \varrho_0 \left( \frac{r_0}{r} - 1 \right) \right]^{1/2} . \tag{27.4}$$

In order to obtain only real values of $r$, it must always be less than $r_0$, which means that only the minus sign on the right of (27.4) gives relevant solutions.

For the solution of (27.4) we introduce a new variable $\zeta$, defined by

$$\cos^2 \zeta = \frac{r}{r_0} . \tag{27.5}$$

Therefore

$$\frac{\dot{r}}{r_0} = -2\dot{\zeta} \cos \zeta \sin \zeta \quad , \quad \frac{r_0}{r} - 1 = \frac{\sin^2 \zeta}{\cos^2 \zeta} \quad , \tag{27.6}$$

and (27.4) gives

$$2\dot{\zeta} \cos^2 \zeta = \left( \frac{8\pi G \varrho_0}{3} \right)^{1/2} . \tag{27.7}$$

With the identity

$$2\dot{\zeta} \cos^2 \zeta = \frac{d}{dt} \left( \zeta + \frac{1}{2} \sin 2\zeta \right) \quad , \tag{27.8}$$

which is easily verified, we can write instead of (27.7) that

$$\zeta + \frac{1}{2} \sin 2\zeta = \left( \frac{8\pi G \varrho_0}{3} \right)^{1/2} t \quad , \tag{27.9}$$

where the integration constant is chosen such that the beginning of the collapse (when $r = r_0$ or $\zeta = 0$) coincides with $t = 0$. It should be noted that $r_0$ no longer explicitly appears in the solution (27.9) and that $\varrho_0 = $ constant. Therefore the solution $\zeta(t)$ is the same for all mass shells. Then, according to (27.6), $r/r_0$ and also $\dot{r}/r_0$ at a given time $t$ are the same for all mass shells. This means that the sphere undergoes a *homologous contraction*. Since $\dot{r}/r_0$ is independent of $r_0$, the relative density variation is independent of $r_0$, and the sphere, which was homogeneous at $t = 0$, remains homogeneous. The time it takes to reach the centre ($r = 0$ or $\zeta = \pi/2$) is the free-fall time

$$t_{\mathrm{ff}} = \left( \frac{3\pi}{32 G \varrho_0} \right)^{1/2} \quad , \tag{27.10}$$

which follows from (27.9) and is the same for all mass shells. With $\varrho_0 = 4 \times 10^{-23}$ g/cm$^3$, corresponding to a slightly enhanced interstellar density, one obtains $t_{\mathrm{ff}} \approx 10^7$ years. It should be noted that expression (27.10) is very similar to the free-fall time $\tau_{\mathrm{ff}}$ for a star we estimated in (2.17), if there $g$ is replaced by $GM/R^2 = 4\pi G \varrho_0 R/3$.

Of course, before the centre is reached the pressure will become relevant as the gas becomes opaque and $T$ increases. Then the free-fall approximation has to be abandoned, and finally the collapse will be stopped.

## 27.2 Collapse onto a Condensed Object

As the collapsing cloud becomes opaque the heating will first start in the central parts, since radiation can escape more easily from gas near the surface. Therefore the collapse will be stopped first in the central region. In order to see what then happens we consider a core which has already reached hydrostatic equilibrium, surrounded by a still-free-falling cloud.

Now let $M$ be the mass of the core. For the sake of simplicity we neglect the self-gravity of the free-falling matter. The simplest case is that for the steady state. This would mean that the core is surrounded by an infinite reservoir of matter from which a steady flow rains down. Then the mass flow with absolute radial velocity $v$,

$$\dot{M} = 4\pi r^2 \varrho v \quad , \tag{27.11}$$

must be constant in space and time. Differentiation of (27.11) with respect to $r$ gives the continuity equation

$$\frac{2}{r} + \frac{1}{\varrho} \frac{d\varrho}{dr} + \frac{1}{v} \frac{dv}{dr} = 0 \quad . \tag{27.12}$$

If for $v$ we take the free-fall velocity $v = v_{ff} = [GM/(2r)]^{1/2}$ and assume $M \approx$ constant, we find

$$\frac{1}{\varrho} \frac{d\varrho}{dr} = -\frac{3}{2r} \quad , \tag{27.13}$$

or

$$\varrho(r) = \frac{\text{constant}}{r^{3/2}} \quad . \tag{27.14}$$

If $R$ is the radius of the core, then at impact the free-falling matter has the velocity $v_{ff}(R) = [GM/(2R)]^{1/2}$.

The matter falling onto the core is stopped at its surface. The kinetic energy is then transformed into heat, part of which is used to heat up the core, the rest being radiated away. If we ignore the heating of the core, the radiation losses are

$$L_{accr} = \frac{1}{2} v_{ff}^2(R)\dot{M} = \frac{1}{4} \frac{GM}{R}\dot{M} \quad . \tag{27.15}$$

$L_{accr}$ is called the *accretion luminosity*. Since for the steady-state solution we have assumed constant $M$ in the expression for $v_{ff}$, (27.15) is only valid if the accretion time-scale

$$\tau_{accr} := M/\dot{M} \tag{27.16}$$

is long compared to the free-fall time $t_{ff}$.

## 27.3 A Collapse Calculation

The collapse of an unstable interstellar cloud can in principle be followed numerically. We will describe the first collapse calculations of an originally homogeneous cloud of one solar mass (LARSON, 1969). The mass fractions of hydrogen, helium, and heavier elements were taken to be $X = 0.651$, $Y = 0.324$, and $Z = 0.025$ respectively. The boundary conditions assumed that the surface of the sphere remained fixed. The equations to be solved are the continuity equation

$$\frac{\partial m}{\partial t} + 4\pi r^2 v \varrho = 0 \tag{27.17}$$

(with the radial velocity $v$ having positive values in outward direction), the equation of motion

$$\frac{\partial v}{\partial t} + v \frac{\partial v}{\partial r} + \frac{Gm}{r^2} + \frac{1}{\varrho} \frac{\partial P}{\partial r} = 0 \quad , \tag{27.18}$$

and the energy equation

$$\frac{\partial u}{\partial t} + P \frac{\partial}{\partial t}\left(\frac{1}{\varrho}\right) + v\left[\frac{\partial u}{\partial r} + P \frac{\partial}{\partial r}\left(\frac{1}{\varrho}\right)\right] + \frac{1}{4\pi \varrho r^2} \frac{\partial l}{\partial r} = 0 \quad , \tag{27.19}$$

where $u$ is the internal energy per unit mass. Here the terms on the left (except for the last one) give the substantial derivative $du/dt + Pd(1/\varrho)/dt$ according to $d/dt = \partial/\partial t + v\partial/\partial r$. In addition we have the relation

$$\frac{\partial m}{\partial r} = 4\pi r^2 \varrho \quad . \tag{27.20}$$

Finally we need an equation which describes the energy transport by radiation. Although the diffusion approximation is certainly not good in those parts of the cloud which are optically thin (see § 5), the equation

$$l = -\frac{16\pi acr^2}{3\kappa\varrho} T^3 \frac{\partial T}{\partial r} \tag{27.21}$$

was used, which is identical with our equation (5.11). The errors introduced do not change the qualitative (and maybe even the quantitative) results too much.

For the absorption properties of a gas at extremely low temperatures, other effects than those for stellar-interior opacities discussed in § 17 have to be considered (GAUSTAD, 1963). As long as they exist, dust grains are the dominant source of opacity. With increasing temperature (above 1000 K) the dust particles evaporate. Then the collapsing material becomes more transparent, the opacity being dominated by molecules.

With (27.17–21), one has five equations for the five unknown variables $m(r,t)$, $v(r,t)$, $P(r,t)$, $T(r,t)$, and $l(r,t)$, while $\varrho$, $\kappa$, and $u$ are given material functions of, say, $P$ and $T$. The equation of state is assumed to be that of an ideal gas (including effects of dissociation and ionization). The numerical solution now has to be determined with one of the methods described in § 11.3. The outer boundary condition at $r = R$ in these calculations is $v(R,t) = 0$. Since the equations show a singularity at the centre, one has to demand as inner boundary condition that the solutions remain regular there. The initial conditions are $v(r,0) = 0$, while $P(r,0)$ and $T(r,0)$ are constant, and therefore $l(r,0) = 0$. The initial values were $T(r,0) = 10$ K, $\varrho(r,0) \approx 10^{-19}$ g/cm$^3$. It should be noted that then almost all hydrogen is in molecular form.

In order to have instability at the beginning, the cloud of one solar mass must be sufficiently dense and, therefore, small. Instability was found numerically for $R < 0.46GM\mu/(\Re T)$. The close resemblance to the critical radius (26.24) for *homologous* collapse should be noted. The calculations began with a slightly compressed cloud with $R = 1.63 \times 10^{17}$ cm. With the density $10^{-19}$ g cm$^{-3}$ the free-fall time according to (27.10) is $6.6 \times 10^{12}$s $\approx 210\,000$ years.

In the following we describe the different phases of the collapse.

## 27.4 The Optically Thin Phase and the Formation of a Hydrostatic Core

In the very first phase the whole collapsing cloud remains optically thin and therefore nearly isothermal with $T \approx 10$ K.

When the instability evolves into the non-linear regime the collapse becomes non-homologous, which is not surprising in view of the outer boundary condition. It holds the outer layers of the sphere at a fixed radius while the inner part is free to collapse. Indeed during collapse the density increases rapidly in the central part, while it remains practically constant in the outer regions. A small central concentration, once formed, will necessarily enhance itself. The free-fall time of a certain mass shell at distance $r$ from the centre is of the order $[G\bar{\varrho}(r)]^{-1/2}$, where $\bar{\varrho}(r)$ is the mean density inside the sphere of radius $r$. If $\bar{\varrho}$ increases towards the centre, then the (local) free-fall time decreases in this direction. Therefore the inner shells fall faster than the outer ones and the central density concentration becomes even more pronounced.

The calculations show that the density distribution – starting from $\varrho =$ constant – approaches the form $\varrho \sim r^{-2}$ over gradually increasing parts of the cloud (see Fig. 27.1). It is not surprising that it does not follow (27.14), since there we have made assumptions (steady state, a free fall determined only by the gravity of a central object) which are not fulfilled here.

The density profiles in Fig. 27.1 can be described as follows. A smaller and smaller homogeneous mass collapses more and more rapidly, continuously releasing matter into the inhomogeneous envelope. There the time-scale of collapse remains much larger because (1) the density is smaller and (2) pressure gradients brake the free fall.

The collapse of the homogeneous central part resembles a free fall as long as the matter can get rid of the released gravitational energy via radiation. The central region becomes opaque once a central density of $10^{-13}$ g cm$^{-3}$ is reached. Now the further increase of density in the centre causes an adiabatic increase of temperature. As a consequence the pressure there increases until the free fall is stopped.

This leads to the formation of a central core in hydrostatic equilibrium surrounded by a still-falling envelope. Immediately after the core has reached hydrostatic equilibrium its mass and radius are $10^{31}$ g and $6 \times 10^{13}$ cm, and the central

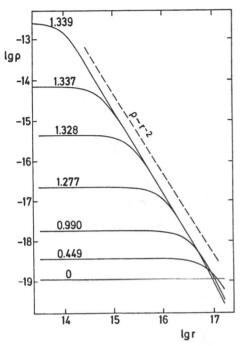

**Fig. 27.1.** The density $\varrho$ (in g cm$^{-3}$) against the distance from the centre $r$ (in cm) in a collapsing cloud. The density distribution is shown by solid lines for different times (labels in $10^{13}$ s after the onset of the collapse). Regions with homologous changes remain homogeneous ($\partial\varrho/\partial r = 0$); regions in free fall approach a distribution with $\varrho \sim r^{-2}$ (i.e. a slope indicated by the dashed line). (After LARSON, 1969)

values are $\varrho_c = 2 \times 10^{-10}$ g cm$^{-3}$, $T_c = 170$ K. The free-fall velocity at the surface of the core is 75 km/s. With increasing core mass and decreasing core radius the velocity of the falling material exceeds the velocity of sound in the core surface regions. Therefore a spherical shock front is formed which separates the supersonic "rain" from the hydrostatic interior. In this shock front the falling matter comes to rest, releasing its kinetic energy. If all the energy released is radiated away (which is approximately the case) the luminosity of the accreting core is given by (27.15).

In certain respects the hydrostatic core resembles a star. But while the surface pressure is virtually zero for a star, here it has to balance the pressure exerted by the infalling material. If $v_e$ and $\varrho_e$ are the velocity relative to the shock front and the density of the falling gas just above it, respectively, and if $P_i$ is the surface pressure, then conservation of momentum demands that

$$P_i = \varrho_e v_e^2 = \varrho_e \frac{GM}{2R} \quad , \tag{27.22}$$

where $M$ and $R$ are the mass and radius of the core. This equation is a special case of the more general condition for shock fronts (see LANDAU, LIFSHITZ, vol. 6, 1959, p. 318) according to which the quantity $P + \varrho v^2$ must have the same values on both sides of the front. In (27.22) $P$ is neglected outside the front, and $v$ inside.

Another difference between an accreting core and a real star is that the accretion energy is released in a thin surface layer, while in a star the energy source is in the deep interior.

At first glance one would expect the whole core to be isothermal. But while matter is raining down on its surface the core is contracting. This has the consequence that $L_{accr}$ as given by (27.15) increases for $\dot{M} \approx$ constant (since $M$ grows and $R$

261

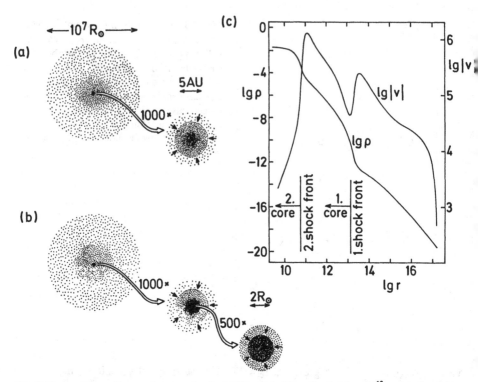

**Fig. 27.2. a–c** The collapse of a gas cloud of $1M_{\odot}$. (a) After about $1.3 \times 10^{13}$ s the cloud has formed an optically thick core. The collapse is stopped there and a shock front develops at the interface between the core, which is in hydrostatic equilibrium, and the still freely falling envelope. (b) When the core has become dynamically unstable owing to dissociation of $H_2$, a second collapse occurs within the core, forming a second shock front at much smaller $r$. (c) Schematic plot of the absolute value of the velocity $v$ (in cm s$^{-1}$) and the density $\varrho$ (in g cm$^{-3}$) against $r$ (in cm), for a time shortly after the formation of a second core within the first one. The regions of the shock fronts are characterized by steep (positive) slopes in the velocity curve

decreases). Since during contraction gravitational energy is released in the deep interior of the core, there must be a finite temperature gradient in order to transport this energy outward. The accreting core in hydrostatic equilibrium is often called a *protostar*. Its diameter is already comparable to the dimensions of the solar system (see Fig. 27.2).

## 27.5 Core Collapse

The accreting protostar heats up in its interior. We have to keep in mind that the gas consists mainly of hydrogen that at low temperatures is in molecular form as $H_2$. When the central temperature reaches about 2000 K, the hydrogen molecules dissociate. The equilibrium between molecular and atomic hydrogen is governed by an equation similar to the Saha equation (see § 14.1). Like ionization, dissociation influences the specific heat, since not all the energy injected into a gas goes into kinetic energy, a fraction being used to break up the molecules into atoms. This decreases $\gamma_{ad}$. For hydrogen molecules there are $f = 5$ degrees of freedom, three belonging to translation and two to rotation around two possible axes. Consequently $\gamma_{ad} = (f+2)/f = 7/5 = 1.40$. This is much closer to the critical value $4/3 = 1.33$ (see §25.3.2) than in the case of a monatomic gas ($\gamma_{ad} = 5/3 = 1.667$). Only a slight reduction of $\gamma_{ad}$ owing to dissociation therefore brings it below the critical value $4/3$. Then the hydrostatic equilibrium becomes dynamically unstable and the protostar starts to collapse again.

In Larson's calculations this happened when the protostar has, compared to the initial values, twice the mass and half the radius. It collapses as long as the gas is partially dissociated. When almost all hydrogen in the central region is in atomic form, $\gamma_{ad}$ increases above $4/3$ (approaching the value $5/3$ for a monatomic gas) and the collapsing protostar forms a dynamically stable subcore in its interior. This core has an initial mass of $1.5 \times 10^{-3} M_\odot$ and an initial radius of $1.3 R_\odot$. Its central density is $2 \times 10^{-2}$ g cm$^{-3}$ and the central temperature is $2 \times 10^4$ K. At the surface of this new core there is another shock front. The situation is illustrated in Fig. 27.2b,c. As a consequence of the second collapse the density below the outer shock front decreases and the outer shock finally disappears.

The evolution of the centre of the $1 M_\odot$ cloud, starting from the original Jeans instability, is given in Fig. 27.3. The curve starts on the left during the isothermal collapse. After the matter has become opaque, $T$ rises adiabatically. The slope is at first 0.4 (corresponding to $\gamma_{ad} = 1.40$ for H$_2$), but then becomes considerably less owing to partial dissociation, and finally reaches 2/3 (corresponding to $\gamma_{ad} = 5/3$ for a monatomic gas).

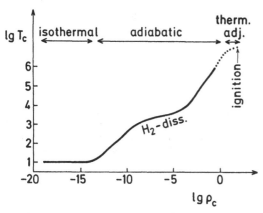

Fig. 27.3. The central evolution of a $1 M_\odot$ cloud from the isothermal collapse to the ignition of nuclear burning. The central temperature $T_c$ (in K) is plotted over the central density $\varrho_c$ (in g cm$^{-3}$). The dotted line is an extrapolation, indicating that after the adiabatic compression a phase of thermally adjusted contraction brings the centre to ignition. (After APPENZELLER, TSCHARNUTER, 1975b)

The central compression is adiabatic as long as the accretion time-scale $\tau_{accr}$ of the core (or of the innermost core, if there are two) is short compared to its Kelvin–Helmholtz time-scale $\tau_{KH}$. But the more the envelope is depleted the more the accretion rate will diminish and consequently $\tau_{accr}$ will grow. When it exceeds $\tau_{KH}$ the core can adjust thermally and the evolution of the central region ceases to be adiabatic. Since then $\dot{M}$ has become very small, the protostar has practically constant mass. We shall discuss its further evolution with constant $M$ in the next section.

## 27.6 Evolution in the Hertzsprung–Russell Diagram

A plot of the evolution of a collapsing cloud in the Hertzsprung–Russell (HR) diagram has to be made with care. The radiation emitted by the core is absorbed in the falling envelope, particularly by dust grains, which heat up and reradiate in the infrared. One can assign an effective temperature to the protostellar models. Defining an effective radius $R$ at the optical depth 2/3 one can derive an effective temperature $T_{eff}$ from $L = 4\pi R^2 \sigma T_{eff}^4$. Evolutionary tracks for initial masses of $1M_\odot$ and $60M_\odot$ are given in Fig. 27.4. To an outside observer the collapsing cloud remains an infrared object as long as the envelope is opaque to visible radiation. The evolutionary track, therefore, starts extremely far to the right in this diagram. This, of course, is no contradiction to the statements about a forbidden region to the right of the Hayashi line (§ 24), since the falling envelope (including the "photosphere") is far from being in hydrostatic equilibrium. Even if we could see the already hydrostatic core, we would not observe a normal star, since its boundary conditions are still perturbed by infalling matter.

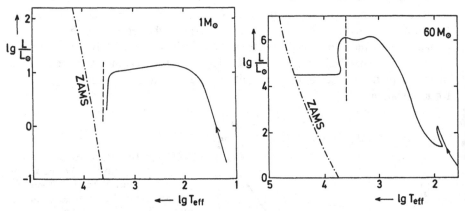

**Fig. 27.4.** Hertzsprung–Russell diagrams with evolutionary tracks for protostars of $1M_\odot$ and $60M_\odot$. The tracks start at the lower right, where the thermal radiation of the clouds is in the infrared, and they finally approach the zero-age main sequence (ZAMS, *dot-dashed*). In the case of $60M_\odot$ part of the mass of the envelope is blown away so that a star of only $17M_\odot$ settles down on the main sequence. The corresponding Hayashi lines are indicated by broken lines. (After APPENZELLER, TSCHARNUTER 1974, 1975)

The thinning out of the envelope has several effects: the first is that it becomes more transparent, and the photosphere ($\tau = 2/3$) moves downwards until it has reached the surface of the hydrostatic core. With decreasing radius of the photosphere, $T_{eff}$ must increase in order to radiate away the energy. In the whole first phase (through the maximum of $L$ in the evolutionary tracks of Fig. 27.4) the luminosity is produced by accretion: $L = L_{accr} \sim \dot{M}$. With decreasing $\dot{M}$, the luminosity $L$ decreases until it is finally provided by contraction of the core. It can even happen that nuclear reactions set in during the accretion phase, although their contribution to $L$ is not important. Another effect is the influence of accretion on the boundary conditions of the core. Strong accretion heats up the surface of the core so much that the core is nearly isothermal and the ram pressure $\varrho_e v_e^2$ is appreciable. With decreasing $\dot{M}$ the boundary conditions become "normal". The core surface cools down, a temperature gradient is built up, and a convection zone develops downwards from the surface.

This convection may or may not penetrate down to the centre. If the object is fully convective, has "normal" boundary conditions, and is already visible, we must see it on the Hayashi line. In any case we have the transition from a protostar to a normal contracting star in hydrostatic, but not yet in thermal, equilibrium.

One should keep in mind that the collapse calculations discussed here are based on simplifying assumptions, encounter unresolved difficulties, and to some extent show unexplained results. (For details see, for instance, TSCHARNUTER, 1985). For example, even slow rotation of the initial cloud may change the (spherically symmetric) results completely. The scenario may also be modified by interstellar magnetic fields frozen in the plasma. The treatment of radiative transfer in the highly extended, non-stationary, partially transparent envelope with uncertain opacity is far from being trivial. Some solutions show a "bounce", where part of the originally collapsing matter is expelled so that the final star has much less mass than the original cloud.

# § 28 Pre-Main-Sequence Contraction

In the last section we left the newly born star while it was still contracting in hydrostatic, but not yet thermal, equilibrium. Essential features of this contraction can already be understood by assuming simple homologous changes. It will turn out that the fate of such a sphere is mainly determined by the equation of state.

## 28.1 Homologous Contraction of a Gaseous Sphere

A star which has not yet reached the temperature for nuclear burning has to supply its energy loss by contraction. This is a consequence of the virial theorem and of energy conservation as discussed in §3.1. We have seen, in particular, that part of the released gravitational energy goes into internal energy, while the rest supplies the luminosity [see. (3.12)]. The characteristic time-scale is $\tau_{KH}$, as shown in §3.3.

In the following we will be concerned with the centre of the star. For this we can use the relations of §20.3, which hold for any mass shell of a homologously contracting star. The equation of state (for fixed chemical composition) was written there as $d\varrho/\varrho = \alpha dP/P - \delta dT/T$. According to (20.34, 38), the variation of the central temperature, $dT_c$, is related to the variation of the central density, $d\varrho_c$, by

$$\frac{dT_c}{T_c} = \frac{4\alpha - 3}{3\delta} \frac{d\varrho_c}{\varrho_c} \quad . \tag{28.1}$$

This defines a field of directions in the $\lg \varrho_c$–$\lg T_c$ plane as displayed in Fig. 28.1. Each arrow there indicates how $T_c$ changes during contraction ($d\varrho_c > 0$). According to (28.1) the slope depends on the equation of state via $\alpha$ and $\delta$. For an ideal gas $\alpha = \delta = 1$ and (28.1) becomes

$$\frac{dT_c}{T_c} = \frac{1}{3} \frac{d\varrho_c}{\varrho_c} \quad . \tag{28.2}$$

Here the slope is 1/3, a contracting ideal gas heats up (the latter conforms with the conclusions drawn from the virial theorem in §3.1). The same slope also holds for non-negligible radiation pressure ($\beta < 1$) as can be seen if (13.16) is introduced into (28.1). In Fig. 28.1 the evolutionary track of a (homologously) contracting ideal gaseous sphere is a straight line with slope 1/3. This necessarily leads closer to the regime of degeneracy, which is separated from that of ideal gas by a line of slope 2/3 [see (16.10) and Fig. 16.1]. The onset of degeneracy changes $\alpha$ and $\delta$ and decreases the slope of the arrows in Fig. 28.1. In the limit of complete non-relativistic degeneracy one has $\alpha \to 3/5$ and $\delta \to 0$. What happens to a sphere

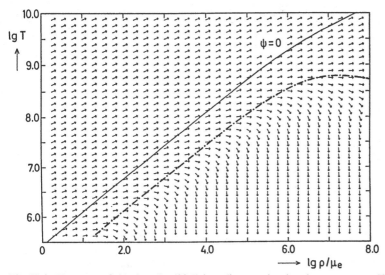

**Fig. 28.1.** The vector field given by (28.1) in a diagram showing the temperature $T$ (in K) over the density $\varrho/\mu_e$ (in g cm$^{-3}$). The arrows indicate the direction in which the centre of a homologously contracting star would evolve. In the upper-left part the equation of state is that of an ideal gas and therefore the arrows have a slope of 1/3. The thin solid line at which the degeneracy parameter $\psi = 0$ indicates roughly the transition from the ideal gas to degeneracy of the electrons. The critical line along which $\alpha = 3/4$ is dot-dashed. On this curve the arrows point horizontally while below it the arrows point downwards

which is contracting and becomes more and more degenerate? Then $\alpha$ will pass the value 3/4 when $\delta$ is still finite and the slope given by (28.1) will change sign. Further contraction leads to cooling: the stronger the degeneracy the steeper will be the then negative slope, until finally the stellar centre tends to cool off at almost constant density. In the case of complete relativistic degeneracy, with $\alpha = 3/4$ and $\delta = 0$, the factor on the right of (28.1) becomes indeterminate. Then the ion gas – although its pressure is negligible compared to that of the degenerate electrons – will determine the slope. A dash-dotted line in Fig. 28.1 connects the points of vanishing slope ($\alpha = 3/4$).

For the sake of simplicity let us first ignore the fact that nuclear reactions set in at certain temperatures. Obviously, the evolutionary track of a contracting gaseous sphere in the lg $\varrho_c$–lg $T_c$ diagram depends very much on the starting point at the left-hand border, as can be seen from Fig. 28.2. If a stellar centre starts there sufficiently low it will reach a maximum temperature and begin to cool again after entering the domain of degeneracy. But if it started on the left at a sufficiently high temperature, it will never be caught by degeneracy and thus will continue to heat up.

Which types of spheres do reach a maximum temperature, and which types have the privilege of heating up forever? This depends on the mass of the sphere. In order to show this we consider two homologous spheres of an ideal gas with masses $M$ and $M' = M/x$ and radii $R$ and $R' = R/z$. Then, according to (20.17), $\varrho_c/\varrho_c' = xz^{-3}$, $P_c/P_c' = x^2 z^{-4}$, and therefore, for an ideal gas, $T_c/T_c' = x/z$. If we now compare states in which the two spheres have the same central density ($xz^{-3} = 1$), we have $T_c/T_c' = x^{2/3} = (M/M')^{2/3}$. This means that in Fig. 28.2 the evolutionary tracks of

267

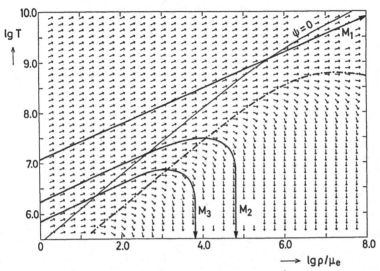

**Fig. 28.2.** Temperature $T$ (in K) over density $\varrho/\mu_e$ (in g cm$^{-3}$) with the vector field and the lines $\psi = 0$ and $\alpha = 3/4$ as in Fig. 28.1. The heavy lines give the "evolutionary tracks" of the centres of three homologously contracting stars of different masses. Mass $M_1$ is so large that the evolution is not remarkably influenced by degeneracy, and the centre continuously heats up during contraction. For mass $M_2 (< M_1)$ degeneracy becomes important in the centre, and consequently a homologous contraction cannot bring the central temperature above a few $10^7$ K (which is not sufficient to start helium burning). Mass $M_3 (< M_2)$ while contracting will start to cool off even before the temperature of hydrogen burning is reached

larger masses are above those of smaller masses. Consequently it is the less massive spheres which will finally be forced by degeneracy to cool off after having reached a maximum central temperature, being smaller the smaller the mass.

This has immediate consequences for the nuclear reactions, which we have ignored up to now. We know that a nuclear burning in a wide range of densities occurs at a characteristic temperature: hydrogen burning near $10^7$ K, helium burning at $10^8$ K. (Since here we are discussing early phases of stellar evolution, we exclude the pycnonuclear reactions, which occur at extremely high densities only; see § 18.4.) One can therefore expect that a contracting sphere below a certain critical mass may never reach the temperature of hydrogen burning, since its central temperature never reaches $10^7$ K.

This important result deduced from simple homology considerations is also manifested in computer calculations of more realistic stellar models. Although the cores formed in the protostar phase do not contract completely homologously, their centres evolve in the $\lg \varrho$–$\lg T$ plane very similarly. Protostars of mass less than about $0.08 M_\odot$ never ignite their hydrogen and thus never become main-sequence stars. Such objects are called *black* or *brown dwarfs*. They are sometimes invoked in order to explain the missing mass in galaxies. Because of their low luminosity they could easily escape detection, and therefore a large amount of matter could be hidden in many such objects. Here we have encountered an evolutionary aspect of the lower end of the main sequence: protostars born with too little mass never reach the state of complete equilibrium by which the main-sequence models are defined. Even if some nuclear reactions have started, they are so slow at these low temperatures that equilibrium abundances (rate of destruction = rate of production) of the involved nuclei are not reached even in the lifetime of the galaxy. But we should note that this, in principle, is a quite different problem from that of the existence of (stable) equilibrium solutions (§23) at the lower end of the main sequence. The critical masses for each are about the same, since they are each caused by degeneracy. Their precise determination suffers from uncertainties of the material functions.

We shall see later that analogous considerations can be used to explain critical masses for the ignition of each higher nuclear burning in contracting cores of evolved stars. And masses above $10M_\odot$ will never be caught by degeneracy in this way (see §34).

## 28.2 Approach to the Zero-Age Main Sequence

We have seen that a contracting star of more than $0.08M_\odot$ ignites hydrogen in its centre and becomes a star on the zero-age main sequence (ZAMS). While the luminosity of the star was originally due to contraction, it now originates from nuclear energy. These two energy sources are quite differently distributed in the star. According to (20.41), $\varepsilon_g \sim T$ is not so much concentrated towards the centre, while hydrogen burning with $\varepsilon_{pp} \sim T^5$ and $\varepsilon_{CNO} \sim T^{18}$ has strong central concentration. Clearly the transition from contraction to hydrogen burning requires a rearrangement of the internal structure. The protostar becomes a zero-age main sequence star with the properties described in §22.

The way in which nuclear reactions take over the energy production is described in detail by IBEN (1965), who calculated the approach to the main sequence of contracting protostars. We first discuss the results for one solar mass. Some reactions of the CNO cycle as given in (18.64) become important before the central temperature has reached that of equilibrium hydrogen burning (where the participating nuclei have equilibrium abundances). At a central temperature of about $10^6$ K, all the $^{12}C$ that had been in the interstellar cloud will burn into $^{14}N$ via the reactions of the first three lines in (18.64). Once switched on, this process will take over the energy generation and stop the contraction. Because of the high temperature sensitivity of $\varepsilon$, the energy is released close to the centre. Consequently the energy flux $l/4\pi r^2$ is large and a convective core that contains 11% of the total mass develops. At the same time, the first reactions of the $pp$ chain become relevant, transforming H into $^3He$ [see the first two lines of (18.62)]. With decreasing $^{12}C$ the $pp$ reactions become more important and $^3He$ can be destroyed by $^3He+^3He$ and $^3He+^4He$ [the two reactions in the third line of (18.62)]. As a consequence the concentration of $^3He$ reaches a maximum at $m = 0.6M$. Outside, the temperature is too low to form $^3He$, while inside, $^3He$ is used up to form $^4He$. With the depletion of $^{12}C$ in the central region the convective core disappears and the $pp$ chain becomes the dominant energy source.

The situation is similar for more massive stars. But then instead of the $pp$ chain, the CNO cycle finally takes over and the abundance of $^{12}C$ becomes that of equilibrium. For stars of $M > 1.5M_\odot$ the effect of pre-main-sequence $^{12}C$ burning can even be seen in the computed evolutionary tracks in the Hertzsprung–Russell diagram: there seems to be another, relatively short-lived main sequence to the right of the ordinary (hydrogen) main sequence. Contracting protostars stay there until their $^{12}$C fuel is used up before they move on to the main sequence. This somewhat prolongs the time a protostar needs to reach the ZAMS.

Iben's calculations were carried out before the results described in §27 were known. He therefore started out with a cool protostar on the Hayashi line and followed the ensuing contraction until the model reached the hydrogen main sequence. But the errors introduced in this way may not be too large and certainly become negligible towards the end of pre-main-sequence contraction when the thermal effects of accretion are forgotten by the star.

This has to do with the fact that, whatever the thermal history of the protostar, its structure has adjusted to thermal equilibrium after a Kelvin–Helmholtz time. Since the main-sequence time-scale (which is relevant for the ensuing evolution) is much longer, the stars settle on the ZAMS quite independently of their past. Whatever their detailed history, tracks of protostars of the same mass (and chemical composition) lead to the same point on the ZAMS.

We now turn to the question of how rapidly stars of different $M$ approach the ZAMS. Decisive for this is the Kelvin–Helmholtz time-scale $\tau_{\text{KH}} \approx c_v \bar{T} M / L$. The mean temperature $\bar{T}$ does not vary too much with $M$, since $T_c$ is anyway just below the ignition temperature of hydrogen. As a rough estimate for $L$ we may take the corresponding ZAMS luminosity, since the evolutionary tracks in their final parts are at about that luminosity (see Fig. 27.4). Then $L \sim M^{3.5}$ and $\tau_{\text{KH}} \sim M^{-2.5}$. This means that massive protostars reach the ZAMS much faster than their low-mass colleagues.

In the Hertzsprung–Russell diagrams of very young stellar clusters (for example NGC 2264 and the Pleiades) one finds that only massive stars are on the main sequence, while the low-mass stars lie to the right of it. It seems that, because of their longer $\tau_{\text{KH}}$, these stars are still in the contraction phase and have not yet begun with nuclear burning. Among them are flare stars (UV Ceti stars) and T Tauri variables. The cause of their (irregular) variability is not yet known.

# § 29 From the Initial to the Present Sun

There is evidence on Earth that the sun has shone for more than 3000 million years with about the same luminosity. From radioactive decay in different materials of the solar system, one nowadays assumes that it was formed $4.65 \times 10^9$ years ago. Since then, the sun has lived on hydrogen burning, predominantly according to the $pp$ chain, and its interior has been appreciably enriched in $^4$He. In the following we show how a model of the present sun can be constructed.

## 29.1 Choosing the Initial Model

While the observations yield information about the mass abundance $Z$ of heavier elements, it is difficult to determine spectroscopically the helium content $Y$ of the solar surface. One therefore uses $Y$ as a free parameter. Furthermore, there is no information about the mixing length $\ell_m$ to be used in the convection theory (see § 7). One normally expresses $\ell_m$ in units of the local pressure scale height $H_P$ and treats the dimensionless quantity $\ell_m/H_P$ as another free parameter. Let us now start the construction of an initial solar model with trial values of $Y$ and $\ell_m/H_P$. Since the model changes only on the (long) nuclear time-scale, it can well be approximated by assuming complete equilibrium. This means that in addition to the inertia term in (9.2) the time derivatives in the energy equation (9.3) can be neglected. The evolution can then be followed numerically until a time of $4.65 \times 10^9$ years after the onset of hydrogen burning has elapsed. During this time interval the molecular weight in the central regions increases owing to the enrichment of helium. Consequently, the luminosity increases slightly, as can be expected from the homology relation (20.20) according to which the luminosity should increase like $\mu^4$. (The fact that the solar evolution is not homologous changes the result only quantitatively.) At the same time, the point in the Hertzsprung–Russell (HR) diagram moves slightly to the left. If our choice of the free parameters were correct, the model after $4.65 \times 10^9$ years should resemble the present sun. But, in general, this will not be the case and the evolutionary track will miss the image point of the present sun. One therefore has to adjust the two free parameters in order to end up with the present sun.

A variation of the mixing length changes the radius slightly, but turns out to have almost no influence on the luminosity. Therefore, while varying $\ell_m$, the initial model will move almost horizontally (Fig. 29.1). If, on the other hand, $Y$ is changed, the mean molecular weight $\mu$ varies. With increasing helium content, $\mu$ also increases, and since the computed models roughly behave as the homologous models of § 20.2.2, the image point of the model moves to the upper left on a line below the main sequence [see the arguments after (20.23)].

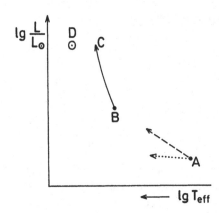

Fig. 29.1. Finding a model that for given values of $Z = 1 - X - Y$ describes the present sun. For arbitrary values of $Y$, $\ell_m$ one obtains a ZAMS model at $A$, from where it shifts along the broken and dotted arrow as a result of independent changes of $Y$ and $\ell_m$ respectively. Based on this, one guesses the values of $Y$, $\ell_m$ that yield the model at $B$. Its evolution is calculated from age zero ($B$) to $t = 4.65 \times 10^9$ years ($C$). The guessed values $Y$, $\ell_m$ are modified until $C$ coincides with $D$ (present sun)

Since small changes in the two parameters do not modify the form of the evolutionary track very much, the whole track makes an approximately parallel shift. Therefore one can find values for $Y$ and $\ell_m/H_P$ for which the end point of the evolutionary track coincides with the point of the (observed) present sun. The procedure is illustrated in Fig. 29.1. A model constructed in this way, and by using the standard assumptions for the input physics, is often called a "solar standard model".

The values of the initial $Y$ and $\ell_m/H_P$, which after $4.65 \times 10^9$ years lead to the present sun, depend sensitively on the details of the computations, for instance the opacities. The solar standard model by BAHCALL et al. (1982) has a rather high initial hydrogen abundance ($X = 0.732$). We shall use it later for discussing the solar-neutrino problem. Here we refer to a model which has been computed by WEISS (1986), who used opacity tables for $Z = 0.021$ and obtained a solar standard model for the hydrogen abundance $X = 1 - Y - Z = 0.695$ and $\ell_m/H_P = 2$. (The difference compared to the results of BAHCALL et al. (1982) is probably due to different mixtures of the elements comprised in $Z$.)

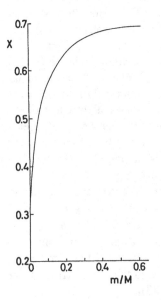

Fig. 29.2. The hydrogen abundance $X$ in the inner parts of a model for the present sun (age $4.65 \times 10^9$ years) as a function of $m/M$. In the homogeneous ZAMS model the hydrogen content was $X = 0.695$ everywhere. (After WEISS, 1986)

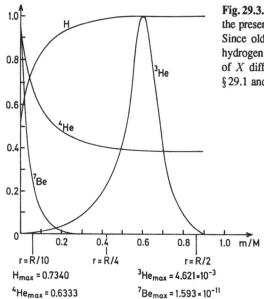

Fig. 29.3. The chemical composition of a model of the present sun computed by BARTENWERFER, 1972. Since older opacity tables were used, the original hydrogen content as well as the present central value of $X$ differ from the model by WEISS described in § 29.1 and in Fig. 29.2

r = R/10
$H_{max} = 0.7340$
$^4He_{max} = 0.6333$

$^3He_{max} = 4.621 \times 10^{-3}$
$^7Be_{max} = 1.593 \times 10^{-11}$

In the central region of the present sun, quite an appreciable percentage of the original hydrogen has already been converted into helium. The hydrogen content $X$ as a function of $m$ is plotted in Fig. 29.2, which shows that the central value of $X$ has dropped to about 0.3. More details about the chemical composition of the present sun are given in Fig. 29.3.

The outer convective zone reaches down to a temperature of $1.9 \times 10^6$ K. The radius of its inner boundary is $r = 0.75\ R$. The temperature gradients $\nabla$, $\nabla_{ad}$, $\nabla_{rad}$ as defined in §§ 5–7 are plotted in Fig. 29.4. In the near-surface regions where $\lg P < 4.9$, one finds $\nabla_{rad} < \nabla_{ad}$ and the layer is stable (Fig. 29.4a). Then convection sets in where $\nabla_{rad}$ exceeds $\nabla_{ad}$. In the outermost part of the convective zone the convection is very ineffective and $\nabla$ is close to $\nabla_{rad}$, according to the considerations in § 7.3. But $\nabla$ does not follow $\nabla_{rad}$ to the extreme values (which at $\lg P = 9$ reach a maximum of $2.5 \times 10^5$). It never exceeds 0.9. Owing to partial ionization of the most abundant elements, $\nabla_{ad}$ is not constant in the outer region of the solar model, as we have already shown in Fig. 14.1b. The deeper inside, the more the actual gradient approaches the adiabatic one, following it up and down (Fig. 29.4a,b). In Fig. 29.4c the convective velocity obtained from $U$, $\nabla_{rad}$, and $\nabla$ according to (7.6,15) is given in units of the (isothermal) velocity of sound $v_s = (\Re T/\mu)^{1/2}$. At the top of the convection zone, $v/v_s$ reaches its maximum of about 0.4.

It is not surprising that one can produce models for the present sun which have the correct position in the HR diagram, since two free parameters, $Y$ and $\ell_m$, can be varied to adjust the two quantities $L$ and $R$. Therefore obtaining a solar model with the right age at the right position in the HR diagram is not much of a test of stellar-evolution theory. At present there are two observational tests to compare the solar interior with model calculations. Both of them seem to require changes of the theoretical stellar models. One of these tests is based on the investigation of

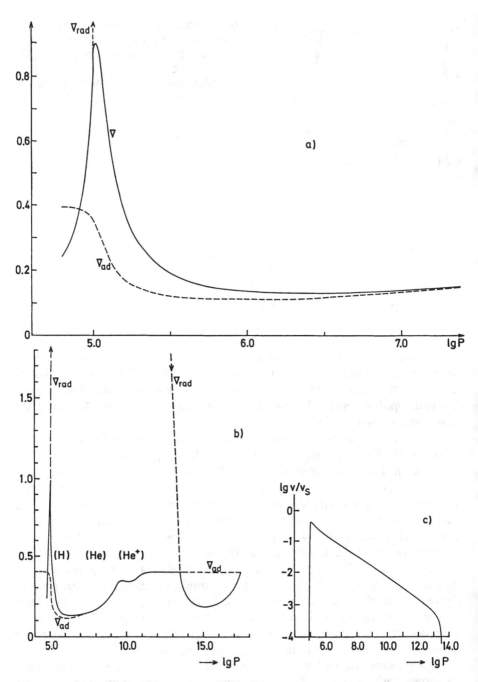

**Fig. 29.4. a–c** Some properties of the model for the present sun described in the text. (a) The temperature gradients in the outer layers, against the pressure $P$ (in dyn cm$^{-2}$). In the outermost layers the actual gradient $\nabla$ (*solid line*) coincides with $\nabla_{rad}$, which then, however, goes up to values above the range of the ordinate. The strong depression of $\nabla_{ad}$ (*lower dashed line*) for $\lg P > 5$ is due to hydrogen ionization. (b) The same curves as in (a) but with compressed scales, such that the whole interior of the model is covered. $\nabla_{rad}$ is still out of the range for almost all of the outer convective zone. The depression of $\nabla_{ad}$ is caused by the ionization of H, He, and He$^+$ (see labels in parenthesis). Note that the centre of the sun is close to convective instability. (c) The convective velocity $v$ in units of the local velocity of sound, $v_s$, in the outer convective zone of the sun

non-radial solar oscillations, commonly called solar seismology. We shall deal with such oscillations later (see § 40.4). The other test is the solar neutrino experiment.

## 29.2 Solar Neutrinos

Some of the nuclear reactions of the $pp$ chain, as well as of the CNO cycle, produce neutrinos (§ 18.6). These reactions were summarized in (18.79). In addition, there are also neutrinos due to the very rare reaction

$$^1H + {}^1H + e^- \rightarrow {}^2H + \nu \quad . \tag{29.1}$$

As already discussed in § 18.6, the neutrinos leave the star practically without interacting with the stellar matter. The energy spectrum of neutrinos from $\beta$ decay is continuous, since the electrons can take part of the energy away, while neutrinos released after an inverse $\beta$ decay are essentially monochromatic. Therefore the first and the third of the reactions of (18.79) have a continuous spectrum, while the second reaction of (18.79) and reaction (29.1) have a line spectrum. Since $^7Be$ can decay into $^7Li$ either in the ground state or in an excited state, the second reaction of (18.79) gives two spectral lines. The neutrino spectrum of the sun as predicted from the reactions of the $pp$ chain, computed for a solar standard model, is given in Fig. 29.5. In order to obtain the neutrino spectrum of the present sun one cannot use the simple (equilibrium) formulae (18.63,65), but must compute the rates of all the single reactions given in (18.62,64) and in addition the reaction (29.1). Consequently one obtains the abundances of all nuclei involved as functions of depth, as shown in Fig. 29.3. For the peak in the abundance of $^3He$ see the discussion in § 28.2.

The chlorine experiment by R. DAVIS (see, for instance, BAHCALL, DAVIS 1976) is sensitive to neutrinos with energies above 0.814 MeV. Therefore, as one can see from Fig. 29.5, only the $^8B$ neutrinos of the sun are counted. The experiment is based on the reaction $^{37}Cl + \nu \rightarrow {}^{37}Ar$, where the decays of radioactive argon nuclei are counted. The rate of neutrino captures is commonly measured in *solar neutrino units* (SNU). One SNU corresponds to $10^{-36}$ captures per second and per target nucleus.

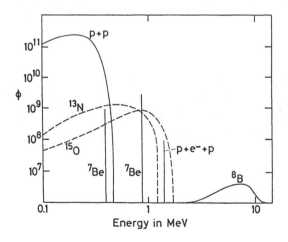

Fig. 29.5. The neutrino spectrum of the sun as predicted from a theoretical solar model. The solid lines belong to reactions of the $pp$ chain while the broken lines are due to reactions of the CNO cycle. The neutrinos from the $pp$ reaction as well as those from $^8B$, $^{13}N$, and $^{15}O$ have continuous spectra, while the monoenergetic neutrinos come from $^7Be$ and from the triple reaction (29.1). The flux $\phi$ for the continuum sources is given in cm$^{-2}$ s$^{-1}$ MeV$^{-1}$ and for the line sources in cm$^{-2}$ s$^{-1}$. (After BAHCALL et al., 1982)

The results of the measurements between 1970 and 1981 gave an $^{37}$Ar production rate of $1.3^{+0.7}_{-0.8}$ SNU, while from theoretical solar models one expects much higher rates. For instance, the solar standard model, after taking into account all possible uncertainties in the input parameters, predicts $7.6 \pm 3.3$ SNU. The measured rate is by roughly a factor 4 less than the predicted one.

The discrepancy has not yet been resolved. The theoretical solar neutrino rate would be less if the helium content in the central region of the models were reduced. The higher the helium content, the higher the central temperature must be in order to produce the solar luminosity. Since the side chain of the $pp$ reactions which contains the $^8$B decay is highly sensitive to temperature, a reduction of the helium content in the central region would decrease the central temperature and therefore the predicted rate of the $^8$B neutrinos. Indeed models in which the helium created since the formation of the sun has been mixed artificially over a larger region of the solar interior (instead of being left where it was formed) can be brought into agreement with the chlorine experiment (IBEN 1969; SCHATZMAN, MAEDER 1981; LAW et al., 1984). However, no convincing mechanism has been found up to now which causes such mixing.

Another possibility for overcoming the discrepancy is the assumption that the sun is not thermally adjusted. If the nuclear burning in the sun should suddenly completely extinguish, then it would take a Kelvin–Helmholtz time (that is, some $10^7$ years) until a change could be recognized at the surface of the sun. But the neutrino flux would immediately become zero. One could therefore imagine that the nuclear energy production may be modulated by some kind of instability in such a way that at present the nuclear energy production (and therefore the neutrino flux) is much smaller than that derived from the present luminosity via thermally adjusted solar models. With this explanation one encounters the difficulty that the sun seems to be stable.

Whether the solar energy production is that predicted by the (thermally adjusted) solar standard model can be checked by another neutrino experiment. The $^8$B neutrinos come from a rather unimportant side line. Its rate may change appreciably without changing the total energy output of the $pp$ chain. This is quite different for the neutrinos of the reaction $p + p$, which is the entrance for *all* branches and a measure for the whole energy production of the chain. Therefore the flux of these neutrinos is proportional to the present nuclear power of the sun. This must be equal to $L_\odot$ for a thermally adjusted model.

The low energy $pp$ neutrinos (Fig. 29.5) can be captured by $^{71}$Ga causing a transition to $^{71}$Ge. The predicted capture rate in this experiment for the solar standard model is $106.4^{+12.8}_{-8.5}$ SNU. Such experiments are presently being prepared.

One should keep in mind that the measurements of neutrino fluxes and their reduction are extremely involved. Even the physics of neutrinos is not yet sufficiently known. It is therefore not clear whether such experiments will tell us more about the sun or more about neutrinos.

# §30 Chemical Evolution on the Main Sequence

## 30.1 Change in the Hydrogen Content

In the main-sequence phase, the large energy losses from a star's surface are compensated by the energy production of hydrogen burning (see §18.5.1). These reactions release nuclear binding energy by converting hydrogen into helium. This chemical evolution of the star concerns primarily its central region, since the energy sources are strongly concentrated towards the centre (§22.2).

Somewhat larger volumes are affected simultaneously if there is a convective core in which the turbulent motions provide a very effective mixing. If the extent of convective regions and the rate of energy production $\varepsilon_H$ for all mass elements are known, the rate of change of the hydrogen content $X_H$ can be calculated according to §8.2.3.

The situation is particularly simple for stars of rather small mass (say $0.1 M_\odot < M \lesssim 1 M_\odot$) that have a radiative core. In the absence of mixing, the change of $X_H$ at any given mass element is proportional to the local value of $\varepsilon_H$. After a small time step $\Delta t$, the change of hydrogen concentration is $\Delta X_H \sim \varepsilon_H \Delta t$ everywhere (with a well-known factor of proportionality). Following the chemical evolution in this way over many consecutive time steps, one obtains "hydrogen profiles" [i.e. functions $X_H(m)$] as shown in Fig. 30.1. At the end of the main-sequence phase, $X_H \to 0$ in the centre.

In more massive stars, the helium production is even more concentrated towards the centre because of the large sensitivity to temperature of the CNO cycle. But the mixing inside the central convective core is so rapid compared to the local production

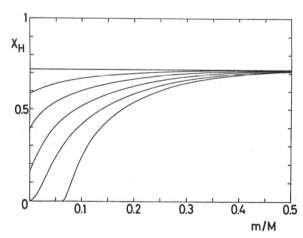

**Fig. 30.1.** Hydrogen profiles showing the gradual exhaustion of hydrogen in a star of $1 M_\odot$. The homogeneous initial model consists of a mixture with $X_H = 0.723$. The hydrogen content $X_H$ over $m/M$ is plotted for six models which correspond to an age of 0.0, 2.2, 4.5, 6.7, 8.9 and 11.4 times $10^9$ years after the onset of hydrogen burning

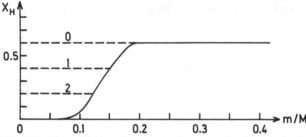

**Fig. 30.2.** The solid line gives the hydrogen profile $X_H(m)$ that is established in a $5M_\odot$ star of extreme population I at the end of hydrogen burning in a shrinking convective core. The dashed lines indicate the hydrogen content in the convective core at the onset (label 0) and at two intermediate phases of central hydrogen burning

of new nuclei that the core is virtually homogeneous at any time. Inside the core, $\Delta X_H \sim \bar{\varepsilon}_H \Delta t$ with an energy production rate $\bar{\varepsilon}_H$ averaged over the whole core. The only difficulty comes from the fact that the border of the convective core may change during the time step $\Delta t$. The numerical calculations show that for stars below $10M_\odot$ the mass $M_c$ of the convective core decreases with progressive hydrogen consumption, which leads to a hydrogen profile $X_H(m)$ as shown in Fig. 30.2 for a $5M_\odot$ star. At the end of central hydrogen burning, one has a helium core with $M_{He} \approx 0.1M$, and the envelope in which $X_H$ still has almost its original value. Similar profiles are established in stars with other values of $M$. The main difference is that with increasing $M$ the hydrogen profile is gradually shifted to larger values of $m/M$, i.e. the relative mass of the produced helium core increases with $M$. The corresponding increase of the convective core with increasing $M$ for zero-age main-sequence models has already been shown in Fig. 22.7.

This simple scenario is seriously complicated, particularly for rather massive stars, by two uncertainties in the theory of convection (convective overshoot and semiconvection). These effects will be indicated separately in § 30.4.

### 30.2 Evolution in the Hertzsprung–Russell Diagram

At the beginning of the main-sequence phase the models are located in the HR diagram on the zero-age main sequence (ZAMS) as described in § 22. Numerical solutions show that their positions change relatively little during the long phase in which hydrogen is exhausted in the central region. A typical evolutionary track (for a $7M_\odot$ star of extreme population I mixture) is given in Fig. 30.3a. Starting from point $A$ on the ZAMS, the luminosity increases by about $\Delta \lg L = 0.192$ to point $B$ and about $\Delta \lg L = 0.074$ from $B$ to $C$. The rise of $L$ is due to the increasing mean molecular weight when $^1H$ is transformed to $^4He$, in accordance with the prediction of the homologous relations [see, for example, (20.20)]. The evolution from $B$ to $C$ is so fast that $\mu$ increases only a very little in this short time interval. From the change of $r$ for different values of $m$ one clearly sees that the star evolves non-homologously, which ultimately is because the chemical composition changes only in the central region. The solutions show that the effective temperature

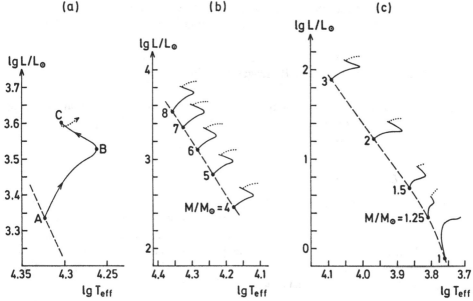

**Fig. 30.3 a–c.** Hertzsprung–Russell diagrams with evolutionary tracks for population I stars during central hydrogen burning (main-sequence phase). The zero-age main sequence is dashed. (a) For stellar mass $M = 7 M_\odot$. Some characteristic models are labelled by $A$ (age zero), $B$ (minimum of $T_{eff}$), and $C$ (exhaustion of central hydrogen). The dotted curve indicates the continuation of the track in the ensuing phase. (b) For stellar masses $M = 4 \ldots 8 M_\odot$. (c) For stellar masses $M = 3 \ldots 1 M_\odot$. (After MATRAKA et al., 1982)

decreases from $A$ to $B$ by $\Delta \lg T_{eff} \approx -0.061$, and then increases again to point $C$ by $\Delta \lg T_{eff} \approx 0.042$. This corresponds to an increase of the radius by $\Delta \lg R \approx 0.218$ ($A$ to $B$), and $\Delta \lg R \approx 0.047$ ($B$ to $C$). Point $B$ is reached after about $2.38 \times 10^7$ years, roughly when the central hydrogen content has dropped to $X_H \approx 0.05$. At point $C$, when $X_H = 0$ in the centre, the age is $2.49 \times 10^7$ years.

The evolutionary tracks are very similar for all stellar masses for which the hydrogen content is exhausted in a convective core of appreciable mass, i.e. on the whole upper part of the main sequence (see Fig. 30.3b). The increments of $\lg L$ from $A$ to $B$ and from $A$ to $C$ become somewhat larger for larger values of $M$, while the changes of $\lg T_{eff}$ remain about the same. The structure of the evolutionary tracks changes for smaller masses which have radiative cores. This can be seen in Fig. 30.3c.

A common feature of all evolutionary tracks described here is that they point in some direction *above* the ZAMS. This is the case only for an evolution producing chemically inhomogeneous models (composed of a helium core and a hydrogen-rich envelope). In an evolution assuming complete mixing of the whole model, $\mu$ would have a constant spatial distribution and would increase in time. Then the star would evolve below the ZAMS, in accordance with the discussion after (20.23). Aside from all details, the observations (e.g. cluster diagrams) show that evolved stars are in fact above and to the right of the ZAMS, i.e. the stars obviously develop chemical inhomogeneities in their interior. This conclusion is very important, in particular, for the theory of stellar rotation. It excludes, for example, a complete mixing by the large-scale currents of rotationally driven meridional circulations (§ 42).

### 30.3 Time-scales for Central Hydrogen Burning

The time $\tau_H$ a star spends on the main sequence while burning its central hydrogen depends on $M$. This is because its luminosity $L$ increases so strongly with $M$. Let us consider this time-scale

$$\tau_H = \frac{E_H}{L} \quad , \tag{30.1}$$

where $E_H$ is the nuclear energy content that can be released by central hydrogen burning. As a rough estimate, we assume that the same fraction of the total mass of hydrogen $M_H$ in the star is consumed in all stars. Then we have $E_H \sim M_H \sim M$. Since $L$ does not vary very much in this phase, we take the $M$–$L$ relation of the zero-age main sequence, $L \sim M^\eta$ [cf. (22.1)]. Introducing these proportionalities into (30.1) we have for the dependence of $\tau_H$ on $M$

$$\tau_H(M) \sim \frac{M}{L} \sim M^{1-\eta} \quad . \tag{30.2}$$

For an average exponent in the $M$–$L$ relation of, say, $\eta = 3.5$ one has $\tau_H \sim M^{-2.5}$, i.e. a strong decrease of $\tau_H$ towards larger values of $M$.

Of course, the numerical results are influenced and modified by a variety of details, some of which are not yet clear. A sequence of calculations for $X_H = 0.602$, $X_{He} = 0.354$ yields $\tau_H/10^7$ years = 8.03, 4.87, 3.32, 2.49, and 1.97, for $M/M_\odot =$ 4, 5, 6, 7, and 8 respectively. In all these cases, by far the largest part of $\tau_H$ was spent in the first phase between points $A$ and $B$, while the last part ($B$–$C$) covered only about 4...5%.

Although the absolute values are very uncertain (§ 30.4), the general trend is clear and has remarkable consequences for the observed HR diagrams of star clusters, by which it is confirmed. Assume that all stars of such a cluster were formed at the same time, i.e. that they now have the same age $\tau_{cluster}$. We must then conclude that all stars with masses larger than a limiting mass $M_0$ have already left the main-sequence region, while stars with $M < M_0$ are still on the main sequence. $M_0$ is given by the condition $\tau_{cluster} = \tau_H(M_0)$. This is the basis for the age determination of such clusters. As mentioned in § 28.2, in extremely young clusters the low-mass stars have not yet even reached the ZAMS.

### 30.4. Complications Connected with Convection

The seemingly nice and clear picture of the main-sequence phase as described above is unfortunately blurred by the notorious problem of convection. Questionable points include the precise determination of those regions in the deep interior in which convective motions occur, and therefore the extent to which the chemical elements are mixed. The mixing influences the later evolution, since the chemical profile, which is established and left behind, is a long-lasting memory. We briefly mention two problems, the first of which concerns all main-sequence stars having convective cores, while the second occurs only in the more massive of these stars.

### 30.4.1 Convective Overshooting

We consider the situation in the surroundings of the outer boundary of a convective core of mass $M_{bc}$, as calculated without allowance for overshooting. This means that here we have defined the boundary to be at the position of neutral stability, i.e. where

$$\nabla_{rad} = \nabla_{ad} \qquad (30.3)$$

according to the classical criterion (6.13). (Without much loss of generality, we may here treat a chemically homogeneous layer, e.g. in the model for a ZAMS star.) Complete mixing and a nearly adiabatic stratification with $\nabla = \nabla_{ad} + \varepsilon$ ($0 < \varepsilon \ll 1$) is assumed in the convective region below $M_{bc}$, while no mixing and $\nabla = \nabla_{rad}$ is assumed for the radiative region above $M_{bc}$ (cf. §6 and §7, in particular §7.3).

This model implies an obvious problem: the boundary between the regimes in which convective motions are present ($v > 0$) and absent ($v = 0$) is determined by the criterion (30.3), which essentially relies on buoyancy forces, and therefore describes the *acceleration* $\dot{v}$ rather than the velocity $v$ (cf. §6.1). Rising elements of convection are accelerated until they have reached $M_{bc}$; the braking starts only beyond this border, which is passed by elements owing to their inertia. The situation is the same as if we were to hope that a car would come to a full stop at the very point where one switches from acceleration to braking. The only way to substantiate this would be to try it (once) right in front of a hard and solid enough wall.

Simple estimates (e.g. SASLAW, SCHWARZSCHILD, 1965) indeed give the impression that there is such a hard wall for elements passing the border $M_{bc}$. We have seen in §7.3 that in the deep interior of the star the elements rise adiabatically such that $\nabla_e = \nabla_{ad}$. From (7.5) we then see that the buoyancy force $k_r$ acting on an element is

$$k_r \sim \nabla - \nabla_{ad} \quad , \qquad (30.4)$$

with a positive factor of proportionality. Below the border, $k_r$ is small and positive (small acceleration) since $\nabla - \nabla_{ad}$ is extremely small, and positive ($\approx 10^{-6}$). In contrast to this, the braking *above* the border is by orders of magnitude more efficient. We have assumed that there $\nabla$ is equal to $\nabla_{rad}$, which drops rapidly below $\nabla_{ad}$ (in Fig. 22.8a by about 0.1 within a scale height). So the force $k_r$ due to $\nabla - \nabla_{ad}$ soon reaches rather large and negative values: an overshooting element can be stopped within a negligible fraction of the pressure scale height.

A significant overshoot, therefore, could result only if the braking were substantially reduced (the "wall" softened). A possibility for this was outlined by SHAVIV, SALPETER (1973), who pointed to the recoupling of the overshoot on the thermal structure of the layer. Consider the temperature excess $DT$ of a moving element ($\nabla_e = \nabla_{ad}$) over the surroundings (gradient $\nabla$). According to (7.4), we have $DT \sim \nabla - \nabla_{ad}$, and $DT$ becomes negative above the border, i.e. the overshooting elements become cooler than the surroundings, which results in a cooling of the upper layers and an increase of the gradient $\nabla$. We may describe it in terms of the convective flux (positive, if it points outwards), which according to (7.3) is

$$F_{con} \sim v \cdot DT \tag{30.5}$$

(with positive factors of proportionality). Above the border, the upwards motion ($v > 0$) of cooler elements ($DT < 0$) represents a negative $F_{con}$. In order to maintain a constant total flux

$$F = F_{con} + F_{rad} = \frac{l}{4\pi r^2} \quad , \tag{30.6}$$

with $F_{con} < 0$, the radiative flux $F_{rad}$ must become larger than the total flux $F$. From (7.1,2) we immediately have

$$\frac{F_{rad}}{F} = \frac{\nabla}{\nabla_{rad}} \quad , \tag{30.7}$$

which shows that $\nabla > \nabla_{rad}$ for $F_{rad} > F$. The increase of $\nabla$, however, reduces the absolute values of $\nabla - \nabla_{ad}$ and of the braking force $k_r$ compared with the situation without overshooting; the elements can penetrate farther into the region of stability than originally estimated, etc.

To find out whether or not this provides an appreciable amount of overshooting is a difficult problem and one that is still far from being solved. In order to find the point where the velocity $v$ vanishes, one needs a self-consistent and detailed solution (including velocities, fluxes, gradients) for the whole convective core. This can only be obtained by using a theory of convection, the uncertainties of which now enter directly into the interior solution of the star. Even if we want to apply the mixing-length theory, the procedure is not clear. Instead of the usual local version of the theory, one needs a non-local treatment. At a given point, for example, the velocity of an element or its temperature excess depends not only on quantities at that point, but on the precise amount of acceleration (and braking) which the element has experienced along its whole previous path. All prescriptions for evaluating this and for averaging quantities like $v$ or $DT$ are as arbitrary as the choice of the mixing length. In fact any detailed modelling of the convective core by a mixing length theory is necessarily ambiguous. For example, it encounters the difficulty that this core extends over less than a scale height of pressure [the local expression of which, $H_P = -dr/d\ln P$, becomes $\infty$ at the centre according to (10.7)]. Different authors using different prescriptions have arrived at answers ranging from virtually no overshoot to rather extensive overshoot; and all of them have been questioned (see RENZINI, 1987). As an example we present results obtained by the treatment of MAEDER (1975). Figure 30.4 shows the typical run of some characteristic functions as obtained from such calculations for $M = 2M_\odot$ and $\alpha = \ell_m/H_P = 1$. Below the "classical" border of stability ($\nabla_{rad} = \nabla_{ad}$), one has typically $\nabla - \nabla_{ad} \approx +10^{-4}$

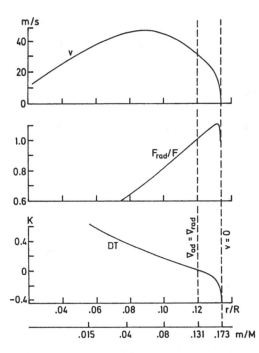

**Fig. 30.4.** Velocity $v$ and temperature excess $DT$ of rising convective elements, and the ratio of the radiation flux $F_{rad}$ relative to the total flux $F$ around the border of stability ($\nabla_{rad} = \nabla_{ad}$) in a star of $2M_\odot$. Overshooting calculated with $\alpha = \ell_m/H_P = 1$ extends to the point where $v = 0$ (after MAEDER, 1975)

which is enough to accelerate the convective elements to $30 \ldots 40\,\mathrm{m\,s^{-1}}$. Above the border, where still $v > 0$, but $DT < 0$, $F_{rad}$ exceeds the total flux $F$ by about 10%, while $\nabla - \nabla_{ad}$ ranges from $-10^{-4}$ to $-10^{-2}$. The overshooting reaches to the point with $v = 0$, which occurs at about 14% of the local scale height $H_P$ above the border, corresponding to an increase of the mass of the convective core $M_c$ of more than 30%. This amount depends on the assumed value of $\alpha$. Figure 30.5a shows the hydrogen profile established during hydrogen burning in a $7M_\odot$ star calculated

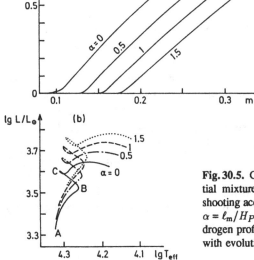

**Fig. 30.5.** Central hydrogen burning for a $7M_\odot$ star (initial mixture $X_H = 0.602$, $X_{rest} = 0.044$) with overshooting according to different assumptions for the ratio $\alpha = \ell_m/H_P$ ($\alpha = 0$ means no overshooting). (a) The hydrogen profile at the end of this phase. (b) HR diagram with evolutionary tracks. (MATRAKA et al., 1982)

with such overshooting for different $\alpha$. (The limit case $\alpha = 0$ is the model calculated without overshooting.) The influence of overshooting on the evolutionary tracks is shown in Fig. 30.5b. The consequences of an increased helium core at the end of this phase are an increased luminosity and an increased age (by about 25% for $\alpha = 1$). However, if such overshooting occurs, its main effect will show up only later, during the phase of helium burning (see §31.4).

But as mentioned before the question of overshooting is quite open and can be settled only by use of a better theory of convection.

### 30.4.2. Semiconvection

Another phenomenon related to convection introduces a large amount of uncertainty in the evolution of rather massive stars, say, for $M > 10 M_\odot$. (This limit depends on the chemical composition; it can even be around $7 M_\odot$ for hydrogen-rich mixtures of extreme population I stars.)

In these stars during central hydrogen burning the convective core retreats, leaving a certain hydrogen profile behind; the radiative gradient $\nabla_{rad}$ outside the core starts to rise and soon exceeds the adiabatic gradient $\nabla_{ad}$. This happens in a region with outwardly increasing hydrogen content (decreasing molecular weight $\mu$); therefore $\nabla_\mu \equiv d \ln \mu / d \ln P > 0$, which makes the layer dynamically stable. Considering the classical criteria for convective stability according to Schwarzschild and Ledoux we find

$$\nabla_{ad} < \nabla_{rad} < \nabla_{ad} + \frac{\varphi}{\delta} \nabla_\mu \ . \tag{30.8}$$

As described in §6.3 the layer is vibrationally unstable ("overstable"). A slightly displaced mass element starts to oscillate with slowly growing amplitude and penetrates more and more into regions of different chemical composition. This gives a rather slow mixing which is called *semiconvection*. The treatment of this process is complicated, one difficulty being that any degree of mixing must have a noticeable reaction on the stratification in the mixed layer.

Suppose that semiconvection occurs in some region of an originally very smooth hydrogen profile (solid line in Fig. 30.6a). The corresponding gradients are schematically sketched in Fig. 30.6b. If the mixing in this semiconvective region were very efficient, we would obtain a "plateau" in the profile (dashed line in Fig. 30.6a). There are obviously two main effects of such a mixing on the gradients. Firstly, any change of profile changes the value of $\nabla_\mu$, which goes to zero in the plateau. Secondly, the mixing increases the hydrogen content $X_H$ in the lower part and decreases $X_H$ in the upper part of the mixed region. In such massive main-sequence stars the opacity is largely dominated by electron scattering, for which $\kappa \sim (1 + X_H)$, [cf. (17.2)]. Since $\nabla_{rad} \sim \kappa$, [cf. (5.28)], the radiative gradient $\nabla_{rad}$ is increased in the lower part and decreased in the upper part of the mixed area. Therefore both these changes (of $\nabla_\mu$ and of $\nabla_{rad}$), which are due to the mixing, will modify the decisive terms entering into (30.8), and as a result some parts can completely change their stability properties (convective – semiconvective – radiative).

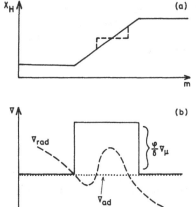

(a)  Fig. 30.6 a, b. Schematic illustration of the example for semiconvection discussed in the text. The solid line in (a) shows a hydrogen profile in which semiconvection occurs. Complete mixing in this layer would lead to the dashed "plateau". The gradients in the same range of $m$ are sketched in (b)

The slow mixing in semiconvective regions can be considered a diffusion process [see, for instance, LANGER et al. (1985)]. The resulting profile will depend on the time-scale $\tau_{\mathrm{diff}}$ of that kind of diffusion and its ratio to the typical time $\tau_*$ in which the stellar properties change (e.g. the composition due to nuclear reactions). For example, a relatively small $\tau_{\mathrm{diff}}$ (large diffusion coefficient) will tend to mix to such an extent that neutrality is nearly reached with $\nabla_{\mathrm{rad}} \approx \nabla_{\mathrm{ad}}$. In general, one should expect a continuous change of the profile, and radiative, semiconvective, and fully convective regions moving slowly through the star. Unfortunately the coefficient of diffusion cannot yet be determined satisfactorily, which is rather serious, since, as in the case of overshooting, the details of the established profile are very decisive for the later evolution of these stars.

Additional complications can arise from the interaction of semiconvection and overshooting. Note that semiconvection can also play a role in later phases, for example if a convective core increases during helium burning and expands into a region of different chemical composition.

## 30.5 The Schönberg–Chandrasekhar Limit

Since the nuclear time-scale for hydrogen burning is large compared with the Kelvin–Helmholtz time-scale, stars can be well represented by models in complete equilibrium throughout this phase. The question is now whether this continues to be valid also for the subsequent evolution. At the end of hydrogen burning, the star is left with a helium core without nuclear energy release surrounded by a hydrogen-rich envelope. At the bottom of this envelope, the temperature is just large enough for further hydrogen burning, which continues at this place in a *shell source*. The problematic part is the possible structure and change of the helium core. A core almost in thermal equilibrium without nuclear energy sources cannot have a considerable luminosity, and hence must be nearly isothermal, since $l \sim dT/dr$.

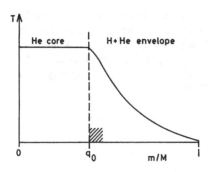

Fig. 30.7. Schematic temperature profile in an equilibrium model having an isothermal helium core of mass $q_0 M$. Hydrogen burns in a shell source (hatched) at the bottom of the envelope

Therefore we consider here equilibrium models consisting of an isothermal helium core of mass $M_c = q_0 M$ and a hydrogen-rich envelope of mass $(1 - q_0)M$ (see Fig. 30.7). For simplicity the chemical composition is taken to change discontinuously at the border of the two regions. The luminosity is supplied by hydrogen shell burning at the bottom of the envelope. In the following, solutions for the core (subscript 0 at its surface $q = q_0$) and solutions for the envelope (subscript e at the lower boundary $q = q_0$) are first discussed separately and then fitted to each other. In view of their importance we will look at the surprising results from different points of view.

### 30.5.1 A Simple Approach – The Virial Theorem and Homology

Important properties of such models can be understood by rather simple considerations, which give at least a qualitatively correct picture. We assume the isothermal core after hydrogen burning to consist of an ideal monatomic gas (molecular weight $\mu_{core}$). To this core, we apply the virial theorem in the form (3.21) which contains a term for the non-vanishing surface pressure $P_0$. Solving for $P_0$, we obtained (26.23), which we here rewrite as

$$P_0 = C_1 \frac{M_c T_0}{R_c^3} - C_2 \frac{M_c^2}{R_c^4} \quad , \tag{30.9}$$

where $C_1$, $C_2$ are positive factors, and $C_1 \sim c_v = 3\Re/(2\mu_{core})$. This describes the resulting surface pressure $P_0$ as the difference between the average interior pressure (first term $\sim \bar{\varrho} T_0$) and the self-gravity term (second term $\sim R_c \bar{g} \bar{\varrho}$), when we use $\bar{\varrho} \sim M_c/R_c^3$ and $\bar{g} \sim M_c/R_c^2$.

For simplicity we assume $T_e$ to be kept at a constant value by the thermostatic action of hydrogen burning. The fitting condition at $q_0$ then requires

$$T_0 = T_e = \text{constant} \quad , \tag{30.10}$$

and $P_0$ depends only on $M_c$ and $R_c$. As explained in § 26.2 the counteraction of the two terms in (30.9), which depend on different powers of $R_c$, has the result that, for $M_c = \text{constant}$, $P_0$ has a *maximum value* $P_{0max}$ at $R_c = R_{cmax}$ [see (26.24)],

$$R_{cmax} = C_3 \frac{M_c}{T_0} \quad , \quad P_{0max} = C_4 \frac{T_0^4}{M_c^2} \quad , \tag{30.11}$$

with some positive constants $C_3$, $C_4$. This can be obtained by solving $\partial P_0/\partial R_c = 0$

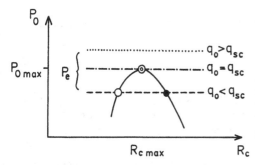

Fig. 30.8. The solid line shows schematically the pressure $P_0$ at the surface of the isothermal core as a function of the core radius $R_c$. Horizontal lines indicate the pressure $P_e$ at the bottom of the envelope for 3 different relative core masses $q_0$. The stable solution is marked by a dot, the unstable solution by an open circle; the solution at $P_{0max}$ is marginally stable

(for constant $T_0$) from (30.9). The function $P_0(R_c)$ for given $M_c$ and $T_0$ is sketched in Fig. 30.8. From (30.11) we see that $P_{0max} \sim M_c^{-2}$, i.e. the *maximum surface pressure of the core decreases strongly with the mass $M_c$ of the core.*

For the functions at the bottom of the envelope we simply assume that all possible envelopes are homologous to each other. Then from (20.17,24) follow $P_e \sim M^2/R^4$ and $T_e \sim M/R$. The latter relation together with (30.10) means that $M/R$ = constant, such that the relation for $P_e$ becomes

$$P_e = C_5 \frac{T_0^4}{M^2} \ . \tag{30.12}$$

We see that $P_e$ is independent of $R_c$, and has the same dependence on $T_0$ as $P_{0max}$, but decreases with $M$ instead of $M_c$. This can lead to difficulties! In Fig. 30.8 the envelope pressure $P_e$ according to (30.12) is given by a horizontal straight line, the height of which depends on $M$.

The remaining fitting conditions for a complete solution of the star require $R_c = r_e$ and $P_0 = P_e$, i.e. we look for an intersection of the two types of curves in Fig. 30.8. Obviously this can be obtained only if $P_e \le P_{0max}$, which together with (30.11) and (30.12) gives the condition

$$q_0 \equiv \frac{M_c}{M} \le q_{SC} \ . \tag{30.13}$$

i.e. the *relative* core mass $q_0$ must not exceed a certain limiting value, which is called the *Schönberg–Chandrasekhar limit* $q_{SC}$.

For $q_0 < q_{SC}$ we have $P_e < P_{0max}$ and there are two intersections in Fig. 30.8. The solution for the smaller value of $R_c$ is thermally unstable, the other one is stable. This can be made plausible by a simple argument. Fig. 30.8 shows that, if we slightly increase the core radius of the stable solution, $P_0$ drops below $P_e$ and the envelope tends to compress the core thus restoring the equilibrium state. The opposite behaviour (further increase of an initial expansion, since $P_0$ exceeds $P_e$) can be seen to result from the perturbation of the unstable equilibrium state and this rough argument is confirmed by a strict eigenvalue analysis.

The solutions merge for $q_0 = q_{SC}(P_e = P_{0max})$ which corresponds to neutral stability. And there are no solutions possible for $q_0 > q_{SC}$, since $P_e$ always exceeds $P_0$. In such a case some basic assumption of our present picture has to be dropped (e.g. equilibrium, or ideal gas). This will be discussed later.

The value of $q_{SC}$ has been computed by SCHÖNBERG, CHANDRASEKHAR (1942). It depends on the ratio of the molecular weights $\mu_{core}/\mu_{env}$, since the envelope pressure depends on $\mu_{env}$, while $P_0$ depends on $\mu_{core}$ via $C_1$. One can write roughly

$$q_{SC} = 0.37 \left( \frac{\mu_{env}}{\mu_{core}} \right)^2 , \qquad (30.14)$$

which means for $\mu_{core} = 4/3$ and a hydrogen-rich envelope $q_{SC} \approx 0.09$. This value is certainly exceeded by the helium cores that are left after central hydrogen burning in stars of the upper main sequence. Stars of somewhat smaller mass may encounter the same difficulty later, when the shell source burns outwards, thus increasing the mass of the helium core above the critical value. The Schönberg–Chandrasekhar limit is therefore quite relevant for the evolution in any phases in which at a first glance one would expect isothermal cores of ideal gas to appear.

### 30.5.2 Integrations for Core and Envelope

More reliable curves in the $P$–$R_c$ diagram (Fig. 30.8) can be easily obtained by numerical integrations for core and envelope (ROTH, 1973).

An envelope solution can be calculated for given $M$ and $M_c$ by requiring the lower boundary conditions $l = 0$, $r = R_c$ to hold at $M = M_c$. The solution gives $P_e$ and $T_e$ at $m = M_c$. By varying $R_c$, one obtains a set of solutions which gives $P_e(R_c)$, $T_e(R_c)$. Two typical envelope curves $P_e(R_c)$ are shown in Fig. 30.9a. It turns out that these curves, in their important parts, are nearly independent of $M_c$ but are raised essentially by a decrease of $M$. [This is qualitatively the same as in the approximation (30.12).] The temperature $T_e$ varies, in fact, very little along such an envelope curve. For later applications (§ 31.1) we briefly mention the surface

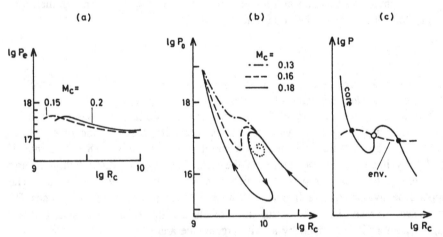

Fig. 30.9 a–c. Some typical curves of the pressure $P$ (in dyn cm$^{-2}$) against the core radius $R_c$ (in cm). (a) The pressure $P_e$ at the lower boundary of the envelope for a stellar mass $M = 2M_\odot$ and two values of the core mass $M_c$ (in $M_\odot$). (b) The pressure $P_0$ at the surface of isothermal cores of different mass $M_c$ (in $M_\odot$). The arrows along the solid curve indicate the direction of increasing central pressure. The dotted spiral is with neglect of degeneracy. (c) Sketch of core and envelope curves for the case of three intersections giving three complete solutions (*filled circles* stable, *open circle* unstable). (After ROTH, 1973)

values of these envelope solutions. Those with large values of $R_c$ are located near the main sequence. With decreasing $R_c$ they move to the right in the HR diagram, and envelopes with the smallest values of $R_c$ are close to the Hayashi line.

The solution for an isothermal core with temperature $T_0$ can be obtained by a straightforward integration starting at the centre with an assumed value of $P = P_c$ and continued until $m = M_c$ is reached. At this point one finds a pair of values $P = P_0$ and $r = R_c$. Many such integrations for different values of the parameter $P_c$ then give the curve $P_0(R_c)$ for the core. The solid line in Fig. 30.9b gives such a curve for cores of mass $M_c = 0.18M_\odot$ and $T_0 = 2.24 \times 10^7$ K. The lower-right part (small $P_0$, large $R_c$) corresponds to small central pressures $P_c$. With increasing $P_c$ the curve leads up to the maximum and decreases again. (This corresponds to the maximum of the core curve in Fig. 30.8, while the horizontal envelope curves there are now replaced by envelope curves like those in Fig. 30.9a.) Then it would follow the dotted spiral, if we artificially suppress the deviation from the ideal-gas approximation in the equation of state. This may be compared with the spiral in the $U$–$V$ plane obtained for an isothermal core in Fig. 21.2. An increasing $P_c$, however, implies an increasing degeneracy of the electron gas. This "unwinds" the spiral and $P_0$ drops, while a gradually increasing fraction of the core becomes degenerate. When degeneracy encompasses practically the whole core, $P_0$ rises again strongly with decreasing $R_c$ (upper-left end of the solid curve in Fig. 30.9b). The dashed and dot-dashed lines demonstrate how the curve changes when $M_c$ is decreased. As predicted by (30.11) the maximum shifts to smaller $R_c$ and larger $P_0$. The main effect, however, is that the minimum is less and less pronounced. This goes so far that finally the maximum, which is decisive for the existence of a Schönberg–Chandrasekhar limit, has disappeared. A similar change of the structure of the curve is obtained if, instead of decreasing $M_c$, we increase the temperature $T_0$.

### 30.5.3 Complete Solutions for Stars with Isothermal Cores

As mentioned, each sequence of envelope solutions yields a relation $T_e = T_e(R_c)$. Assume now that along a corresponding sequence of isothermal-core solutions $T_0$ is varied such that $T_0(R_c) = T_e(R_c)$ for all $R_c$. This deforms a core curve in Fig. 30.9b only slightly. Any intersection of this new core curve with a corresponding envelope curve gives a complete solution, since we then have at $m = M_c$

$$r_e = R_c \quad , \quad P_e = P_0 \quad , \quad T_e = T_0 \quad , \quad l_e = l_0 = 0 \quad , \tag{30.15}$$

i.e. continuity of all variables.

Suppose that the core curve has a pronounced maximum. We can then obviously expect to have up to 3 solutions (see Fig. 30.9c), one with an ideal gas (largest $R_c$), the second with partial degeneracy (intermediate $R_c$), and the third with large degeneracy (smallest $R_c$) in the core. If the envelope curve passes below the minimum or above the maximum of the core curve, there will be only one solution. And there can also be only one solution with a monotonic core curve.

The resulting solutions for different values of $M$ and $q_0 = M_c/M$ can best be reviewed by representing them as linear series of models (Fig 30.10) in which $q_0$

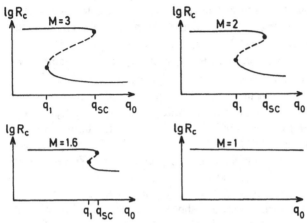

**Fig. 30.10.** Linear series of complete equilibrium solutions for four different stellar masses $M$ (in $M_\odot$) having an isothermal core of mass $M_c = q_0 M$. Each solution here is characterized by its core radius $R_c$ and its relative core mass $q_0$. Branches with thermally stable solutions are shown by solid lines, branches with unstable solutions by dashed lines. The turning point at $q_0 = q_{SC}$ defines the Schönberg–Chandrasekhar limit. (After ROTH, 1973)

varies as parameter while $M$ is fixed. Each model is represented here by its core radius $R_c$ in order to give an easy connection with the foregoing fitting procedure. Note that these sequences are a continuation of those discussed in § 23.4 (see Fig. 23.9) towards small core masses $M_c$ without helium burning.

Figure 30.10 shows that for larger $M$ the linear series consist of three branches. Two of them contain thermally stable models (solid lines), the other unstable models (dashed). On the upper and lower stable branch, the isothermal cores have no or strong degeneracy respectively. The branches are connected by two turning points (at $q_1$, and $q_{SC}$) where the models have marginal stability (zero eigenvalues) and where the local uniqueness of the solution is violated (cf. § 12). The turning point with the larger $q_0$ defines the Schönberg–Chandrasekhar limit. Its value $q_{SC}$ turns out to be nearly independent of $M$. For $q_1 < q_0 < q_{SC}$ there are three solutions, otherwise one solution. When going to gradually smaller $M$, we see that $q_1$ approaches $q_{SC}$, until both turning points merge and finally disappear for $M < 1.4 M_\odot$. For such small $M$, therefore, one has only one (stable) branch and no Schönberg–Chandrasekhar limit. This agrees with what one expects from the core curves given in Fig. 30.9b. It shows, for example, that the curves are already monotonic for $M = 1.3 M_\odot$ and $q_0 \le 0.1$ (i.e. $M_c \le 0.13 M_\odot$).

Instead of $R_c$, we might have plotted the stellar radius $R$ over the parameter $q_0$. As mentioned above, small $R_c$ corresponds to large $R$ and vice versa. The sequences for large enough $M$ would then exhibit a stable dwarf branch for $q_0 < q_{SC}$, a stable giant branch for $q_0 > q_{SC}$ and an unstable intermediate branch.

In evolutionary models one will encounter a smooth profile rather than a discontinuity of the chemical composition. In such a case it is profitable to define the effective core mass by the point of maximum shell source burning, and the Schönberg–Chandrasekhar limit by the transition from thermal stability to instability derived from an eigenvalue analysis. One then finds again that $q_{SC} \approx 0.1$.

# VI  Post-Main-Sequence Evolution

# § 31 Evolution Through Helium Burning – Massive Stars

## 31.1 Crossing the Hertzsprung Gap

After central hydrogen burning, the star has a helium core, which in the absence of energy sources tends to become isothermal. Indeed thermal equilibrium would require that the models consist of an isothermal helium core (of mass $M_c = q_0 M$, radius $R_c$), surrounded by a hydrogen-rich envelope [of mass $(1 - q_0)M$] with hydrogen burning in a shell source at its bottom. Such models were discussed in detail in § 30.5. We now once more consider the case of $M = 3M_\odot$, which is typical for stars on the upper part of the main sequence (say $M > 2.5M_\odot$). The possible solutions were comprised in a linear series of models consisting of 3 branches. This is shown in the first graph of Fig. 30.10, and again in Fig. 31.1, which also gives the position in the HR diagram.

Suppose that the relative mass of the core $q_0$ has not yet reached the Schönberg–Chandrasekhar limit $q_{SC}(\approx 0.1)$ at the end of central hydrogen burning. The model then can easily settle into a state contained in the uppermost branch of Fig. 31.1a, which consists of stars close to the main sequence (Fig. 31.1b). Let us imagine a "quasi-evolution" of this simple model by assuming that $M_c$ grows because of shell burning while complete equilibrium is maintained. The result is that the model is shifted to the right in Fig. 31.1a. This proceeds continuously until the model reaches the Schönberg–Chandrasekhar limit, represented by the turning point which terminates the uppermost branch of the linear series. Further increase of $M_c$ would require the model to jump discontinuously onto the lower branch in Fig. 31.1a. This decrease of $R_c$ (i.e. compression of the core) would be accompanied by a large jump in the HR diagram, from the main sequence to the region of the Hayashi line (Fig. 31.1b). This means that such equilibrium stars have to become giants because the main-sequence solutions (which the stars had selected owing to their history) cease to exist, while the red-giant solutions (which have coexisted for a long time) are still available. In Fig. 31.1a and b the quasi-evolution of increasing $M_c$ is indicated by solid lines, while those parts of the linear series which can obviously not be reached are broken. We will see that basic features of this jump in the simple quasi-evolution (particularly the compression of the core together with an expansion of the envelope to a red-giant stage) are recovered in the real evolution which, of course, leads through non-equilibrium models. In any case, a phase of thermal non-equilibrium must follow after central hydrogen burning, since a continuation via suitable equilibrium models would involve a discontinuity.

As an example for the real evolution we may take numerical solutions obtained for upper-main-sequence stars with a fairly hydrogen-rich initial mixture ($X_H =$

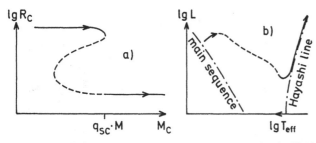

Fig. 31.1. (a) The same linear series for $M = 3M_\odot$ as in Fig. 30.10. The core radius $R_c$ is plotted against the core mass $M_c$. In a quasi-evolution with increasing mass $M_c$ of the isothermal helium core the model shifts along the solid lines, as indicated by the arrows. (b) The corresponding position in the HR diagram

0.602, $X_{He} = 0.354$, $X_{rest} = 0.044$). The transition from central to shell burning can be seen from Fig. 31.2a. Such graphs are very useful for illustrating in a concise way the occurrence of certain properties as functions of $m$ and $t$. Any line parallel to the ordinate indicates what one would encounter in different layers when moving along the radius of the star at that moment of the evolution. Figure 31.2b gives the corresponding evolutionary track in the HR diagram. The first part of Fig. 31.2a (from $A$ to $C$) shows the phase of central hydrogen burning which exhausts $^1H$ in the core within about $5.6 \times 10^7$ years for $5M_\odot$. With hydrogen being depleted there, the burning together with the convection ceases rather abruptly in the central region. At the same time, hydrogen burning starts in an initially rather broad shell around the core, i.e. in the mass range of the outwards-increasing hydrogen content left by the shrinking convective core (cf. Fig. 30.2a). Later this shell source narrows remarkably in mass-scale, particularly when it has consumed the lower tail of the hydrogen profile. After phase $C$ the evolution is so much accelerated that the abscissa has had to be expanded. The models are no longer in thermal equilibrium, i.e. the time derivatives ($\varepsilon_g = -T\partial s/\partial t$) in the energy equation are not negligible [cf. (4.27,28)].

The radial motion of different mass elements in this phase is shown in Fig. 31.3 for a star of $7M_\odot$. After a short resettling at the end of central hydrogen burning (point $C$) we see that core and envelope change in opposite directions: an expansion of the layers above the shell source (at $m \approx 0.15M$) is accompanied by a contraction of the layers below. The fact that $\dot{r}$ changes sign at the maximum of a shell source is a pattern very characteristic for models with strongly burning shell sources; it can occur in quite different phases of evolution, for contracting or expanding cores, for one or two shell sources. Such shell sources seem to represent sorts of mirror in the pattern of contraction and expansion inside a star ("mirror principle" of radial motion).

The $\varepsilon_g$ term also changes sign at the maximum of the shell source. One finds that $\varepsilon_g > 0$ in the contracting core, and $\varepsilon_g < 0$ in the expanding envelope. The energy released in the contracting core must flow outwards, which prevents the core from becoming isothermal. Such a massive star starts on the main sequence with relatively low central density (cf. Fig. 22.5) and therefore remains non-degenerate during the present contraction, which then leads to heating. When the central temperature has

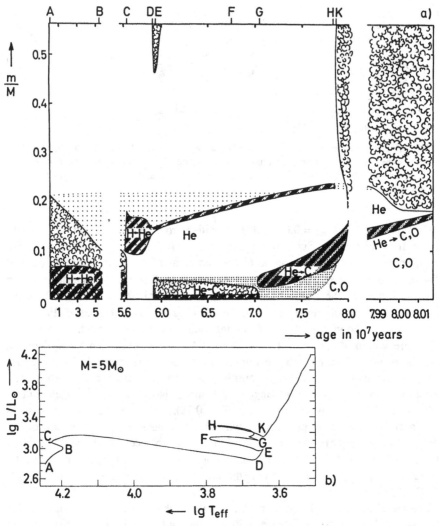

**Fig. 31.2. (a)** The evolution of the internal structure of a star of $5M_\odot$ of extreme population I. The abscissa gives the age after the ignition of hydrogen in units of $10^7$ years; each vertical line corresponds to a model at a given time. The different layers are characterized by their values of $m/M$. "Cloudy" regions indicate convective areas. Heavily hatched regions indicate where the nuclear energy generation ($\varepsilon_H$ or $\varepsilon_{He}$) exceeds $10^3$ erg g$^{-1}$ s$^{-1}$. Regions of variable chemical composition are dotted. The letters $A \ldots K$ above the upper abscissa indicate the corresponding points in the evolutionary track, which is plotted in Fig. 31.2 **(b)**. (After KIPPENHAHN et al., 1965)

reached about $10^8$ K, helium is ignited. The core has thus tapped a large new energy source which stops its rapid contraction, and the star again reaches a stage of complete (thermal and hydrostatic) equilibrium. The whole core contraction from $C$ to $D$ has proceeded roughly on the Kelvin–Helmholtz time-scale of the core (in less than $3 \times 10^6$ years for $5M_\odot$ and about $5 \times 10^5$ years for $7M_\odot$). In the same time, the outer layers have rapidly expanded and the stellar radius is increased appreciably (roughly by a factor 25 in Fig. 31.3).

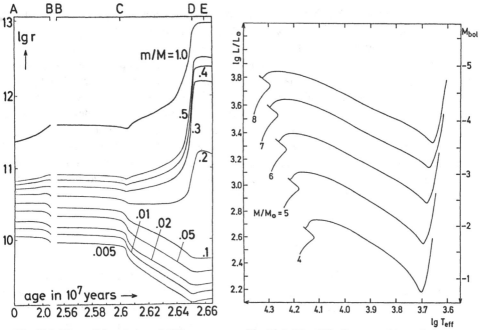

**Fig. 31.3.** The radial variation of different mass shells (characterized by their $m/M$ values) in the post-main-sequence phase of a star of $7 M_\odot$. The letters $A \ldots E$ correspond to the evolutionary phases labelled in the two Figs. 31.2 for a star of $5 M_\odot$. (After HOFMEISTER et al., 1964)

**Fig. 31.4.** The HR diagram with evolutionary tracks from the zero-age main sequence to the onset of helium burning for stars with different masses $M$ (from $4 M_\odot$ to $8 M_\odot$) and for an initial composition with $X_H = 0.602$, $X_{He} = 0.352$. (After MATRAKA et al., 1982)

The evolutionary path in the HR diagram for the $5 M_\odot$ star is shown in Fig. 31.2b. The expansion transforms the star into a red giant so rapidly that there is little chance of observing it during this short phase of evolution. This explains the existence of the well-known *Hertzsprung gap*, an area between main sequence and red giants with a striking deficiency of observed stars.

The evolution is qualitatively similar for all stars in which helium burning is ignited before the core becomes degenerate, and in which possible complications due to semi-convection cannot prevent the star from moving close to the Hayashi line. This includes stellar masses of, say, $2.5 M_\odot < M < 10 M_\odot$. A set of evolutionary tracks in this phase for different $M$ is shown in Fig. 31.4.

Let us briefly come back to the linear series of equilibrium models with isothermal cores discussed at the beginning of this paragraph. We have seen that the real non-equilibrium evolution can lead to quite different equilibrium stages (with helium burning). Therefore we should extend the linear series such that it includes *all* possible equilibrium solutions for the given parameters ($M$, chemical composition), i.e. solutions with isothermal cores *and* solutions with helium burning. This completion will be done in detail later (§ 32.8) when we are familiar with the transition between the two types of models, though a simple general conclusion may be drawn immediately. For certain ranges of the parameters the linear series will again have the 3 branches (2 stable, 1 unstable) with isothermal helium cores as discussed above. In addition there is a fourth branch of (stable) models with helium burning cores. The general results described in § 12.5 then require the existence of at least one further (unstable) branch, since they can be "added" only in pairs of opposite stability properties. Consequently we know that there must exist at least 5 branches yielding up to 5 solutions (3 stable, 2 unstable) for given parameters in a certain range. To which of these different equilibrium solutions a real evolution will finally tend depends on the initial stage and on the details of an intermediate non-equilibrium evolution. For example, if the helium core becomes sufficiently dense by its contraction after central hydrogen burning, then the temperature for helium burning is not reached and the evolution leads to an equilibrium stage with an isothermal, degenerate core (as will be seen in § 32 to happen with less-massive stars).

### 31.2 Central Helium Burning

As a consequence of the rapid contraction and heating of the core, central helium burning sets in (at the age of $5.9 \times 10^7$ years for $5M_\odot$). The star is then in the red-giant region of the HR diagram, close to the Hayashi line ($D$–$E$ in Fig. 31.2b). Correspondingly it has a very deep outer convection zone, which can be seen in Fig. 31.2a to reach down to $m/M \approx 0.46$. The larger $M$, the deeper the convection zone penetrates. For $7M_\odot$ it extends down to $m/M < 0.3$, into layers in which the composition was already slightly modified by the earlier hydrogen burning. Therefore some products of this burning are now dredged up by the convection and distributed all over the envelope. We here encounter one of the mechanisms by which nuclear species produced in the very deep interior can be lifted to the stellar surface.

The high temperature sensitivity of helium burning provides a strong concentration of the energy release towards the centre and therefore the existence of a convective core. The core contains roughly 5% of $M$, i.e. much less mass than during hydrogen burning.

At first the dominant reaction is $3\alpha \rightarrow {}^{12}$C (cf. § 18.5.2). With increasing abundance of ${}^{12}$C the reaction ${}^{12}$C $+\alpha \rightarrow {}^{16}$O gradually takes over. When ${}^4$He has already become rather rare the depletion of ${}^{12}$C on account of ${}^{16}$O is larger than the production of ${}^{12}$C by the $3\alpha$ reaction, and ${}^{12}$C decreases again after having reached a maximum abundance. This is explained by the fact that the production of ${}^{12}$C is proportional to $X_\alpha^3$, while its depletion is proportional to $X_\alpha X_{12}$. The change of the abundances can be seen from Fig. 31.5, which shows the final composition for such stars to be ${}^{12}$C and ${}^{16}$O in roughly equal amounts with only a very small admixture of ${}^{20}$Ne. Note, however, that the final ratios ${}^{16}$O/${}^{12}$C and ${}^{20}$Ne/${}^{16}$O are much larger when calculated with more modern reaction rates (ARNETT, THIELEMANN, 1985). Anyway, they increase with increasing stellar mass, since $T$ increases.

The phase of central helium burning lasts roughly $1.1 \times 10^7$ years, which is about 20% of the duration of the main-sequence phase. This fraction seems to be surprisingly large in view of the facts that now $L$ is somewhat higher, the exhausted core much smaller, and the specific gain of energy (per unit of mass of the fuel) is only 1/10, as compared with hydrogen burning. The simple reason is that most of

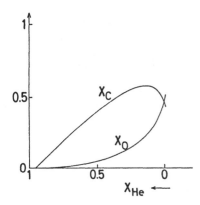

**Fig. 31.5.** Variation of the abundance of $^{12}$C and $^{16}$O during the depletion of $^4$He in the centre of a $5M_\odot$ star. (After MEYER-HOFMEISTER, 1967)

the total energy output in this phase comes from hydrogen shell source burning. For a star of $5M_\odot$ helium burning contributes only about 6%, 20%, and 48% at points $E$, $F$, and $G$ respectively: a rather small release of nuclear energy inside the core is obviously sufficient to prevent it from contraction and to bring the whole star nearly into thermal equilibrium. The luminosity $L_{He}$ produced between points $E$ and $F$ by helium burning in a helium core of mass $M_{He}$ is roughly equal to the luminosity a pure helium star of $M = M_{He}$ would have on the helium main sequence (cf. § 23.1). In fact the helium-burning core resembles in several respects a star on the helium main sequence with $M = M_{He}$. For later applications we note that the radius $R_{He}$ of the core changes rather little during most of this phase. It increases very slowly until the central helium content has dropped to $X_{He} \approx 0.3$. It is only in the final phase of central helium burning ($X_{He} < 0.1$) that the core contracts and $R_{He}$ drops appreciably. It should be mentioned that the evolution can be affected by convective overshooting, which enlarges the convective core during central helium burning.

Let us now look at the HR diagram in Fig. 31.2b. After point $E$ the star goes at first down along the Hayashi line, then leaves this line and moves far to the left. The "bluest" point $F$, for $5M_\odot$, is reached after $8.3 \times 10^6$ years (75% of the helium-burning phase) when the central helium content is down to about $X_{He} \approx 0.25$. The track then leads back towards point $G$ in the vicinity of the Hayashi line. The further evolution in which another loop may occur will be discussed in § 31.5.

The extension of the loops, i.e. the distance of their bluest points from the Hayashi line, depends on the stellar mass $M$. We limit the discussion to a range of not too large masses, say $M < 10M_\odot$, where the situation is relatively simple and clear. Large loops are obtained for stars with large $M$. With decreasing $M$ the loops become gradually smaller and finally degenerate to a mere down and up along the Hayashi line. This can be seen in Fig. 31.6, which gives the evolutionary tracks for a comparable set of computations. The loops for different stellar masses cover a roughly wedge-shaped area which is bordered by the Hayashi line and the connection of the bluest points of the loops (i.e. points $F$ where $T_{eff}$ has a maximum). The duration of characteristic phases as obtained from these calculations can been seen from Table 31.1. Point $E'$ corresponds to the minimum of $L$ after $E$, where the leftwards motion starts, while $G'$ indicates the end of the central helium burning. [As with most numerical values obtained up to now from evolutionary calculations,

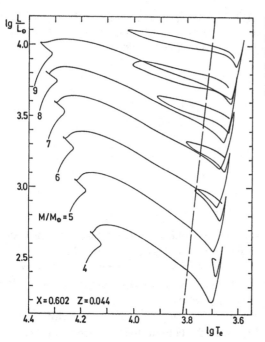

**Fig. 31.6.** Hertzsprung–Russell diagram with evolutionary tracks for stars in the mass range from $4M_\odot$ to $9M_\odot$ from the main sequence through helium burning (after MATRAKA et al., 1982). The broken line indicates the Cepheid strip

**Table 31.1** Characteristic points and the time elapsed after the zero-age main-sequence stage in the evolutionary tracks of models of different masses (after MATRAKA et al., 1982). The meaning of the points $E$, $E'$, $F$, $G'$ is explained in the text

|  |  | $t$ (in $10^6$a) | $\lg L/L_\odot$ | $\lg T_e$ |
|---|---|---|---|---|
| $4M_\odot$ | $E$ | 87.857 | 2.591 | 3.659 |
|  | $E'$ | 98.839 | 2.376 | 3.685 |
|  | $F$ | 116.58 | 2.495 | 3.700 |
|  | $G'$ | 123.83 | 2.497 | 3.684 |
| $5M_\odot$ | $E$ | 51.435 | 2.943 | 3.645 |
|  | $E'$ | 57.228 | 2.770 | 3.669 |
|  | $F$ | 65.663 | 2.987 | 3.768 |
|  | $G'$ | 69.552 | 2.904 | 3.663 |
| $6M_\odot$ | $E$ | 34.478 | 3.253 | 3.632 |
|  | $E'$ | 38.381 | 3.116 | 3.653 |
|  | $F$ | 42.428 | 3.302 | 3.788 |
|  | $G'$ | 44.756 | 3.223 | 3.650 |
| $7M_\odot$ | $E$ | 25.912 | 3.541 | 3.615 |
|  | $E'$ | 28.592 | 3.392 | 3.640 |
|  | $F$ | 30.641 | 3.637 | 3.912 |
|  | $G'$ | 32.228 | 3.500 | 3.637 |
| $8M_\odot$ | $E$ | 20.013 | 3.787 | 3.605 |
|  | $E'$ | 21.460 | 3.593 | 3.631 |
|  | $F$ | 22.345 | 3.861 | 4.005 |
|  | $G'$ | 24.308 | 3.713 | 3.635 |

these data should be taken as an indication of typical relative properties, rather than as absolutely reliable. For other data see, e.g., the review article by IBEN, (1974).]

The situation is much more complicated for still larger masses, where the loops do not continue to grow with $M$ and the tracks remain well separated from the Hayashi line. Unfortunately this depends on the uncertain details of the mixing during the earlier main-sequence phase (compare § 30.4, § 31.4).

The importance of the loops comes from the fact that they occur during a nuclear, slow phase of evolution in which the star has a sufficient chance of being observed (contrary to the foregoing phase of core contraction). We therefore expect to find helium-burning stars as red giants in the area of the HR diagram covered by the loops. This is in fact the case, as can be seen from HR diagrams of open clusters (see Fig. 31.7). They often show a more or less extended giant branch, which is clearly separated from the main sequence by the Hertzsprung gap, and which sets out nicely the range of loops for the corresponding values of $M$. However, the observed even distribution of stars along the giant branch is hard to explain theoretically. The calculations rather predict a strong concentration towards the Hayashi line (corresponding to the long duration of the phase $E–E'$ in Table 31.1).

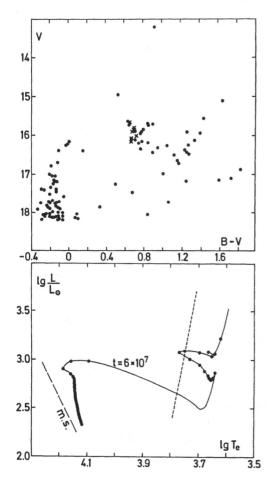

Fig. 31.7. (*above*) Equivalent of an HR diagram (magnitude $V$ against colour index $B - V$) for the cluster NGC 1866 (after ARP, THACKERAY, 1967). Crosses indicate Cepheid variables. (*below*) The isochrone (positions of stars of different mass but the same age) for the age $t = 6 \times 10^7$ years (after MEYER-HOFMEISTER, 1969). The density of points on the line indicates the expected star density. The straight broken line is the zero-age main sequence, the dotted line the central line of the Cepheid strip

299

## 31.3 The Cepheid Phase

It is of particular significance that the loops are necessary for explaining the observed $\delta$ Cephei variables. The observations show that these stars are giants, located in the HR diagram in a narrow strip roughly parallel to the Hayashi line and a few $10^2$ K wide (cf. Fig. 31.6). Indeed the theory of stellar pulsations which will be described in § 39 predicts that a star is vibrationally unstable if it is located in the "instability strip" of the HR diagram, where the observed Cepheids are found. This is a consequence of the way in which the outer stellar envelope (particularly the helium ionization zone) reacts on small perturbations. When a stellar model has evolved into the instability strip, the oscillation will grow to finite, observable amplitudes. This phenomenon does not show up in the normal evolutionary calculations which are carried out by neglecting the inertia terms in the equation of motion, since these terms are necessary to obtain an oscillation at all. The calculated evolution therefore gives only the unperturbed solution.

The evolutionary tracks discussed above cross the instability strip up to three times. For all stars a first crossing occurs in the short phase of core contraction when the star moves from $C$ to $D$ (Fig. 31.6). This passage is so rapid that there is scarcely a chance for observing a star as a Cepheid in this phase. So we are left with the much slower second and third passages, which occur only for sufficiently large loops. According to Fig. 31.6 this is roughly the case for all stars with $M \gtrsim 5M_\odot$. This lower mass limit for Cepheids depends of course on all the uncertainties of the the loops in the computed evolutionary tracks.

The theory of stellar pulsations (§§ 38,39) also gives the period $\Pi$ of the oscillation. For the evolutionary models the theory in fact yields values of $\Pi$ comparable with the observed Cepheid periods. In a first approximation, $\Pi$ is shown to depend only on the mean density $\bar{\varrho}$ of the whole star as

$$\Pi\sqrt{\bar{\varrho}} = \text{constant} \quad , \quad \bar{\varrho} \sim M/R^3 \quad . \tag{31.1}$$

Indeed $\Pi$ is of the order of the hydrostatic time-scale $\tau_{\text{hydr}}$ introduced in (2.19).

During a passage through the Cepheid strip from right to left ($E \rightarrow F$), the radius $R$ decreases, which means that $\Pi$ must also decrease according to (31.1). During a passage in the opposite direction ($F \rightarrow G$), the period $\Pi$ will increase. The calculated evolutionary velocity in the passage yields changes of the period which should be in a range accessible to measurements. (Note that the period of a Cepheid can be measured with very high precision by observations covering many periods.)

Since the Cepheid strip is rather narrow, each passage defines reasonably well a pair of average values of $L$ and $R$, and (31.1) then gives the corresponding period $\Pi$. When going from the lowest to the highest passages in Fig. 31.6, we find that $\Pi$ increases, since its variation is dominated by the increase of $R$, which enters into $\bar{\varrho}$ with the third power. In fact this, together with the properties of the instability strip in § 39, will be shown to lead to the famous $\Pi$–$L$ relation of Cepheids, which is the basic standard for the determination of extragalactic distances.

The duration of a passage $\tau_{\text{cep}}$ increases strongly towards lower values of $L$ (i.e. of $\Pi$). For an assumed width of $\Delta \lg T_{\text{eff}} = 0.025$ for the strip, the crossing on the way from $E$ to $F$ takes $\tau_{\text{cep}} = 9.1 \times 10^3$ years for $8 M_\odot$ and $\tau_{\text{cep}} = 2.3 \times 10^6$ years for $5 M_\odot$. From $\tau_{\text{cep}}$ one can draw conclusions on the number of Cepheids to be expected. It turns out that this number should increase substantially towards smaller values of $\Pi$, reach a maximum (at a period of a few days), and then drop steeply, since the loops no longer reach the Cepheid strip. This is at least qualitatively in agreement with the observations.

A less favourable result concerns the masses of the Cepheids. One value, called the "evolutionary mass" $M_{\text{ev}}$, can be obtained with the help of evolutionary calculations essentially by comparing the luminosities. On the other hand, non-linear pulsation calculations show that the form of the light curves should depend on $M$, and a comparison with observed light curves gives a "pulsational mass" $M_{\text{pul}}$. Now one finds that $M_{\text{ev}}$ notoriously exceeds $M_{\text{pul}}$ by 20 – 40%. This problem is amply discussed in the literature (compare COX, 1980). It seems as if a major role is played by the difficulties involved in the transformation from measured quantities into physically relevant values (such as $L$ and $R$).

If there were an appreciable amount of convective overshooting during the earlier main-sequence phase (§ 30.4.1), then the stars have a larger core mass $M_{\text{He}}$, causing a higher luminosity during helium burning (cf. Fig. 30.5b). This obviously reduces the derived values of $M_{\text{ev}}$ (bringing it closer to $M_{\text{pul}}$). On the other hand, too large values of $M_{\text{He}}$ owing to overshooting would suppress the loops completely (cf. § 31.4).

We have dwelt at length on this short phase of evolution, since the Cepheids are important and offer a major fraction of those rare cases which, at least in principle, allow a quantitative test of the theory.


## 31.4 To Loop or Not to Loop ...

In § 31.3 we saw how important it is to find evolutionary tracks looping through the red-giant region during central helium burning. It was all the more noteworthy when one learned that the loops depend critically on some uncertain input parameters (e.g. $\kappa$, $\varepsilon$, treatment of convection, composition) used in the calculations. A detailed classification of all influences, including their mutual interaction, is far too involved. Rather we point out a few characteristic properties of the models which allow a phenomenological prediction on the looping. [We here follow the discussion of LAUTERBORN et al. (1971a,b); for other descriptions see ROBERTSON (1971), FRICKE, STRITTMATTER (1972)].

For not too large masses (say, $M \lesssim 7 M_\odot$), the evolution through the loops is so slow that the $\varepsilon_g$ terms scarcely play a role. So we can reproduce the loops sufficiently well by models in complete equilibrium. Let us again consider solutions for the helium core (mass $M_c$, radius $R_c$, luminosity $l_0$) and for the hydrogen-rich envelope separately before fitting them to a full solution. The core luminosity $l_0$ is supplied by central helium burning; hydrogen shell burning at the bottom of the envelope gives the additional luminosity $L-l_0$.

For given chemical composition a solution for the envelope can be obtained after specifying a pair of values $R_c$, $l_0$ as inner boundary conditions at $m = M_c$.

(This is quite analogous to the usual central conditions $r = l = 0$ at $m = 0$.) Any solution gives a point in the HR diagram as well as pressure and temperature at $M_c$, i.e. values for $L$, $T_{\text{eff}}$, $P_0$, $T_0$. For the first part of the loop, helium burning contributes relatively little to $L$. Consequently we may approximate the envelope by setting $l_0 = 0$. (This can be done, of course, only for the calculation of the envelope, which is dominated by hydrogen burning; in the core, $l_0$ cannot be neglected since it represents the whole local luminosity of this region.) The envelope solutions then form a two-parameter set in which we treat $M_c$, $R_c$ as free parameters.

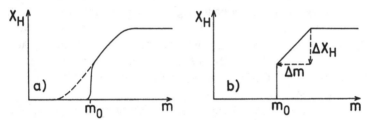

**Fig. 31.8.** The hydrogen abundance in an evolved star. (a) The convective core has left a fairly smooth profile (*dashed line*) which afterwards is steepened by shell burning. The shell is centred at $m_0$. Consequently $X_H = 0$ for $m < m_0$. For $m > m_0$ there is still a region in which $X_H$ is not constant. (b) Schematic description of the chemical profile given by the solid line in (a)

Next we look for a simple description of the chemical composition in the envelope. Figure 31.8a shows a typical hydrogen profile. A rather moderate increase of $X_H$ is the relic of hydrogen consumption in the convective core during the main-sequence phase. The very narrow shell source has already eaten away the lower part of this profile (dashed) and produced a steep increase of $X_H$ above the momentary helium core. We idealize this by a profile described by the parameters $\Delta m$ and $\Delta X$, as shown in Fig. 31.8b. The further shell burning will obviously increase $M_c$ and decrease $\Delta m$ and $\Delta X$.

Now the envelope solution and its position in the HR diagram depend on the 4 parameters $M_c$, $R_c$ $\Delta m$, $\Delta X$. We would like to have a simple function of these parameters which can serve as a measure for the separation from the Hayashi line. The back-and-forth motion in the loop would then correspond to a non-monotonic variation of this function. A hint for a suitable procedure can be found in Fig. 31.1. The envelopes there shift monotonically to the right in the HR diagram, while the cores move through all three branches of the linear series with increasing ratio $M_c/R_c$. This is essentially the surface potential of the core and plays a decisive rôle in many descriptions of radial expansion and contraction during the evolution. So we consider an "effective core potential"

$$\varphi := h\frac{M_c}{R_c} \quad , \tag{31.2}$$

where we count $M_c$, $R_c$ in solar units. The function $h = h(\Delta m, \Delta X)$ takes account of the influence of the chemical profile. We normalize it by setting $h = 1$ for a simple step profile ($\Delta m = \Delta X = 0$) and specify it later for other profiles. For a step profile and for $M = 5M_\odot$ five sequences of envelope solutions with constant $M_c$

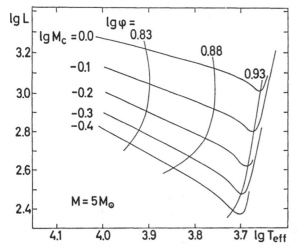

Fig. 31.9. Envelope solutions for $M = 5M_\odot$ with homogeneous composition down to $M_c(h = 1)$ for different values of the core mass $M_c$ (in $M_\odot$). Lines of constant core potential $\varphi$ are indicated. (After LAUTERBORN et al., 1971)

but varying $R_c$ are shown in Fig. 31.9. The plotted lines $\varphi =$ constant illustrate that $\varphi$ may indeed serve as an indicator of the distance from the Hayashi line. In particular we can find a critical value $\varphi_{cr}$ such that all envelopes with

$$\varphi > \varphi_{cr} \tag{31.3}$$

are close to, and move upwards along, the Hayashi line with increasing $\varphi$. The line $\varphi = \varphi_{cr}$ may therefore roughly connect the minima of the envelope curves, and from Fig. 31.9 we see that $\lg \varphi_{cr} = 0.93$ for $5M_\odot$. For $M = 3M_\odot$ and $7M_\odot$, it is 0.83 and 0.99 respectively.

The function $h$ is defined so that models with different profiles but equal distances from the Hayashi line have the same $\varphi$. Numerical experiments with different profiles have shown that the simple approximation

$$h = e^{\text{constant}\cdot\Delta m\cdot\Delta X} \tag{31.4}$$

is sufficient. Here $h$ depends only on the product $\Delta m \cdot \Delta X$, that is to say on the amount of helium in and just above the shell source. The profile influences the envelope mainly through a hydrostatic effect.

Finally, relations between $M_c$ and $R_c$ have to be derived from solutions for the core. Each solution of an envelope of given $M_c$, $R_c$ yields a pair of values $P_0$, $T_0$. For each $M_c$ we vary $R_c$ and get the functions $P_0(R_c)$ and $T_0(R_c)$, which are taken as outer boundary conditions for the core. For a specified composition and different $M_c$ the core solutions then give the required relation $R_c(M_c)$, which is quite different for $\varphi$ larger or smaller than $\varphi_{cr}$, namely $M_c/R_c \sim M_c^{0.4}$ ($\varphi < \varphi_{cr}$) and $\sim M_c^{0.25}$ ($\varphi > \varphi_{cr}$). Therefore this factor tends to increase $\varphi$ when the shell source burns outwards. We then have, in addition, the influence of the chemical evolution of the core on $R_c$. As mentioned earlier, an appreciable effect occurs only after the central helium content has dropped below, say, 0.1. The following rapid decrease of $R_c$ tends to increase $\varphi$. Both these effects (the increase of $M_c$ and the decrease of $R_c$) tend to shift the model to the right in the HR diagram and may therefore finish a loop, but they can never start it.

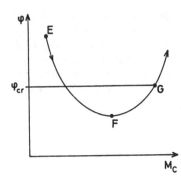

Fig. 31.10. Sketch of the effective potential $\varphi$ as a function of the core mass $M_c$ for an evolution through a loop. The points $E$, $F$, $G$ refer to those in Fig. 31.2

Obviously the responsibility for the onset of a loop rests with the function $h$. In fact, when the shell source burns farther into the profile, $\Delta m$ and $\Delta X$ (cf. Fig. 31.8) become smaller and $h$ decreases according to (31.4). This outweighs the increase of $M_c/R_c$ in the first phase after $E$, and $\varphi$ becomes smaller (Fig. 31.10). Sooner or later, however, the factor $M_c/R_c$ takes over, since it continues to grow steadily, while $h$ will level off near its maximum $h = 1$ when the shell source has "crunched up" almost the entire profile. Therefore $\varphi$ reaches a minimum $\varphi_{min}$ and then increases again. The turning of $\varphi$ at $\varphi_{min}$ can be caused either by the growth of $M_c$ or by the drop of $R_c$ due to helium depletion. Which of these effects occurs earlier will depend on the ratio of the time-scales for shell source and central burning.

So we have found a non-monotonic variation of $\varphi$. Whether this results in a loop, and if so the length of the loop, will depend on $\varphi_{cr}$ and the starting value $\varphi(E)$ by which we denote the value of $\varphi$ at point $E$. For small $M$, $\varphi(E)$ exceeds $\varphi_{cr}$ by so much that even $\varphi_{min}$ remains above $\varphi_{cr}$, and no loop occurs. The variation of $\varphi$ then is reflected only in a motion down-and-up near the Hayashi line (Fig. 31.6, for $M \lesssim 4M_\odot$). High values of $M$ bring $\varphi(E)$ close to $\varphi_{cr}$, and therefore in the further evolution $\varphi$ goes below $\varphi_{cr}$. A case with $\varphi_{min} < \varphi_{cr}$ is illustrated in Fig. 31.10. When $\varphi$ drops below $\varphi_{cr}$ the model detaches from the Hayashi line and starts looping to the left. The turn to the right begins at point $F$ when $\varphi = \varphi_{min}$.

Now it is obvious that many factors can modify the loops. For example, all properties changing the ratio of the time-scales for central helium burning and shell burning can shift $\varphi_{min}$ and thus the bluest point of the loop. In particular, we have to mention all the uncertainties concerning convection. Appreciable overshooting on the main sequence shifts the whole profile outwards. This can increase $M_c$ and consequently $\varphi(E) - \varphi_{cr}$ such that the loop becomes smaller, if it is not completely suppressed. Other factors affect the decisive upper part of the hydrogen profile. Aside from careless integrations during the main-sequence phase there are also physical uncertainties which can leave faulty profiles in the models. An example is the mixing by the outer convection zone during its deepest penetration, which in turn depends on the chosen mixing length in the superadiabatic layer. A similar problem causes the semi-convective region in main-sequence stars of large $M$ (cf. § 30.4.2). The assumption that this region is fully mixed leads to a plateau in the calculated pro-file with a discontinuous drop of $X_H$ at its bottom. The presence or absence of this plateau must strongly influence the function $h$. Correspondingly the literature presents quite different evolutionary tracks for massive stars during helium burning

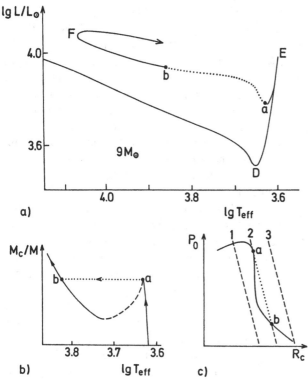

**Fig. 31.11.** (a) Evolutionary track in the HR diagram for $M = 9M_\odot$, from the post-main-sequence contraction through central helium burning. Capital letters $D$, $E$, $F$ refer to characteristic phases as in Fig. 31.2. Corresponding equilibrium models jump discontinuously from $a$ to $b$ (see text). (b) Linear series of equilibrium models for $M = 9M_\odot$ during central helium burning. The parameter $M_c$ (core mass) is plotted against the effective temperature. There are two thermally stable branches (*solid lines*) and an unstable one (*dashed*). Arrows indicate a quasi-evolution with increasing $M_c$. (c) Pressure at the bottom of the envelope of $9M_\odot$ equilibrium models (*solid line*), against the core radius $R_c$. The dashed and dotted lines indicate schematically the pressure at the surface of the core for 3 consecutive phases (1), (2), and (3) after point $E$. Complete models exist for the intersections (dots). The points labelled $a$ and $b$ correspond to those in Fig. 31.11(a, b). (After LAUTERBORN, et al., 1971)

(some with loops near the Hayashi line, others more to the left and completely detached from this line) for different assumptions on the semi-convective mixing. We see that details, which have originated from different regions and from earlier phases when the effects were scarcely recognizable, can now pop up and modify the evolution appreciably. The present phase is a sort of magnifying glass, revealing relentlessly the faults of calculations of earlier phases.

We now have to describe a modification occurring for stars of larger masses, say $M > 7M_\odot$ (depending on the chosen mixing length). This is illustrated in Fig. 31.11a for $M = 9M_\odot$. Models in complete equilibrium roughly approximate the evolution in most parts of helium burning, except for an intermediate phase (dotted) where the models jump discontinuously (from $a$ to $b$). This has to be expected from the structure of the linear series of these equilibrium models shown in Fig. 31.11b. The value of $M_c$ is taken as the parameter (which in contrast to our earlier plots of linear series is used here as the ordinate). Each solution is characterized by its value of $T_{eff}$, which facilitates the comparison with Fig. 31.11a. We see that there are 3 branches connected via turning points in which local uniqueness is violated and the Henyey determinant $|H|$ vanishes (see § 12). On the dashed branch, we have $|H| < 0$ and the models are thermally unstable, while the other branches consist of stable models with $|H| > 0$. The equilibrium evolution starts on the right branch, which it goes up along until the first turning point is reached. The slightest further increase of $M_c$ due to shell source burning forces the model to jump discontinuously onto the left stable branch, i.e. from a to b in Fig. 31.11a. The occurrence of these different branches can be traced back to a peculiarity of the function $P_0(R_c)$ of the envelope solutions (see above). For larger $M$ this function has the form sketched in Fig. 31.11c. The left/right part belongs to envelopes near/off the Hayashi line respectively. The point here is that the transition between the two parts is steeper than the function $P_0(R_c)$ for the core (illustrated for three consecutive phases 1, 2, and 3 by the straight dashed and dotted lines). Note that the lines for core and envelope with increasing $M_c$ shift in this diagram, but only their relative position is important. We start with phase 1, where only one solution is possible, namely the fit given by an intersection on the upper part of the envelope curve. In the ensuing evolution, this intersection shifts to point $a$, after which it has to jump to point $b$ on the lower part of the envelope curve. This is connected with a sudden increase of $R_c$, i.e. a decrease of $\varphi$, which makes the model jump to the left in the HR diagram. Of course, the real evolution replaces the jump by a rapid evolution through thermal non-equilibrium stages. For smaller $M$, the envelope curve is everywhere less steep than the core curve, and this problem does not occur except for small values of the mixing length.

## 31.5 After Central Helium Burning

In the central core, helium burning terminates when $^4$He is completely processed to $^{12}$C, $^{16}$O, and $^{20}$Ne (in various ratios, depending on the temperatures, i.e. on the stellar mass, and on the reaction rates used). The burning continues in a concentric shell surrounding the exhausted core and the formation of this shell source for $5M_\odot$ can be seen in Fig. 31.2a. While the helium shell burns outwards, the C–O core increases in mass and contracts. Obviously the situation resembles that before central helium burning. Now, however, the star has two shell sources, since the hydrogen shell is still burning at the bottom of the hydrogen-rich envelope. Again the type of mirror principle for radial motions as described in § 31.1 seems to act: the core contracts, the helium region between the two shell sources expands, the envelope contracts. In the HR diagram of Fig. 31.2b the model moves to the left from $G$ to $H$. Then the temperature in the hydrogen shell source drops so far that hydrogen burning ceases. The outer of the two "mirrors" has disappeared, and core contraction is now accompanied by expansion of all layers above the helium-burning shell. In the HR diagram the model moves to the right ($H \rightarrow K$) and then upwards. The luminosity increases strongly (more than by a factor 10 in Fig. 31.2b) with increasing mass of the C–O core. These two quantities are in fact correlated, as we shall see in

§ 32 and § 33. Whether or not the calculations yield a second loop ($G \to H \to K$ in Fig. 31.2b) depends on the input values used, e.g. burning rates, $\kappa$, $M$.

From Fig. 31.2a we see that the outer convective envelope gradually reaches further down until it contains more than 80% of the stellar mass. Its lower boundary clearly penetrates into a range of mass through which the hydrogen shell source has burned during the preceding $\sim 10^7$ years, processing all $^1$H to $^4$He, and nearly all $^{12}$C and $^{16}$O to $^{14}$N. These nuclei are now dredged up by the outer convection zone and can appear at the surface. This is usually called the phase of the *second dredge up*.

With the inward motion of the lower border of convection, the H–He discontinuity has come rather close to the helium shell source where $T \approx 2 \times 10^8$ K. The approach of this hot helium shell heats up the hydrogen until it is ignited – the hydrogen shell source is reactivated. Before we continue discussing these massive stars in § 33, we have in the next section to describe the evolution of low-mass stars through central helium burning.

# § 32 Evolution Through Helium Burning – Low-Mass Stars

## 32.1 Post-Main-Sequence Evolution

Compared to massive stars, those of lower masses (typically $M < 2.3 M_\odot$) evolve in a qualitatively different way after the exhaustion of hydrogen in their central regions. There are several reasons for this difference. Low-mass main-sequence stars have small, or no, convective cores and degeneracy is important, if not on the main sequence, then shortly afterwards. In addition they start at a point on the main sequence much closer to the Hayashi line than the starting points of massive stars.

For example, if hydrogen is consumed in a well-mixed convective core, there will be a helium core of appreciable mass at the very end of central hydrogen burning. However, stars of around $1 M_\odot$ have no convective cores; they consume hydrogen as illustrated in Fig. 30.1. Consequently they produce a growing helium core starting at zero mass. Therefore there is a smooth transition from central to shell burning. These stars start with such large central densities ($\gtrsim 10^2$ g cm$^{-3}$) that the electron gas is at the border of degeneracy, which has several consequences. The Schönberg–Chandrasekhar limit (§ 30.5) is not important: initially, the core mass $M_c$ is below $0.1 M$. When, however, with outward burning shell source $M_c > 0.1 M$, the core contraction has produced sufficient degeneracy, making this limit irrelevant. The stars can then well exist in thermal equilibrium with a degenerate, isothermal helium core. This means that there is no "need" for a *rapid* core contraction as described in § 31.1 and no equivalent of the Hertzsprung gap. Another consequence of degeneracy is that core contraction is not connected with heating. This is in contrast to the pre-main-sequence contraction (§ 28.1) and to post-main-sequence core contraction, which leads to helium ignition in massive stars.

At least in the first phases to be discussed here, the growth of the core mass is slow (since the productivity of the shell source is low), and the whole core settles at the temperature of the surrounding hydrogen-burning shell. This means that the core temperature is far from that of the ignition of helium ($\approx 10^8$ K). In low-mass stars, helium burning will be seen to start much later owing to secondary effects, after the core mass has grown up to a certain limit. Therefore the shell-burning phase between the central hydrogen and helium burning is a nuclear, slow phase and one can expect to find many such stars in the sky.

The contraction of the core is (as in the case of larger $M$) accompanied by an expansion of the hydrogen-rich envelope outside the shell source. However, as long as the luminosity does not change drastically the expansion cannot carry the star far away from its starting point on the main sequence. The reason is that this point is already close to the Hayashi line, which cannot be crossed (§ 24).

Any further expansion of the envelope is only possible if the luminosity increases. In fact the calculations show that $L$ now increases by more than a factor $10^2$ while $M_c$ grows.

Surprisingly enough it turns out that $L$ soon depends on the properties of the core only and is practically independent of the mass $M - M_c$ of the envelope (and therefore of $M$). In this phase the models can be well described analytically by a generalized form of homology.

## 32.2 Shell-Source Homology

Consider a model in complete equilibrium consisting of a degenerate helium core (mass $M_c$, radius $R_c$) surrounded by an extended envelope of hydrogen with abundance $X_H$. The core mass $M_c$ grows owing to hydrogen-shell burning, which provides the luminosity $L$:

$$\dot{M_c} = \frac{L}{X_H E_H} \tag{32.1}$$

(where $E_H$ is the energy gain per unit mass of hydrogen). This equation could easily be integrated if $L$ were constant. However, while evolution proceeds, $L$ grows too, since there is a relation between $L$ and $M_c$. The properties of the shell (and therefore $L$) are mainly determined by $M_c$ and $R_c$, while they are almost independent of the properties of the envelope. This can be understood from the fact that the core is highly concentrated and the gravity at its surface is very large. Then, according to hydrostatic equilibrium, $|dP/dm| \sim m/r^4$ is very large and $P$ drops by powers of 10 within a thin mass shell just above the core surface. In other words, the extended envelope above this layer is nearly weightless and has no influence on the burning shell.

We now present an analytic approach of REFSDAL, WEIGERT (1970) giving relations between the properties of the core and the physical variables in the hydrogen-burning shell. For this purpose we will generalize the homology considerations of §20 and use again the power approximations for $\kappa$ and $\varepsilon$

$$\kappa = \kappa_0 P^a T^b \quad , \quad \varepsilon = \varepsilon_0 \varrho^{n-1} T^\nu \quad . \tag{32.2}$$

Here we have replaced the exponent $\lambda$ used in §20 by $n - 1$.

For the gas pressure we will use the ideal-gas equation

$$P = \frac{\Re}{\mu} \varrho T \quad , \tag{32.3}$$

since we only want to apply it to regions outside the core, where the gas is not degenerate. We also neglect radiation pressure since it is not important for low-mass stars. In §33 we shall apply the relations derived here to massive stars and then take radiation pressure into account.

We now assume for the density, temperature, pressure, and local luminosity in the region of the hydrogen-burning shell that there exists a simple dependency on

$M_c$ and $R_c$:

$$\varrho(r/R_c) \sim M_c^{\varphi_1} R_c^{\varphi_2} \quad , \tag{32.4}$$

$$T(r/R_c) \sim M_c^{\psi_1} R_c^{\psi_2} \quad , \tag{32.5}$$

$$P(r/R_c) \sim M_c^{\tau_1} R_c^{\tau_2} \quad , \tag{32.6}$$

$$l(r/R_c) \sim M_c^{\sigma_1} R_c^{\sigma_2} \quad . \tag{32.7}$$

These homology-type relations have the following meaning: we compare two stellar models of different core masses $M_c$ and $M_c'$ and core radii $R_c$ and $R_c'$. We define homologous points, $r$ and $r'$, in the two models by

$$\frac{r}{R_c} = \frac{r'}{R_c'} \quad ; \tag{32.8}$$

the physical quantities at homologous points in the two models shall then be connected by relations (32.4–7). This indeed is very similar to the considerations of § 20.1, though there the homologous points were defined with respect to the total radius $R$, whereas we here define them with respect to the core radius $R_c$. While there, for example in (20.17), the physical quantities vary like powers of $M$ and $R$; they here vary like powers of $M_c$ and $R_c$. For example, with our new concept of homology (20.17) is replaced by (32.4,6), which are written explicitly as

$$\frac{\varrho}{\varrho'} = \left(\frac{M_c}{M_c'}\right)^{\varphi_1} \left(\frac{R_c}{R_c'}\right)^{\varphi_2} \quad , \tag{32.9}$$

$$\frac{P}{P'} = \left(\frac{M_c}{M_c'}\right)^{\tau_1} \left(\frac{R_c}{R_c'}\right)^{\tau_2} \quad . \tag{32.10}$$

We now introduce relations (32.4–7) into the stellar-structure equations in order to determine the exponents. We therefore write (2.4), (5.11), and (4.22) in the form

$$dP \sim M_c \varrho \, d(1/r) \quad , \tag{32.11}$$

$$d(T^4) \sim \kappa \varrho l \, d(1/r) = \kappa_0 \varrho P^a T^b l \, d(1/r) \quad , \tag{32.12}$$

$$dl \sim \varepsilon \varrho \, d(r^3) = \varepsilon_0 \varrho^n T^\nu \, d(r^3) \quad , \tag{32.13}$$

with positive factors of proportionality. In (32.11) we have assumed that $m \approx M_c =$ constant, which is a sufficient approximation in the region in which $P$ drops to negligible values. This assumption yields decisive differences from the relations discussed in § 20. Introducing (32.4,5,6) into (32.3) we easily obtain for the exponents

$$\tau_1 = \varphi_1 + \psi_1 \quad , \quad \tau_2 = \varphi_2 + \psi_2 \quad . \tag{32.14}$$

We now integrate (32.11,12,13) over the shell, starting with (32.11): we choose a radius $r_0$ sufficiently larger than $R_c$ that $P(r_0/R_c) \ll P(r/R_c)$, and find from (32.11) that

$$P(r/R_c) = P(r_0/R_c) + \int_{1/r_0}^{1/r} G M_c \varrho \, d(1/r) \approx \frac{G M_c}{R_c} \int_{x_0}^{x} \varrho \, dx \quad , \tag{32.15}$$

with $x = R_c/r$. If we do the same for another model with $M_c'$, $R_c'$, we find for the pressure at the homologous radius $r'$

$$P'(r'/R'_c) \approx \frac{GM'_c}{R'_c} \int_{x_0}^{x} \varrho' \, dx = \frac{GM'_c}{R'_c} \left(\frac{M'_c}{M_c}\right)^{\varphi_1} \left(\frac{R'_c}{R_c}\right)^{\varphi_2} \int_{x_0}^{x} \varrho \, dx \quad , \qquad (32.16)$$

where (32.9) has been introduced into the integral. Comparing (32.16) with (32.15) yields

$$P(r/R_c) \sim M_c^{\varphi_1+1} R_c^{\varphi_2-1} \quad , \qquad (32.17)$$

and if we compare this with (32.6) we find

$$\tau_1 = \varphi_1 + 1 \quad , \quad \tau_2 = \varphi_2 - 1 \quad . \qquad (32.18)$$

The same procedure can be carried out using equations (32.12,13). For the integration in the first case we again choose $r_0$ sufficiently far outside, where the temperature is small compared to its values in the shell; for the integration of (32.13) we take $r_0 = R_c$, where the local luminosity vanishes. We then obtain

$$(4 - b)\psi_1 = \varphi_1 + a\tau_1 + \sigma_1 \quad , \qquad (32.19)$$
$$(4 - b)\psi_2 = \varphi_2 + a\tau_2 + \sigma_2 - 1 \quad , \qquad (32.20)$$
$$\sigma_1 = n\varphi_1 + \nu\psi_1 \quad , \quad \sigma_2 = n\varphi_2 + \nu\psi_2 + 3 \quad . \qquad (32.21)$$

Equations (32.14,18,19,20,21) are 8 linear inhomogeneous algebraic equations for the 8 exponents in (32.4–7). The solutions are

$$\varphi_1 = -\frac{\nu - 4 + a + b}{N} \quad , \quad \varphi_2 = \frac{\nu - 6 + a + b}{N} \quad , \quad \psi_1 = 1 \quad , \quad \psi_2 = -1 \quad ,$$
$$\tau_1 = 1 + \varphi_1 \quad , \quad \tau_2 = \varphi_2 - 1 \quad , \quad \sigma_1 = \nu + n\varphi_1 \quad , \quad \sigma_2 = 3 - \nu + n\varphi_2 \quad ,$$

$$(32.22)$$

with

$$N = 1 + n + a \quad . \qquad (32.23)$$

Equations (32.22) allow us to determine the variations of the physical quantitites from one model (characterized by $M_c$, $R_c$) to another (characterized by $M'_c$, $R'_c$). The temperature and the local luminosity at homologous points vary as

$$T \sim M_c^{\psi_1} R_c^{\psi_2} = M_c/R_c \quad , \qquad (32.24)$$

$$l \sim M_c^{\nu+n\varphi_1} R_c^{3-\nu+n\varphi_2} \quad . \qquad (32.25)$$

This holds for all homologous points, also for those at the upper border of the range of integration where $l = L$. Therefore the luminosity of these shell source models depends on $M_c$ (rather than on $M$) and on the mode of energy generation (in striking contrast to main-sequence type models, cf. § 20). As an illustration we assume $a = b = 0$ (electron scattering, see § 17.1) and $\nu = 13$, $n = 2$ (CNO cycle, see § 18.5.1). Then $\varphi_1 = -3$, $\varphi_2 = 7/3$, and we find

$$L \sim M_c^7 R_c^{-16/3} \ . \tag{32.26}$$

We have obtained relations $T(M_c, R_c)$ and $L(M_c, R_c)$ independent of $M$. In order to see how $T$ and $L$ vary along an evolutionary sequence of models with increasing $M_c$, one has to know how $R_c$ varies with $M_c$. Since the cores in the evolution under consideration are degenerate, they resemble white dwarfs whose radii decrease with increasing mass (see § 19.6, § 35). We therefore can expect from (32.24) that the temperature in the shell source increases with $M_c$; and according to (32.26) the luminosity increases strongly with $M_c$ even with $R_c$ = constant (this increase being much steeper than the $L(M)$ relation for main-sequence stars).

We now need a relation $R_c(M_c)$. The classical mass–radius relation for white dwarfs (§ 35) is, of course, not directly applicable to these cores. Below the shell there must be a transition from complete through partial to no degeneracy. Compared to the outer layers of white dwarfs, this transition region is very hot (like the shell source) and may occupy an appreciable fraction of the core volume. (For a discussion of this problem see REFSDAL, WEIGERT, 1970.) Nevertheless, as a simple example for $R_c(M_c)$ we here take the relation for the cold white dwarfs of Table 35.1, yielding $d \ln R_c / d \ln M_c$ for different values of $M_c$. This can be used in

$$\frac{d \ln L}{d \ln M_c} = \sigma_1 + \sigma_2 \frac{d \ln R_c}{d \ln M_c} \ , \tag{32.27}$$

which follows from (32.7). The coefficients $\sigma_1$ and $\sigma_2$ are determined by (32.22). For $a = b = 0$, $n = 2$, $\nu = 14$, one finds $d \ln L / d \ln M_c \approx 8 \ldots 10$. We can also integrate (32.27) numerically when starting from a correctly computed model, which gives an initial value $L$ for a given $M_c$. The results of such an integration are shown in Fig. 32.1 by the left part of the solid curve where radiation pressure can be neglected ($\beta \approx 1$).

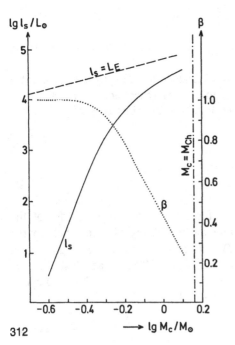

Fig. 32.1. The luminosity $l_s$ (*solid curve, left ordinate*) at the top of the hydrogen-burning shell around a degenerate helium core of mass $M_c$. The dotted line indicates the importance of the radiation pressure, the value of $\beta(= P_{gas}/P_{total})$ being given by the ordinate at the right. When $M_c$ approaches the Chandrasekhar mass $M_{Ch}$ (*dot-dashed vertical line*) the luminosity curve has the tendency to approach the Eddington luminosity $L_E$ (*dashed line*) for which gravity equals the radiation-pressure gradient (for an opacity dominated by electron scattering)

For the temperature at homologous points, say at the bottom of the hydrogen-burning shell, instead of (32.27) we obtain from (32.24)

$$\frac{d\ln T}{d\ln M_c} = 1 - \frac{d\ln R_c}{d\ln M_c} \quad , \tag{32.28}$$

and we get $d\ln T/d\ln M_c$ somewhat larger than 1. Since the cores are assumed to be isothermal, this also gives the increase of the central temperature $T_c$. We see that in this way $T_c$ can be raised to helium ignition even by models in complete equilibrium.

## 32.3 Evolution to the Helium Flash

In the following we describe the evolution of a star of $1.3 M_\odot$ as calculated by THOMAS (1967). The chemical composition of the initial model on the ZAMS is $X_H = 0.9$, $X_{He} = 0.099$, $Z = 0.001$, which at that time seemed to be the appropriate mixture for a star of population II. The essential results, however, do not depend too much on the chosen chemical composition. The initial model has $L = 1.91 L_\odot$, $T_{eff} = 6760$ K. Nuclear energy is released in the central region at $T_c = 1.48 \times 10^7$ K via the $pp$ chain. There is a small convective core containing 4.3% of the total mass, which disappears long before the exhaustion of hydrogen in the centre. There is also an outer convective zone, which reaches inwards from the photosphere to about $r \approx 0.95 R$.

The evolutionary track in the HR diagram is shown in Fig. 32.2, while the internal evolution is illustrated by Fig. 32.3. In the HR diagram the image point of the model first moves upwards and then to the right. At the same time, the model switches from central nuclear burning to shell burning, as can be seen in Fig. 32.3. We have already learned from the shell-source homology of § 32.2 that the luminosity must grow with increasing core mass. The calculated evolution confirms these predictions once the core is sufficiently compressed. The track is very close to the Hayashi line, leading up along the "ascending giant branch" to higher luminosities and correspondingly larger radii. The neighbourhood of the line of fully convective stars can also be seen from the internal structure of the models. Figure 32.3 shows that the outer convective zone penetrates deeply inwards until more than 70% of the total mass is convective. It then reaches into layers which are already contaminated by products of nuclear reactions (see dotted area in Fig. 32.3). The processed material is distributed over the whole convective region and therefore also brought to the surface. This type of partial mixing, which we have already encountered for massive stars in § 31, is called *first dredge up*.

The monotonic increase of the luminosity is interrupted when the hydrogen-burning shell reaches the layer down to which the outer convective zone has mixed at the moment of deepest penetration. At this point the mixing has produced a discontinuity in molecular weight between the homogeneous hydrogen-rich outer layer and the helium-enriched layers below. When the shell source reaches the discontinuity, the molecular weight of the shell material becomes smaller. This causes the drop of luminosity at $L \approx 100 L_\odot$ (see Fig. 32.2) as can easily be understood.

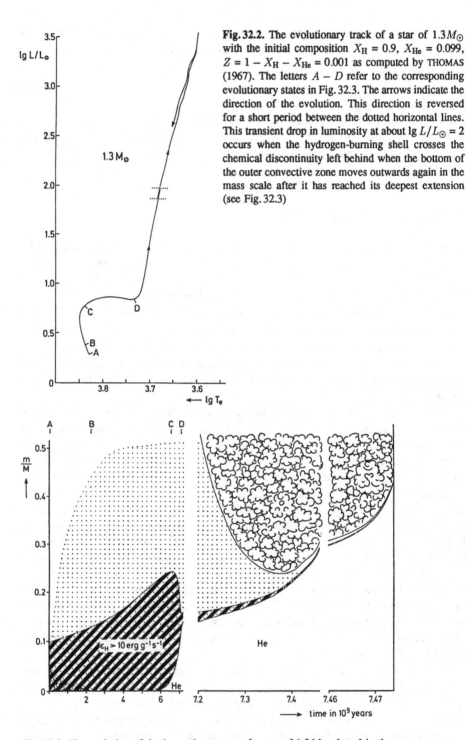

**Fig. 32.2.** The evolutionary track of a star of $1.3 M_\odot$ with the initial composition $X_H = 0.9$, $X_{He} = 0.099$, $Z = 1 - X_H - X_{He} = 0.001$ as computed by THOMAS (1967). The letters $A - D$ refer to the corresponding evolutionary states in Fig. 32.3. The arrows indicate the direction of the evolution. This direction is reversed for a short period between the dotted horizontal lines. This transient drop in luminosity at about $\lg L/L_\odot = 2$ occurs when the hydrogen-burning shell crosses the chemical discontinuity left behind when the bottom of the outer convective zone moves outwards again in the mass scale after it has reached its deepest extension (see Fig. 32.3)

**Fig. 32.3.** The evolution of the internal structure of a star of $1.3 M_\odot$ plotted in the same manner as in Fig. 31.2(a). The main region of hydrogen burning is hatched, "cloudy" areas indicate convection. Regions of variable hydrogen content are dotted. (After THOMAS, 1967)

314

For this purpose we follow the considerations of §32.2, but this time we vary the molecular weight $\mu$ at homologous points while keeping $M_c$, $R_c$ and all other parameters constant. Analogously to (32.4–7) we write

$$\varrho(r/R_c) \sim \mu^{\varphi_3} \quad , \tag{32.29}$$

$$T(r/R_c) \sim \mu^{\psi_3} \quad , \tag{32.30}$$

$$P(r/R_c) \sim \mu^{\tau_3} \quad , \tag{32.31}$$

$$l(r/R_c) \sim \mu^{\sigma_3} \quad , \tag{32.32}$$

and with the same procedure as in §32.2 we find

$$\varphi_3 = \frac{4-b-\nu}{N} \quad , \quad \psi_3 = 1 \quad , \quad \tau_3 = \varphi_3 \quad , \quad \sigma_3 = \nu + n\varphi_3 \quad , \tag{32.33}$$

with $N = 1+n+a$. For example, using again the values $\nu = 13$, $n = 2$, $a = b = 0$ as in §32.2, we see that (32.32) becomes $l \sim \mu^7$. Therefore the luminosity decreases with decreasing $\mu$, which explains the transient reduction of $L$. After the shell source has passed the discontinuity, $\mu$ remains at its reduced value and the luminosity grows again with increasing core mass.

Evolutionary calculations for somewhat different total masses $M$ yield similar results. Near the main sequence the tracks are shifted relative to each other according to their different starting points on the ZAMS. When approaching the Hayashi line the tracks merge. (This is not exactly true, since different total masses have slightly different Hayashi lines.) After the cores are sufficiently condensed they are virtually independent of the envelope (and therefore of the total mass $M$). However, they determine the total luminosity according to the $L(M_c)$ relation. Consequently stars of different $M$ but the same $M_c$ have the same $L$ and are practically at the same point in the HR diagram.

The same convergence of the evolution for different $M$ must occur for all properties of the shell source and the core. For example, the central values of density and temperature converge to the same evolutionary track in the $\varrho_c$–$T_c$ plane.

Numerical calculations show that with growing core mass the temperature in the core rises. This is due to two effects which are of approximately the same order. The first is the increase of the temperature in the surrounding shell source where $T \sim M_c/R_c$ after (32.24). While this effect already occurs in models of complete equilibrium, there is an additional effect due to non-stationary terms. With growing $M_c$ the core contracts, releasing energy. If this occurs rapidly enough, it heats up the transition layer below the shell, and therefore the whole core. An inward-directed temperature gradient is built up in the transition region, such that the energy released by $\varepsilon_g$ terms is carried away. All this is enhanced by increasing $L$: the rate $\dot{M}_c$ is proportional to $L$, which in turn increases by a high power of $M_c$, and the process speeds up more and more. Both these effects, controlled by the growth of $M_c$, finally increase the core temperature to $\approx 10^8$ K at which helium is ignited. This happens when $M_c \approx 0.45 M_\odot$, independently of $M$. The matter in the core is highly degenerate and the nuclear burning is unstable. The resulting thermal runaway terminates the slow and quiet evolution along the ascending giant branch.

## 32.4 The Helium Flash

We start with some analytic considerations and assume that helium is ignited in the centre, where the electron gas is assumed to be non-relativistic and degenerate. In §25.3.5 we have discussed the secular stability of nuclear burning in a small central sphere of mass $m_s$, "luminosity" $l_s = \varepsilon m_s$ and gravothermal specific heat $c^*$. Assuming a homologous reaction of the layers above, a small relative temperature perturbation $\vartheta (= dT_c/T_c)$ was shown in (25.35) to evolve according to

$$\dot{\vartheta} = \frac{l_s}{c_p m_s T_c} (\varepsilon_T + \kappa_T - 4) \vartheta \quad , \tag{32.34}$$

where we have set $\delta = 0$ and therefore $c^* = c_p$ according to (25.29). For helium burning we have $\varepsilon_T > 19$ (see §18.5.2), which certainly dominates the other terms in the parenthesis which thus is positive: the onset of helium burning in the degenerate core is unstable and results in a thermal runaway. The time-scale of the thermal runaway is of the order $c_p m_s T_c/l_s = c_p T_c/\varepsilon$, i.e. of the order of the thermal time-scale of the helium-burning region.

The homologous linear approximation which yielded (32.34) can only give a very rough picture of the events after helium ignition. Nevertheless we can try to discuss the consequences which follow from our simple formalism. From (25.25,26) one obtains

$$\frac{d\varrho_c}{\varrho_c} = \frac{3\delta}{4\alpha - 3} \vartheta \quad , \tag{32.35}$$

and for $\alpha = 3/5$, $\delta = 0$ (completely degenerate non-relativistic gas) we find $d\varrho_c = 0$. Therefore, while during the thermal runaway the central temperature is rising, the matter neither expands nor contracts. The central density remains constant, and in the $\lg \varrho_c - \lg T_c$ diagram, the centre evolves vertically upwards as indicated in Fig. 32.4. The reason is that in the (fully) degenerate gas the pressure does not depend on temperature and therefore remains constant during the thermal runaway. But only an increase of pressure could lift the weight of the mass above and cause an expansion. Since the $Pdv$ work is zero, all nuclear power goes into internal energy. During

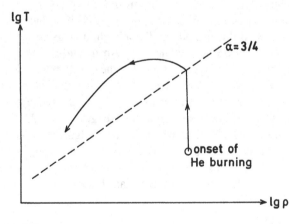

Fig. 32.4. Schematic sketch of the changes of temperature and density during the helium flash. After the ignition temperature is reached in the regime of degeneracy the temperature rises almost without a change of density until degeneracy is removed near the broken line. Then a phase of almost isothermal expansion ensues followed by a phase of stable helium burning in the non-degenerate regime

the thermal runaway there is an enormous overproduction of nuclear energy. The local luminosity $l$ at maximum comes to $10^{11} L_\odot$, about that of a whole galaxy, but only for a few seconds. (The expression "helium flash" is quite appropriate indeed!) However, almost nothing of it reaches the surface, since it is absorbed by expansion of the non-degenerate layers above.

With increasing temperature at constant density, the degeneracy is finally removed. This happens roughly when in Fig. 32.4 the border ($\alpha = 3/4$) between degeneracy and ideal gas is crossed. Then with further increase of $T$ the core expands. With the removal of degeneracy the gravothermal specific heat becomes negative again and central helium burning becomes stable; the expansion stops the increase of temperature. The overproduction then is gradually removed by cooling until the temperature has dropped to "normal" values for quiet (stable) helium burning. In the $\lg \varrho_c - \lg T_c$ plane the core settles near the image point of a homogeneous helium star of mass $M_c$, which is of the order of $0.45 M_\odot$.

There is another prediction we can make for the changes in the HR diagram. Until the onset of helium burning the total luminosity of the star (which is just the power produced in the shell) increases with increasing core mass as expected from (32.26). After degeneracy is removed in the central region, the core expands and $R_c$ increases. During the short phase of the flash, $M_c$ remains practically unchanged. From (32.26) we therefore expect the luminosity to be appreciably reduced after the flash phase, and this indeed can be seen from Fig. 32.2.

## 32.5 Numerical Results for the Helium Flash

In § 32.4 we have tacitly assumed that the maximum temperature is in the centre. This, however, is not the case if neutrinos are created in the very interior of the core and provide an energy sink there, since they leave the star without noticeable interaction. Then the maximum of temperature is not in the centre but at a finite value of $m$ (see Fig. 32.5). From there, energy flows outwards ($l > 0$) and inwards

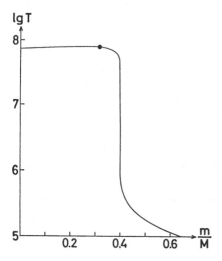

Fig. 32.5. The temperature $T$ (in K) as a function of the mass variable $m$ in the $1.3 M_\odot$ model described in Figs. 32.2,3 shortly before the onset of (unstable) helium burning. Owing to neutrino losses the maximum temperature does not occur in the centre but near $m/M = 0.3$ (*dot*). (After THOMAS, 1967)

($l < 0$). This energy is released by core contraction in the transition zone below the burning shell as mentioned in § 32.3. The transport mechanisms are radiation and conduction. The inward-going energy is carried away by neutrinos. Then the ignition of helium and the flash will not take place in the centre but in the concentric shell of maximum temperature. This is near $m/M = 0.3$ according to Fig. 32.5. [Note that the calculations discussed here assume an unusually low value of $\mu$ in the envelope. Therefore, according to (32.30,32) $T$ in the shell source and $L$ are smaller for the same $M_c$ and $R_c$, and helium ignites at correspondingly larger $M_c$.]

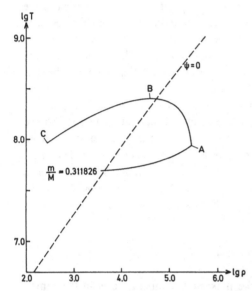

Fig. 32.6. Temperature $T$ (in K) versus density $\varrho$ (in g cm$^{-3}$) for the mass shell at which helium ignites in the $1.3 M_\odot$ model. The letters $A - C$ refer to the corresponding evolutionary states in Figs. 32.2,3. The dashed line (degeneracy parameter $\psi = 0$ for $\mu_e = 2$) roughly separates the regimes of degeneracy and non-degeneracy of the electron gas. (After THOMAS, 1967)

In Fig. 32.6, the evolution is shown in a $\lg \varrho - \lg T$ diagram for the shell in which helium is ignited. We see that the shell behaves roughly as predicted in Fig. 32.4 for the centre. When the temperature of helium burning is reached at point $A$ the core matter heats up. After degeneracy is removed near point $B$, the core expands and a non-degenerate phase follows with stable helium burning, roughly at the same temperature at which the flash phase had started but at much lower densities. The internal structure of the model after the ignition of helium is indicated in Fig. 32.7.

The calculations discussed here were carried out with neutrino rates which have turned out to be too high. In later calculations for $1.3 M_\odot$, with more realistic neutrino rates, the igniting shell was at $m/M = 0.11$. For stellar masses in the range $0.7 \leq M/M_\odot \leq 2.2$, model calculations up to the ignition of helium have been carried out with the improved neutrino rates (SWEIGART, GROSS, 1978). It turned out that helium ignites at $m/M \approx 0.17$ for $M = 0.7 M_\odot$, while with increasing total mass the shell of ignition moves closer to the centre.

Although the properties of the regions in which the flash occurs can change drastically within a few seconds, it seems as if inertia terms can be neglected even in the most violent phases of the flash; however, this is a question that is not completely settled. It is also not clear how convection behaves during the rapid evolution of the helium flash and whether the ignition of helium and the flash in a shell proceeds in strict spherical symmetry.

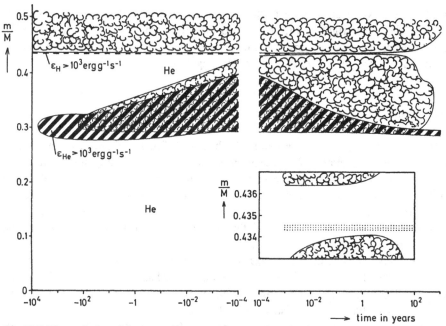

**Fig. 32.7.** The evolution of the internal structure of a star of $1.3 M_\odot$ during the helium flash. The zero point of the abscissa corresponds to the age $7.474 \times 10^9$ years of the abscissa of Fig. 32.3. The main regions of nuclear energy release are hatched; the hydrogen-burning shell is, in the mass scale of the ordinate, so narrow that it appears as a broken line. It extinguishes at $t \approx 10^{-3}$ years. "Cloudy" areas indicate convection. The close approach of the outer convective envelope and the convective region above the helium-burning shell is shown with a strongly enlarged ordinate in a window at the lower right. There the dotted area indicates the transition region of the chemical composition left by the (then extinguished) hydrogen-burning shell

If helium is ignited off centre, then the burning forms a shell enriched in carbon and oxygen which surrounds a helium sphere. But if the molecular weight decreases in the direction of gravity, the layer is secularly unstable: a mass element pushed down so slowly that it could adjust its pressure and temperature to that of the new surroundings ($DP = 0$, $DT = 0$, in the terms of § 6) would have a higher density ($D\varrho > 0$, because $D\mu > 0$) and would sink deeper. This corresponds to the "salt finger instability" discussed in § 6.5. In the case discussed here it will cause mixing between the shell in which carbon and oxygen are produced, and the helium region below. The linear stability analysis is rather easy, though it is difficult to follow the instability into the non-linear regime and, for instance, to determine the characteristic time for this mixing process. Simple assumptions about the flow pattern suggest that mixing due to the inwardly decreasing molecular weight is slow compared to the nuclear time-scale and can therefore be neglected (KIPPENHAHN et al., 1980).

More spectacular mixing than in the case just discussed would occur if the convective shell, forming above the helium-burning shell during the flash, were to merge with the outer convective layer. Then hydrogen-rich matter would be mixed down to regions with temperatures where helium burning gives rise to quite unusual nuclear reactions. Although the boundaries between the two convective zones come very close to each other, they never merge. This can be seen in the detailed picture on the lower right of Fig. 32.7. Also the model calculations carried out with different parameters never give mixing between the hydrogen-rich envelope and the convective layer just above the helium-burning shell.


## 32.6 Evolution after the Helium Flash

After the violent phase of the helium flash there follows a phase of quiet burning in non-degenerate matter. The transition to this is not particularly well covered by calculations. Most authors preferred to start with models that belong to a later state in which they already resemble the horizontal-branch stars of globular clusters.

Although during the flash helium is ignited in a shell, it will also burn in the central region after some time, and the stars can be approximated by models on generalized main sequences (cf. § 23.4). For example, a $0.9 M_\odot$ star, having a helium core of $0.45 M_\odot$ after the flash, corresponds to the generalized main sequence for $q_0 = 0.5$. Then from Fig. 23.8 we expect that the model should lie in the HR diagram near the Hayashi line at a luminosity of about $L \approx 100 L_\odot$, appreciably lower than just before the flash. This is also what we had expected from the analytic discussion at the end of § 32.4, and the evolutionary track in Fig. 32.2 in fact points downwards in the right direction. When in the subsequent phase $q_0$ increases with growing $M_c$, the model should cross over to generalized main sequences of larger $q_0$, i.e. move to the left with slightly increasing luminosity.

Detailed calculations, carried out in order to reproduce the horizontal branch of globular clusters, show that the models after the helium flash depend not only on $q_0$ but also on the chemical composition (FAULKNER, 1966). Let us discuss models of different masses $M$ in complete equilibrium at the onset of quiet helium burning in a core of $M_c = 0.5 M_\odot$ with a hydrogen-burning shell at the bottom of the envelope. The composition of the envelope $X_H = 0.65$ is kept fixed while $X_{CNO}$ (the elements participating in the CNO cycle) is varied over a large range. In Fig. 32.8 we see that for $X_{CNO} = 10^{-2}$, models in the range $M = 1.25 \ldots 0.75 M_\odot$ are close to the Hayashi line. Only models with even smaller mass are located considerably to the left: for $M = 0.6 M_\odot$ the effective temperature is almost 9000 K. In order to cover the whole observed horizontal branch with such models for a fixed value of $X_{CNO}$, one would have to assume that the models differ in mass. Suppose that during the slow evolution before the helium flash the stars lose an appreciable, but from star to star different, amount of mass from their surfaces. Then the stars would start their evolution after the flash with the same core masses but different envelope masses: those which have lost more mass would lie on the left, while those which have lost only little mass would lie in the red region. If we therefore identify the observed horizontal branches with (theoretical) *zero-age horizontal branches*, we must assume that low-mass stars, ascending in the HR diagram into the region of red giants, undergo appreciable mass losses of various amounts.

However, it could also be that the observed horizontal branches reflect the evolution of stars after their appearance on the zero-age branch. When their cores grow owing to shell hydrogen burning, and the helium is consumed in their central part,

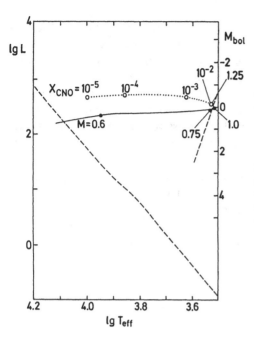

Fig. 32.8. The position in the HR diagram of models with helium cores for different values of the mass $M$ and of the abundance $X_{CNO}$. For all models the hydrogen abundance in the envelope was assumed to be $X_H = 0.65$, and the abundance of all elements heavier than helium was taken to be $2X_{CNO}$. The solid line gives a sequence of models for constant $X_{CNO} = 10^{-2}$ and different masses $M$, ranging from $0.6 M_\odot$ to $1.25 M_\odot$. The dotted line indicates a sequence of models with constant mass ($M = 1.25 M_\odot$) but for different values of $X_{CNO}$ ranging from $10^{-5}$ to $10^{-2}$ (after FAULKNER, 1966). The dashed lines indicate the main sequence and the Hayashi line for $1.25 M_\odot$

it could well be that their evolutionary tracks loop back and forth, populating the horizontal branch. Then the observed branch would not be the locus of zero-age models. We will come to the further evolution in § 32.8.

The results plotted in Fig. 32.8 reveal another important property of zero-age models. If one keeps the total mass constant but decreases $X_{CNO}$, then the image points of the models in the HR diagram move to the left. Thus one could in principle populate the horizontal branch by zero-age models of constant mass but varying $X_{CNO}$, but it is not reasonable to assume large variations of the composition within one cluster. However, the dependence of the models on $X_{CNO}$ helps us to understand an observed correlation between horizontal-branch characteristics of different globular clusters and their composition: the concentration of stars on the horizontal branch shifts from left to right with increasing contents of heavier elements. This is in obvious accordance with Fig. 32.8. [For attempts to explain the horizontal branches of different clusters, see ROOD (1973).] Since the horizontal branch crosses the instability strip (see § 39) we can expect pulsating horizontal-branch stars. Indeed there one finds the RR Lyrae variables.

Faulkner's result also indicated that an envelope composition with $X_H = 0.65$ fits the observations better than one with $X_H = 0.9$, which at that time was considered typical for Pop. II.

## 32.7 Evolution from the Zero-Age Horizontal Branch

A so-called zero-age horizontal branch model has a homogeneous non-degenerate helium core of mass $M_c \approx 0.45 M_\odot$, surrounded by a hydrogen-rich envelope of mass $M - M_c$. The total luminosity consists of a contribution from (quiet) central helium burning and one from the hydrogen-burning shell.

A complication occurs during the following evolution of these models. The stars have a central convective core which becomes enriched in carbon and oxygen during helium burning. The opacity in this temperature–density range is dominated by free–free transitions. However, the free–free opacity increases with increasing carbon and oxygen abundance as can be seen from the factor $B$ in (17.5,6). As a consequence a semi-convective layer is formed above the convective central region (CASTELLANI et al., 1971). The situation is very similar to that in massive stars on the main sequence (see § 30.4.2) where the opacity is governed by electron scattering and increases with increasing helium abundance.

The mass of the helium core grows owing to hydrogen-shell burning, while in the convective core, helium is consumed and carbon and oxygen are produced. After some time a carbon–oxygen core will be formed in the central region of the helium core. Then nuclear burning takes place in two shells (hydrogen and helium burning) and in the subsequent phases of evolution, the masses within these shells will grow.

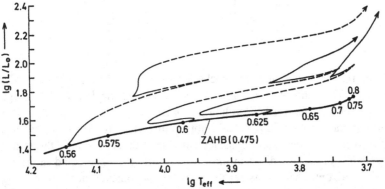

**Fig. 32.9.** Hertzsprung–Russell diagram with the zero-age horizontal branch and the evolution afterwards. The thick line labelled with ZAHB is the zero-age horizontal branch for models with a helium core of $M_c = 0.475 M_\odot$ and a hydrogen-rich envelope ($X_H = 0.699$, $X_{He} = 0.3$) with different masses $M - M_c$. The total masses $M$ (in $M_\odot$) are indicated for a few points. For 3 of these models the ensuing evolution is shown by the thin lines. Phases of slow evolution are given by solid lines, those of rapid evolution by broken lines. The models evolve from the ZAHB first in the slow phase of central helium burning with a hydrogen-burning shell. This phase, which lasts for some $10^7$ years, is followed by a phase of rapid evolution during which the models go from helium burning in the centre to shell burning. After that a slow phase of double shell burning occurs. (After STROM et al., 1970. See also IBEN, ROOD, 1970)

Some results of evolutionary calculations for different parameters are shown in Fig. 32.9. The evolutionary tracks start on the zero-age horizontal branch and, after some back and forth, approach the Hayashi line when the central helium becomes exhausted. They lead upwards with increasing core mass, and the corresponding branch in the HR diagram is called the *asymptotic giant branch* (AGB). It has to be distinguished from the giant branch (GB), along which the image points in the HR diagram move upwards *before* ignition of helium. The models of the post-horizontal-branch evolution occupy a region above the horizontal branch. During their evolution some of them cross the instability strip (see § 39), where one finds the pulsating W Virginis stars (compare the sketch in Fig. 32.10).

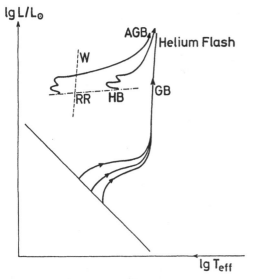

lg L/L⊙

W

AGB

Helium Flash

RR    HB    GB

lg T_eff

Fig. 32.10. Sketch of the evolution of low-mass stars in the HR diagram. For three slightly different masses the evolutionary tracks in the post-main-sequence merge in the giant branch (GB). After the helium flash they appear on the zero-age horizontal branch (HB), evolve towards the upper right, and merge in the asymptotic giant branch (AGB). The broken line indicates the positions of the variable RR Lyr stars (RR) and of the W Vir stars (W)

After the hydrogen shell has burned outwards for some time, the temperature in this shell drops, and hydrogen-shell burning extinguishes. The layer of transition between the hydrogen-rich envelope and the region of helium stays now at a fixed value of $m$. But there is still the active helium-burning shell moving to higher values of $m$ and therefore approaching the bottom of the hydrogen-rich envelope. Since helium burning proceeds at a temperature of $\gtrsim 10^8$ K, which is about ten times the temperature of hydrogen ignition, hydrogen burning starts again, and once more there are two shell sources. In this phase, shell burning becomes secularly unstable, resulting in a thermal runaway. This leads to a cyclic phenomenon (reoccurring here within some $10^5$ years) known as *thermal pulses*. Their general properties will be discussed in § 33.3 in connection with their appearance in massive stars where the unstable shells are in the deep interior, and the response of the surface is moderate. In the case of low-mass stars, the luminosity and the surface temperature can vary appreciably with each pulse. This is the more pronounced the less mass is left above the unstable shells. If a thermal pulse occurs in certain critical phases (with neither too much nor too little mass above the shells) the models can even move rapidly through large regions of the HR diagram (KIPPENHAHN et al., 1968, SCHÖNBERNER, 1979). The evolution displayed in Fig. 32.11 goes through 11 pulses, the onsets of which are indicated by heavy dots. The variation of the surface values is not too large, since the envelope mass is either still too large (pulses 1...10) or too small (pulse 11).

The pulses are more or less an envelope phenomenon and are of no influence on the core. The inner part of the C–O core resembles more and more a white dwarf. Only the hydrogen-rich envelope, small in mass but thick in radius, at first gives the star the appearance of a red giant. After the envelope mass has dropped below, say, one per cent the envelope starts to shrink. With decreasing envelope mass the star moves within a few times $10^4$ years to the left of the main sequence (see Fig. 32.11). Then shell burning extinguishes and the star becomes a white dwarf.

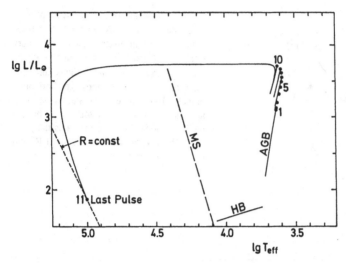

**Fig. 32.11.** The evolutionary track of a star of $0.6 M_\odot$ ($X_H = 0.749$, $X_{He} = 0.25$) for the phases after central helium burning. The model moves upwards along the asymptotic giant branch (AGB) until thermal pulses occur (indicated by full circles). The changes during a pulse are shown only for pulse 9 and pulse 10. Before the last pulse the track has reached the white-dwarf area of the HR diagram. The main sequence (MS), the horizontal branch (HB), and a line of constant radius in the white-dwarf region are indicated (after IBEN, RENZINI, 1983)

It is clear that the mass in the envelope is diminished by two effects: the hydrogen burning at the bottom and a mass loss from the surface. Therefore the stage at which the star leaves the asymptotic branch, turning to the left, is sensitive also to the amount of mass loss in the red-giant phase. This influences the mass of the final white dwarf (cf. § 33.4). Unfortunately details of the mass loss are not well known.

## 32.8 Equilibrium Models with Helium Cores – Continued

The foregoing paragraphs have shown the importance of equilibrium (or near-equilibrium) models with helium cores. Certain sequences of such models were discussed in § 23.4 as the "generalized main sequences". We are now ready to complete this discussion by including also unstable models and those with isothermal cores.

The models to be discussed are in complete equilibrium and consist of a helium core of mass $M_{He} = q_0 M$ and a hydrogen-rich envelope of mass $M_e = (1 - q_0)M$ (Fig. 23.7). Let us consider the following sequences of such models: along one sequence the stellar mass $M$ is fixed, while $q_0$ is varied from $q_0 = 0$ (a star on the normal hydrogen main sequence) to $q_0 = 1$ (a star on the helium main sequence). In § 23.4 we have already presented those parts of the sequences which start on the helium main sequence and belong to cores with $M_{He} \gtrsim 0.30 M_\odot$ (necessary for stable central helium burning).

We now have all the means to understand how these sequences can be continued into the normal main sequence. Extensive model sequences of the type discussed

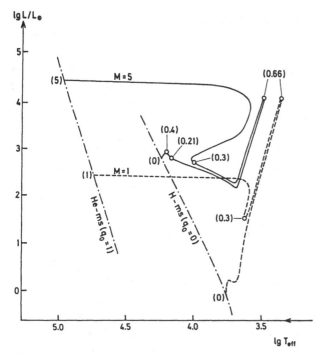

**Fig. 32.12.** Hertzsprung–Russell diagram with sequences of complete-equilibrium models having a hydrogen-rich envelope and a helium core of mass $M_{He} = q_0 M$, for stellar masses $M = 5M_\odot$ (*solid line*) and $M = 1M_\odot$ (*broken line*). A few characteristic core masses $M_{He}$ (in $M_\odot$) are given in parentheses. The sequences terminate at the helium main sequence ($q_0 = 1$, $M_{He} = M$) and the hydrogen main sequence ($q_0 = 0$, $M_{He} = 0$). Open circles indicate turning points of the corresponding linear series (see Fig. 32.13) where $M_{He}$ has an extremum. (After KOSLOWSKI, PACZYŃSKI, 1975)

here are calculated by KOSZLOWSKI, PACZYŃSKI (1975). The sequences for $M = 5$ and $1M_\odot$ are shown in the HR diagram (Fig. 32.12). (We also include here the discussion of the more massive models in order to show the differences.) We start with the fairly simple sequence for $M = 1M_\odot$ (broken line) and with $q_0 = 0$ (on the hydrogen main sequence). With increasing helium core the sequence goes somewhat upwards, then to the right, and then far up along the Hayashi line. This nearly follows the evolutionary track in the post-main-sequence phase for low-mass stars (Fig. 32.2). On this part of the sequence the models have a degenerate helium core which is isothermal; it has the temperature of the surrounding hydrogen shell source that produces the luminosity $L$. Such models are well described by the shell-homology relations of § 32.2 giving, e.g., the increase of $L$ as caused by the increase of $M_{He}$. According to (32.24) the temperature $T_s$ in the shell source (and therefore the central temperature $T_c$) also increases with $M_{He}$. In this way $T_c$ has reached the ignition point for helium burning when $M_{He} \approx 0.66M_\odot$ and the model is at the top of our sequence. Nuclear burning in a degenerate core is unstable, and we have here a transition from stable to unstable models. Obviously this must be a turning point in the corresponding linear series of equilibrium models (see § 12.4). It corresponds to the onset of the helium flash in the *evolutionary* sequences. There, however, the core is additionally heated by the increasing release of gravitational energy, which builds up a $T$ gradient inside the core, thus providing an earlier ignition (at $M_{He} \approx 0.45 M_\odot$ instead of $0.66 M_\odot$).

The fact that there is a turning point for the model with $M_{He} = 0.66 M_\odot$ means that $M_{He}$ has reached a local maximum (and of course also $L$, which is determined by the core mass). Then $M_{He}$ decreases again on an unstable branch. This leads down to another turning point with a local minimum of $M_{He}$ at the value of $0.3 M_\odot$ where stable helium burning sets in. Here we have reached the part of the sequences discussed earlier (Fig. 23.9), along which $M_{He}$ increases monotonically to $M_{He} = M$ ($q_0 = 1$, on the helium main sequence). The transition from unstable to stable helium burning at $M_{He} \approx 0.3 M_\odot$ could be expected; the helium cores behave very much like helium stars of the same mass on the helium main sequence, the stable branch of which extends only down to $\approx 0.3 M_\odot$ (Fig. 23.8).

The equilibrium sequences for more massive stars are more complicated, as shown for $M = 5 M_\odot$ in Fig. 32.12 (solid line). Starting again on the normal main sequence ($q_0 = 0$), we encounter two turning points (local extrema of $M_{He}$, change of stability) not present in the low-mass case. The local maximum with $M_{He} \approx 0.40 M_\odot$ defines a well-known critical point of stars with non-degenerate isothermal He-cores: the Schönberg–Chandrasekhar limit ($M_{He} \approx 0.08 M_\odot$). The corresponding maxima of $M_{He}$ can be read off the curves in Fig. 30.12. When $M_{He}$ increases again after the turning point at $M_{He} = 0.21 M_\odot$, the cores become degenerate, the sequence approaches the Hayashi line and moves up with increasing $M_{He}$. As in the low-mass case, the helium flash occurs at the top of this branch, where $M_{He} \approx 0.66 M_\odot$. The following unstable branch goes down, then to the left and ends at the turning point at $M_{He} \approx 0.3 M_\odot$, where the branch with stable helium burning cores starts, which was shown in Fig. 23.9.

The different turning points and the stability properties are much better displayed if we plot linear series of these equilibrium models. This is done in Fig. 32.13, which gives the luminosity $L$ for fixed $M$ as a function of the parameter $M_{He}$. The turning points are recognizable here as the local maxima and minima of $M_{He}$. Since they define a zero eigenvalue of the thermal stability problem (cf. § 12.4), a stable branch (solid line) and an unstable one (broken line) merge at these points.

Following the linear series for $M = 1 M_\odot$ and starting from small $M_{He}$, we encounter the first turning point at $M_{He} \approx 0.66 M_\odot$, i.e. at the helium flash (F). The following unstable branch leads to the next turning point at $M_{He} \approx 0.3 M_\odot$ where we have the transition to the stable helium-burning branch that ends on the helium main sequence. For larger $M$ (in Fig. 32.13 the curves for $M = 2, 3$, and $5 M_\odot$) recognize the additional turning point marking the Schönberg–Chandrasekhar limit (SC) as a local maximum of $M_{He}$. All these sequences for different $M$ merge into the same line leading to the helium flash, since $L$ depends only on $M_{He}$. Only the sequences for $M = 1$ and $5 M_\odot$ are plotted after point F. The following unstable branch for $M = 5 M_\odot$ is more complicated than that for $M = 1 M_\odot$, since it loops upwards before reaching the final stable helium-burning branch at $M_{He} \approx 0.3 M_\odot$. The diagram shows nicely that these last stable branches always start at nearly the same value of $M_{He}$, the value at which the stable helium main sequence ends in a turning point (minimum of $M_{He}$).

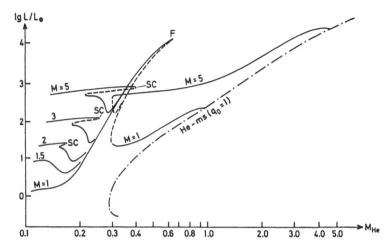

**Fig. 32.13.** Linear series of equilibrium models with hydrogen-rich envelopes and helium cores, for a few values of the stellar mass $M$ (in $M_\odot$). The luminosity is plotted against core mass $M_{He}$. Thermally stable and unstable branches are shown by solid and broken lines respectively. The labelled turning points correspond to the Schönberg–Chandrasekhar limit (SC) and the helium flash (F). The sequences end on the helium main sequence (*dot-dashed*) where $M_{He} = M(q_0 = 1)$. (After KOSZLOWSKI, PACZYŃSKI, 1975)

Figure 32.13 also clearly shows the existence of multiple solutions for given values of $M$ and $M_{He}$ (i.e. given parameters $M$ and chemical composition). In the case of $M = 1 M_\odot$ we find three solutions (2 stable, 1 unstable) for $0.3 \lesssim M_{He}/M_\odot \lesssim 0.66$, and one stable solution outside this range. For $M = 5 M_\odot$ there are also ranges of $M_{He}$ in which 1 (stable) or 3 (2 stable, 1 unstable) solutions exist; but from $M_{He} \approx 0.3 M_\odot$ to $M_{He} \approx 0.4 M_\odot$ (the SC limit), there are 5 solutions (3 stable, 2 unstable) for given $M_{He}$. The existence of so many solutions for fixed parameters was initially shown via other arguments (ROTH, WEIGERT, 1972). For still larger values of $M$ the linear series acquire two additional turning points, which give rise to two more solutions. And then there must exist other branches (also for lower $M$) which are not shown here and which are not connected with the plotted branches. (Such additional branches can start from the other, unstable branches of the hydrogen and helium main sequence.) From the general description of linear series in § 12.5 it is clear that these additional branches will always come in pairs (1 stable, 1 unstable), such that also the number of additional solutions for given $M$ and $M_{He}$ increases by an even number. As mentioned earlier, the problem of which of the many stable equilibrium solutions can be reached in the evolution can only be decided by evolutionary calculations based on thermally unadjusted models.

# § 33 Later Phases

### 33.1 Nuclear Cycles

The stellar evolution described above may seem to be rather complicated, at least where the changes of the surface layers are concerned, for example, in the case of evolutionary tracks in the HR diagram. The processes appear much simpler and even become qualitatively predictable if we concentrate only on the central evolution. Extrapolating from central hydrogen and helium burning of sufficiently massive stars, we can imagine that the central region continues to pass through cycles of nuclear evolution which are represented by the following simple scheme:

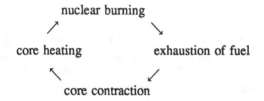

The momentary burning will gradually consume all nuclei inside the convective core that serve as "fuel". The exhausted core then contracts. This raises the central temperature until the next higher burning is ignited etc.

As long as this scheme works, gradually heavier elements are built up near the centre from cycle to cycle. The new elements are evenly distributed in convective cores which usually become smaller with each step. For example, in the first cycle (hydrogen burning) the star develops a massive helium core, inside which a much smaller C–O core is produced in the next cycle (helium burning), and so on.

We have also seen that after the core is exhausted the burning usually continues in a concentric shell at the hottest place where the fuel is still present. A shell source can survive several of the succeeding nuclear cycles, each of which generates a new shell source, such that several of them can simultaneously burn outwards through the star. They are separated by mass shells of different chemical composition; gradually heavier elements are encountered when going inwards from shell to shell. One then speaks of an "onion skin model". A schematical cross-section of such a model is shown in Fig. 33.1. The shell structure of the chemical composition can in fact become more complicated than that, since some shell sources bring forth a convective (or semi-convective) subshell, inside which the newly processed material is completely (or partially) mixed. This can be recognized in Fig. 34.7, which shows the interior composition of a model for a $25 M_\odot$ star in a very advanced stage (just

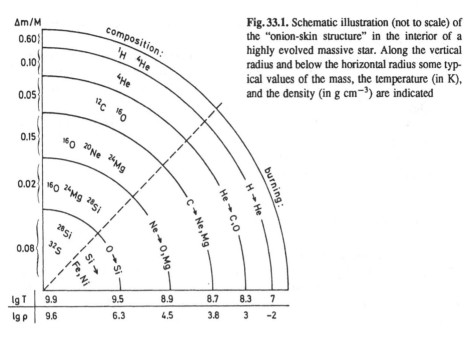

Fig. 33.1. Schematic illustration (not to scale) of the "onion-skin structure" in the interior of a highly evolved massive star. Along the vertical radius and below the horizontal radius some typical values of the mass, the temperature (in K), and the density (in g cm$^{-3}$) are indicated

| $\lg T$ | 9.9 | | 9.5 | | 8.9 | | 8.7 | 8.3 | 7 |
|---|---|---|---|---|---|---|---|---|---|
| $\lg \varrho$ | 9.6 | | 6.3 | | 4.5 | | 3.8 | 3 | -2 |

before core collapse, see § 34). We have also seen that, depending on the change of $T$ in certain regions, a shell source may stop burning for some time and be reignited later.

The simple evolution through nuclear cycles as sketched above can obviously be interrupted, either temporarily or for good. From the discussion of the nuclear reactions in § 18 we know that the cycles must come to a termination, at the latest, when the innermost core consists of $^{56}$Fe (or neighbouring nuclei) and no further exothermic fusions are possible. However, it is easily seen that the sequence of cycles can be interrupted much earlier by another effect. Each contraction between consecutive burnings increases the central density $\varrho_c$. Assuming homology for the contracting core and ignoring the influence of the rest of the star, we obtain from (28.1) the change of the central temperature $T_c$

$$\frac{dT_c}{T_c} = \left(\frac{4\alpha - 3}{3\delta}\right)\frac{d\varrho_c}{\varrho_c} \ . \tag{33.1}$$

The decisive factor, in parenthesis on the right-hand side, depends critically on the equation of state which is written as $\varrho \sim P^\alpha T^{-\delta}$. For an ideal gas with $\alpha = \delta = 1$ we have $dT_c/T_c = (1/3)(d\varrho_c/\varrho_c)$. This means that each contraction of the central region increases the temperature, as well as the degeneracy parameter $\psi$ of the electron gas [$\psi$ = constant for $dT/T = (2/3)(d\varrho/\varrho)$ (cf. § 15.4, 16.2)]. With increasing degeneracy the exponents $\alpha$ and $\delta$ become smaller. When the critical value $\alpha = 3/4$ is reached ($\delta$ is then still $> 0$), the contraction ($d\varrho_c > 0$) no longer leads to a further increase of $T_c$ according to (33.1). The degeneracy in the central region has obviously decoupled the thermal from the mechanical evolution, and the cycle of consecutive nuclear burnings is interrupted. In this case the next burning can be ignited only via more complicated secondary effects, which originate, for example, in the evolution of the surrounding shell source (cf. § 32.2).

Other complications may arise if the central region of a star suffers an appreciable loss of energy by strong neutrino emission (cf. § 18.6). We have already seen (§ 32.5) that this can decrease the central temperature and, therefore, influence the onset of a burning.

In any case, the nuclear cycles tend to develop central regions with increasing density and with heavier elements. We should note, however, that the later nuclear burnings are not capable of stabilizing the star long enough for us to observe many stars in such phases (as is the case with central hydrogen burning and helium burning).

## 33.2 Shell Sources and Their Stability

As mentioned above, a sufficiently evolved star can have several active shell sources. Their productivity may change considerably and even go to zero for some time. Neighbouring shell sources can influence each other, since each type of burning requires a separate range of temperature. For example, if a helium shell source operating at roughly $2 \times 10^8$ K approaches a hydrogen-rich layer, we can expect an enormous increase of hydrogen burning, which usually proceeds at $T \lesssim 3 \times 10^7$ K. It is also clear that different shell sources will generally move with different "velocities" $\dot{m}_i$ through the mass, unless their contributions $L_i$ to the total luminosity are in certain ratios. If $X_i$ denotes the mass concentration of the reacting element ahead of the shell source, and $q_i$ the energy released by the fusion of one unit of mass, then $\dot{m}_i = L_i/(q_i X_i)$. For example, the relative motion of the hydrogen and helium shell sources through the mass is given by the ratio

$$\frac{\dot{m}_H}{\dot{m}_{He}} = \frac{L_H}{L_{He}} \frac{q_{He}}{q_H} \frac{X_{He}}{X_H} . \tag{33.2}$$

This gives a stationary situation with roughly equal velocities only if $L_H \approx 7 L_{He}$, since typically $X_H \approx 0.7$, $X_{He} \approx 1$, and $q_H/q_{He} \approx 10$. Otherwise the two shell sources approach each other or the inner one falls behind.

Shell-source models for several evolutionary phases can be approximated well by solutions obtained by assuming complete equilibrium. While burning outwards, a shell source has the tendency to concentrate the reactions over steadily decreasing mass ranges. One then has to deal with rather short *local* nuclear time-scales, defined as those time intervals in which the burning shifts the very steep chemical profile over a range comparable to its own extension. This can require computations with unreasonably short time steps, which are usually avoided by using special techniques.

All changes become much more rapid and the assumption of complete equilibrium certainly has to be dropped if the shell source is thermally unstable. The reasons for such instabilities will be made plausible by considering a very simple model for the shell source and its perturbation. The procedure is completely analogous to that used in § 25.3.5 for the stability of a central nuclear burning. The only difference between the two cases is that the burning regions are geometrically different and the density reacts differently to an expansion.

a)

b)

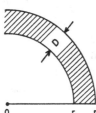

0          r          0          $r_0$    r

**Fig. 33.2.** The main region of nuclear energy production (*hatched*) in the cases of (a) central burning and (b) shell source burning

Let us compare the two cases of a central burning and a shell-source burning in Fig. 33.2. In the central case, the mass of the burning region is $m \sim \varrho r^3$, and an expansion $dr > 0$ with $dm = 0$ requires a relative change of the density [compare with (25.25)]

$$\frac{d\varrho}{\varrho} = -3\frac{dr}{r} \quad . \tag{33.3}$$

In the case of a shell source of thickness $D$, we write the upper boundary of the burning region as $r = r_0 + D$ (cf. Fig. 33.2b). For relatively small $D$ the mass in the burning shell is $m \sim \varrho r_0^2 D$. If the burning region expands with roughly $r_0 = $ constant as a reaction to an energy perturbation, we have $dr = dD$, and the condition $dm = 0$ now leads to

$$\frac{d\varrho}{\varrho} = -\frac{dD}{D} = -\frac{r}{D}\frac{dr}{r} \quad . \tag{33.4}$$

We now assume that the mass outside $r_0 + D$ expands or contracts homologously. Then for the pressure in the shell we can use the relation $dP/P = -4dr/r$ as in (25.25). When comparing (33.4) with (33.3) we see that we only have to replace the factor 3 by the factor $r/D$ when going from the central case to that of a shell source. This can be done directly in expression (25.29) for the gravothermal heat capacity $c^*$. For simplicity we neglect the perturbation of the flux $dl_s$ and have from (25.30)

$$c^*\frac{dT}{dt} = d\varepsilon \quad ; \quad c^* = c_P\left(1 - \nabla_{\text{ad}}\frac{4\delta}{4\alpha - r/D}\right) \quad . \tag{33.5}$$

(Note that the time derivative $dT/dt$ represents a differential perturbation; it could be replaced by $d(dT/dt)$ since $T = T_0 + dt$ with time-independent $T_0$.) If $c^*$ is positive, then the shell source is unstable, since an additional energy input ($d\varepsilon > 0$) leads to higher $T$ and further increased burning.

We first recover the well-known flash instability in the case of strong degeneracy of the electron gas with $\delta \to 0$. Indeed we have seen in §32 that the helium flash occurs in a shell rather than in the centre if the central part is cooled by neutrino emission.

In addition, (33.5) shows that there is a new instability which can occur even for an ideal monatomic gas ($\alpha = \delta = 1$, $\nabla_{\text{ad}} = 2/5$) and which has no counterpart in the case of central burning. It depends only on the geometrical thickness $D$ of the

331

shell source. If $D/r$ is small enough (in our simple representation smaller than 1/4), $c^*$ is positive and the shell source is secularly unstable. This instability of a shell source is called *pulse instability* for reasons which will become obvious very soon.

It is amazing that such a simple geometrical property can cause a thermal instability, though it becomes more plausible if we consider the change of the pressure in the shell source as a hydrostatic reaction to the lifting of the layers above (for which we simply assume homology). Suppose that the shell tries to get rid of the perturbation energy by expansion. A substantial relative increase of the thickness $dD/D > 0$ gives the same absolute value for the relative decrease of the density $d\varrho/\varrho < 0$, but only a very small relative increase $dr/r$, if $D/r \ll 1$ [cf. (33.4) and Fig. 33.2b]. This means that the layers above are scarcely lifted, so that their weight remains about constant and hydrostatic equilibrium requires $dP/P \approx 0$. In fact with the homology relation $dP/P = -4dr/r$ and (33.4) we find the connection between $dP$ and $d\varrho$ to be

$$\frac{dP}{P} = 4\frac{D}{r}\frac{d\varrho}{\varrho} \quad . \tag{33.6}$$

Considering the equation of state

$$\frac{d\varrho}{\varrho} = \alpha\frac{dP}{P} - \delta\frac{dT}{T} \quad , \tag{33.7}$$

we see that expansion ($d\varrho/\varrho < 0$) necessarily leads to an *increase* of the temperature ($dT/T > 0$), since $dP/P \to 0$ for $D/r \to 0$:

$$\frac{d\varrho}{\varrho} = -\delta\frac{dT}{T} \quad . \tag{33.8}$$

Therefore the expansion of a thin shell source does not stabilize it, but rather enforces the liberation of energy by heating. This means that the shell source reacts just as if the equation of state were $\varrho \sim 1/T$, which, of course, gives instability [cf. (33.5) with $\alpha = 0$ and $\delta = 1$].

While the foregoing discussion provides the main points correctly, it can easily be completed by also considering the perturbation of the local luminosity. Then some of the surplus energy can flow away, and instability requires, in addition, that the temperature sensitivity of the burning exceeds a certain limit, which is usually fulfilled. The eigenvalue analysis of such stellar models has shown that they are indeed thermally unstable and that the unstable modes are complex (HÄRM, SCHWARZSCHILD, 1972).

The pulse instability was first found (SCHWARZSCHILD, HÄRM, 1965) for a helium shell source in calculations for a $1M_\odot$ star. The same type of instability was encountered independently in a two-shell source model for $5M_\odot$, and here it turned out that the instability leads to nearly periodic relaxation oscillations, which were called *thermal pulses*, as described below (WEIGERT, 1966). They are now known to be quite a common phenomenon with sufficiently evolved stars.

## 33.3 Thermal Pulses of a Shell Source

Thermal pulses occur in models containing one or more shell sources, and in stars of different masses and compositions. We start by describing their properties according to the calculation of the first 6 pulses found in the $5M_\odot$ model, whose foregoing evolution was described in §31. The instability occurs in the helium shell source after it has reached $m/M \approx 0.1597$. It then contributes only a little to the surface luminosity $L$, which is almost completely supplied by the nearby hydrogen shell source located at $m/M \approx 0.1603$.

The instability results immediately in a thermal runaway: the shell source reacts to the surplus energy with an increase in $T$, which enhances the release of nuclear energy etc. The increase of $T$ is connected with an expansion according to (33.8). This can be seen from Fig. 33.3a and b which give $T$ and $\varrho$ at maximum $\varepsilon_{He}$ in the unstable shell source as functions of time. (Note that the thermal runaway in a *flash* instability would proceed with $\varrho$ = constant.) Since helium burning has an extreme temperature sensitivity, the increase of $T$ strongly enhances the productivity $L_{He}$ of the shell source, in later pulses even to many times the surface value $L$. But most of this energy is used up by expansion of the layers above, and this expansion reduces considerably the temperature in the hydrogen shell source, such that $L_H$ decreases significantly. After starting rather slowly the thermal runaway accelerates more and more until reaching a sharp peak within a few years. The helium shell source is now widely expanded and is therefore no longer unstable. The whole region then starts to contract again, which heats up the hydrogen shell source so that it regains its large

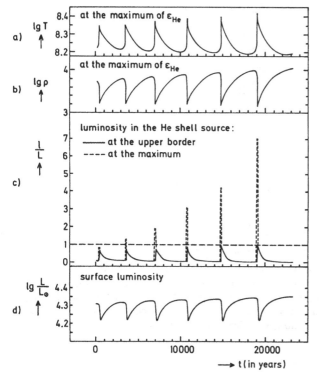

Fig. 33.3. Thermal pulses of the helium shell source in a $5M_\odot$ star after central helium burning. For the first 6 pulses, some characteristic functions are plotted against time from the onset of the first pulse. $T$ is in K, $\varrho$ in g cm$^{-3}$. (After WEIGERT, 1966)

productivity. Within a time of a few $10^3$ years the whole region has asymptotically recovered its original overall structure, the helium shell source becomes unstable again and the next pulse starts. Figure 33.3 shows that the amplitude of the pulses and the time between consecutive pulses grows (in these calculations from 3200 to 4300 years). The reason for these changes is that the chemical composition around the shells changes considerably from pulse to pulse. Later calculations (for a review see IBEN, RENZINI, 1983) showed that a nearly periodic behaviour is usually reached after roughly 20 pulses. The amplitude of a pulse has then become so large that during the maximum $L_{He}$ exceeds $L$ by orders of magnitude. The changes of the chemical composition still provide a small deviation from periodicity. Otherwise we would expect strictly periodic relaxation oscillations, i.e. the solution would have reached a limit cycle.

The surface luminosity (Fig. 33.3d) drops in each pulse by typcially $\Delta \lg L \approx$ 0.1...0.2 for models with rather massive outer envelopes. The visible reaction of the surface is much more pronounced if the pulses occur in a shell source close to the surface. Such models can move quite spectacularly through the HR diagram (compare with §32.7).

We now turn to the change of the chemical composition by a combination of burning and convection. Figure 33.4 shows (with expanded scales) $m$ against $t$ during the peak of two pulses. The high fluxes near the maximum of helium burning create a short-lived convective shell (CS), which, in the later pulses, comes very close to the H–He discontinuity. For a short time, almost the entire matter between the two

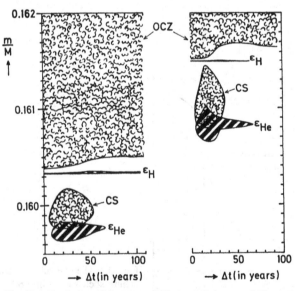

**Fig. 33.4.** Evolution of the mass shells around the two shell sources in a $5M_\odot$ star near the maxima of the first and sixth thermal pulses of the helium shell source (compare Fig. 34.3). The mass variable $m$ is plotted against time, starting from an arbitrary zero point. Note the strongly expanded scales on both axes. "Cloudy" areas indicate the convective shell (CS) and the outer convective zone (OCZ); striped areas show the regions of strongest nuclear energy production ($\varepsilon > 3 \times 10^7$ erg g$^{-1}$s$^{-1}$). (After WEIGERT, 1966)

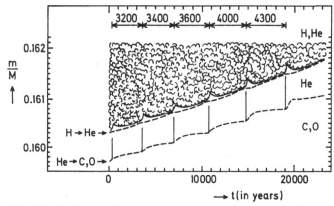

**Fig. 33.5.** Evolution of the mass elements around the two shell sources (*broken lines*) during the first 6 thermal pulses in a $5M_\odot$ star (compare Fig. 33.3). The "cloudy" area represents the outer convective zone (OCZ). The convection in the inter-shell region (CS in Fig. 33.4) at the maximum of each pulse is so short-lived that it appears here as a vertical spike. The time (in years) between consecutive pulses is indicated at the top. (After WEIGERT, 1966)

shells is mixed into the helium-burning shell, the products of which are spread over the intershell region. The outer convection zone (OCZ), which extends to the surface, can be seen to reach down nearly to the hydrogen shell source. The lower boundary of the OCZ moves during each pulse at first somewhat outwards, and then back again (compare also with Fig. 33.5, where the $t$ axis is more compressed). According to other calculations the lower border of the OCZ can even descend beyond the former location of the H–He discontinuity into the intershell region. Hydrogen-rich material is then transported downwards, while intershell material is dredged up by the OCZ and distributed over the whole outer envelope (*third dredge up*). Nuclei processed in the very hot helium shell source during the last pulse can thus be lifted to the surface. (For details of these problems and references, see IBEN, RENZINI, 1983.)

Helium burning transforms $^4$He into $^{12}$C and $^{16}$O, and the hydrogen shell source converts $^{16}$O and $^{12}$C into $^{14}$N, which is left behind when the shell burns outwards between two pulses. The CS of the next pulse sweeps these $^{14}$N nuclei down into the helium shell source where they are burned in the chain $^{14}$N $(\alpha, \gamma)$ $^{18}$F $(\beta^+\nu)$ $^{18}$O $(\alpha, \gamma)$ $^{22}$Ne. During a pulse in fairly massive stars, the helium shell source attains a temperature so high that $^{22}$Ne is also burned, in the reaction $^{22}$Ne $(\alpha, n)$ $^{25}$Mg. This can provide a neutron source sufficiently strong to build up elements beyond the iron peak in the so-called s process (i.e. with neutron captures being *slow* compared with beta decay) (IBEN, 1975; TRURAN, IBEN, 1977). In other cases a corresponding neutron source may be provided by $^{13}$C nuclei, which are burned via the chain $^{12}$C $(p, \gamma)$ $^{13}$N $(\beta^+\nu)$ $^{13}$C $(\alpha, n)$ $^{16}$O in the helium shell. The problem is how to bring a sufficient amount of $^{13}$C into the helium shell. This could happen if hydrogen-rich material is directly swept up during a pulse, or if it diffuses into the $^{12}$C-rich region between two pulses and is then processed to $^{13}$C. Other modifications of the surface composition, particularly of the ratio of $^{12}$C and $^{14}$N, can occur if a burning starts at the lower boundary of the OCZ. The details of all these processes and their results are still hard to foresee, since they depend critically on the precise extensions of the

335

two convective zones involved (the OCZ and the CS) and on their uncomfortably rapid changes.

The properties of the thermal pulses depend on the type of star in which they occur. The cycle time $\tau_p$ (between the peaks of two consecutive pulses) becomes smaller with increasing mass $M_c$ of the degenerate C–O core inside the helium shell source. From a large sequence of calculations it follows roughly (PACZYŃSKI, 1975) that

$$\lg \left( \frac{\tau_p}{1 \text{ year}} \right) \approx 3.05 + 4.50 \left( 1 - \frac{M_c}{M_\odot} \right) \quad . \tag{33.9}$$

For $M_c \approx 0.5 M_\odot$ the cycle time is of the order of $10^5$ years, while near the limit mass $M_c \approx 1.4 M_\odot$ it would be of the order of 10 years only. We now consider the number of pulses that can occur until $M_c$ has reached 1.4 $M_\odot$. Suppose that the hydrogen shell source moves outwards by $\Delta m$ per cycle time and produces most of the energy $L\tau_p$. Although $L \sim M_c$ (cf. § 33.5), $\Delta m$ decreases strongly with growing $M_c$ owing to the decrease of $\tau_p$. One can estimate that, depending on the details of the model, the total number of pulses (determined mainly by the very small $\tau_p$ in the last phases) must be 8000 ... 10000 before $M_c \approx 1.4 M_\odot$. Of course, the shell source cannot burn further than to within a few $10^{-3} M$ from the surface. Therefore the total number of pulses will be much smaller if the stellar mass is well below $1.4 M_\odot$, either originally or owing to mass loss. In low-mass stars one can expect only 10 pulses or so, as seen, for instance, in Fig. 32.11. These, however, occur very close to the surface and can affect the observable values certainly much more than pulses of a shell source in the deep interior.

During a thermal pulse, the star changes quite rapidly, particularly in the layers of the shell sources. Consequently the calculations have to use short time steps (often of the order of 1 year), and the number of models to be computed per pulse is large (up to $10^3$). It is clear that one cannot hope to compute straightforwardly through the whole phase of about $10^4$ pulses in medium-mass and massive stars. Indeed one may try to suppress the pulses artificially by neglecting the time-dependent terms ($\varepsilon_g$) in the energy equation and computing models in complete equilibrium. This gives (hopefully) an average evolution which might suffice in order to describe the evolution of the central core, and therefore of the final fate of the star. For stars of small mass (originally or by mass loss) the situation is better. One can certainly calculate through all of the relatively few pulses that occur before such a star becomes a white dwarf.


## 33.4 Evolution of the Central Region

The description of the nuclear cycles in § 33.1 has already given a rough outline of the central evolution of a star. We recognize it easily in Fig. 33.6, where the evolution of the centre is plotted in the $\lg \varrho_c$–$\lg T_c$ plane according to evolutionary calculations for very different stellar masses $M$. We see that $T_c$ indeed rises roughly $\sim \varrho_c^{1/3}$ [cf. (33.1)] as long as the central region remains non-degenerate. Of course, the details of the central evolution are much more complicated than predicted by the simple

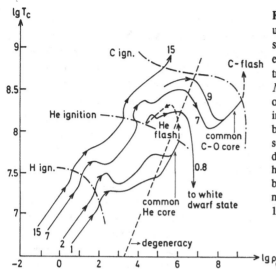

**Fig. 33.6.** Evolution of the central values of temperature $T_c$ (in K) and density $\varrho_c$ (in g cm$^{-3}$) for stars of different masses (from $0.8 M_\odot$ to $15 M_\odot$). The tracks are labelled with the stellar mass $M$ (in $M_\odot$). The conditions for ignition of hydrogen, helium, and carbon burning are indicated by dot-dashed lines. The broken straight line shows roughly the separation between non-degeneracy and degeneracy of the electron gas at not too high temperatures. The star of $0.8 M_\odot$ has been computed with the assumption of mass loss from the surface. (After IBEN, 1974)

vector field in Fig. 28.1. During the burnings the curves bulge out to the upper left. This is not surprising, since then the changes are far from homologous [which is assumed in (33.1) and for Fig. 28.1], for example owing to the restratification from a radiative to a convective core. After these interludes of burning, the evolution returns more or less to the normal slope. A parallel shift of the track from one to the next contraction is to be expected, since the contracting region (the core) will in general have a larger molecular weight, but a smaller mass.

We have already mentioned in § 28.1 and in § 33.1 the important fact that each contraction with $T_c \sim \varrho_c^{1/3}$ brings the centre closer to the regime of electron degeneracy. The degree of non-relativistic degeneracy is constant on the steeper lines $T \sim \varrho^{2/3}$. Such a line of constant degeneracy parameter $\psi$ is plotted in Fig. 33.6 for constant $\mu_e$, say $\mu_e = 2$. It is thus valid for the evolution after central hydrogen burning. Before this phase there is a hydrogen-rich mixture with lower $\mu_e$ and the line of the same $\psi$ has to be shifted to the left. Once the central region has reached a certain degree of degeneracy (where $\alpha = 3/4$ in the simple model of § 28.1), $T_c$ no longer increases, and the next burning is not reached in this way (if at all). This happens the earlier in the nuclear history, the closer to degeneracy a star has been at the beginning, i.e. the smaller $M$ is (cf. Fig. 33.6 and § 22.2). Therefore which nuclear cycle is completed before the star develops a degenerate core depends on the stellar mass $M$.

If the evolution were to proceed with complete mixing, we would only have to consider homogeneous stars of various $M$ and different compositions, and to see whether their contraction leads to ignition ($M > \widetilde{M_0}$) of a certain burning or to a degenerate core ($M < \widetilde{M_0}$). These limits for reaching the burning of H, He, and C are $\widetilde{M_0} \approx 0.08$, 0.3, and $0.8 M_\odot$, respectively.

We know that the evolution lies far from the case of complete mixing, and only the innermost core of a star is processed by a burning. But for sufficiently concentrated cores, the central contraction proceeds independently of the conditions at its boundary, i.e. independently of the non-contracting envelope. Therefore the above values $\widetilde{M}_0$ give roughly the limits for the masses of the corresponding cores.

Standard evolutionary calculations (assuming a typical initial composition, no convective overshooting, and no mass loss) give the following characteristic ranges of $M$. After central hydrogen burning, *low-mass stars* with $M < M_1(\text{He}) \approx 2.3 M_\odot$ develop degenerate He cores. After central helium burning, *medium-mass stars* with $M < M_1(\text{C–O}) \approx 9 M_\odot$ develop a degenerate C–O core. And in *massive stars* with $M > M_1(\text{C–O})$ even the C–O core remains non-degenerate while contracting for the ignition of the next burning. The precise values of the limiting masses $M_1$ depend, for example, on the assumed initial composition and on the rates for the neutrino losses, which can raise $M_1(\text{C–O})$ appreciably. Another important influence is the downwards penetration of the outer convection zone after central helium burning (in the second dredge-up phase). This lowers the mass of the core and therefore encourages the evolution into stronger degeneracy, i.e. it lowers $M_1$.

After a star has developed a strongly degenerate core it has not necessarily reached the very end of its nuclear history. This is only the case if the shell-source burning cannot sufficiently increase the mass of the degenerate core. However, the next burning is only delayed, and it will be ignited later in a "flash" if the shell source is able to increase the mass of the core to a certain limit $M_c'$. We have seen in §32.3 that the critical mass for ignition of helium in a degenerate core is $M_c'(\text{He}) \approx 0.45 M_\odot$. The corresponding critical mass of a degenerate C–O core is $M_c'(\text{C–O}) \approx 1.4 M_\odot$ as we shall see immediately. Note that these limits are appreciably larger than the corresponding lower limits $(\widetilde{M}_0)$ for reaching a burning by non-degenerate contraction, as described above. This indicates the possibility that the evolution depends discontinuously on $M$ around the limits $M_1(\text{He})$ and $M_1(\text{C–O})$. For example, stars with $M = M_1(\text{He}) - \Delta M$ ignite helium via a flash in a degenerate core of mass $0.45 M_\odot$, while stars with $M = M_1(\text{He}) + \Delta M$ can ignite helium burning via core contraction in (nearly) non-degenerate cores of about $0.3 M_\odot$ (cf. the idealized scheme in Fig. 33.7). Here one could imagine a bifurcation at $M = M_1$, where fluctuations would decide into which of the two regimes the star turns. In reality the limit will be "softened up" (a little bit of degeneracy leading to a baby flash, etc.).

The evolution of degenerate C–O cores is similar to that of degenerate helium cores in low-mass stars (§32.3,4). The structure of the core is more or less independent of the details of the envelope. Therefore the evolution of the central values converges for stars of different $M$ as long as the core mass is the same (cf. Fig. 33.6).

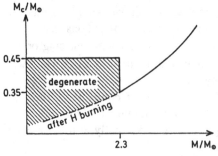

Fig. 33.7. The solid line shows schematically the mass $M_c$ of the helium core at the onset of helium burning as a function of the stellar mass $M$. The broken line shows the core mass at the end of hydrogen burning in low-mass stars, before the electron gas in the core becomes degenerate

While the mechanical structure of such a core is determined by its mass $M_c$, its thermal properties depend on the surrounding shell source and on the neutrino losses. If the shell source were extinguished, the core would simply cool down with $\varrho_c =$ constant (on a vertical line in Fig. 33.6) to the white-dwarf state. The continuous burning of the shell source increases $M_c$, which in turn increases the temperature in the shell source (cf. § 32.2). It also increases the central density, as we know from the discussion of the structure of degenerate configurations (§ 19.6), i.e. the evolution goes to the right in Fig. 33.6. The contraction due to this effect releases a large amount of gravitational energy, which, in the absence of energy losses (by conduction or neutrinos), would heat the core adiabatically.

However, there are strong neutrinos losses $\varepsilon_\nu$ in this part of the $T$-$\varrho$ diagram (cf. Fig. 18.9), which modify the whole situation. Since $\varepsilon_\nu$ increases appreciably with $T$, we should first make sure that there is no thermal runaway in the degenerate core (a "neutrino flash"), in analogy to a flash at the onset of a burning. This can be easily shown by the stability consideration presented in § 25, where we analysed the reaction of the central region on an assumed increase $d\varepsilon$ of the energy *release*. This led to (25.30) with gravothermal heat capacity $c^*$ (25.29). Now we replace $d\varepsilon$ by the small energy *loss* $-d\varepsilon_\nu$. If we neglect the perturbation of the flux ($dl_s = 0$) for simplicity, (25.29,30) become

$$c^* \frac{dT}{dt} = -d\varepsilon_\nu \quad , \quad c^* = c_P \left( 1 - \nabla_{ad} \frac{4\delta}{4\alpha - 3} \right) \quad . \tag{33.10}$$

Obviously the reversal of the sign of the right-hand side in the first equation (33.10) has reversed the conditions for stability. An ideal gas with $\alpha = \delta = 1$ has the gravothermal heat capacity $c^* < 0$, and neutrino losses are unstable since $\dot{T} > 0$ (a thermal runaway with ever increasing neutrino losses). Degenerate cores with $\alpha \to 3/5$, $\delta \to 0$ have $c^* > 0$, i.e. $\dot{T} < 0$, and these cores are stable: a small additional energy loss reduces $T$ and $\varepsilon_\nu$ such that the core returns to a stable balance. In the following scheme we summarize the different properties of thermal stability we have encountered:

|  | Burning ($\varepsilon > 0$) | Neutrinos ($-\varepsilon_\nu < 0$) |
|---|---|---|
| Ideal gas | stable | unstable |
| Degeneracy | unstable | stable |

According to § 33.3 the scheme also holds for burning in shell sources, where we have in addition the pulse instability for thin shells.

Numerical calculations approve the above conclusions: instead of leading to a thermal runaway, the neutrino losses cool the central region of a degenerate core such that $\varepsilon_\nu$ remains moderate. Typical "neutrino luminosities" $L_\nu$ (= total neutrino energy loss of the star per second) remain only a fraction of the normal "photon

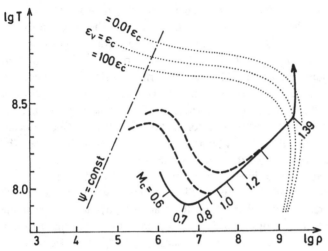

**Fig. 33.8.** Temperature $T$ (in K) and density $\varrho$ (in g cm$^{-3}$) in the C–O core of a $3M_\odot$ star after central helium burning. The solid line gives the evolution of the centre with increasing core mass $M_c$ (in $M_\odot$). The carbon flash starts at about $M_c = 1.39 M_\odot$ when the energy production by carbon burning ($\varepsilon_C$) exceeds the neutrino losses ($\varepsilon_\nu$). Some lines of constant ratio $\varepsilon_\nu/\varepsilon_C$ are dotted. The broken lines show the $T$ stratification in the core for two consecutive stages; neutrino losses have produced a maximum of the temperature outside the centre. (After PACZYŃSKI, 1971)

luminosity" $L$. The temperature profiles inside the cores of two different $M_c$ are shown in Fig. 33.8 by the broken S-shaped curves. They follow roughly lines of $\varepsilon_\nu = $ constant. With increasing $M_c$ the point for the centre moves along the solid line to the right, and extremely high values of $\varrho_c$ would necessarily occur if $M_c$ could go to the Chandrasekhar limit of $1.44M_\odot$. Shortly before this limit, at $M_c \approx 1.4M_\odot$, the central values reach the dotted line $\varepsilon_\nu = \varepsilon_C$, to the right of which pycnonuclear carbon burning dominates over the neutrino losses, $\varepsilon_C > \varepsilon_\nu$. Now carbon burning starts with a thermal runaway. If this happens in the centre, then explosive carbon burning will finally disrupt the whole star, such that one should expect a supernova outburst that does not leave a remnant (a neutron star); compare this also with § 34. This could be different if the reaction rates had to be changed such that the first ignition occurred at the maximum of $T$, i.e. in a shell rather than in the centre. With improved reaction rates one has in fact found that O is ignited before C; but the principal story remains that the degenerate C–O core is ignited when its mass $M_c \approx 1.4M_\odot$.

The just-described central evolution is the same for all stars that are able to develop a degenerate C–O core of $M_c \approx 1.4M_\odot$. The obvious condition for this is that the stellar mass $M$ is larger than that limit. For $\dot{M} = 0$ this would include all stars in the range $1.4M_\odot < M < 9M_\odot$, i.e. the intermediate-mass stars ($M \approx 2.3 \ldots 9M_\odot$) and the low-mass stars with $M > 1.4M_\odot$. More precisely the stellar mass $M$ must be larger than $1.4M_\odot$ *at the moment of ignition* (which does not occur before $M_c \approx 1.4M_\odot$). This can require that the initial stellar mass $M_i$ (on the main sequence) was much larger than $1.4M_\odot$, if $M$ has been reduced in the meantime by a strong mass loss.

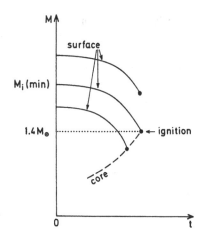

**Fig. 33.9.** For 3 different initial masses $M_i$ the solid lines show schematically the decrease of the stellar mass $M$ due to mass loss, while the mass of their degenerate C–O cores (*dashed line*) increases owing to helium shell burning. Carbon burning is ignited when the core mass reaches about $1.4 M_\odot$. This never occurs for $M_i < M_i(\min)$, since then the surface reaches the core before it can grow to $1.4 M_\odot$

Obviously there are two competing effects, the increase of $M_c$ due to shell-source burning and the simultaneous decrease of the stellar mass $M$ due to mass loss. Their changes in time are schematically shown in Fig. 33.9, and the outcome of this race decides the final stage of the star. The two values ($M$ and $M_c$) reach their goal at $1.4 M_\odot$ simultaneously if the initial mass has the critical value $M_i(\min)$. Stars with $M_i > M_i(\min)$ will ignite the C–O core, since $M_c$ can reach $1.4 M_\odot$. For stars with initial masses $M_i < M_i(\min)$, the mass loss will win and $M_c$ never reaches $1.4 M_\odot$. Such stars will finally cool down to the white-dwarf state after the shell source has died out near the surface (cf. § 32.7). Unfortunately the total loss of mass during the evolution is not well known. Rough guesses indicate a total mass loss of $\Delta M \approx 2.5 \ldots 3.5 M_\odot$, which would mean a critical initial mass in the range $M_i(\min) \approx 4 \ldots 5 M_\odot$. Of course, if the mass loss were so large that even stars with $M_i \approx 9 M_\odot$ were reduced to $M < 1.4 M_\odot$ before carbon ignition, then all intermediate stars (developing a degenerate C–O core) would become white dwarfs. In any case, there are drastic differences between the final stages (white dwarfs or explosions) to be expected for stars in a narrow range of $M_i$ near $M_i(\min)$.

It is clear that we have the same competition between $\dot{M}_c > 0$ and $\dot{M} < 0$ in the analogous problem of determining initial masses for which the degenerate helium cores are ignited (at $M_c \approx 0.45 M_\odot$). In this case the bifurcation of the evolution concerns mainly the composition of the final white dwarfs (He or C–O).

Finally, we have to consider the stars with $M > 9 M_\odot$, in which the C–O core does not become degenerate during the contraction after central helium burning. Therefore $T_c$ rises sufficiently during this contraction to start the (non-explosive) carbon burning. Here the neutrino losses can become very large, carrying away most of the energy released by carbon burning. In the later burnings, massive stars can have neutrino luminosities up to $10^6$ times larger than $L$; but these stages are very shortlived: for example silicon burning lasts just a few days (see Table 33.1).

These massive stars will go all the way through the nuclear burnings until Fe and Ni are produced in their central core. (Such a case is illustrated in the onion skin model in Fig. 33.1.) After the core has become unstable and collapses, electron captures by these nuclei transform the core into a neutron star, while the envelope is blown away by a supernova explosion (see § 34).

**Table 33.1** The ratio $L_\nu/L$ (neutrino to photon luminosity) at ignition, and the duration $\tau$ of late burnings (after WEAVER, ZIMMERMANN, WOOSLEY, 1978)

| Burning: | $M = 15M_\odot$: $(L \approx 10^4 L_\odot)$ | | $M = 25M_\odot$: $(L \approx 3 \times 10^5 L_\odot)$ | |
|---|---|---|---|---|
| | $L_\nu/L$ | $\tau$(years) | $L_\nu/L$ | $\tau$(years) |
| C | 1.0 | $6.3 \times 10^3$ | 8.3 | $1.7 \times 10^2$ |
| Ne | $1.8 \times 10^3$ | 7 | $6.5 \times 10^3$ | 1.2 |
| O | $2.1 \times 10^4$ | 1.7 | $1.9 \times 10^4$ | 0.51 |
| Si | $9.2 \times 10^5$ | 0.017 | $3.2 \times 10^6$ | 0.004 |

## 33.5 The Core-Mass–Luminosity Relation for Large Core Masses

We have seen that medium-mass stars, after central helium burning, develop a degenerate C–O core which is separated from the hydrogen-rich envelope by a thin helium layer. At its bottom there is helium-shell burning, which contributes only, say, 10 per cent to $L$. Most of the luminosity is produced in a hydrogen shell source at the bottom of the envelope. It is not too bad an approximation if we simply assume $L \approx L_H$, the hydrogen luminosity generated above a condensed core of mass $M_c$ and radius $R_c$. We also have seen that $L$ increases with increasing $M_c$ (giving the upwards motion along the asymptotic branch) and here face the same situation as for low-mass stars on the ascending giant branch. One can again derive the dependence of the properties of the shell on $M_c$ and $R_c$ by homology relations as in §32.2, assuming the simple power laws (32.2) for $\kappa$ and $\varepsilon$. But since we are dealing with rather massive cores and high temperatures here, the radiation pressure cannot be neglected. We therefore have to replace (32.3) by

$$P = \frac{\Re}{\mu}\varrho T + \frac{a}{3}T^4 = \frac{1}{\beta}\frac{\Re}{\mu}\varrho T \quad . \tag{33.11}$$

If again we write in the neighbourhood of given $P$ and $T$ the equation of state as a power law, $\varrho \sim P^\alpha T^{-\delta}$, we know from (13.16) that $\alpha = 1/\beta$, $\delta = (4 - 3\beta)/\beta$. Therefore we have as equation of state

$$P \sim \varrho^\beta T^{4-3\beta} \quad . \tag{33.12}$$

As in (32.4–7), we write the quantities $\varrho$, $T$, $P$, and $l$ in the shell as powers of $M_c$ and $R_c$. By the same procedure as in §32.2 we can derive equations for the exponents. For the sake of simplicity we restrict ourselves to the case $a = b = 0$ and obtain, instead of (32.22),

$$\varphi_1 = \frac{4 - \nu}{N} \quad , \quad \varphi_2 = \frac{\nu - 12 + 6\beta}{N} \quad ,$$

$$\psi_1 = \frac{1+n}{N} \quad , \quad \psi_2 = \frac{2\beta - n - 3}{N} \quad ,$$

$$\tau_1 = \beta\varphi_1 + (4 - 3\beta)\psi_1 \quad , \quad \tau_2 = \beta\varphi_2 + (4 - 3\beta)\psi_2 \quad ,$$

$$\sigma_1 = \frac{4n + \nu}{N} \quad , \quad \sigma_2 = \frac{3 - \nu - 3n}{N}\,\beta \qquad (33.13)$$

with

$$N = (4 - 3\beta)(1 + n) + (1 - \beta)(\nu - 4) \quad . \qquad (33.14)$$

For $\beta = 1$ the relations (33.13,14) agree with (32.22,23) for $a = b = 0$.

With increasing core mass, $\beta$ in the shell must decrease strongly, as can be seen from the following considerations. From (33.11,12) we have

$$\beta \sim \frac{\varrho T}{P} \sim \varrho^{1-\beta} T^{-3(1-\beta)} \quad . \qquad (33.15)$$

If we here replace $\varrho$, $T$ by (32.4,5), then the dependence of $\beta$ on $M_c$, $R_c$ is given by

$$\frac{d\ln\beta}{d\ln M_c} = (1 - \beta)\left[(\varphi_1 - 3\psi_1) + (\varphi_2 - 3\psi_2)\frac{d\ln R_c}{d\ln M_c}\right] \quad . \qquad (33.16)$$

One may start from an initial model that has been computed by solving the stellar structure equations numerically. This gives initial values for $M_c$, $R_c$, $L$, and $\beta$. Starting from these initial values we want to integrate (33.16). For simplicity, let us take for the derivative on the right-hand side of (33.16) Chandrasekhar's mass-radius relation of white dwarfs, and for the exponents in the energy generation $n = 2$, $\nu = 14$. The result of such an integration is shown by a dotted line in Fig. 32.1. In the same way, (32.27) can be integrated with $\sigma_1$, $\sigma_2$ from (33.13) and $\beta(M_c)$ as derived from the solution of (33.16). This gives the solid curve in Fig. 32.1. In spite of all approximations used, the integrated curves illustrate clearly the essential points.

For small core masses, $\beta \approx 1$ and the relation (32.25) holds, giving a steep increase of $L$ with $M_c$ [$L \sim M_c^7$ after (32.26)]. For larger $M_c$, radiation pressure becomes more and more important and $\beta$ decreases. This gives a much smaller slope of the $L(M_c)$ curve. Indeed in the limit $\beta = 0$ (33.13) gives $\sigma_1 = 1$, $\sigma_2 = 0$, independent of $n$ and $\nu$:

$$L \sim M_c \quad . \qquad (33.17)$$

The $L$-$M_c$ relation has become extremely simple, and we do not have to worry about the correct $R_c$-$M_c$ relation. Indeed from numerical models PACZYŃSKI (1970) derived

$$\frac{L}{L_\odot} = 5.92 \times 10^4 \left(\frac{M_c}{M_\odot} - 0.52\right) \qquad (33.18)$$

as an interpolation formula for sufficiently large $M_c$.

# § 34 Final Explosions and Collapse

We have seen that stars can evolve to the white-dwarf stage through a sequence of consecutive hydrostatic states if they develop a degenerate core and have final masses less than the Chandrasekhar limit $M_{\text{Ch}}$. It is not well known, however, how much mass the stars can have initially (on the main sequence) in order to end this way. Sometimes an upper limit of $4M_\odot$ is quoted, but it may be even up to $10M_\odot$. The main uncertainty here is the total amount of mass lost by stellar winds.

Other stars certainly undergo explosions, ejecting a large part of their mass, if not disrupting completely. In the case where a neutron star is left as a remnant the core must have undergone a collapse, since it cannot reach the neutron-star stage by a hydrostatic sequence. Collapse and explosions are connected with supernova events, but as yet there is no fully developed theory that could explain for sure the mechanisms responsible for the different observed phenomena. The appearance of SN 1987A and the large amount of new information resulting has made the situation even more complicated. In this section we only discuss some effects which certainly play an important role in late phases of more massive stars, and that will probably be part of future theories of supernovae.

Connected with supernovae is the interesting problem of nucleosynthesis in stars, a topic beyond the scope of this book. For a review see HILLEBRANDT (1986), WOOSLEY (1986).

## 34.1 The Evolution of the C–O Core

After central helium burning, further evolution depends critically on the question whether or not the C–O core becomes degenerate in the ensuing contraction phase. Clearly this will depend on the mass of the core. Since its contraction is practically independent of the envelope, the core can be considered as if it were a contracting gaseous sphere with zero surface pressure, as discussed in § 28.

We first estimate the critical core mass that separates the case where the contraction leads to increasing temperatures from the case where degeneracy prevents further heating. For this purpose we replace the equation of state by an interpolation formula between different asymptotic behaviours. In the cores of evolved stars the molecular weight per electron is $\mu_e \approx 2$, while that per ion is $\mu_0 \geq 12$, and therefore the pressure of non-degenerate electrons ($\sim 1/\mu_e$) dominates the ion pressure ($\sim 1/\mu_0$). This holds even more so if the electrons are degenerate. For simplicity we here neglect radiation pressure, as well as the creation of electron–positron pairs, which can also lead to partial degeneracy at very high temperatures and low densities (see §34.3.5). We then approximate the equation of state by the simple form

$$P \approx P_e = \frac{\Re}{\mu_e} \varrho T + K_\gamma \left(\frac{\varrho}{\mu_e}\right)^\gamma . \tag{34.1}$$

In the second term the exponent $\gamma$ is not a constant, allowing for non-relativistic and relativistic degeneracy. It varies from $\gamma = 5/3$ for $\varrho \ll 10^6$ g cm$^{-3}$ to $\gamma = 4/3$ for $\varrho \gg 10^6$ g cm$^{-3}$, while $K_\gamma$ varies from the constant in (15.23) to that in (15.26).

The equation of hydrostatic equilibrium (2.4) yields as a rough estimate for the central values (which we denote by subscript 0):

$$P_0 \approx \frac{GM_c\bar{\varrho}}{R_c} = f\, GM_c^{2/3} \varrho_0^{4/3} . \tag{34.2}$$

Here we have used the fact that $P_0$ is almost given by the weight of the core material alone and $\bar{\varrho} = 3M_c/(4\pi R_c^3)$ is assumed to be proportional to $\varrho_0$. The dimensionless factor $f$, containing, for example, the ratio $\bar{\varrho}/\varrho_0$, is kept constant in this consideration. Using (34.1) for the centre and eliminating $P_0$ from (34.2) yields

$$\frac{\Re}{\mu_e} T_0 = f\, GM_c^{2/3} \varrho_0^{1/3} - K_\gamma \varrho_0^{\gamma-1} \mu_e^{-\gamma} . \tag{34.3}$$

On the right-hand side, the first term dominates in the non-degenerate case, while the two terms are about equal for high degeneracy.

For a given mass $M_c$, (34.3) gives an evolutionary track in the lg $\varrho_0$-lg $T_0$ plane in Fig. 34.1, similar to the tracks shown in Fig. 28.2. Starting with rather small $\varrho_0$ and $\gamma = 5/3$, the central temperature $T_0$ grows with $\varrho_0$ and has a maximum at $\varrho_{0max}$, after which $T_0$ decreases again until $T_0 = 0$ is reached at a density of $8\varrho_{0max}$. The behaviour of these evolutionary tracks is the same as that discussed in §28, if there $M$ is replaced by $M_c$. (The way we have made our estimate here, keeping $f$ constant during contraction, is equivalent to the assumption of homology there.) For example, in the non-degenerate case [first term on the right of (34.3) dominant] the slope of

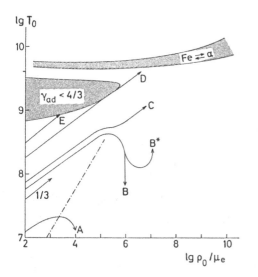

Fig. 34.1. Schematic evolution of the central values $T_0$ (in K) and $\varrho_0$ (in g cm$^{-3}$) for different core masses. The dot-dashed line corresponds to the left-hand part of the dot-dashed line in Figs 28.1,2. Five evolutionary tracks are plotted which illustrate the different cases discussed in the text: $A$ and $B$ correspond to case 1. $B^*$ illustrates case 2, where the core gains mass after it has become degenerate and undergoes a carbon flash. The curves $C$, $D$ correspond to case 3, while curve $E$ corresponds to case 4

345

the tracks is 1/3 as indicated on the left hand side of Fig. 34.1, and the tracks for different $M_c$ are shifted at the same values of $\varrho_0$ like $T_0 \sim M_c^{2/3}$, in analogy to §28.1.

With sufficiently growing central density, relativistic degeneracy becomes important, and $\gamma \to 4/3$, $K_\gamma \to K_{4/3}$. If we now write $\gamma = 4/3 + \chi$ (where $\chi \to 0$ for $\varrho/\mu_e > 10^7$ g cm$^{-3}$), we can replace (34.3) by

$$\frac{\Re}{\mu_e} T_0 = \varrho_0^{1/3} \left( f\, GM_c^{2/3} - K_{(4/3+\chi)}\mu_e^{-(4/3+\chi)} \varrho_0^\chi \right) \quad . \tag{34.4}$$

This shows that with increasing $\varrho_0$ the temperature $T_0$ does not become zero, but rises again $\sim \varrho^{1/3}$ if

$$M_c > M_{crit} = \left( \frac{K_{4/3}}{fG} \right)^{3/2} \mu_e^{-2} \quad . \tag{34.5}$$

Obviously the critical value of $M_c$ obtained in (34.5) is of the order of the Chandrasekhar mass $M_{Ch}$ as in (19.29,30). [Note that a comparison of (34.1) with (19.3) shows that $K_{4/3} = K\mu_e^{4/3}$.] In fact if $M_c = M_{crit}$ as defined here, then the core at zero temperature is fully relativistic, degenerate, and in hydrostatic equilibrium, which requires $M_c = M_{Ch}$.

We can therefore say that during contraction of a core with $M_c \lesssim M_{Ch}$ the central temperature reaches a maximum and afterwards decreases because of degeneracy, while for $M_c \gtrsim M_{Ch}$ the temperature continues to increase, roughly proportionally to $\varrho_0^{1/3}$.

We consider next the maximum temperature an evolutionary track reaches for $M_c < M_{crit}$ in the non-relativistic regime. We simply set $\gamma = 5/3$, $K_\gamma = K_{5/3}$ in (34.3) and introduce $M_{crit}$ from (34.5), obtaining

$$\Re T_0 = K_{4/3} \left( \frac{M_c}{M_{crit}} \right)^{2/3} \left( \frac{\varrho_0}{\mu_e} \right)^{1/3} - K_{5/3} \left( \frac{\varrho_0}{\mu_e} \right)^{2/3} \quad . \tag{34.6}$$

This gives a maximum temperature $T_{0max}$ for

$$\frac{\varrho_{0max}}{\mu_e} = \frac{1}{8} \left( \frac{K_{4/3}}{K_{5/3}} \right)^3 \left( \frac{M_c}{M_{crit}} \right)^2 \approx 2.38 \times 10^5 \text{g cm}^{-3} \left( \frac{M_c}{M_{crit}} \right)^2 \quad , \tag{34.7}$$

with the value

$$T_{0max} = \frac{1}{4\Re} \frac{K_{4/3}^2}{K_{5/3}} \left( \frac{M_c}{M_{crit}} \right)^{4/3} \approx 0.5 \times 10^9 \text{ K} \left( \frac{M_c}{M_{crit}} \right)^{4/3} \quad . \tag{34.8}$$

(Note that $K_{4/3}$ and $K_{5/3}$ have different dimensions.) For cores with $M_c \lesssim M_{crit}$, therefore, $T_0$ cannot exceed $\approx 0.5 \times 10^9$ K. This is in rough agreement with the "summit" of the dotted line in Fig. 28.1.

The events in the following stages depend sensitively on details of the material functions, the initial models, and the numerical calculations. These factors can decide, for example, whether core collapse is followed by an explosion, whether a remnant is left, etc. In view of the uncertainties involved and the many complications which can occur, it is not surprising that the present picture is not too clear. Nevertheless we will tentatively classify the different evolutionary scenarios according to the core mass $M_c$ after helium burning. As can be seen, for example, from (34.3), the tracks for lower mass are below those for higher mass. We distinguish four cases, each of which is represented by one or more schematic evolutionary tracks in Fig. 34.1.

*Case 1*: If $M_c < M_{crit} \approx M_{Ch}$, and if there is no sufficiently massive envelope (due either to the original mass or to mass loss), so that $M_c$ cannot approach $M_{Ch}$ during the shell burning phase, then $T_0$ grows in the non-degenerate regime until a maximum is reached. Then the core becomes degenerate, starts to cool, and the star must become a white dwarf. Only if it is a member of a binary system and accretes sufficient mass at certain rates can carbon finally be ignited in a flash. From the shell in which the flash occurs, two detonation waves (see § 34.2.4) can start, a helium detonation front moving outwards and a carbon detonation front moving inwards. In this double-detonation model the star will finally be disrupted (for a summary see, for instance, HILLEBRANDT, 1986). Such explosions in binary systems are nowadays believed to be the cause of type I supernovae.

*Case 2*: If initially $M_c < M_{crit}$, but if there remains an envelope sufficiently massive, so that because of shell burning, $M_c$ can grow to $M_{Ch}$, the core becomes degenerate and cools after having reached a maximum temperature. But $\varrho_0$ increases with $M_c$, and finally carbon burning begins (for example by pycnonuclear reactions; compare with § 33.4). It starts in a highly degenerate state and is therefore explosive. This carbon flash can occur in stars that have started on the main sequence in the range $4 \lesssim M/M_\odot < 8$, if their mass loss has not been too strong. We will discuss the carbon flash in § 34.2.

*Case 3*: If $M_{crit} < M_c \lesssim 40 M_\odot$, the evolutionary track misses the non-relativistic region of degeneracy. The core heats up, reaching successively higher nuclear reactions. For $M_c \lesssim 4 M_\odot$, electron captures by Ne and Mg reduce the pressure and start a central collapse. For $M_c \gtrsim 4 M_\odot$, photodisintegration of the nuclei brings $\gamma_{ad}$ below 4/3 and triggers a collapse. The collapse may lead to neutron-star formation and to ejection of the envelope (see § 34.3). This mechanism is assumed to cause type II supernovae.

*Case 4*: If $M_c \gtrsim 40 M_\odot$, the cores also reach the carbon burning in a non-degenerate state as in Case 3. But afterwards their evolutionary tracks in Fig. 34.1 cross the region of pair creation, which also reduces $\gamma_{ad}$. If $\gamma_{ad} < 4/3$ in an appreciable fraction of the core, say within 40% of its mass, then the core collapses adiabatically until the temperature of oxygen burning is reached. This may stop the collapse and make the star explode; if not, the collapse would lead into the region of instability because of photodisintegration, and the events would be as in Case 3. We will discuss this in § 34.3.5.

## 34.2 Carbon Burning in Degenerate Cores

Consider stars starting with masses in the range $4 \lesssim M/M_\odot \lesssim 8$ and having not too large a mass loss. After helium burning they will form a C–O core that is degenerate, and in the subsequent evolution $M_c$ grows owing to shell burning until it comes close to $M_{Ch}$. During this phase the central density increases with $M_c$ (similar to a sequence of white dwarfs with increasing mass). The energy released in the core during this contraction is transported by electron conduction in the direction of the centre, where the temperature is smaller and neutrino losses (see § 18.6) carry away the energy. The increase of the central density or of the temperature at the place of its maximum finally ignites carbon burning.

### 34.2.1 The Carbon Flash

The ignition of carbon in degenerate C–O cores of mass $M_c \approx M_{Ch}$ has already been discussed in § 33.4. As described there, the ignition of carbon may occur in the centre or in the shell of maximum temperature. The general properties of the flash are the same in both cases. We discuss here the central ignition. In Fig. 34.2 the $\lg \varrho_0$-$\lg T_0$ plane is shown again with an evolutionary path of the centre. The stability behaviour of the degenerate core depends critically on the question whether

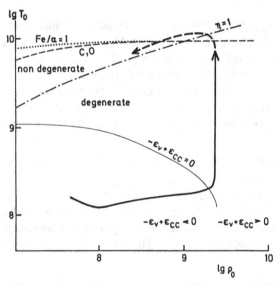

**Fig. 34.2.** Schematic evolution of the central region during and after the carbon flash (*heavy*). It corresponds to the evolution of type $B^*$ in Fig. 34.1. The flash starts when the central density $\varrho_0$ (in g cm$^{-3}$) or the central temperature $T_0$ (in K) is so high that the neutrino losses do not overcome the energy generation by carbon burning. The temperature then rises almost at constant density until degeneracy is removed. The dot-dashed line labelled $\eta = 1$ indicates where the gas pressure is twice the (degenerate) pressure at temperature zero; it roughly separates the regions of degeneracy and non-degeneracy. The broken line labelled C, O gives the temperature reached if all the energy released by carbon burning is used to increase the internal energy. The dotted line labelled Fe/$\alpha$ = 1 shows the points for which statistical equilibrium gives equal abundances of iron and helium

the energy balance is dominated by neutrino losses ($\varepsilon_{CC} - \varepsilon_\nu < 0$: stable) or by carbon burning ($\varepsilon_{CC} - \varepsilon_\nu > 0$: unstable). The borderline $\varepsilon_{CC} - \varepsilon_\nu = 0$ bends down at a few $10^9$ g cm$^{-3}$, since $\varepsilon_{CC}$ here increases mainly with increasing density (pycnonuclear reactions, see §18.4). Numerical calculations indicate that C–O cores reach the critical border $\varepsilon_{CC} - \varepsilon_\nu = 0$ between stability and instability at a density of $2 \times 10^9$ g cm$^{-3}$.

The slightest increase in temperature now makes $\varepsilon_{CC} - \varepsilon_\nu > 0$. Because of degeneracy the pressure does not increase and there is no consumption of energy through expansion. Therefore the temperature rises even more: a violent flash occurs. As in the case of the helium flash (see §32.4) the involved matter heats up at constant density until degeneracy is removed. Then it expands.

### 34.2.2 Nuclear Statistical Equilibrium

How violent the carbon flash can become is seen from a simple estimate. In a mixture of equal parts of C and O the carbon burning can release $2.5 \times 10^{17}$ erg/g and the subsequent oxygen burning twice this amount. If all this energy is used to heat the material, it can reach the temperatures indicated by the dashed line labelled C, O in Fig. 34.2. This line is somewhat curved since the specific heat depends slightly on the density. At these temperatures of nearly $10^{10}$ K the energy of the photons exceeds the binding energy of the nuclei, which are thus disintegrated. Photodisintegration, for example of Ne nuclei

$$^{20}\text{Ne} + \gamma \rightarrow\, ^{16}\text{O} + \alpha \quad , \tag{34.9}$$

was discussed in §18.5.3. The inverse reaction of (34.9) can also occur and the photon generated by this process can disintegrate another Ne nucleus. The processes are very similar to ionization and recombination of atoms. In statistical equilibrium the abundances of O, Ne, and $\alpha$ particles can be derived from a set of equations similar to the Saha equation (14.11):

$$\frac{n_O n_\alpha}{n_{Ne}} = \frac{1}{h^3} \left( \frac{2\pi m_O m_\alpha kT}{m_{Ne}} \right)^{3/2} \frac{G_O G_\alpha}{G_{Ne}} e^{-Q/kT} \quad , \tag{34.10}$$

where $G_O$, $G_\alpha$ and $G_{Ne}$ are the statistical weights, while $Q$ is the difference of binding energies

$$Q = (m_O + m_\alpha - m_{Ne}) c^2 \quad . \tag{34.11}$$

In addition to (34.10) there are two other conditions, one of which relates the particle numbers to the density, the other one describing the initial composition, since (34.9) and its inverse cannot change $n_O - n_\alpha$. Of course, one cannot consider a single reaction only, but has to take into account all reactions that can take place simultaneously. For example, $\alpha$ particles generated by (34.9) can also be captured by $^{12}\text{C}$ or $^{20}\text{Ne}$. (The problem is similar when ionization of different elements takes place simultaneously. They are not independent of each other, since all of them produce electrons which influence all recombination rates.)

If the temperatures are sufficiently high, many nuclei are disintegrated by photons and their fragments react again. The abundances of the different elements are then determined by a set of "Saha formulae" of the type (34.10). The nucleus $^{56}_{26}$Fe as the most stable one plays a crucial role in this statistical equilibrium. It can be disintegrated by photons into $\alpha$ particles and neutrons:

$$\gamma +^{56}_{26}\mathrm{Fe} \rightleftarrows 13\alpha + 4\mathrm{n} \quad . \tag{34.12}$$

In order to determine the ratio $n_{\mathrm{Fe}}/n_\alpha$ we consider quite general reactions of the type

$$\gamma + (Z, A) \rightleftarrows (Z - 2, A - 4) + \alpha \quad , \tag{34.13}$$

$$\gamma + (Z, A) \rightleftarrows (Z, A - 1) + \mathrm{n} \quad . \tag{34.14}$$

We start with the nucleus $(26,56) = {}^{56}\mathrm{Fe}$ and consider 13 reactions of type (34.13) and four of type (34.14). Then the abundance ratios are all given by equations like (34.10), and they can be combined to

$$\frac{n_\alpha^{13} n_\mathrm{n}^4}{n_{\mathrm{Fe}}} = \frac{G_\alpha^{13} G_\mathrm{n}^4}{G_{\mathrm{Fe}}} \left(\frac{2\pi kT}{h^2}\right)^{24} \left(\frac{m_\alpha^{13} m_\mathrm{n}^4}{m_{\mathrm{Fe}}}\right)^{3/2} e^{-Q/kT} \quad , \tag{34.15}$$

with

$$Q = (13m_\alpha + 4m_\mathrm{n} - m_{\mathrm{Fe}}) c^2 \quad . \tag{34.16}$$

If one assumes that the numbers of protons to neutrons (independently of whether they are free or in nuclei) have a ratio $n_\mathrm{p}/n_\mathrm{n} = 13/15$, as it is in the nucleus $^{56}\mathrm{Fe}$, then

$$n_\mathrm{n} = \frac{4}{13} n_\alpha \quad . \tag{34.17}$$

This, for instance, would be approximately the case in a mixture in which $^{56}\mathrm{Fe}$ is by far the most abundant heavy nucleus and its disintegration yields almost all neutrons and $\alpha$ particles. Then the left-hand side of (34.15) can be replaced by

$$\left(\frac{4}{13}\right)^4 \frac{n_\alpha^{17}}{n_{\mathrm{Fe}}} \quad . \tag{34.18}$$

Ignoring the binding energies, we can write the density as

$$\varrho = (56n_{\mathrm{Fe}} + 4n_\alpha + n_\mathrm{n}) m_\mathrm{u} \quad , \tag{34.19}$$

where $m_\mathrm{u}$ is the atomic mass unit. For given values of $\varrho$, $T$, and the ratio $n_\mathrm{n}/n_\alpha$ [corresponding to (34.17)] with (34.15, 18,19) we have two equations for $n_{\mathrm{Fe}}$ and $n_\alpha$.

Suppose again that the ratio of protons to neutrons per unit volume, normally called $\overline{Z}/\overline{N}$, is 13/15. Then equilibrium demands that all matter goes into $^{56}\mathrm{Fe}$ (the nucleus of the highest binding energy per nucleon) for temperatures that are not too high, and into $^4\mathrm{He}$ for high temperatures (see Fig. 34.3a). However, if we assume

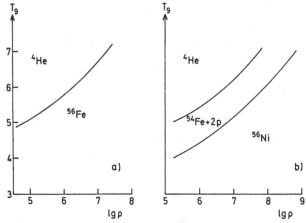

**Fig. 34.3.** (a) In the temperature–density diagram ($T$ in $10^9$ K, $\varrho$ in g cm$^{-3}$) the curve separates the regions in which equilibrium demands matter to be in the form of $^4$He and $^{56}$Fe respectively, for the case of $\overline{Z}/\overline{N} = 13/15$. (b) the corresponding equilibrium regions for $\overline{Z}/\overline{N} = 1$

$\overline{Z}/\overline{N} = 1$, then for the former temperatures $^{56}_{28}$Ni is the dominant nucleus, since it has the highest binding energy per nucleon of all nuclei with $Z = N$. With increasing temperature the equilibrium shifts from $^{56}$Ni to $^{54}$Fe+2p and finally to 14 $^4$He (see Fig. 34.3b).

The value $\overline{Z}/\overline{N}$ at the occurrence of photodisintegration depends on the weak interaction processes ($\beta$ decays) during the nuclear history of the stellar matter. In any case, in equilibrium at moderate temperatures one expects nuclei of the iron group, which with increasing temperature disintegrate into $\alpha$ particles and protons and neutrons.

### 34.2.3 Hydrostatic and Convective Adjustment

Even during the rapid helium flash the star remains very nearly in hydrostatic equilibrium, and convection can carry away all the released nuclear energy without becoming appreciably superadiabatic. The situation is completely different if unstable carbon burning proceeds in a degenerate core on a time-scale of milliseconds.

Consider the events after the onset of the carbon flash in the centre. The rapid rise of the central temperature is sufficient for immediately starting higher nuclear reactions, such as oxygen burning, which release additional energy. In one single runaway the central temperature rises so much that statistical equilibrium between Fe and He is reached, and eventually degeneracy is removed (see Fig. 34.2). Then the pressure increases and the central region starts to expand. This will occur roughly on a time-scale $\tau_\varepsilon$, in which the central temperature and the internal energy $u$ rise. Since $\dot{T}/T \approx \varepsilon_{CC}/u$, we have

$$\tau_\varepsilon = \frac{c_P T}{\varepsilon_{CC}} \ . \tag{34.20}$$

The other regions of the core react on the central expansion on the hydrostatic

351

time-scale $\tau_{\text{hydr}} \approx (G\bar{\varrho})^{-1/2}$ [compare with (2.19)], where $\bar{\varrho}$ is the mean density of the core. As long as $\zeta := \tau_\varepsilon/\tau_{\text{hydr}} \gg 1$ the core follows the central expansion quasi-hydrostatically. If, however, $\zeta \ll 1$, then the layers above cannot react rapidly enough, and a compression wave will move outwards with the speed of sound. If the push by the suddenly expanding burning region is sufficiently strong, an outwards travelling shock wave may develop.

Owing to the energy release in the flash, a central convective core will form, which has two effects. Part of the surplus energy is carried away (reducing the intensity of the flash), and new nuclear fuel is brought to the region of carbon burning (enhancing the flash). A characteristic time-scale for convection is $\tau_{\text{conv}} \approx \ell_{\text{m}}/v_{\text{s}}$, where $\ell_{\text{m}}$ is the mixing length and $v_{\text{s}}$ the local velocity of sound. Indeed turbulent elements will scarcely move faster than $v_{\text{s}}$, since otherwise shock waves would strongly damp the motion. If $\xi := \tau_\varepsilon/\tau_{\text{conv}} \gg 1$, convection is able to carry away all the nuclear energy released. If, however, $\xi \ll 1$, then convection cannot carry away the released energy.

The time-scales $\tau_{\text{hydr}}$ and $\tau_{\text{conv}}$ are very short indeed. For the central parts of the core with $\varrho > 10^8$ g cm$^{-3}$, one finds typically $\tau_{\text{hydr}} \approx 0.1$ s, and $\tau_{\text{conv}}$ is of the same order. However, for $T = 2\ldots3 \times 10^9$ K the local time-scale $\tau_\varepsilon$ for the flash is of the order of $10^{-6}$ s. Therefore $\zeta$ and $\xi$ are both $\ll 1$. This means that, instead of hydrostatic adjustment, a compression wave will start outwards and that "convective blocking" prevents a rapid spread of released energy in the core. The changes caused by the flash in one mass element propagate comparatively slowly to other parts.

### 34.2.4 Combustion Fronts

The local nuclear time-scale $\tau_\varepsilon$ at the onset of the flash is rather short. If a flash is started somewhere in a degenerate C–O core, the burning proceeds at such high rates that the fuel in this mass element is used up almost instantaneously. To be more precise, the consumption is completed locally before the layers above can adjust. Only then is the unburnt material ahead heated to ignition (either by compression or by energy transport), and the flash proceeds outwards. But the burning is always confined to a layer of (practically) zero thickness. We have an outward-moving *combustion front*, which can be of two different types.

We have seen that a shock wave develops. Matter in front penetrates the discontinuity with supersonic velocity and is compressed and heated. If this suffices to ignite the fuel, then the combustion front coincides with the shock front moving outwards supersonically. This is called a *detonation front*.

If the compression in the shock does not ignite the fuel, then the ignition temperature is reached owing to energy transport (convection or conduction). This gives a slower, subsonic motion for the burning front and contains a discontinuity in which density and pressure drop. This is a *deflagration front*.

Obviously the speed of a deflagration front is controlled by that of energy transport. This in turn depends on the conductivity (thermal or convective) and on the temperature difference between the deflagration front and the material ahead.

In both cases the deviations from hydrostatic equilibrium are mainly confined to a thin shell across which the pressure is discontinuous and all nuclear energy is released. The momentum of the matter approaching a detonation front supersonically is balanced by the higher pressure behind the front; the momentum of the matter approaching a deflagration front subsonically is balanced by the recoil of the matter moving away from it behind the front.

For an account of the theory of the two types of combustion fronts see COURANT, FRIEDRICHS (1948). As with normal shock waves, the theoretical results follow from the conservation of mass, momentum, and energy of the matter going through the discontinuity. For energy conservation, however, it also has to be taken into account that energy is released at the discontinuity. This makes the *two* types of solutions (detonation and deflagration waves) possible, while the theory of normal shock waves allows only that solution in which the density of matter going through the discontinuity increases.

In principle, detonation fronts as well as deflagration fronts can occur in stars. Which of the two will develop depends on the details of the transport mechanism, which determines the motion of a deflagration front and of the preceeding shock. Therefore numerical calculations have to decide which of the two types of combustion fronts will be built up.

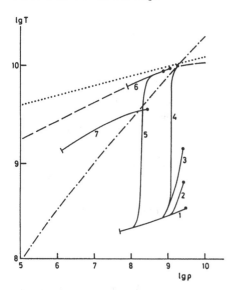

Fig. 34.4. The evolution through thermally unstable carbon burning. The structure of a stellar core at the onset of carbon burning is labelled 1. The centre is indicated by a thick point in the subsequent models 1 ... 7. After ignition the centre heats up at almost constant density until in model 4 degeneracy is removed. (The dot-dashed line gives the locations at which the gas pressure is 1.3 times the (degenerate) pressure at temperature zero.) Then the density decreases. Until model 5 the surface of the core remains unchanged, since the detonation wave triggered by carbon burning has not yet reached the surface of the core. The broken line as in Fig. 34.2 gives the temperature accessible with the energy of carbon burning. The dotted line as in Fig. 34.2 corresponds to equilibrium between Fe and He with Fe/$\alpha$ = 1. (After ARNETT, 1969)

Let us first assume that the central flash leads to a detonation front. The evolution of the central core is sketched in Fig. 34.4, giving the stratification for 7 consecutive stages. In particular stages 4 and 5 show clearly that $T$ and $\varrho$ increase if one crosses the front in an inward direction, and that the outer layers of the core do not react until the front has arrived. This is because the detonation front moves supersonically. Oxygen burning follows immediately, and its contribution has to be included into the energy release in the detonation front. Then as long as there is no expansion or heat leakage the matter that has gone through the shock has a temperature given by the dashed line in Fig. 34.4. When the shock wave reaches the surface of the core

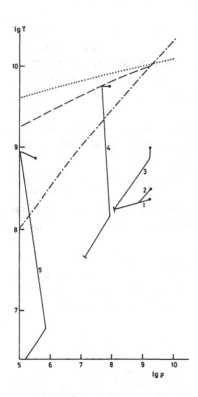

**Fig. 34.5.** The evolution of a stellar core during carbon deflagration. Ignition starts with model 1. Then the centre moves as in the case of detonation. But after model 3 the outer layers of the core are also involved. At the same time, a deflagration front develops. Note that the density decreases in the inward direction in the front (after NOMOTO et al., 1976). The dot-dashed and dotted lines correspond to those of Fig. 34.4

(stage 6 in Fig. 34.4) all of the core mass is at a temperature of about $5 \times 10^9$ K. Then the iron peak elements are formed in statistical equilibrium.

The corresponding evolution of the core in the case of a deflagration front is shown in Fig. 34.5. One can see that the layers ahead expand long before the front arrives, a sign of the subsonic motion of the deflagration front. The increase of $T$ in the front is accompanied by a decrease of $\varrho$. A basic difference to the result of a detonation front is that only the innermost part of the core is heated to $T \approx 5 \times 10^9$ K, where iron peak elements can be formed. Because of the expansion these high temperatures are no longer reached when the front has moved a bit further outwards.

### 34.2.5 Numerical Solutions

Numerical computations suggest that stars prefer deflagration rather than detonation. The reason is illustrated by a simple consideration: numerical calculations show that $^{12}C$ is ignited at a density of $2 \times 10^9$ g/cm$^3$. Assuming complete relativistic degeneracy and $\mu_e = 2$, one finds from (15.26) $P_e = 1.24 \times 10^{27}$ dyn/cm$^2$ and with (15.27) $u_e = U_e/\varrho = 3P_e/\varrho = 1.87 \times 10^{18}$ erg/g. Carbon burning, followed by oxygen burning, adds to this $\approx 5 \times 10^{17}$ erg/g, i.e. only 27 per cent. Correspondingly the overpressure is not very large, and therefore the shock is not extremely strong.

A critical point for computing a deflagration front is that a method of dealing with time-dependent convection is needed. In the simplest case one may just assume a velocity of the front, for example (NOMOTO et al., 1976)

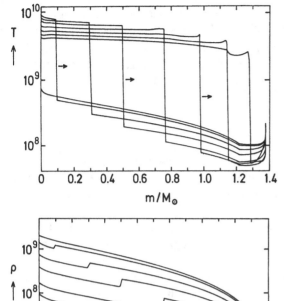

**Fig. 34.6.** The temperature and density distribution during the propagation of a carbon deflagration wave (after NO-MOTO et al., 1984). Note that in the outgoing front the density drops in the inward direction. The eight stages plotted in both diagrams correspond to 0, 0.6, 0.79, 0.91, 1.03, 1.12, 1.18, 1.24, and 3.22 s after the onset of carbon

$$v_D = \alpha \left( \frac{Gm}{4r^2} \, \ell_m \Delta \lg \varrho \right) \quad , \tag{34.21}$$

where $\alpha$ is a free parameter and $\ell_m$ the mixing length, while $\Delta \lg \varrho$ refers to the difference in density ahead and behind the deflagration front.

Calculations by NOMOTO et al. (1984) used the model of UNNO (1967) for time-dependent convection. Their results are displayed in Fig. 34.6, showing $T$ and $\varrho$ in the core for 8 consecutive stages of evolution. Note that between the first (onset of burning) and the last only 3.22 s have elapsed. One sees nicely the sharp changes of $\varrho$ and of $T$ in the discontinuity moving outwards in mass. The drop of $\varrho$ in the whole core from stage to stage reflects the expansion of the core.

Although the deflagration front moves outwards subsonically, the core is normally destroyed by the carbon flash. A rough estimate can make it plausible that there is enough energy for disrupting the core. If the matter were completely relativistic, $\gamma_{ad}$ would be 4/3 and the total energy $W$ would be zero (see the remark at the end of § 19.9). Therefore one can expect that the total energy is very small: $|W| = |E_g + E_i| \ll |E_g|$. From (19.50) with $n = 3$ we obtain for the gravitational energy of the core

$$|E_g| \approx \frac{3}{2} \frac{G M_c^2}{R_c} \quad . \tag{34.22}$$

If we take $M_c = 0.7 M_\odot$ and for $R_c$ the corresponding white-dwarf radius ($R \approx 2.2 \times 10^9$ cm), we find $|W| \ll |E_g| \approx 9 \times 10^{49}$ erg. On the other hand, the energy released by carbon burning (if half of the core mass is $^{12}$C) is $2.5 \times 10^{17}$ erg/g $\cdot M_c = 3.5 \times 10^{50}$ erg. Therefore carbon burning releases enough energy for a disruption of the whole core, which indeed has been found from numerical computations.

### 34.2.6 Carbon Burning in Accreting White Dwarfs

Rather similar phenomena to those described above for C–O cores of single stars can occur in C–O white dwarfs which are members of binary systems. They can receive appreciable amounts of matter from their companions. The accreted matter is compressed and heated, and its ignition can give rise to various phenomena.

For example, if helium is accreted with relatively low rates (about $10^{-8} M_\odot$ /year), a helium flash will be ignited in a shell of high density. The result can be a double detonation wave: a helium detonation front running outwards and a carbon detonation front going to the centre. As a result the white dwarf will be disrupted.

For higher accretion rates the new material can burn quietly near the surface, thus simply increasing the mass of the C–O white dwarf. When it approaches $M_{Ch}$, the density in the inner parts becomes so large that carbon burning starts either in the centre, or in the shell of maximum temperature. This results in a flash, and a deflagration (or detonation) front starts, as discussed above for single stars. The white dwarf will also be disrupted. It is this mechanism which at present is generally believed to cause the Type I supernovae. Note that the binary scenario had to be invoked, since the spectra of these supernovae show no hydrogen, and because evolving single stars of $M < 10 M_\odot$ may lose so much mass that their C–O core can never come close to $M_{Ch}$.

Carbon deflagration and the detonation of helium shells in accreting white dwarfs have been investigated by NOMOTO et al. (1985).

## 34.3 Collapse of Cores of Massive Stars

According to Fig. 34.1 one can expect that the cores of massive stars will not cool, because of non-relativistic degeneracy, but will heat up during core contraction until the next type of nuclear fuel is ignited. The core then is either non-degenerate (larger core mass $M_c$), or degenerate but to the upper right of the "summit" of the line $\alpha = 3/4$ in Fig. 28.1. In both cases the gravothermal heat capacity is negative, and the burning is self-controlled. In the following we discuss stars with core masses in the range $M_{Ch} < M_c < 40 M_\odot$. The evolutionary paths of these stars will avoid the region of $\alpha < 3/4$, where in Fig. 28.1 the arrows point downwards.

After going through several cycles of nuclear burning and contraction, the core will heat up to silicon burning. Nuclear burning in several shell sources has produced layers of different chemical composition, as shown in Fig. 34.7. Finally the central region of the core reaches a temperature at which the abundances are determined by nuclear statistical equilibrium. In this stage the core is in a peculiar state in several

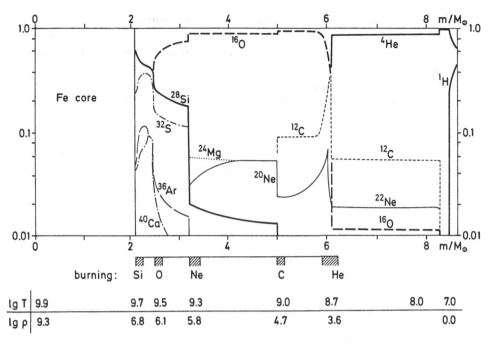

| burning: | Si | O | Ne | | C | He | | |
|----------|----|----|----|----|----|----|----|----|

| lg T | 9.9 | | 9.7 | 9.5 | 9.3 | 9.0 | 8.7 | 8.0 | 7.0 |
|------|-----|----|-----|-----|-----|-----|-----|-----|-----|
| lg ρ | 9.3 | | 6.8 | 6.1 | 5.8 | 4.7 | 3.6 | | 0.0 |

**Fig. 34.7.** The chemical composition in the interior of a highly evolved model of a $25 M_\odot$ star of population I. The mass concentrations of a few important elements are plotted against the mass variable $m$. Below the abscissa the location of shell sources and typical values of temperature (in K) and density (in g $cm^{-3}$) are indicated. (After WOOSLEY, WEAVER, 1986)

respects. Since the electron gas dominates the pressure, and since at temperatures of $T_9 \approx 10$ the electrons are relativistic ($kT \approx 1.7 m_e c^2$), the adiabatic exponent $\gamma_{ad}$ is close to 4/3. In the more massive stars photodisintegration of heavy nuclei reduces $\gamma_{ad}$ even more (like partial ionization). In addition general relativistic effects increase the critical value of $\gamma$ above 4/3, and the core becomes dynamically unstable. As a consequence core collapse sets in. For less massive stars the relativistic electrons are degenerate with high Fermi energies. Then electron captures by heavy nuclei reduce the pressure and start the collapse. For this stage we now discuss a simple solution.

### 34.3.1 Simple Collapse Solutions

Suppose we have a core at the onset of collapse, say, with central values $\varrho_0 = 10^{10}$ g $cm^{-3}$, $T_0 \approx 10^{10}$ K. The electrons are relativistically degenerate. Then the equation of state is polytropic and can be written as

$$P = K' \varrho^{4/3} , \qquad (34.23)$$

where $K' = K_{4/3}/\mu_e^{4/3}$ [compare with (15.26)]. Therefore the core can be described by a polytrope of index 3. We have already discussed the collapse of such a polytrope in § 19.11. As we have seen there, the parameter $\lambda$ appearing in the modified Emden equation (19.81) is a measure of the deviation from hydrostatic equilibrium, which corresponds to the value $\lambda = 0$. Solutions with finite radius are possible only for values $0 < \lambda < \lambda_m = 6.544 \times 10^{-3}$, where $\lambda = \lambda_m$ corresponds to the strongest deviation from equilibrium. For $\lambda > \lambda_m$ no homologous collapse of a polytrope of $n = 3$ is possible.

We now adapt the formalism of §19.11 for application to the collapse of stellar cores. The solution of the spatial structure is given by the function $w(z)$, which obeys (19.81). We denote the value of $z$ at the surface of the collapsing core by $z_3$, so that $w(z_3) = 0$; for $\lambda = 0$ one has $z_3 = 6.897$. It increases with $\lambda$ and reaches the maximum value 9.889 for $\lambda = \lambda_m$. The limit $\lambda = \lambda_m$ is reached when the surface of the core collapses with the acceleration of free fall.

If we apply (19.75) to the surface we have

$$z_3 \ddot{a} = -\frac{4}{3} \lambda \frac{(K')^{3/2}}{\sqrt{\pi G}} \frac{z_3}{a^2} \quad . \tag{34.24}$$

If this is equal to the free-fall acceleration $-GM_c/(az_3)^2$, then

$$\lambda = \lambda_m = \frac{3}{4} \sqrt{\frac{\pi G}{K'^3}} \frac{GM_c}{z_3^3} \quad . \tag{34.25}$$

On the other hand, (19.67,81) give

$$\frac{\varrho}{\varrho_0} = w^3 = \lambda - \frac{1}{z^2} \frac{d}{dz} \left( z^2 \frac{dw}{dz} \right) \quad , \tag{34.26}$$

and therefore with $r = az$, $R_c = az_3$, and

$$\bar{\varrho} = \frac{3}{R_c^3} \int_0^{R_c} \varrho r^2 dr \quad , \tag{34.27}$$

after some manipulation we find

$$\frac{\bar{\varrho}}{\varrho_0} = \lambda - \left[ \frac{3}{z} \left( \frac{dw}{dz} \right) \right]_{z=z_3} \quad . \tag{34.28}$$

If we apply this to the limit case $\lambda = \lambda_m$ in which $dw/dz$ vanishes at the surface (compare with Fig. 19.3), we find $\bar{\varrho}/\varrho_0 = \lambda_m$.

The core may start out from the (marginally stable) equilibrium for which $\lambda = 0$. Here the actual acceleration at the surface is zero, since gravity and pressure gradient cancel each other. But if the pressure is slightly decreased, the core will start to collapse ($\lambda > 0$). The numerical integration of (19.81) for different values of $\lambda$ in the range $0 \leq \lambda \leq \lambda_m$ gives values for $z_3$ and $\bar{\varrho}/\varrho_0$ in the ranges $6.897 \leq z_3 \leq 9.889$ and $0.01846 \leq \bar{\varrho}/\varrho_c \leq 0.0654$ (GOLDREICH, WEBER, 1980). If we determine the masses for different collapsing polytropes, we can use the expression

$$M_c \equiv \frac{4\pi a^3 z_3^3 \varrho_0}{3} \frac{\bar{\varrho}}{\varrho_0} = \frac{4\pi z_3^3}{3} \left( \frac{K'}{\pi G} \right)^{3/2} \frac{\bar{\varrho}}{\varrho_0} \quad , \tag{34.29}$$

which has been derived with the help of (19.67). Equation (34.29) for $\lambda = 0$ gives the Chandrasekhar mass $M_{\text{Ch}}$, as can be seen from (19.29,30) and (34.28). In fact all masses obtained for different values of $\lambda$ in the narrow interval $0 \le \lambda \le \lambda_m$ are close to the Chandrasekhar mass, namely $M_{\text{Ch}} \le M_c \le 1.0499 M_{\text{Ch}}$.

Only core masses in this small interval can collapse homologously. Now we know that $M_{\text{Ch}} \sim \mu_e^{-2}$. Electron captures during the collapse increase $\mu_e$ and reduce $M_{\text{Ch}}$. Therefore the upper bound for $M_c$ for homologous collapse decreases. If initially $\mu_e = \mu_{e0}$ and $M_{\text{Ch}} = M_{\text{Ch}0}$, then after some time not more than the mass

$$M_c = 1.0499 \left( \frac{\mu_{e0}}{\mu_e} \right)^2 M_{\text{Ch}0} \sim \mu_e^{-2} \tag{34.30}$$

can collapse homologously. (Note that, strictly speaking, the whole formalism should be repeated for a time-dependent $K'$.) Numerical integrations in fact indicate that during collapse the mass of the homologously collapsing part of the core decreases with increasing $\mu_e$ as given by (34.30).

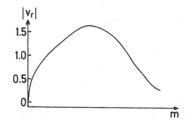

Fig. 34.8. Schematic picture of the velocity distribution in a collapsing stellar core originally of $1.4 M_\odot$ after numerical calculations (VAN RIPER, 1978). Note the two regimes: on the left $|v_r|$ (in units of $10^9$ cm s$^{-1}$) increases in the outward direction. It corresponds to a (roughly) homologously collapsing part, while on the right $|v_r|$ decreases with m. This corresponds to the free-fall regime

Figure 34.8 shows the infall velocity as a function of $m$ as obtained from numerical computations. The maximum separates the homologously collapsing inner core (left) from the free-falling outer part of the core (right). During collapse the boundary between the two regimes is not fixed but moves to smaller $m$ values: mass from the inner core is released into the free-fall regime. This corresponds to the decrease of $M_{\text{Ch}}$ with increasing $\mu_e$ as discussed above.

The collapse is extremely short-lived; it takes a time which is of the order of the free-fall time. If the core starts with an initial density of $10^{10}$ g cm$^{-3}$ one obtains $\tau_{\text{ff}} \approx (G\bar{\varrho})^{-1/2} \approx 40$ ms at the onset of collapse, while it is 0.4 ms for $\bar{\varrho} = 10^{14}$ g cm$^{-3}$.

### 34.3.2 The Reflection of the Infall

Because of the collapse, the density finally approaches that of neutron stars (nuclear densities of the order $10^{14}$ g cm$^{-3}$). Then the equation of state becomes "stiff", i.e. the matter becomes almost incompressible. This terminates the collapse.

If the whole process were completely elastic, then the kinetic energy of the collapsing matter would be sufficient to bring it back after reflection to the state just before the collapse began. This energy can be estimated roughly from

$$E \approx GM_c^2 \left( \frac{1}{R_n} - \frac{1}{R_{\text{wd}}} \right) \approx \frac{GM_c^2}{R_n} \approx 3 \times 10^{53} \text{erg} \quad , \tag{34.31}$$

where $M_c$ is the mass of the collapsing core, while $R_n$ and $R_{wd}$ are the typical radii of a neutron star and of a white dwarf. We compare this with the energy $E_e$ necessary to expel the envelope, which had no time to follow the core collapse,

$$E_e = \int_{M_{wd}}^{M} \frac{Gm\,dm}{r} \ll \frac{GM^2}{R_{wd}} \approx 3 \times 10^{52} \text{erg} \qquad (34.32)$$

for $M = 10M_\odot$. Realistic estimates bring $E_e$ down to $10^{50}$ erg, and therefore only a small fraction of the energy involved in the collapse of the core is sufficient to blow away the envelope. In predicting what happens after the bounce, one has to find out what (small) fraction of the energy of the collapse can be transformed into kinetic energy of outward motion. Remember that the energy estimated in (34.31) would suffice only to bring back the whole collapse to its original position – and no energy would be left for expelling the envelope. But if a remnant (neutron star) of mass $M_n$ remains in the condensed state, the energy of its collapse is available. The question is how this can be used for accelerating the rest of the material outwards.

A possible mechanism would be a shock wave moving outwards. The remnant is somewhat compressed by inertia beyond its equilibrium state and afterwards, acting like a spring, it expands, pushing back the infalling matter above. This creates a pressure wave, steepening when it travels into regions of lower density. The kinetic energy stored in such a wave may be sufficient to lift the envelope into space. However, the following problem arises. One can imagine that the neutron star formed has a mass of the order of the final Chandrasekhar mass $M_{ChF}$. The rest of the collapsing matter still consists mainly of iron. When, after rebounce, this region is passed by the shock wave, almost all of its energy is used up to disintegrate the iron into free nucleons. Therefore only a small fraction of the initial kinetic energy remains in the shock wave and is available for lifting the envelope.

### 34.3.3 Effects of Neutrinos

Before collapse, neutrinos were created by the processes described in § 18.6, and their energy is of the order of the thermal energy of the electrons. During collapse, neutrino production by neutronization becomes dominant. As soon as the density approaches values of $10^{12}$ g cm$^{-3}$, inverse $\beta$ decay becomes more pronounced, and the neutron-enriched nuclei decay. During this neutronization neutrinos are released. In connection with the recent supernova SN 1987A, neutrinos have been observed – manifest evidence that core collapse is indeed connected with the supernova phenomenon. The typical energy of the neutrinos released during collapse is of the order of the Fermi energy of the (relativistic) electrons. Therefore when using the relation $\varrho = \mu_e n_e m_u$ and (15.11,15) one finds

$$\frac{E_\nu}{m_e c^2} \approx \frac{E_F}{m_e c^2} = \frac{p_F}{m_e c}$$

$$= \left(\frac{3}{8\pi m_u}\right)^{1/3} \frac{h}{m_e c} \left(\frac{\varrho}{\mu_e}\right)^{1/3} \approx 10^{-2} \left(\frac{\varrho}{\mu_e}\right)^{1/3} . \qquad (34.33)$$

Here and in the following formulae (34.36,37) $\varrho$ is in g cm$^{-3}$.

If heavy nuclei are present, the neutrinos interact predominantly through the so-called "coherent" scattering (rather than scattering by free nucleons):

$$\nu + (Z, A) \rightarrow \nu + (Z, A) \quad . \tag{34.34}$$

The cross-section is of the order of

$$\sigma_\nu \approx \left(\frac{E_\nu}{m_e c^2}\right)^2 A^2 10^{-45} \text{cm}^2 \quad , \tag{34.35}$$

which with (34.33) gives

$$\sigma_\nu \approx A^2 \left(\frac{\varrho}{\mu_e}\right)^{2/3} 10^{-49} \text{cm}^2 \quad . \tag{34.36}$$

This allows an estimate of the mean-free-path $\ell_\nu$ of neutrinos in the collapsing core. If $n = \varrho/(A m_u)$ is the number density of nuclei, then with (34.36)

$$\ell_\nu \approx \frac{1}{n \sigma_\nu} = \frac{1}{\mu_e A} \left(\frac{\varrho}{\mu_e}\right)^{-5/3} 1.7 \times 10^{25} \text{cm} \quad . \tag{34.37}$$

Can $\ell_\nu$ become comparable with the dimension of the collapsing core, say $10^7$ cm? With $\mu_e = 2$, $A \approx 100$, we obtain from (34.37) $\ell_\nu = 10^7$ cm for $(\varrho/\mu_e) = 3.6 \times 10^9$ g cm$^{-3}$. Obviously we cannot simply assume that the neutrinos escape without interaction. The more the density rises, the smaller $\ell_\nu$, and the collapsing core becomes opaque for neutrinos. Then they can only diffuse through the matter via many scattering processes. For sufficiently high density the diffusion velocity becomes even smaller than the velocity of the collapse. Calculations show that the neutrinos cannot escape by diffusion within the free-fall time $\tau_{ff}$ of the core if $\varrho \gtrsim 3 \times 10^{11}$ g cm$^{-3}$: the neutrinos are then trapped.

In the schematic picture of the core structure (Fig. 34.9), the place where the infall velocity of matter equals the velocity of outward neutrino diffusion is indicated

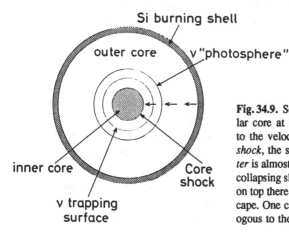

Si burning shell

outer core    ν "photosphere"

inner core

v trapping surface

Core shock

Fig. 34.9. Schematic picture of a collapsing stellar core at bounce. The short arrows correspond to the velocity field. At the sphere labelled *core shock*, the shock is formed inside which the *matter* is almost at rest. Above the shock there is a still collapsing shell in which neutrinos are trapped. But on top there is a shell from which neutrinos can escape. One can define a neutrino photosphere analogous to the photosphere in a stellar atmosphere

361

as the "neutrino trapping surface". Below it the neutrinos are trapped; above it they diffuse outwards until reaching the so-called "neutrino photosphere". This provides the boundary of the opaque part of the core and is located one mean free path $\ell_\nu$ beneath the surface. From here the neutrinos leave the core almost without further interaction.

In detailed calculations a rather complicated transfer problem is encountered. In particular one has to consider and use the distribution function of the neutrinos (rather than their average energy). This is obvious, since the cross-section as given in (34.35) depends on the energy of the neutrinos: those with low energy can escape more easily than those of high energy.

The congestion of the neutrinos, resulting from the opaqueness of the core, influences the further neutronization. With increasing density the neutrinos become degenerate with a high Fermi energy. Electron capture becomes less probable, since the new neutrinos have to be raised to the top of the Fermi sea. When a density of $3 \times 10^{12}$ g cm$^{-3}$ is reached, neutronization stops and $\gamma_{ad}$ has increased to the value 4/3, which corresponds to relativistic degeneracy. The collapse continues until $\varrho > 10^{14}$ g cm$^{-3}$. The dissociation of heavy nuclei into free neutrons and protons make the equation of state stiff ($\gamma_{ad} > 5/3$) and the collapse is stopped. Further neutronization can proceed only as far as the neutrinos diffuse outwards. Most of this takes place in the neutronization shell between trapping surface and neutrino photosphere (Fig. 34.9) where the density is several $10^{11}$ g cm$^{-3}$.

### 34.3.4 Numerical Results

Many authors have followed up the hydrodynamical evolution of the collapsing core numerically, using different initial models and different equations of state. As an example, Fig. 34.10a,b shows calculations for the collapse of stars of 20 and $25 M_\odot$. In most cases the core bounce does not expel the envelope. Quite generally a violent core bounce is needed to mimic a supernova explosion. This requires a rather soft equation of state, say $\gamma_{ad}$ slightly above 4/3. Then the central region compresses elastically and expands again, converting the infall energy into kinetic energy of outward motion. If, however, the equation of state is very stiff ($\gamma_{ad}$ appreciably above 4/3), the infalling matter is stopped at a more or less stationary shock front. Here the energy of infall is converted into internal energy. The interior core represents an accreting neutron star.

### 34.3.5 Pair-Creation Instability

From Fig. 34.1 one can see that evolutionary tracks for cores of sufficient mass enter a region on the left-hand side of the diagram where also $\gamma_{ad} < 4/3$ (FOWLER, HOYLE, 1964). In this region many photons have an energy exceeding the rest-mass energy of two electrons, $h\nu \geq 2m_e c^2$. Therefore electron–positron pairs can be spontaneously formed out of photons in the fields of nuclei. Admittedly the pairs do annihilate, creating photons again, but there is always an equilibrium number of pairs present. The *mean* energy of the photons $h\nu \approx kT$ equals the rest energy of the electron–positron pair only at a temperature of $1.2 \times 10^{10}$ K, but even at $10^9$ K

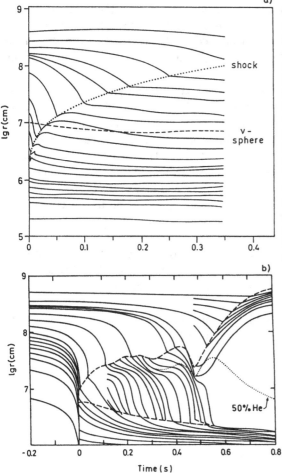

a)

Fig. 34.10. Two numerical simulations of core collapse. (a) Radius versus time for selected zones of a $20M_\odot$ model. The positions of the shock front and neutrino photosphere are indicated by dotted and broken lines. In this model the envelope is not shed (after HILLEBRANDT, 1987). (b) Radius versus time for selected zones of a $25M_\odot$ model. The shock front is given by the upper dashed line, the neutrino photosphere by the lower one. The dotted line gives the points at which the helium content is 50%. In this model the envelope is blown into space. Note that the model separates into a collapsing core and an outgoing envelope at mass $m = 0.1665M_\odot$ and time $t = 0.44$ s. (After WILSON, 1985)

b)

Time (s)

appreciable pair creation occurs because of the high-energy photons of the Planck distribution.

For an account of the thermodynamic effects of pair creation see, for example, COX, GIULI (1968). In many respects pair creation can be considered in analogy to ionization or dissociation (a photon being "ionized" or "dissociated" into a pair $e^-$, $e^+$). Regarding the stability of massive cores, the crucial point is that the pair creation reduces $\gamma_{ad}$, as incomplete ionization or photodisintegration does. Indeed, if the gas is compressed, not all the energy is used to increase the temperature, but part of it is used to create pairs. Other reductions of $\gamma_{ad}$ are due to high radiation pressure according to (13.16,21,24) and to relativistic electrons. All these effects bring $\gamma_{ad}$ below the critical value 4/3 for dynamical instability.

The total number of electrons consists of those from pairs and those from normal ionization of atoms. With increasing $\varrho$ the Fermi energy rises. This diminishes the possibility for pair creation, since newly created electrons now need an energy exceeding the Fermi energy. Correspondingly the instability region in Fig. 34.1 is limited to the right at a density of $5 \times 10^5$ g cm$^{-3}$.

The pairs created are not relativistic, having $\gamma_{ad} = 5/3$. (Note that a photon with $h\nu = m_e c^2$ can only create a pair with zero kinetic energy!) For higher temperatures there are so many pairs that they dominate and bring $\gamma_{ad}$ of the whole gas–radiation mixture slightly above 4/3, which limits the instability region towards high temperatures.

For the evolution of cores into the region of pair instability, radiation pressure is important, and therefore one cannot use our simple formulae of § 34.1. Furthermore, for a core instability it is not sufficient that the evolutionary track of the star's *centre* moves through the area with $\gamma_{ad} < 4/3$. Since in reality a mean value of $\gamma$ over the whole core decides upon its dynamical stability (§ 38.1), an appreciable fraction of the core mass must lie in that density–temperature range. According to numerical results this happens to cores of masses of $30 M_{crit}$ and more, where $M_{crit}$ is defined in (34.5). The corresponding main-sequence masses depend on the uncertain mass loss, but a realistic guess seems to be that stars initially with $M > 80 M_\odot$ later develop pair-unstable cores.

Numerical calculations indicate that, in a collapsing core of this type, oxygen is ignited explosively and the core runs into the (unstable) region of photodisintegration, though not very much is known of its final fate. In most numerical calculations, the pair instability causes a disruption of the core.

There is also the possibility of violent pulsations which lead to explosive mass loss, but no total disruption. This situation is not yet fully investigated, but may occur in the final evolution of stars with initial masses near $80 M_\odot$ (EL EID, LANGER, 1986).

# VII  Compact Objects

Stellar evolution can lead to somewhat extreme final stages. We have seen in §32 and §33 that the evolution tends to produce central regions of very high density. On the other hand it is known that stellar matter can be ejected. The mechanisms are only partly (if at all) understood, but they do exist according to observations (normal mass loss, planetary nebulae, explosions). It may be that in certain cases the whole star explodes without any remnant left (see §34). Often enough, however, only the widely expanded envelope is removed, leaving the condensed core as a *compact object*. Relative to "normal stars" these objects are characterized by small radii, high densities, and strong surface gravity.

There are 3 types of compact objects, distinguished by the "degree of compactness": white dwarfs (WD), neutron stars (NS), and black holes (BH). Typical values for WD are $R \approx 10^{-2} R_\odot$, $\varrho \approx 10^6$ g cm$^{-3}$, escape velocity $v_E \approx 0.02c$; their configuration is supported against the large gravity by the pressure of highly degenerate electrons (instead of the "thermal pressure", which dominates in the case of normal stars). For NS one has typically $R \approx 10$ km, $\varrho \approx 10^{14}$ g cm$^{-3}$, $v_E \approx c/3$; their pressure support is provided by densely packed, partially degenerate neutrons. This is the dominant species of particles since normal nuclei do not exist above a certain density. Indeed a NS represents very roughly a huge "nucleus" of $10^{57}$ baryons.

As a simple illustration, suppose that in both cases (WD and NS) ideal, non-relativistic degenerate fermions (of mass $m_e$ or $m_n$) provide the pressure balancing the gravity. The stars then are polytropes of index $n = 3/2$. With a mass–radius relation (19.28), where the constant of proportionality can be seen to be $\sim K \sim 1/m_{fermion}$, we have $R \sim 1/m_{fermion}$. The ratio of $m_n$ to $m_e$ then provides the ratio of typical radii for WD and NS of the same mass. The pressure–gravity balance by degenerate neutrons can only be maintained up to limiting masses corresponding to about $2 \times 10^{57}$ fermions.

Clearly for objects with gravity fields like those in NS general relativity becomes important. It will be the dominant feature for the last group of compact objects, namely BH with $R \approx 1$ km and $v_E = c$.

The first WD was detected long before theoreticians were able to explain it, whereas NS were predicted theoretically before they were, accidentally, discovered in the sky. And up to now BH are found with certainty in books only.

The physics of compact objects is interesting and complex enough to fill special textbooks (e.g., SHAPIRO, TEUKOLSKY, 1983). We refer to these for details and limit ourselves to indicating a few main characteristics.

# §35 White Dwarfs

It is characteristic for configurations involving degenerate matter that mechanical and thermal properties are more or less decoupled from each other. Correspondingly we will discuss these two aspects separately. When dealing with the mechanical problem (including the $P$ and $\varrho$ stratification, the $M$–$R$ relation, etc.) one may even go to the limit $T \to 0$. Of course, such cold matter can not radiate at all and it is more appropriate to denote these objects as "black dwarfs". The thermal properties, on the other hand, are responsible for the radiation and the further evolution of white dwarfs. The evolution indeed leads from a white dwarf (WD) to a black dwarf, since it is – roughly speaking – the consumption of fossil heat stored in the WD which we see at present. (Concerning the evolution *to* the white-dwarf stage see §§ 33,34.)

### 35.1 Chandrasekhar's Theory

This theory treats the mechanical structure of WD under the following assumptions. The pressure is produced only by the ideal (non-interacting) degenerate electrons, while the non-degenerate ions provide the mass. The electrons are supposed to be fully degenerate, but they may have an arbitrary degree of relativity $x = p_F/m_e c$, which varies as $\varrho^{1/3}$. Therefore we no longer have a polytrope as we had in the limiting cases $x \to 0$ and $x \to \infty$. The equation of state can be written as

$$P = C_1 f(x) \quad , \quad \varrho = C_2 x^3 \quad ; \quad x = p_F/m_e c \quad , \tag{35.1}$$

according to (15.13,15), which also define the constants $C_1$ and $C_2$, while (15.14) gives $f(x)$.

In order to describe hydrostatic stratification we start with Poisson's equation (19.2), in which we eliminate $d\Phi/dr$ by (19.1) and substitute $P$ and $\varrho$ from (35.1) obtaining

$$\frac{C_1}{C_2} \frac{1}{r^2} \frac{d}{dr} \left( \frac{r^2}{x^3} \frac{df(x)}{dr} \right) = -4\pi G C_2 x^3 \quad . \tag{35.2}$$

Differentiating the left-hand side of (15.12) with respect to $x$, one obtains an expression for $df(x)/dx$ which shows that

$$\frac{1}{x^3} \frac{df(x)}{dr} = 8 \frac{d}{dr} \left[ (x^2 + 1)^{1/2} \right] = 8 \frac{dz}{dr} \quad , \tag{35.3}$$

with

$$z^2 := x^2 + 1 \quad .\qquad (35.4)$$

Therefore (35.2) becomes

$$\frac{1}{r^2}\frac{d}{dr}\left(r^2\frac{dz}{dr}\right) = -\frac{\pi G C_2^2}{2C_1}(z^2-1)^{3/2} \quad ,\qquad (35.5)$$

and as in §19.2 we replace $r$ and $z$ by dimensionless variables $\zeta$ and $\varphi$:

$$\zeta := \frac{r}{\alpha} \quad , \quad \alpha = \sqrt{\frac{2C_1}{\pi G}}\frac{1}{C_2 z_c} \quad ,\qquad (35.6)$$

$$\varphi := \frac{z}{z_c} \quad ,\qquad (35.6)$$

where $z_c$ is the central value of $z$, characterizing the central density. Then from (35.5)

$$\frac{1}{\zeta^2}\frac{d}{d\zeta}\left(\zeta^2\frac{d\varphi}{d\zeta}\right) = -\left(\varphi^2 - \frac{1}{z_c^2}\right)^{3/2} \quad .\qquad (35.7)$$

This is Chandrasekhar's differential equation for the structure of WD. We write it in the form

$$\frac{d^2\varphi}{d\zeta^2} + \frac{2}{\zeta}\frac{d\varphi}{d\zeta} + \left(\varphi^2 - \frac{1}{z_c^2}\right)^{3/2} = 0 \quad ,\qquad (35.8)$$

and see that it is very similar (differing only in the parenthesis) to the Emden equation (19.10) for polytropes. In fact (35.8) becomes the Emden equation for indices $n = 3$ and $n = 3/2$ if we go to the limits $z \to \infty$ (i.e. $x \to \infty$) and $z \to 1$ (i.e. $x \to 0$) respectively. The central conditions are now

$$\zeta = 0 \ : \ \varphi = 1 \quad , \quad \varphi' = 0 \quad .\qquad (35.9)$$

Starting with these values, (35.8) can be integrated outwards for any given value of $z_c$. The density stratification is found if $\mu_e$ (which enters via $C_2$) is also specified:

$$\varrho = C_2 x^3 = C_2(z^2-1)^{3/2} = C_2 z_c^3 \left(\varphi^2 - \frac{1}{z_c^2}\right)^{3/2} \quad .\qquad (35.10)$$

The surface is reached at $\zeta = \zeta_1$, where $\varrho$ becomes zero, i.e. after (35.1,4,6)

$$\zeta = \zeta_1 \ : \ x_1 = 0 \quad , \quad z_1 = 1 \quad , \quad \varphi_1 = 1/z_c \quad .\qquad (35.11)$$

The value of $R$ is

$$R = \alpha\zeta_1 = \sqrt{\frac{2C_1}{\pi G}}\frac{1}{C_2 z_c}\zeta_1 \quad ,\qquad (35.12)$$

and $M$ can be found if we replace $r$ and $\varphi$ by (35.6) and (35.10):

$$M = \int_0^R 4\pi r^2 \varrho \, dr$$

$$= 4\pi\alpha^3 C_2 z_c^3 \int_0^{\varphi_1} \zeta^2 \left(\varphi^2 - \frac{1}{z_c^2}\right)^{3/2} d\zeta$$

$$= 4\pi\alpha^3 C_2 z_c^3 \left(-\zeta^2 \frac{d\varphi}{d\zeta}\right)_1$$

$$= \frac{4\pi}{C_2^2} \left(\frac{2C_1}{\pi G}\right)^{3/2} \left(-\zeta^2 \frac{d\varphi}{d\zeta}\right)_1 . \tag{35.13}$$

The integrand in the second equation (35.13) was simply replaced by the derivative on the left-hand side of (35.7).

Table 35.1 Numerical results of Chandrasekhar's theory of white dwarfs. Subscripts c and 1 refer to centre and surface, respectively. (After COX, GIULI, 1968)

| $1/z_c^2$ | $x_c$ | $\zeta_1$ | $(-\zeta^2 d\varphi/d\zeta)_1$ | $\varrho_c/\mu_e$ (g cm$^{-3}$) | $\mu_e^2 M$ ($M_\odot$) | $\mu_e R$ (km) |
|---|---|---|---|---|---|---|
| 0 | $\infty$ | 6.8968 | 2.0182 | $\infty$ | 5.84 | 0 |
| 0.01 | 9.95 | 5.3571 | 1.9321 | $9.48 \times 10^8$ | 5.60 | 4.170 |
| 0.02 | 7 | 4.9857 | 1.8652 | $3.31 \times 10^8$ | 5.41 | 5.500 |
| 0.05 | 4.36 | 4.4601 | 1.7096 | $7.98 \times 10^7$ | 4.95 | 7.760 |
| 0.1 | 3 | 4.0690 | 1.5186 | $2.59 \times 10^7$ | 4.40 | 10.000 |
| 0.2 | 2 | 3.7271 | 1.2430 | $7.70 \times 10^6$ | 3.60 | 13.000 |
| 0.3 | 1.53 | 3.5803 | 1.0337 | $3.43 \times 10^6$ | 2.99 | 16.000 |
| 0.5 | 1 | 3.5330 | 0.7070 | $9.63 \times 10^5$ | 2.04 | 19.500 |
| 0.8 | 0.5 | 4.0446 | 0.3091 | $1.21 \times 10^5$ | 0.89 | 28.200 |
| 1.0 | 0 | $\infty$ | 0 | 0 | 0 | $\infty$ |

Table 35.1 gives the results of integrations for different values of $z_c$ from $\infty$ to 1, i.e. from $x_c = \infty$ (fully relativistic) to $x_c = 0$ (non-relativistic), with the resulting $M$–$R$ relation being plotted in Fig. 35.1. As in the simple case of polytropes (§ 19.6), we find an $M$–$R$ relation with $dR/dM < 0$, but the exponent of $M$ is no longer constant as it is in (19.28). The stellar mass $M$ cannot exceed the Chandrasekhar limit $M_{\text{Ch}}$ as given by (19.30),

$$M_{\text{Ch}} = \left(\frac{2}{\mu_e}\right)^2 \times 1.459 M_\odot , \tag{35.14}$$

since this limit case ($z_c \to \infty$) coincides with a polytropic structure of index $n = 3$.

These characteristics certainly call for a simple explanation, since they contradict the everyday experience that spheres of given material (say iron) become larger with increasing mass. This experience is not only obtained by handling small iron spheres, but also by measurements of planets.

Let us consider rough averages (taken over the whole star) of the basic equation of hydrostatic equilibrium (9.16). Replacing there the absolute value of $dP/dm$ by

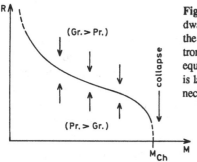

Fig. 35.1. Sketch of the classical mass–radius relation of white dwarfs according to Chandrasekhar's theory (assuming that the pressure is provided only by an ideal, degenerate electron gas). The arrows indicate the direction into which a non-equilibrium configuration is pushed if the gravitational force is larger or smaller than the pressure gradient. Corrections are necessary at both ends of the curve (*dashed*)

$P/M$ and $m/r^4$ by $M/R^4$, we obtain

$$\frac{P}{M} \approx \frac{GM}{4\pi R^4} \tag{35.15}$$

where $P$ is some average value. We replace it by the average density $\varrho \sim M/R^3$, using a degenerate equation of state,

$$P \sim \varrho^\gamma \sim \left(\frac{M}{R^3}\right)^\gamma . \tag{35.16}$$

The pressure term $f_p$, i.e. the left-hand side of (35.15), and the gravity term $f_g$, on the right-hand side, are then

$$f_p \sim \frac{M^{\gamma-1}}{R^{3\gamma}} \quad ; \quad f_g \sim \frac{M}{R^4} . \tag{35.17}$$

Their ratio $f$ must be unity for hydrostatic equilibrium:

$$f := \frac{f_g}{f_p} \sim M^{2-\gamma} R^{3\gamma-4} = \begin{cases} M^{1/3}R , & \text{for } \gamma = 5/3 \\ M^{2/3} , & \text{for } \gamma = 4/3 \end{cases} . \tag{35.18}$$

Suppose we have a given stellar mass $M < M_{Ch}$ and non-relativistic electrons with $\gamma = 5/3$. Then the star can easily find an equilibrium by adjusting $R$ such that $f = 1$. If we now slightly increase $M$, then $f > 1$ (gravity exceeds the pressure force), and $R$ must *decrease* in order to regain equilibrium ($f = 1$). This explains the structure of the $R$–$M$ relation (cf. Fig. 35.1).

However, if the electrons are relativistic ($\gamma = 4/3$), then $f$ is independent of $R$. Equilibrium can be achieved only by adjusting $M$ to a certain value $M_{Ch}$. If $M < M_{Ch}$, then $f < 1$, i.e. the dominant pressure term makes the star expand. This makes the electrons less relativistic and increases $\gamma$ above the critical value $4/3$. For $M > M_{Ch}$, $f > 1$, and the dominant gravity term makes the star contract; but this does not help either, and the star must collapse without finding an equilibrium. So $M_{Ch}$ is quite obviously a mass limit for these equilibrium configurations.

## 35.2 The Corrected Mechanical Structure

The admirable lucidity of the theory of §35.1 is based completely on the simplicity of the equation of state for an ideal, fully degenerate electron gas used there [cf. (35.1)]. It certainly requires corrections near both ends of the mass range. For cold (or nearly cold) configurations of $M \to 0$ we should get the behaviour $R \to 0$, $\varrho \approx$ constant as for planets (or even smaller spheres) instead of $R \to \infty$, $\varrho \to 0$ (as we have already explained above). At least there should be the possibility for a smooth transition to the planets, which in this connection can well be considered cold bodies. The corrections to be applied here are due to the electrostatic interaction. Near the limiting mass, on the other hand, we have encountered very high densities, with the simple theory yielding $\varrho \to \infty$ for $M \to M_{\text{Ch}}$. In this domain we have to allow for effects of the weak interaction (inverse $\beta$ decay) and the possibility of pycnonuclear reactions. Some influences on the equation of state have already been indicated in §4 and §5.

Let us first treat the main effects of *electrostatic interaction* in a cold plasma with nuclei of type $(Z, A)$ and electrons of density $n_e$. We have seen in §16.4 that matter in WD can be crystallized, and we will come back later to the condition for this. Let us suppose that the ions form a regular lattice and the electrons are evenly distributed. For the density encountered in WD the Wigner–Seitz approximation is not too bad, and so we divide the lattice into neutral Wigner–Seitz spheres of radius $R' = Z^{1/3} r_e a_0$ ($r_e$ = average separation of the electrons in units of the Bohr radius $a_0$). Each sphere contains one ion (point charge $+Z$ in the centre) and $Z$ electrons (a uniformly distributed charge $-Z$). In order to find the Coulomb energy $Z E_C$ of the sphere we take concentric shells of radius $y$ and charge $-3Z e y^2 dy/R'^3$ and remove them to infinity, thereby overcoming the potential difference $Ze(1 - y^3/R'^3)$. An integration over the whole sphere gives the energy per electron as

$$-E_C = \frac{9}{10} \frac{Z e^2}{R'} = \frac{9}{10} \frac{Z^{2/3} e^2}{r_e a_0} \approx 2 \frac{Z}{A^{1/3}} \varrho_6^{1/3} \text{ keV} \quad , \tag{35.19}$$

with $\varrho_6 = \varrho/10^6$ g cm$^{-3}$. Even for $T \to 0$ the ions cannot sit at rest precisely on their points in the lattice. Instead, the ions of mass $m_0 = A m_u$ and density $n_0$ oscillate around their positions with some ion plasma frequency $\omega_E$ (with $\omega_E^2 \sim Z^2 e^2 n_0/m_0$) such that the *zero-point energy* is $Z E_{\text{zp}} = 3\hbar\omega_E/2$ per ion. With $\varrho = n_0 A m_u$ we have per electron

$$E_{\text{zp}} = \frac{3}{2} \sqrt{\frac{4\pi}{3} \frac{\hbar e}{A m_u}} \varrho^{1/2} \approx \frac{0.6}{A} \varrho_6^{1/2} \text{ keV} \quad . \tag{35.20}$$

For $^{12}$C ($Z = 6$, $A = 12$) and $\varrho = 10^6$ g cm$^{-3}$, the energies are $-E_C \approx 5.2$ keV and $E_{\text{zp}} \approx 0.05$ keV $\ll -E_C$. The ratio $-E_C/E_{\text{zp}} \sim Z A^{2/3} \varrho^{-1/6}$ varies only very little with $\varrho$ and increases towards heavier elements.

Therefore cold configurations ("black dwarfs") are crystallized. The ions form a regular lattice which minimizes the energy; they perform low-energy oscillations around their average positions, where they are kept by mutual repulsive forces.

The energy per electron is now

$$E = E_0 + E_C + E_{zp} \approx E_0 + E_C < E_0 \quad , \tag{35.21}$$

where $E_0$ is the mean energy of an electron in an ideal Fermi gas. The influence of $E_C$ on the pressure is seen from

$$P \equiv -\frac{\partial E}{\partial(1/n)} \approx -\frac{\partial E_0}{\partial(1/n)} - \frac{\partial E_C}{\partial(1/n)} < P_0 \quad , \tag{35.22}$$

where the derivatives are taken for constant entropy, and $P_0$ is the pressure of the ideal Fermi gas. The lowering of $E$ and $P$ due to $E_C < 0$ comes from the concentration of all positive charges into the nucleus, while the negative charges are much more uniformly distributed. The average electron–electron distance is thus larger than the average electron–nucleus distance, and the repulsion is smaller than the attraction. A few calculated values of the ratio $P/P_0$ for different $Z$ and relativity parameter $x$ are given in Table 35.2. As expected the reduction of $P$ increases with the charge $Z$ and with decreasing $\varrho$ (decreasing Fermi energy). It will therefore be the dominant correction at small $M$, providing there the described reduction of $R$. The above approximation breaks down, of course, when it yields $P \lesssim 0$.

**Table 35.2** Values of $P/P_0$, where $P$ includes the Coulomb interaction and $P_0$ is for an ideal Fermi gas. $x$ is the relativity parameter; $\varrho$ is in g cm$^{-3}$. (After SALPETER, 1961)

| $x$ | $\varrho \cdot 2/\mu_e$ | $P/P_0$ $(Z=2)$ | $P/P_0$ $(Z=6)$ | $P/P_0$ $(Z=26)$ |
|---|---|---|---|---|
| 0.05 | $2.44 \times 10^2$ | 0.760 | 0.564 | $-0.063$ |
| 0.1 | $1.95 \times 10^3$ | 0.880 | 0.782 | 0.467 |
| 1 | $1.95 \times 10^6$ | 0.988 | 0.975 | 0.933 |

For the very high densities occurring near the upper end of the mass range, *pycnonuclear reactions* have to be considered (cf. §18.4). These were defined as nuclear reactions which depend mainly on $\varrho$ (instead of $T$, as in the case of thermonuclear reactions). They can occur even at $T \to 0$ as a consequence of the small oscillations of the nuclei in the lattice with energy $E_{zp}$, combined with the tunnel effect. Reactions set in rather abruptly at a certain density limit $\varrho_{pyc}$ and use up all fuel within a short time (say $10^5$ years) once $\varrho \gtrsim \varrho_{pyc}$. The limits $\varrho_{pyc}$ for the different reactions are not wellknown, since the relevant cross-sections are very uncertain. The values of $\varrho_{pyc}$ increase towards heavier elements; the orders of magnitude are $\varrho_{pyc} \approx 10^6$, $10^9$, and $10^{10}$ g cm$^{-3}$ for burning of $^1$H, $^4$He, and $^{12}$C respectively.

*Inverse $\beta$ decay* becomes important at high densities. Consider a nucleus $(Z - 1, A)$ which is $\beta$-unstable and decays under normal conditions to the stable nucleus $(Z, A) + e^- + \bar{\nu}$ (we always drop the subscript 'e' for the neutrinos), the decay energy being $E_d$. If $(Z, A)$ is surrounded by a degenerate electron gas with a kinetic energy at the Fermi border

$$E_F = m_e c^2 \left[ (1 + x^2)^{1/2} - 1 \right] \quad , \tag{35.23}$$

such that $E_F > E_d$, then $(Z, A)$ becomes unstable against electron capture, i.e. we have the inverse $\beta$ decay

$$(Z, A) + e^- \rightarrow (Z - 1, A) + \nu \quad . \tag{35.24}$$

In general we have to deal with the particularly stable even–even nuclei $(Z, A)$ and then $E_d(Z - 1, A) < E_d(Z, A)$. If $E_F > E_d(Z, A)$, then also $E_F > E_d(Z - 1, A)$, and the inverse $\beta$ decay proceeds further to $(Z - 2, A)$. The new nuclei are now stabilized by the Fermi sea, i.e. they cannot eject an electron with $E_d(< E_F)$, since it would not find a free place in phase space. $E_F$ increases with $\varrho$. Therefore for each type of nucleus $(Z, A)$ there is a threshold $\varrho_n$ of the density above which neutronization occurs. For $^1$H and $^4$He ($\varrho_n = 1.2 \times 10^7$ and $1.4 \times 10^{11}$ g cm$^{-3}$) this is of no interest, since clearly $\varrho_n \gg \varrho_{pyc}$ such that pycnonuclear burning will set in before neutronization can occur. Even for the decay $^{12}_{6}$C $\rightarrow$ $^{12}_{5}$B $\rightarrow$ $^{12}_{4}$Be one has $\varrho_n = 3.9 \times 10^{10}$ g cm$^{-3}$ $> \varrho_{pyc}$, though this is reversed for heavy nuclei. The decay $^{56}_{26}$Fe $\rightarrow$ $^{56}_{25}$Mn $\rightarrow$ $^{56}_{24}$Cr, for example, has a threshold $\varrho_n = 1.14 \times 10^9$ g cm$^{-3}$ $< \varrho_{pyc}$.

In "normal" stars we were used to imposing the chemical composition as an arbitrary free parameter. This was reasonable, since the usual transformation of the elements by thermonuclear reactions takes a sufficiently long time, and configurations with a momentary (non-equilibrium) composition are astronomically relevant. This may be different for very high densities, at which pycnonuclear reactions or inverse $\beta$ decay can transform the nuclei in relatively short time-scales. The other extreme, then, is to impose only the baryon number per volume and ask for the corresponding *equilibrium composition*. In reality the approach to nuclear equilibrium may be too slow to be accomplished. But one can imagine having reached it after an artificial acceleration by suitable catalysts, leading to the expression "cold catalyzed matter". Because of their history, WD will scarcely have reached that stage of equilibrium (they usually consist of $^4$He, or $^{12}$C and $^{16}$O, instead of $^{56}$Fe, etc.). But in order to see the connection between different types of objects, we briefly describe a few characteristics of equilibrium matter.

The equilibrium composition can be found by starting with a certain type of nucleus $(Z, A)$, and varying $Z$ and $A$ until the minimum of energy is obtained. For isolated nuclei the counteraction of attracting nuclear and repelling Coulomb forces gives a maximum binding of the nucleons at $^{56}$Fe (cf. § 18.1). Therefore $^{56}$Fe will be the equilibrium composition for small $\varrho$ ($< 8 \times 10^6$ g cm$^{-3}$). With increasing $\varrho$ this balance is shifted to heavier and neutron-enriched nuclei, since replacing a proton by a neutron decreases the repulsive Coulomb force inside the nucleus; and the $\beta$ decay, which would then result in isolated nuclei, is here prohibited by the filled Fermi sea of the surrounding electrons. Another influence comes from the lattice energy (35.19), which gives only a small correction to $P$ at high $\varrho$, but reduces the Coulomb energy at the surface of the nucleus. The sequence of equilibrium nuclei is (the maximum density in g cm$^{-3}$ is shown in parenthesis): $^{56}$Fe($8 \times 10^6$), $^{62}$Ni($2.8 \times 10^8$), $^{64}$Ni($1.3 \times 10^9$), ..., $^{120}$Sr($3.6 \times 10^{11}$), $^{122}$Sr($3.8 \times 10^{11}$), $^{118}$Kr($4.4 \times 10^{11}$). For $\varrho > 4 \times 10^{11}$ g cm$^{-3}$ it is energetically more favourable that further neutrons are free rather than bound in the nucleus: the "neutron drip" sets in. The composition consists of two phases: the lattice of nuclei (with sufficient electrons for neutrality) plus free neutrons. Their number increases with $\varrho$, and at

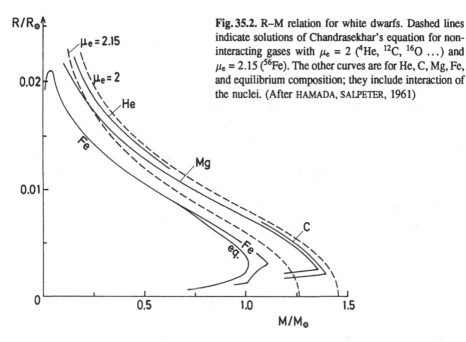

Fig. 35.2. R–M relation for white dwarfs. Dashed lines indicate solutions of Chandrasekhar's equation for non-interacting gases with $\mu_e = 2$ ($^4$He, $^{12}$C, $^{16}$O ...) and $\mu_e = 2.15$ ($^{56}$Fe). The other curves are for He, C, Mg, Fe, and equilibrium composition; they include interaction of the nuclei. (After HAMADA, SALPETER, 1961)

$\varrho \approx 4 \times 10^{12}$ g cm$^{-3}$ their pressure $P_n$ even exceeds $P_e$. At $2 \times 10^{14}$ g cm$^{-3}$ the nuclei are dissolved, leaving a degenerate neutron gas with a small admixture of protons and electrons (see § 36.1). The $P$–$\varrho$ relation can be calculated, giving the equation of state as shown in Fig. 16.2.

Once an equation of state is given, one can easily integrate the mechanical equations outwards, starting from a variety of values for the central pressure which leads to a pair of values $M$, $R$. The $M - R$ relations obtained in this way by HAMADA, SALPETER (1961) are plotted as solid curves in Fig. 35.2 for different compositions (He, C, Mg, Fe, and equilibrium composition). For comparison the relations for an ideal Fermi gas (Chandrasekhar's theory) are plotted for $\mu_e = 2$ (for example $^4_2$He, $^{12}_6$C, $^{24}_{12}$Mg) and $\mu_e = 2.15$ ($^{56}_{26}$Fe); in the latter case, the mass limit is already lowered to $M_{\text{Ch}} \approx 1.25 M_\odot$. Relative to these classical models there is a clear reduction of $R$, particularly at small $M$, owing to the Coulomb interaction reducing $P$. This effect increases with $Z$. The curve for $^{56}$Fe shows a maximum of $R$ beyond which it decreases for $M \to 0$. In fact such a maximum of $R$ ($\approx 0.02$, $0.05$, $0.12 R_\odot$ for Fe, He, H, respectively) occurs for all compositions at values of $M$ between a few $10^{-3}$ to $10^{-2} M_\odot$. In this regime the equation of state is not well known; it is certainly completely dominated by Coulomb effects, and the inhomogeneous distribution of the electrons has to be considered. In any case, we find here the natural transition between WD ($dR/dM < 0$) and planets ($dR/dM > 0$). (Note that Jupiter with $R \approx 0.1 R_\odot$ and $M \approx 10^{-3} M_\odot$ is not far from this border; in fact its radius is far above $R_{\text{max}}$ for He and close to that of H, so that it must consist essentially of H.)

Towards large $M$ the curves for C, Mg, and Fe show kinks at the mass limit. These are due to a phase transition in the centre, since $\varrho_c$ reaches one of the limits described above. For $^{12}$C we find here $\varrho_c = \varrho_{\text{pyc}}$, and pycnonuclear reactions then transform $^{12}$C $\to$ $^{24}$Mg, which by inverse $\beta$ decay becomes $^{24}$Ne. Models on the lower branch beyond the kink consist of Ne cores and C envelopes. The curve for $^{24}$Mg reaches $M_{max}$ when $\varrho_c = \varrho_n$, and inverse $\beta$ decay gives central cores of $^{24}$Ne. For $^{56}$Fe we see the result of the inverse $\beta$ decay to $^{56}$Cr at $M_{max}$, and to $^{56}$Ti at the following second kink (beyond which the models consist of $^{56}$Ti cores, $^{56}$Cr shells, and $^{56}$Fe envelopes). The curve for equilibrium composition, which coincides with $^{56}$Fe for $\varrho \lesssim 8 \times 10^6$ g cm$^{-3}$, is below and to the left of all other curves; it always has the largest average $\mu_e$. At the maximum $M$ ($\approx 1.0 M_\odot$) one finds $\varrho_c \approx 2 \times 10^9$ g cm$^{-3}$, with $^{66}_{28}$Ni nuclei giving a relatively large $\mu_e$. Towards the end of the plotted equilibrium curve, $^{120}$Kr is reached and the first neutrons are freed. (From here follows the sequence of equilibrium configurations which leads to neutron stars, see § 36.) The whole curve appears fairly smooth, since the change of the composition here proceeds in small steps via neighbouring nuclei, while the transit of a non-catalyzed composition to equilibrium is first delayed by large thresholds and then occurs in a big jump.

Concerning inhomogeneous models of WD with nonequilibrium composition, we briefly mention the case of a low-mass envelope of light elements (particularly $^1$H) being placed on a WD of $^4$He, or $^{12}$C and $^{16}$O. This may happen by mass exchange in close binary systems. Aside from possible instabilities during the onset of nuclear burning (which can lead to the ejection of a nova shell), there is a strong influence on the equilibrium radius described by $d \lg R / d \lg M_H$ of the order 10 $\ldots 10^2$. This means that the addition of a $^1$H envelope of only 1% of $M$ increases $R$ by about 50% and more. In fact the white dwarf will scarcely be recognizable as such.

The connection with other types of configurations is seen in Fig. 36.2, which gives the $M$–$R$ relation for cold catalyzed matter (equilibrium composition). When going along the curve in the direction of increasing $\varrho_c$, one encounters extrema of $M$ (open circles) in which the stability properties change. An example is the point at $M = M_{max}$ for the white-dwarf sequence, beyond which a branch of unstable models follows (see the discussion of § 36.2).


### 35.3 Thermal Properties and Evolution of White Dwarfs

In the very interior of a WD, the degenerate electrons provide a high thermal conductivity. This, together with the small $L$, does not allow large temperature gradients. The situation is different when going to the outermost layers. With decreasing $\varrho$ the matter is less and less degenerate, and the dominant heat transfer becomes that by radiation (or convection), which is much less effective. Therefore we expect to find a non-degenerate outer layer in which $T$ can drop appreciably and which isolates the degenerate, isothermal interior from outer space.

We simplify matters by assuming a discontinuous transition from degeneracy to non-degeneracy (ideal gas) at a certain point (subscript 0). For the envelope we use the radiative solution (10.23) for a Kramers opacity ($\kappa = \kappa_0 P T^{-4.5}$) and a zero constant of integration:

$$T^{8.5} = BP^2 \quad ; \quad B = 4.25 \frac{3\kappa_0}{16\pi acG} \frac{L}{M} \quad . \tag{35.25}$$

Replacing $P$ by $\Re \varrho T / \mu$ and solving for $\varrho$ here, we have

$$\varrho = B^{-1/2} \frac{\mu}{\Re} T^{3.25} \quad . \tag{35.26}$$

The transition point is assumed to be where the degenerate electron pressure equals the pressure of an ideal gas, i.e. according to (16.6)

$$\varrho_0 = C_1^{-3/2} T_0^{3/2} \quad ; \quad C_1 = 1.207 \times 10^5 \frac{\mu}{\mu_e^{5/3}} \, \text{cgs} \quad . \tag{35.27}$$

This density $\varrho_0$ is reached according to (35.26) at a temperature $T = T_0$ given by

$$T_0^{3.5} = \frac{B}{C_1^3} \left( \frac{\Re}{\mu} \right)^2 = \vartheta \frac{L/L_\odot}{M/M_\odot} \quad , \tag{35.28}$$

where all factors are comprised in $\vartheta$. For typical compositions and values of $\kappa_0$, one has roughly

$$T_0 \approx \vartheta^{2/7} \left( \frac{L/L_\odot}{M/M_\odot} \right)^{2/7} \approx \left( \frac{L/L_\odot}{M/M_\odot} \right)^{2/7} 5.9 \times 10^7 \, \text{K} \quad . \tag{35.29}$$

For $M = M_\odot$ and the range $L/L_\odot = 10^{-4} \ldots 10^{-2}$ this yields $T \approx 4.2 \ldots 16 \times 10^6$ K, which is, by assumption, also the temperature in the whole (isothermal) interior. Typical values for the density at the transition point are then, according to (35.27), of the order of $\varrho_0 \approx 10^3$ g cm$^{-3}$ (i.e. $\ll \varrho_c$).

An idea of the radial extension $R - r_0$ of the non-degenerate envelope is easily obtained from (10.32). We can neglect $T_{\text{eff}}$ ($\approx 10^4$ K) against $T_0$ and get

$$\frac{R - r_0}{r_0} \approx \frac{\Re T_0}{\mu \nabla} \frac{R}{GM} \approx 0.82 \frac{R/R_\odot}{M/M_\odot} \frac{T_0}{10^7 \text{K}} \quad . \tag{35.30}$$

(The numerical factor is given for $\mu = 4/3$, $\nabla = 0.4$.) The relative radial extension of the non-degenerate envelope then is typically 1% or less. This means that the radius of a WD is well approximated by the integrations which assume complete degeneracy throughout.

The rather high internal temperatures of $10^6 \ldots 10^7$ K set a limit to the possible hydrogen content in the interior. If hydrogen were present with a mass concentration $X_H$, we would expect hydrogen burning via the $pp$ chain. For average values $T = 5 \times 10^6$ K, $\varrho = 10^6$ g cm$^{-3}$, (18.63) gives $\varepsilon_{pp} \approx 5 \times 10^4 X_H^2$ erg g$^{-1}$s$^{-1}$ and the luminosity for $M = 1 M_\odot$ would be

$$L/L_\odot \approx \frac{M_\odot}{L_\odot} \varepsilon_{pp} \approx 2.5 \times 10^4 X_H^2 \quad , \tag{35.31}$$

such that the observed $L \leq 10^{-3} L_\odot$ allows only $X_H \lesssim 2 \times 10^{-4}$. Stability considerations (§ 25.3.5) indeed rule out that the luminosity of normal WD is generated by thermonuclear reactions, which was first pointed out by MESTEL (1952). A stable burning could only be expected in nearly cold configurations that produce their extremely small $L$ ("black" or "brown" dwarfs) by pycnonuclear reactions near $T = 0$.

If there are no thermonuclear reactions, then which reservoirs of energy are involved when a normal WD loses energy by radiation? The means for obtaining the answer are provided in §3.1. For a configuration in hydrostatic equilibrium the virial theorem (3.9) requires $\zeta \dot{E_i} + \dot{E_g} = 0$.

The potential energy in the gravitational field $E_g(< 0)$ is given by (3.3). The total internal energy of the star $E_i = E_e + E_{ion}$ consists of the contributions from electrons and ions. By $\zeta$ we mean an average of the quantity $\zeta'$, defined by the relation

$$\zeta' u = 3\frac{P}{\varrho} \quad , \qquad\qquad (35.32)$$

where $u$ is the internal energy per unit mass. For highly degenerate electrons, $\zeta'$ varies from $\zeta' = 2$ (non-relativistic) to $\zeta' = 1$ (relativistic case). For the ions, $\zeta' = 2$ if they are an ideal gas cf. (3.5)]. If there is crystallization, the contributions $u_C$ of Coulomb energy and $u_p$ of lattice oscillations (phonons) have to be considered. For the static Coulomb part we note that $u_C = n_e E_C/\varrho$, with $E_C \sim \varrho^{1/3}$ according to (35.19). Then one finds from (35.22) that $P_C/\varrho = u_C/3$, i.e. $\zeta' = 1$. The situation is more difficult with $u_p$, but this contributes relatively little.

Summing up all effects, the average over the whole WD will obviously be somewhere in the range $1 < \zeta < 2$. As in "normal" stars we have a simple relation between $E_i$ and $E_g$, the absolute values of both being of the same order.

The total energy is $W = E_i + E_g$. The energy equation requires $L = -\dot{W}$, which together with the virial theorem gives [compare with (3.12)]

$$L = -\dot{W} = -\frac{\zeta - 1}{\zeta}\dot{E_g} = (\zeta - 1)\dot{E_i} \quad . \qquad\qquad (35.33)$$

Therefore $L > 0$ requires a contraction ($\dot{E_g} < 0$) and an increase of the internal energy ($\dot{E_i} > 0$). So far, it is the same as with normal, non-degenerate stars. The crucial question is how $E_i$ is distributed between electrons ($E_e$) and ions ($E_{ion}$).

We recall the situation for a normal star with both electrons and ions being non-degenerate. Then there is equipartition with $E_{ion} \sim E_e \sim T$, such that also $E_i = E_{ion} + E_e \sim T$; $\dot{E_i} > 0$ means $\dot{T} > 0$. Thus the $loss$ of energy ($L > 0$) leads to a $heating$ ($\dot{T} > 0$). This was expressed in §25.3.4 by saying that the star has negative gravothermal specific heat, $c^* < 0$.

For demonstrating the behaviour of a WD let us simply assume that the electrons are non-relativistic degenerate and the ions form an ideal gas. Then $\zeta = 2$ and $L = -\dot{E_g}/2$, i.e. the star must contract, releasing twice the energy lost by radiation. Since $-E_g \sim 1/R \sim \varrho^{1/3}$, we have $\dot{E_g}/E_g = (1/3)\dot{\varrho}/\varrho$. (Here $\varrho$ is some average value.) The compression, however, increases the Fermi energy $E_F$ of the electrons. Their internal energy is $E_e \approx E_F \sim p_F^2 \sim \varrho^{2/3}$, such that $\dot{E_e}/E_e = (2/3)\dot{\varrho}/\varrho$. So we have a simple relation between $\dot{E_g}$ and $\dot{E_e}$:

$$\dot{E_e} \approx 2\frac{E_e}{E_g}\dot{E_g} = -\frac{E_e}{E_i}\dot{E_g} \quad . \qquad\qquad (35.34)$$

Here $E_i$ is introduced via the virial theorem in the form $E_g = -2E_i$.

If the WD is already cool, then $E_{\text{ion}} \ll E_{\text{e}}$ and $E_{\text{i}} = E_{\text{ion}} + E_{\text{e}} \approx E_{\text{e}}$. This means $\dot{E}_{\text{e}} \approx -\dot{E}_{\text{g}} = 2L$, and nearly as much energy as released by contraction is used up by raising the Fermi energy of the electrons. With $\dot{E}_{\text{e}} \approx -\dot{E}_{\text{g}}$, the energy balance $L = -\dot{E}_{\text{ion}} - \dot{E}_{\text{e}} - \dot{E}_{\text{g}}$ becomes

$$L \approx -\dot{E}_{\text{ion}} \sim -\dot{T} \quad . \tag{35.35}$$

Therefore, the ions release about as much energy by cooling as the WD loses by radiation. The contraction is then seen to be the consequence of the decreasing ion pressure (even though $P_{\text{ion}}$ is only a small part of $P$). In spite of the decreasing ion energy, the whole internal energy rises, since $\dot{E}_{\text{ion}} + \dot{E}_{\text{e}} \approx L$. This evolution tends finally to a cold black dwarf; then the contraction has stopped and all of the internal energy is in the form of Fermi energy.

Of course, the relations just derived should have somewhat different numerical factors, since $\zeta$ is not exactly 2 (a certain degree of relativity in the central part, the ion gas not being ideal, etc.). But the essence of the story remains the same.

The foregoing discussion opens the possibility of arriving at a very simple *theory of the cooling of WD*. We start with the energy equation (4.28), setting there

$$\varepsilon_{\text{g}} = -c_v \dot{T} + \frac{T}{\varrho^2} \left( \frac{\partial P}{\partial T} \right)_v \dot{\varrho} \quad , \tag{35.36}$$

which follows from the first equation (4.17). We now integrate (4.28) over the whole star, taking not only $\varepsilon_{\text{n}} = \varepsilon_\nu = 0$, but also neglecting the compression term in (35.36),

$$-L \approx \int_0^M c_v \dot{T} \, dm \approx c_v \dot{T}_0 M \quad , \tag{35.37}$$

where an isothermal interior is assumed with $T = T_0$. If the ions are an ideal gas, then

$$c_v^{\text{ion}} = \frac{3}{2} \frac{k}{A m_{\text{u}}} \quad . \tag{35.38}$$

For the specific heat of the degenerate electrons one can derive (CHANDRASEKHAR, 1939, p. 394)

$$\begin{aligned} c_v^{\text{el}} &= \frac{\pi^2 k^2}{m_{\text{e}} c^2} \frac{Z}{A m_{\text{u}}} \frac{\sqrt{1+x^2}}{x^2} T \qquad [x = p_{\text{F}}/m_{\text{e}} c] \\ &\approx \frac{\pi^2 k}{2} \frac{Z}{A m_{\text{u}}} \frac{kT}{E_{\text{F}}} \quad , \quad \text{for } x \ll 1 \quad . \end{aligned} \tag{35.39}$$

The ratio (for $x \ll 1$)

$$\frac{c_v^{\text{el}}}{c_v^{\text{ion}}} = \frac{\pi^2}{3} Z \frac{kT}{E_{\text{F}}} \tag{35.40}$$

is small for small $kT/E_{\text{F}}$ and not too large $Z$. In the numerical examples below we will take $c_v = c_v^{\text{ion}}$. Then (35.37) describes $L$ as given by the change of the internal energy of the ions.

In (35.28) we eliminate $L$ with (35.37) and obtain a differential equation for $T$ (where we drop the subscript 0 for the interior):

$$\dot{T} = -\frac{L_\odot}{M_\odot} \frac{1}{c_v \vartheta} T^{7/2} \quad . \tag{35.41}$$

This can be rewritten with (35.29) as $-\dot{L} \sim L^{12/7}$, which together with $R \approx$ constant describes the motion in the HR diagram. Equation (35.41) is easily integrated from $t = 0$ when the temperature was much larger than it is now, to the present time $t = \tau$. The result gives the *cooling time*

$$\tau = \frac{2}{5} \frac{M_\odot}{L_\odot} c_v \vartheta T^{-5/2} = \frac{2}{5} c_v \frac{MT}{L}$$

$$= \frac{2}{5} \left(\frac{M_\odot}{L_\odot} \vartheta\right)^{2/7} c_v \left(\frac{M}{L}\right)^{5/7} \approx \frac{4.7 \times 10^7 \text{years}}{A} \left(\frac{M/M_\odot}{L/L_\odot}\right)^{5/7} \quad . \tag{35.42}$$

Here we have used (35.28,29). For $A = 4$, $M = M_\odot$ and $L/L_\odot = 10^{-3}$ one has $\tau \approx 10^9$ years.

The specific heat $c_v$ is obviously very important. Larger values of $c_v$ give a slower cooling ($\dot{T} \sim 1/c_v$), i.e. a larger cooling time ($\tau \sim c_v$). The simplest assumption would be $c_v = c_v^{\text{ion}} = 3k/(2Am_u)$, but this requires several corrections. For small $M$ (i.e. moderate $\bar{\varrho}$) and larger $T$ and $Z$, one cannot neglect the contribution of the electrons. From (35.40) we have $c_v^{\text{el}} \approx 0.25 c_v^{\text{ion}}$ for $T = 10^7$K, $M = 0.5M_\odot$ and a C–O mixture.

For small $T$ the ions dominate completely: $c_v = c_v^{\text{ion}}$, but their specific heat is influenced by crystallization. We indicate only a few aspects of the rather involved theory for these processes.

The properties of the ions depend critically on two dimensionless quantities, $\Gamma_c$ and $T/\Theta$. The ratio $\Gamma_c$ of Coulomb energy to kinetic energy of the ions is defined in (16.25). For $\Gamma_c \approx 10^2$ a heated crystal will melt (or a cooling plasma will crystallize), which determines the melting temperature $T_m$ given in (16.26). For $\Gamma_c < 1$ the thermal motion does not allow any correlation between the positions of the ions, no lattice is possible, and the ions behave as a gas.

The other ratio, $T/\Theta$, contains a characteristic temperature $\Theta$ which is essentially the Debye temperature and is defined by

$$k\Theta = \hbar\Omega_p \quad , \quad \Omega_p = \frac{2Ze}{Am_u} (\pi\varrho)^{1/2} \quad , \tag{35.43}$$

with $\Omega_p$ being the ion plasma frequency [cf. the zero-point energy (35.20) where we used $\omega_E = \Omega_p/3$]. This gives

$$\Theta = \frac{he}{km_u\sqrt{\pi}} \frac{Z}{A} \varrho^{1/2} \approx 7.8 \times 10^3 \text{K} \cdot \frac{Z}{A} \varrho^{1/2} \tag{35.44}$$

($\varrho$ in g cm$^{-3}$). $k\Theta$ is a characteristic energy of the lattice oscillations, which cannot be excited for $T/\Theta < 1$. For typical WD composed of C, O, or heavier elements, one has $\Theta < T_m$.

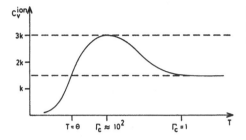

Fig. 35.3. Schematic variation of the specific heat per ion with the temperature $T$ in white-dwarf matter

Figure 35.3 shows how the specific heat $C_v$ per ion changes with $T$. Starting at very large $T$ ($\Gamma_c \ll 1$), the ions form an ideal gas. Each degree of freedom contributes $kT/2$ to the energy (i.e. $k/2$ to $C_v$), and $C_v = 3k/2$. With decreasing $T$ one finds an increasing correlation of the ion positions owing to the growing importance of Coulomb forces in the range $\Gamma_c \approx 1 \ldots 10$. This gives additional degrees of freedom, since energy can go into lattice oscillations, and $C_v$ increases above $3k/2$, with the maximum of $C_v = 3k$ being reached when the plasma crystallizes at $T = T_{\mathrm{m}}$. With further decreasing $T$ gradually fewer oscillations are excited, and the specific ion heat $C_v$ even drops below $3k/2$ around $T = \Theta$. For $T \to 0$ finally, $C_v \sim T^3$.

These large variations of $C_v$ (increase by a factor 2, then decrease to zero) of course influence the cooling times [cf. (35.42)]. In addition there is the release of the latent heat of about $kT$ per ion when the material crystallizes. Attempts have been made to connect these changes of $\tau$ with the observed number of WD as a function of $L$.

Of course, one can easily improve the simple theory of cooling by taking into account all terms in the energy equation. This includes, for example, the fact that $\varepsilon_\nu \neq 0$, i.e. there is an additional cooling by neutrino losses, particularly from very hot WD. Another point of correction is that the transport of energy in the outer layers can be due to convection [while the solution (35.25) assumes purely radiative transfer].

# §36 Neutron Stars

As early as 1934 Baade and Zwicky correctly predicted the birth of the strange objects neutron stars in supernova explosions (BAADE, ZWICKY, 1934). The first models were calculated by OPPENHEIMER, VOLKOFF (1939), and the stage was then left for the next 28 years to particle physicists who struggled with the problem of matter at extreme densities (a struggle not yet finished). Radio astronomers accidentally found the first pulsar in 1967; it was interpreted soon after as a rapidly rotating neutron star (GOLD, 1968). Everything is extreme with neutron stars, their interior state (simulating a huge nucleus), the velocity of sound (not far from $c$), their rotation (frequencies 1 ... 1000 Hz), and their magnetic fields (up to $10^{12}$ gauss). One is far from really understanding them. So we content ourselves here with a few remarks on the state of matter and the resulting models.

## 36.1 Cold Matter Beyond Neutron Drip

Neutron stars (NS) are born hot ($T > 10^{10}$ K) in the collapse of a highly evolved star (see §34). But the interior temperature drops rapidly because of neutrino emission: after a day, temperatures of $10^9$ K are reached; after 100 years, maybe $10^8$ K. And this ($kT \approx 10$ keV) can be considered cold in view of the degenerate nearly relativistic neutrons ($E_F \approx 1000$ MeV). The equation of state is essentially the same as for $T \approx 0$. We refer to the descriptions of high-density matter in §35.2 and of the equation of state in §16.

With increasing density the rising Fermi energy of the electrons provides an increasing neutronization by electron captures. The neutron-rich equilibrium nuclei ($^{118}$Kr) begin to release free neutrons at $\varrho_{dr} \approx 4 \times 10^{11}$ g cm$^{-3}$. This is called the *neutron drip*. The matter consists of nuclei (usually arranged in a lattice) plus sufficient electrons for charge neutrality, and free neutrons. Their number $n_n$ increases with $\varrho$, and so does their pressure $P_n$. While $P \approx P_e \gg P_n$ still at $\varrho = \varrho_{dr}$, we have $P_n = P/2$ at $\varrho \approx 4 \times 10^{12}$ g cm$^{-3}$, and $P_n > 0.8P$ for $\varrho \gtrsim 1.5 \times 10^{13}$ g cm$^{-3}$, and finally $P_n \approx P$. Note that all characteristic densities quoted here and in the following depend in general on the model assumed for the particles and their interaction. The higher the values of $\varrho$, the more uncertain are the details (see below).

With progressing neutron drip the number of nuclei is diminished by fusion. The nuclei more or less touch each other at a density $\varrho_{nuc} \approx 2.4 \times 10^{14}$ g cm$^{-3}$, and hence they merge and dissolve, leaving a degenerate gas (or liquid) of neutrons plus a small admixture of $e^-$ and $p$. The concentrations of these particles can be calculated as an equilibrium between back and forth exchanges in the reaction $n \rightleftharpoons p + e^-$. (The neutrinos leave the system immediately and can be left out of the considerations.) The conditions are that the Fermi energies fulfil $E_F^n = E_F^p + E_F^e$, and that $n_e = n_p$ for neutrality. This gives that $n_p$ is about 1% (or less) of $n_n$ for a wide range of $\varrho$ up to $\varrho_{nuc}$. Increasing relativity of the neutrons raises this ratio slowly, until at an infinite relativity parameter one finds the limiting ratio $n_n : n_p : n_e = 8 : 1 : 1$. When $\varrho$ exceeds $10^{15}$ g cm$^{-3}$, the Fermi energy of the neutrons, $E_F = [(p_F c)^2 + (m_n c^2)^2]^{1/2}$, will gradually exceed the rest masses of the hyperons of lowest mass (such as $\Lambda$, $\Sigma$, $\Delta$, ...). These particles will then appear, i.e. a "hyperonization" begins. Finally even free quarks can occur.

We now come to the *equation of state*, in particular the dependence of $P$ on $\varrho$. For $\varrho$ up to $\varrho_{drip}$ the pressure is dominated by the relativistic, degenerate electrons, and $P \approx P_e \sim \varrho^{4/3}$ [cf. (15.26)].

The onset of the neutron drip ($\varrho = \varrho_{drip}$) has severe consequences for the equation of state. An increase $d\varrho$ mainly increases $n_n$ at the expense of $n_e$ (which yields the pressure), such that the increment $dP$ is small. Therefore the gas becomes more compressible, which is described as a "softening" of the equation of state (in the opposite case one speaks of "stiffening"). In other terms the adiabatic index $\gamma_{ad} = (d \ln P / d \ln \varrho)_{ad}$ drops appreciably below the critical value 4/3 (cf. §25.3.2), and only when $P_n$ contributes sufficiently to $P$ will $\gamma_{ad}$ again rise above 4/3 at $\varrho \approx 7 \times 10^{12}$ g cm$^{-3}$.

When the neutron pressure $P_n$ dominates one may tentatively consider the approximation that the gas consists of *ideal* (non-interacting), *fully degenerate* neutrons. These are fermions like the electrons, and they obey the same statistics, so that the same relations hold as derived in §15.2, if there $m_e$ is replaced by $m_n$ and $\mu_e$ by 1 (since we now have one nucleon per fermion). Instead of (15.23) and (15.26) we can write

$$P_n = K_{\gamma'} \varrho_0^{\gamma'} \qquad (36.1)$$

with the non-relativistic and relativistic limit cases (for $\varrho_0 \ll 6 \times 10^{15}$ and $\varrho_0 \gg 6 \times 10^{15}$ g cm$^{-3}$ respectively)

$$\gamma' = \frac{5}{3} \quad , \quad K_{5/3} = \frac{1}{20} \left( \frac{3}{\pi} \right)^{2/3} \frac{h^2}{m_n^{8/3}} \quad ,$$

$$\gamma' = \frac{4}{3} \quad , \quad K_{4/3} = \frac{1}{8} \left( \frac{3}{\pi} \right)^{1/3} \frac{hc}{m_n^{4/3}} \quad , \qquad (36.2)$$

with $m_u \approx m_n$. In (36.1) we have used the rest-mass density $\varrho_0 = n_n m_n$. For relativistic configurations instead of $\varrho_0$ one has to use the total mass–energy density $\varrho = \varrho_0 + u/c^2$. This distinction was not necessary for the electron gas, where $\varrho_0$ (coming mainly from the non-degenerate nucleons) was always large compared with the energy density $u/c^2$ coming from the degenerate electron gas. Now both $\varrho_0$ and $u/c^2$ are provided by the degenerate neutrons. For non-relativistic neutrons, $\varrho_0 \gg u/c^2$ and $\varrho \approx \varrho_0$; for relativistic neutrons, $\varrho_0 \ll u/c^2$ and $\varrho \approx u/c^2$. For relativistic particles, however, we know that $P = u/3$, i.e. $P = \varrho c^2/3$. So we can write

$$P_n \sim \varrho^\kappa \quad ,$$

$$\kappa = 5/3 \quad \text{(non-relativistic)} \quad ,$$

$$\kappa = 1 \quad \text{(relativistic)} \quad . \tag{36.3}$$

The distinction between $\varrho$ and $\varrho_0$ will be seen to be important for NS models. The relation $P = \varrho c^2/3$ also yields the velocity of sound directly as $v_s^2 = (dP/d\varrho)_{ad} = c^2/3$, i.e. $v_s = 0.577c$.

Of course, with the densities considered here the *interaction between nucleons* is far from being negligible. It dominates the behaviour long before the limit $6 \times 10^{15}\,\mathrm{g\,cm^{-3}}$, where $p_F = m_n c$, is reached. In order to calculate its influence on the equation of state, one faces two problems. The first is the determination of a reasonable potential. In the absence of a rigorous theory and of experiments at such high densities, one has to use a model of the interacting particles that meets the results of low-energy scattering, the properties of saturation of nuclear forces, etc. It is not surprising that such models yield large uncertainties when extrapolated and applied to the densities found in NS. The qualitative influence of some effects on the equation of state is quite obvious. For example, the interaction between two nucleons depends (aside from spin and isospin properties) on their distance. When approaching each other they first feel an attraction, which turns to repulsion below a critical distance (in the extreme: at an inner hard core). Attraction (dominant at not too high $\varrho$) reduces $P$ and gives a softer equation of state. Repulsion (dominant at very high $\varrho$ and small average particle distances) increases $P$ and thus stiffens the equation of state. Obviously details of the potential can shift the border appreciably between these two regimes.

Other uncertainties are connected with the appearance of new particles when $\varrho$ increases. For example, if hyperons of some type occur in sufficient number, they contribute to $\varrho$, but scarcely to $P$, since their creation lowers the Fermi sea of the neutrons. Therefore "hyperonization" makes the gas more compressible. At ultra-high densities (say $\approx 10\varrho_{nuc}$) so many new resonances appear that, in the extreme, attempts have been made to describe their number in a certain energy range only by statistics (which leads, e.g., to the rather soft Hagedorn equation of state). But if the nucleons almost touch each other, one might have to consider something like quark interaction. The question was even discussed whether this might lead to quark matter and possibly to quark stars[1]. As early as $\varrho \gtrsim 2\varrho_{nuc}$ the possibility of the reaction $n \rightarrow p + \pi^-$ (if $E_n \geq E_p + E_{\pi^-}$) gives the possibility of having a Bose–Einstein condensate of the cold $\pi^-$ bosons in momentum space with zero momentum, i.e. no contribution to $P$, but to $\varrho$.

The second quite general problem for determining the equation of state is that, even if the potential were known exactly, one would not know how to solve convincingly the many-body problem. Several attempts use different assumptions and yield different results.

To resume, we must stress that the equation of state is highly uncertain for two independent reasons (concerning the potential and the many-body problem). In

---

[1] The full beauty of this term can be savoured only in German, where the term "quark" means a popular, soft white cheese or, in slang, complete nonsense.

fact particle physics cannot yet decide which of the available equations of state is correct, but the softest ones now seem to be ruled out by observation of neutron stars (see below). In Fig. 16.2 just one of them is plotted and we do not claim to have chosen the best.

We conclude by mentioning some other consequences of the interaction that scarcely influence the $P$-$\varrho$ relation, but might have influence on the further evolution of NS. The neutron liquid becomes *superfluid* if the neutrons are paired. In the range of attraction they can be correlated such that there are pairs with opposite momentum (on top of the Fermi sea) and opposite spins, thus forming bosons with spin 0. The pairing lowers the energy, i.e. the disruption of a pair requires the input of the latent heat $\Delta$. This quantity is quite appreciable, namely $\Delta \approx 1 \ldots 2$ MeV. While superfluidity of helium occurs only below temperatures of a few K, the neutron liquid should be superfluid if $kT$ is below the latent heat, which means if $T < 10^{10}$ K. The corresponding pairing of the charged protons gives *superconductivity*. Another possibility arises from the repulsive forces between neutrons at very small distances. If these forces turned out to be sufficiently dominant, then the neutrons could be forced to settle in a regular lattice to minimize the energy of this interaction, though it is unclear whether such a *solidification* of the neutron matter will happen. Superfluidity and solidification can possibly influence the rotation of NS. Superfluidity also affects the heat capacity (i.e. the cooling), while superconductivity is important for the magnetic fields.

## 36.2 Models of Neutron Stars

For a given equation of state of the form $P = P(\varrho)$ it is easy to obtain the corresponding hydrostatic models of NS. One has only to integrate the relativistic equation of hydrostatic equilibrium (2.31) (the TOV equation) together with (2.30), starting at $r = 0$ with a chosen central density $\varrho_c$. Since the equation of state is independent of $T$, these two equations suffice for obtaining the mechanical structure. This is seen after replacing $P$ by $\varrho$ in (2.31), so that there are two equations for the variables $\varrho$ and $m$. When the integration comes to $\varrho = P = 0$, the surface is reached, i.e. we have found $R = r$ and $M = m(R)$. (We do not have to worry about the obvious failure of the equation of state for $P \to 0$. The transition region to the non-degenerate atmosphere, and even the whole atmosphere, are negligibly thin so that the error made is small.)

Repeating this integration for a variety of starting values $\varrho_c$, one can produce a sequence of models for the chosen equation of state. They give, in particular, the relations $M = M(\varrho_c)$, $R = R(\varrho_c)$, and by elimination of $\varrho_c$ also $R = R(M)$ (cf. Fig. 36.1).

The resulting relations $M(\varrho_c)$ and $R(M)$ change considerably if we replace the equation of state by another one, as can be seen in Fig. 36.1, where the results are plotted for six well-known equations of state. The persisting common feature is that all relations $M(\varrho_c)$ show a minimum and a maximum of $M$, although at quite different values. One can easily understand the qualitative changes which occur when a soft equation of state is replaced by a stiffer one. The matter is then less

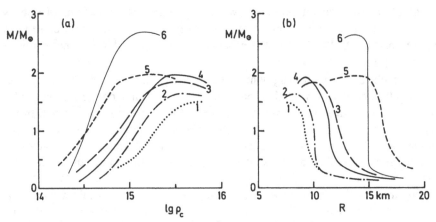

**Fig. 36.1.** The relations $M$ against $\varrho_c$ (in g cm$^{-3}$) and $M$ against $R$ of neutron-star models calculated using 6 different equations of state (*labels* 1 ... 6). (After BAYM, PETHICK, 1979)

compressible; for given $M$ one expects a larger $R$ and a smaller $\varrho_c$. For given $\varrho_c$ one can put more mass on top until reaching the surface with $\varrho = 0$. This lowers the gravity inside the model, and $M_{max}$ is higher. A particularly soft equation of state is that for the ideal degenerate neutron gas in (36.3), since the repulsive forces at small particle distances are completely neglected. Correspondingly OPPENHEIMER, VOLKOFF (1939) obtained for this equation of state a maximum mass of only $M_{max} \approx 0.72M_\odot$. Normally the maxima range roughly between $1M_\odot$ and $3M_\odot$. We have stressed in § 36.1 that particle physics cannot yet supply the correct equation of state. All the more interesting is the binary pulsar PSR 1913 + 16, for which the masses could be determined quite well when details of the orbital motion were interpreted as general relativistic effects. The result for the NS was $M \approx 1.42M_\odot$, which rules out all equations of state so soft that their $M_{max}$ is below $1.42M_\odot$. Here seems to be one of the very few cases where astrophysical measurements set a discriminating limit to particle physics.

The *maximum mass* for NS is very important, not only in connection with evolutionary considerations, but also in the attempt to identify compact objects with $M > M_{max}$ as black holes. If our ignorance of the equation of state does not yet allow the determination of $M_{max}$ to better than the interval 1.5 ... $3M_\odot$, we should at least understand that such a maximum mass (well below $5M_\odot$) must exist.

In order to make this plausible, we neglect effects of *general* relativity, i.e. consider the usual equation of hydrostatic equilibrium but keep those of *special* relativity as allowed for in (36.3). Let us consider some averages of $P$ and $\varrho$ over the whole star. As in (35.15) the normal hydrostatic equation then yields the estimate $P \sim M^2/R^4$. Here we eliminate $R$ by $\varrho \sim M/R^3$ and obtain $P \sim M^{2/3}\varrho^{4/3}$, introduce $\varrho \sim P^{1/\kappa}$ from the equation of state (36.3), and then solve for $M$ and find

$$M \sim \varrho^{3(\kappa-4/3)/2} \quad . \tag{36.4}$$

In the non-relativistic limit, $\kappa = 5/3$, giving $M \sim \varrho^{1/2}$ and $dM/d\varrho > 0$. The extreme relativistic case requires $\kappa = 1$, which gives $M \sim \varrho^{-1/2}$ and $dM/d\varrho < 0$.

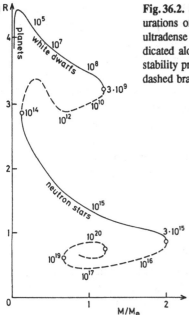

Fig. 36.2. Schematic mass–radius relation ($R$ in km) for configurations of cold catalyzed matter, from the planetary regime to ultradense neutron stars. Some values of $\varrho_c$ (in g cm$^{-3}$) are indicated along the curve. At the extrema of $M$ (*open circles*) the stability problem has a zero eigenvalue. Solid branches are stable, dashed branches are unstable

Somewhere on the border between the two regimes we expect $dM/d\varrho = 0$, i.e. the maximum mass. (The average $\varrho$ treated here will be a sufficient measure for $\varrho_c$ too.) Therefore the maximum of $M$ must occur when the neutrons *start* to become relativistic and the energy density $u/c^2$ begins to overtake the rest-mass density $\varrho_0$. Only by neglecting $u/c^2$ in $\varrho$ [taking (36.1) instead of (36.3)] could we obtain the Chandrasekhar mass of $M_{Ch} = 5.73 M_\odot$ as the mass limit for an *infinite* relativity parameter ($\gamma' = 4/3$). Clearly, therefore, $M_{max} < M_{Ch}$. The here neglected influence of general relativity [i.e. the description of hydrostatic equilibrium by the TOV quation (2.31)] tends to decrease $M_{max}$ even more (see below).

Closely connected with the extrema of $M$ are the *stability properties*. The relation $M = M(\varrho_c)$ can be considered to represent a linear series of equilibrium models with the parameter $\varrho_c$ (cf. § 12). Figure 36.2 shows a schematic overview of the resulting $M$–$R$ relation for cold catalyzed matter from the regime of planets to that of ultra-dense NS. Starting from planets, $\varrho_c$ increases monotonically along the curve (compare with typical values of $\varrho_c$ indicated in Fig. 36.2). There are extrema of $R$ which may be interesting in other connections but are not important for the sequence $M(\varrho_c)$. However, one also encounters extrema of $M$ (open circles). The most important are $M_{min}$ and $M_{max}$ for NS, as well as the maximum $M$ for white dwarfs. These are critical points of the linear series (turning points) where the stability problem has a zero eigenvalue, and where a stable and a (dynamically) unstable branch of the linear series merge[2]. The stable branches are those with $dM/d\varrho_c > 0$, i.e. the branch of NS with $M_{min} < M < M_{max}$ (and the white-dwarf and planetary

---

[2] Note that a thermal stability problem does not exist for these idealized cold configurations, and the whole problem reduces to that of dynamical stability.

branch with $M <$ maximum mass for white dwarfs). So one could as well find the extrema for $M$ by looking for $\omega^2 = 0$ in the spectrum of eigenvalues of the dynamical stability problem. In fact it is found that $\omega_0^2 = 0$ (zero eigenvalue of the fundamental) at $M = M_{max}$. For further increasing $\varrho_c$ there follows an infinite number of maxima and minima of $M$. Correspondingly the curve $R = R(M)$ spirals into a limiting point, which is reached for $\varrho_c \to \infty$. All of these branches are unstable, since the further extrema only indicate that additional harmonics become unstable. The stability analysis can also be made for general relativistic configurations. In the Newtonian limit one has the well-known result that $\omega_0^2 = 0$ when an average of the exponent $\gamma_{ad}$ is $\gamma_{cr} = 4/3$ (see § 25.3.2), and in addition it can be shown (see SHAPIRO, TEUKOLSKY, 1983) that small effects of general relativity ($GM/Rc^2 \ll 1$) change the critical value from 4/3 to

$$\gamma_{cr} = \frac{4}{3} + \Lambda \frac{GM}{Rc^2} \quad , \tag{36.5}$$

where $\Lambda$ is a positive quantity of the order of unity. Therefore general relativity increases $\gamma_{cr}$, making the star more unstable, since stability requires $\bar{\gamma}_{ad} > \gamma_{cr}$. For $M = 1M_\odot$, $R = 10$ km the correction term in (36.5) is about 0.15, i.e. far from being negligible. $\gamma_{cr}$ can be raised well above 5/3 (even above 2 for certain models near $M_{max}$) such that all but the stiffest equations of state would give instability. This increase of $\gamma_{cr}$ is an important factor in determining the value of $M_{max}$ (together with the lowering of $\bar{\gamma}_{ad}$).

A very stiff equation of state, for example, gives $M_{max} = 2.7M_\odot$, with $R = 13.5$ km and $\varrho_c = 1.5 \times 10^{15}$ g cm$^{-3}$, while a softer one yields $M_{max} = 2M_\odot$, with $R = 9$ km and $\varrho_c = 3.3 \times 10^{15}$ g cm$^{-3}$ (curves with labels 6 and 4 in Fig. 36.1). At present there is no equation of state that can be considered realistic and that would give $M_{max}$ above $3M_\odot$.

The model is also marginally stable ($\omega_0^2 = 0$) at the minumum mass $M_{min}$, where the unstable branch begins leading to the white dwarfs. This instability is essentially caused by the lowering of $\gamma'$ in connection with the neutron drip (see § 36.1). We have seen that the release of free neutrons from nuclei results in $\gamma' \lesssim 4/3$ in the range $\varrho \approx 4 \times 10^{11} \dots 7 \times 10^{12}$ g cm$^{-3}$. Typical models for the minimum mass of stable neutron stars give $M_{min} \approx 0.09M_\odot$, $R \approx 160$ km, $\varrho_c \approx 1.5 \times 10^{14}$ g cm$^{-3}$. The average density is, of course, much smaller ($\approx 10^{10}$ g cm$^{-3}$), and the averaged $\gamma_{ad}$ becomes just equal to $\gamma_{cr}$ (which is here close to 4/3).

Let us dwell briefly on the meaning of the mass values quoted for NS. The stellar mass $M$ is here always the "gravitational mass", which is the value measurable for an outside observer [cf. the comments in § 2.6 after (2.29)]. $M$ differs from the proper mass $M_0 = Nm_0$, given by the total number $N$ of nucleons with a rest mass $m_0$, since in relativity the total binding energy $W$ of the configuration appears as a mass $\Delta M = W/c^2$, such that

$$M = M_0 + \frac{W}{c^2} = M_0 + \Delta M \quad . \tag{36.6}$$

In the Newtonian limit (for weak fields) we were used to identifying particularly the internal energy $E_i$ (from motion and interaction of particles) and the potential energy

$E_g$ in the gravitational field. Then for a static, stable configuration, $W = E_i + E_g < 0$, since $E_g < 0$ and $-E_g > E_i$. (In the Newtonian limit $E_g$ and $E_i$ were related by the virial theorem, cf. §3.) Correspondingly we may now say that the mass of a NS is increased by the internal energy and decreased by the (negative) potential energy, and the latter term wins. Therefore $W < 0$, and we have a mass defect $\Delta M < 0$. Depending on the precise model, $|\Delta M|$ can go up to $10 \dots 25\%$ of $M$ near $M_{max}$. Formally $M$ is given as an integral over $4\pi r^2 \varrho dr$, where $\varrho$ is the total mass–energy density ($\varrho_0 + u/c^2$) and $4\pi r^2 dr$ is *not* the volume element. This is rather given by $dV = 4\pi r^2 e^{\lambda/2} dr$ with $e^{\lambda/2}$ being a component of the metric tensor (cf. §2.6). Then simply

$$\Delta M \equiv M - M_0 = \int_0^R (4\pi r^2 \varrho dr - \varrho_0 dV)$$

$$= \int_0^R 4\pi r^2 \varrho \left( 1 - e^{\lambda/2} \frac{\varrho_0}{\varrho} \right) dr \quad . \tag{36.7}$$

Here $\varrho_0/\varrho < 1$, but $e^{\lambda/2} > 1$, and the product of both is $>1$, such that $\Delta M < 0$. So if we find a NS with mass $M$ we know that it started off as a more massive configuration. The mass defect $|\Delta M|$ was radiated away in the course of evolution by photons, neutrinos, or gravitational radiation. In that sense the original Kelvin–Helmholtz hypothesis that contraction supplies the radiated energy has turned out to be correct. The mass defect reaches a maximum at $M = M_{max}$ and then decreases again towards models with still larger $\varrho_c$.

The maximum mass for NS is scarcely influenced by rotation. Except for the very few most rapidly spinning pulsars, centrifugal forces play practically no role in NS, since the overwhelming gravitational forces dominate completely.

Now we turn to describe the *stratification of matter inside a NS model*. At the very outer part there must be an atmosphere of "normal" non-degenerate matter. Going inwards, we come to gradually larger densities and encounter all characteristic changes of high-density matter as described in §36.1

The *atmosphere* of a NS is very hot and incredibly compressed. Typical temperatures are of the order of $10^6$ K (see below). The extension is very small owing to the high surface gravity $g_0 \approx 1.3 \times 10^{14}$ cm s$^{-2}$. (For comparison, $g_0 = 2.7 \times 10^4$ cm s$^{-2}$ for the sun and $\approx 10^8$ cm s$^{-2}$ for white dwarfs.) This gives a pressure scale height of the order of 1 cm only. In the surface layers (say $\varrho \lesssim 10^6$ g cm$^{-3}$) the behaviour of the matter is still influenced by the temperature and also by strong magnetic fields.

Not far below the surface, the densities will be in and above the range typical for the interior of white dwarfs ($\gtrsim 10^6$ g cm$^{-3}$). As an example we discuss the model for a NS of $M = 1.4 M_\odot$ (see Fig. 36.3), calculated by using an equation of state of moderate stiffness (label 4 in Fig. 36.1) which gives $M_{max} \approx 2 M_\odot$. The radius of the $1.4 M_\odot$ model is 10.6 km.

Below the surface there is a solid *crust* ($10^6 \lesssim \varrho \lesssim 2.4 \times 10^{14}$ g cm$^{-3}$) of thickness $\Delta r \approx 0.9$ km. The matter in the crust contains nuclei, which are mainly Fe near the surface (cf. the equilibrium composition as a function of $\varrho$ described in §35.2). These nuclei will form a lattice, thus minimizing the energy of Coulomb interaction as in crystallized white dwarfs. The outer crust consists only of these

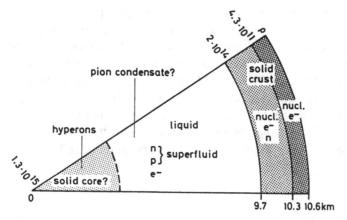

**Fig. 36.3.** Illustration of the interior structure of a neutron-star model with $M = 1.4 M_\odot$ calculated with the same equation of state as the sequence labelled 4 in Fig. 36.1. A few characteristic values of the density (in g cm$^{-3}$) are indicated along the upper radius. (After PINES, 1979)

nuclei plus a degenerate electron gas, though this changes over a depth of $\Delta r \approx 0.3$ km to where $\varrho = \varrho_{dr} \approx 4 \times 10^{11}$ g cm$^{-3}$ is reached. In the subsequent inner crust ($4 \times 10^{11} \lesssim \varrho \lesssim 2 \times 10^{14}$ g cm$^{-3}$), a liquid of free neutrons exists in addition to the nuclei (still arranged in a lattice) and the electrons. With decreasing $r$ the neutrons become more and more abundant at the expense of the nuclei, and the lattice disappears with the nuclei, until all nuclei are dissolved at $\varrho = \varrho_{nuc} \approx 2.4 \times 10^{14}$ g cm$^{-3}$, which therefore defines the lower boundary of the solid crust, at a depth of 0.9 km.

Below the crust there is the *interior neutron liquid* ($\varrho \gtrsim 2.4 \times 10^{14}$ g cm$^{-3}$) consisting mainly of interacting neutrons in equilibrium with a few protons and electrons. The neutrons will be superfluid, the protons superconductive.

It is unclear whether there is finally a central *solid core* in which the neutrons form a solid owing to their repulsive forces at small particle distances. The central density of this model is $\varrho_c \approx 1.3 \times 10^{15}$ g cm$^{-3}$.

The superfluidity of the neutron and proton liquids and the solid parts (crust and possible core) play a role in the attempts to explain the observed "glitches" of pulsars. These are sudden spin-ups, interrupting from time to time the normal, regular spin-down (decrease of the rotation frequency $\Omega$). There is a hypothesis according to which a glitch is due originally to a "starquake", decreasing suddenly the moment of inertia $I_c$ of the crust. Conservation of angular momentum requires a corresponding increase of $\Omega$. The relaxation to the normal state depends critically on the coupling of the rotating crust and the rotating interior liquid (and possible solid core). The charged components could be coupled magnetically, while the superfluid matter may couple via vortices. This coupling is the basis of another model of the glitches: the superfluid neutron liquid in the interior and in the inner crust is considered to rotate with an angular velocity slightly different from that of the lattice of nuclei in the crust. The coupling is provided by vortices in the liquid and is thought to break down suddenly when the crust has been decelerated sufficiently by the pulsar mechanism on the outside. The vortices can contain an appreciable fraction of the star's angular momentum and their distortion induces immediate changes of the observed rotation.

The thermal properties (except for the earliest stages) in principle follow once the mechanical models are given. One can then calculate the thermal conductivity, which, together with a given outward flux of energy, determines the $T$ gradient at any point. It turns out that like white dwarfs (§ 35.3) the NS have a nearly isothermal interior because of the high thermal conductivity. Only in the outermost layers does $T$ drop, by typically a factor of $10^2$, to the surface temperature. Particularly in the first, hot phases the cooling will be very rapid because of strong neutrino losses.

# § 37 Black Holes

Black holes (BH) represent the ultimate degree of compactness to which a stellar configuration can evolve. Having already called the neutron star a strange object, one cannot help labelling BH as weird. From the many fascinating aspects that are accessible via the full mathematical procedure (cf. MISNER, THORNE, WHEELER, 1973; SHAPIRO, TEUKOLSKY, 1983; CHANDRASEKHAR, 1983) we will indicate only a few points, showing that this is really a final stage of evolution, not just another late late phase. We limit the description to non-rotating BH without charge.

The theoretical description to be applied is that of general relativity (see, e.g., LANDAU, LIFSHITZ, vol. 2, 1965). We consider the gravitational field surrounding a very condensed mass concentration $M$ with spherical symmetry. The vacuum solution of Einstein's field equations (2.24) for this case was found as early as 1916 by K. Schwarzschild. It gives the line element $ds$, i.e. the distance between neighbouring events in 4-dimensional space–time as

$$\begin{aligned}
ds^2 &= g_{ij}dx^i dx^j \\
&= \left(1 - \frac{r_s}{r}\right) c^2 dt^2 - \left(1 - \frac{r_s}{r}\right)^{-1} dr^2 - r^2 d\vartheta^2 - r^2 \sin^2 \vartheta \, d\varphi^2 \\
&= \left(1 - \frac{r_s}{r}\right) c^2 dt^2 - d\sigma^2 \quad ,
\end{aligned} \tag{37.1}$$

where one has to sum from 0 to 3 over the indices $i$ and $j$, and where the usual spherical coordinates $r$, $\vartheta$, $\varphi$ are taken as the spatial coordinates $x^1$, $x^2$, $x^3$, and $x^0 = ct$. The critical parameter $r_s$ in (37.1) is the *Schwarzschild radius*

$$r_s = \frac{2GM}{c^2} \quad , \tag{37.2}$$

which has the value $r_s = 2.95$ km for $M = M_\odot$. The second component of the metric tensor, $(1 - r_s/r)^{-1}$, becomes singular at $r = r_s$, but one can show that this is a non-physical singularity disappearing when other suitable coordinates are used.

The proper time $\tau$, as measured by an observer carrying a standard clock, is related to the line element $ds$ along his world line by

$$d\tau = \frac{1}{c} ds \quad . \tag{37.3}$$

For a stationary observer ($dr = d\vartheta = d\varphi = 0$) at infinity ($r \to \infty$) the proper time $\tau_\infty$ coincides with $t$ according to (37.1). Consider two stationary observers, one at $r$, $\vartheta$, $\varphi$, the other at infinity. Their proper times $\tau$ and $\tau_\infty$ are related to each other by

$$\frac{d\tau}{d\tau_\infty} = \left(1 - \frac{r_s}{r}\right)^{1/2} \quad . \tag{37.4}$$

Suppose that the first of them operates a light source emitting signals at regular intervals $d\tau$, for example an atom emitting with the frequency $\nu_0 = 1/d\tau$. The other one receives the signals and measures the intervals in his own proper time as $d\tau_\infty$, i.e. he measures another frequency $\nu = 1/d\tau_\infty$. The resulting red shift due to the gravitational field is therefore

$$z \equiv \frac{\nu_0 - \nu}{\nu} = \frac{\nu_0}{\nu} - 1 = \frac{d\tau_\infty}{d\tau} - 1 = \left(1 - \frac{r_s}{r}\right)^{-1/2} - 1 \quad , \tag{37.5}$$

which gives $z \to \infty$ for $r \to r_s$.

The metric components in equation (37.1) show that the 4-dimensional space–time $(x^0, \ldots, x^3)$ is curved, and this holds also for the 3-dimensional space $(x^1, x^2, x^3)$. At the surface of a mass configuration of mass $M$ and radius $R$, the Gaussian curvature $K$ of position space can be written as

$$K = -\frac{GM}{c^2 R^3} = -\frac{1}{2} \frac{r_s}{R} \frac{1}{R^2} \quad . \tag{37.6}$$

This is usually very small compared with the curvature $R^{-2}$ of the 2-dimensional surface. For example, $-K \approx 2 \times 10^{-6} R^{-2}$ at the surface of the sun. But one already has $-K \approx 0.15 R^{-2}$ for a neutron star, and the two curvatures are comparable at the surface of a BH with $R = r_s$.

Consider a test particle small enough for the gravitational field not to be disturbed which moves freely in the field from point $A$ to $B$. Its world line in 4-dimensional space–time is then a *geodesic*, i.e. the length $s_{AB}$ is an extremum. This is to say, any infinitesimal variation does not change the length:

$$\delta s_{AB} \equiv \delta \int_A^B ds = 0 \quad . \tag{37.7}$$

If the test particle moves locally with a velocity $v$ over a spatial distance $d\sigma$, then the proper time interval $d\tau$ will be the smaller, the larger $v$. It becomes [cf. (37.1)]

$$d\tau = ds = 0 \quad , \quad \text{for } v = c \quad , \tag{37.8}$$

i.e. for photons or other particles of zero rest mass: they move along null geodesics. For material particles the requirement $v < c$ of special relativity (which is locally valid) means $d\tau^2$ and $ds^2 > 0$. Such separations are called *time-like*. World lines of material particles must be time-like. Separations with $ds^2 < 0$ (or $d\tau^2 < 0$) would require $v > c$; they are called *space-like*. For example, the distance between two simultaneous events ($dt = 0$) is space-like.

The null geodesics ($ds^2 = 0$), giving the propagation of photons, describe hyper-cones in space–time which are called *light cones*. In order to also see their properties near $r = r_s$, we introduce a new time coordinate $\bar{t}$ given by

$$\bar{t} = t + \frac{r_s}{c} \ln \left| \frac{r}{r_s} - 1 \right| \quad , \tag{37.9}$$

which transforms (37.1) to

$$ds^2 = \left(1 - \frac{r_s}{r}\right) c^2 d\bar{t}^2 - 2\frac{r_s}{r} c \, dr \, d\bar{t}$$
$$- \left(1 + \frac{r_s}{r}\right) dr^2 - r^2 d\vartheta^2 - r^2 \sin^2\vartheta \, d\varphi^2 \quad , \tag{37.10}$$

which is non-singular at $r = r_s$. We consider only the radial boundaries of the light cones, i.e. the path of radially ($d\vartheta = d\varphi = 0$) emitted photons. Then (37.10) yields for $ds^2 = 0$, after division by $c^2 dr^2$, the quadratic equation

$$\left(1 - \frac{r_s}{r}\right) \left(\frac{d\bar{t}}{dr}\right)^2 - \frac{2r_s}{cr}\frac{d\bar{t}}{dr} - \frac{1}{c^2}\left(1 + \frac{r_s}{r}\right) = 0 \quad , \tag{37.11}$$

which has the solutions

$$\left(\frac{d\bar{t}}{dr}\right)_1 = -\frac{1}{c} \quad , \quad \left(\frac{d\bar{t}}{dr}\right)_2 = \frac{1}{c}\frac{1 + r_s/r}{1 - r_s/r} \quad . \tag{37.12}$$

These derivatives are inclinations of the two radial boundaries of the light cone in an $r - \bar{t}$ plane (see Fig. 37.1). The first always corresponds to an inward motion with the same velocity $c$. The second derivative changes sign at $r = r_s$, being positive for $r > r_s$, where photons can be emitted outwards ($dr > 0$). With decreasing $r$, $(d\bar{t}/dr)_2$ becomes larger so that the light cone narrows and its axis turns to the left in Fig. 37.1. At $r = r_s$ the light cone is such that no photon can be emitted to the outside ($dr > 0$). This is the reason for calling a configuration with $R = r_s$ a "black hole", and for speaking of the Schwarzschild radius $r_s$ as the radius of a BH of mass $M$. For $r < r_s$ both solutions (37.12) are negative and the whole light cone is turned inwards. Therefore inside $r_s$ all radiation (together with all material particles, which can move only inside the light cone) is drawn inexorably towards the centre. This means also that no static solution ($dr = d\vartheta = d\varphi = 0$) is possible inside $r_s$, since it would require a motion vertically upwards in Fig. 37.1, i.e. outside the light cone.

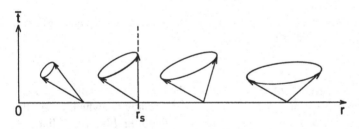

**Fig. 37.1.** Illustration of light cones at different distances $r$ from the central singularity, inside and outside the Schwarzschild radius $r_s$

In order to describe the motion of a material particle, we consider all variables to depend on the parameter $\tau$, the proper time, varying monotonically along the world line: $d\tau = ds/c$. Dots denote derivatives with respect to $\tau$. For example, $\dot{x}^\alpha = dx^\alpha/d\tau$ is the $\alpha$ component of a 4-velocity. Introducing $dx^\alpha = \dot{x}^\alpha d\tau$ into (37.1) gives the useful identity

392

$$c^2 = g_{ij}\dot{x}^i\dot{x}^j = c^2\left(1 - \frac{r_s}{r}\right)\dot{t}^2$$
$$- \left(1 - \frac{r_s}{r}\right)^{-1}\dot{r}^2 - r^2\left(\dot{\vartheta}^2 + \sin^2\vartheta\,\dot{\varphi}^2\right) \quad . \tag{37.13}$$

The condition that the world line be a geodesic means that the variation $\delta s = \delta\tau = 0$, which yields the Euler–Lagrange equations

$$\frac{d}{d\tau}\left(\frac{\partial L}{\partial \dot{x}^\alpha}\right) - \frac{\partial L}{\partial x^\alpha} = 0 \quad , \tag{37.14}$$

with the Lagrangian $L$ given by

$$2cL = \left[g_{ij}\dot{x}^i\dot{x}^j\right]^{1/2}$$
$$= \left[c^2\left(1 - \frac{r_s}{r}\right)\dot{t}^2 - \left(1 - \frac{r_s}{r}\right)^{-1}\dot{r}^2 - r^2\left(\dot{\vartheta}^2 + \sin^2\vartheta\dot{\varphi}^2\right)\right]^{1/2} \quad . \tag{37.15}$$

From (37.13,15) follows the value $L = 1/2$. For $x^0 = ct$, (37.14) becomes simply

$$\frac{d}{d\tau}\left[\left(1 - \frac{r_s}{r}\right)\dot{t}\right] = 0 \quad , \quad \left(1 - \frac{r_s}{r}\right)\dot{t} = \text{constant} \equiv A \quad . \tag{37.16}$$

We confine ourselves to the discussion of a radial infall ($\dot{\vartheta} = \dot{\varphi} = 0$) starting at $\tau = 0$ with zero velocity at the distance $r_0$. Instead of also deriving the equation of motion for $x^1 = r$ from (37.14), we simply introduce the second equation (37.16) into (37.13) and solve it for $\dot{r}$:

$$\dot{r} = c\left[A^2 - 1 + \frac{r_s}{r}\right]^{1/2} \quad . \tag{37.17}$$

For our purposes we set $A^2 - 1 = -r_s/r_0$. According to (37.17) this means that the particle starts with zero velocity at $r = r_0$. The integration of (37.17) then yields

$$\tau = \frac{1}{2}\frac{r_0}{c}\sqrt{\frac{r_0}{r_s}}\,(\sin\eta + \eta) \quad , \tag{37.18}$$

with the parameter $\eta = \text{arc}\cos(2r/r_0 - 1)$, as can be verified by differentiation. This function $\tau = \tau(r)$ is shown in Fig. 37.2 for $r_0 = 5r_s$. Again, nothing special happens in the proper time when the particle reaches $r = r_s$. The total proper time for reaching $r = 0$ is

$$\tau_0 = \frac{\pi}{2}\frac{r_s}{c}\left(\frac{r_0}{r_s}\right)^{3/2} \quad . \tag{37.19}$$

For $r_0 = 10\,r_s$ and $5\,r_s$ we have $\tau_0 = 49.67\,r_s/c$ and $17.56\,r_s/c$, respectively. These are very short times indeed, since for $M = M_\odot$ the characteristic time is only $r_s/c = 9.84 \times 10^{-6}$s.

The motion in terms of the coordinate time $t$ of an observer at infinity is quite different. The relation between $t$ and $\tau$ is given by (37.16) as $d\tau/dt = (1 - r_s/r)/A$, which goes to zero when $r \to r_s$. By this relation and (37.17) one obtains a differential equation for $t(r)$, which is integrated to give

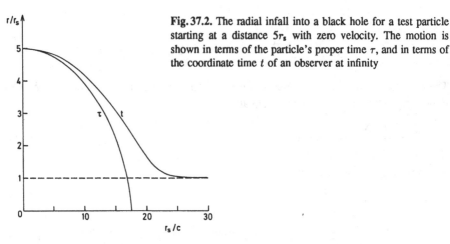

Fig. 37.2. The radial infall into a black hole for a test particle starting at a distance $5r_s$ with zero velocity. The motion is shown in terms of the particle's proper time $\tau$, and in terms of the coordinate time $t$ of an observer at infinity

$$\frac{t}{r_s/c} = \ln\left|\frac{\xi + \mathrm{tg}\,\eta/2}{\xi - \mathrm{tg}\,\eta/2}\right| + \xi\left[\eta + \frac{r_0}{2r_s}(\eta + \sin\eta)\right] \quad , \tag{37.20}$$

with $\eta$ as in (37.18) and $\xi = (r_0/r_s - 1)^{1/2}$. The curve $t = t(r)$ is also shown in Fig. 37.2 for $r_0 = 5r_s$. The fact that the observer sees the $\tau$ clock of the particle slowing down completely for $r \to r_s$ has the result that $t = t(r)$ approaches $r = r_s$ only asymptotically for $t \to \infty$. Events inside $r = r_s$ are completely shielded for the distant observer by the coordinate singularity at the Schwarzschild radius acting as an "event horizon".

These few considerations may suffice to illustrate some important properties of configurations which collapse into a BH. [Note that the Schwarzschild metric (37.1) is a vacuum solution, which is not valid inside the mass configuration, but holds from the surface outwards.]

As observed from the infalling surface (proper time $\tau$) the collapse proceeds fairly rapidly and in particular quite smoothly through the Schwarzschild radius $r = r_s$. Once the surface is inside $r_s$ a static configuration is no longer possible and the final collapse into the central singularity within a very short time is unavoidable. This is shown by the fact that material particles have world lines only inside the local light cone, and this is open only towards $r = 0$ (even radiation falls to $r = 0$). Note that it would not help to invoke an extreme pressure exerted by unknown physical effects, since the pressure would also contribute to the gravitating energy. The singularity at $r = 0$ is an essential one (as opposed to the mere coordinate singularity at $r = r_s$) with infinite gravity, though the physical conditions there are not yet clear. Quantum effects should have to be included and one can speculate whether they might remove the singularity.

The collapse of a star will present itself quite differently for an astronomer who is (we hope) very far away. In his coordinate time $t$ he will see that the collapse of the stellar surface slows down more and more, the closer it comes to $r_s$. In fact he will find that this critical point is not reached within finite time $t$; for him the collapsing surface seems to become stationary there. Of course, the approach of the surface to $r_s$ strongly affects the light received by the distant observer. He receives photons in ever increasing intervals and with ever decreasing energy, since the redshift $z \rightarrow \infty$ according to (37.5). Thus the collapsing star will finally "go out" for the distant observer. Only a strong gravitational field is left. This may be detected either through radiation emitted in the vicinity of the BH by infalling matter or (better) by the motion of a visible companion forming a double star with the BH. It is in this latter way that one hopes to prove some day the existence of BH. The necessary steps would be to ascertain that there is a double star the invisible component of which is a compact object and has a mass larger than the maximum mass for neutron stars. At present there are candidates for such objects (like the X-ray source Cyg X–1), but no proven cases.

It should be mentioned that aside from the Schwarzschild solution for non-rotating, uncharged BH there exist solutions which describe a rotating BH (Kerr metric) and a charged BH (Newman metric), the combination of these covering the full generality of possible properties of a BH: it is fully defined by mass, angular momentum, and charge. This surprising scantiness of properties left after the final collapse was summarized by Wheeler: "a black hole has no hair".

# VIII Pulsating Stars

# §38 Adiabatic Spherical Pulsations

## 38.1 The Eigenvalue Problem

The functions $P_0(m)$, $r_0(m)$, and $\varrho_0(m)$ are supposed to belong to a solution of the stellar-structure equations (9.1–4) for the case of complete equilibrium. Let us assume that we perturb the hydrostatic equilibrium, say by compressing the star slightly and releasing it again suddenly. It will expand and owing to inertia overshoot the equilibrium state: the star starts to oscillate. The analogy to the oscillating piston model (see §6.6) is obvious. More precisely we assume the initial displacement of the mass elements to be only radially directed ($d\vartheta = d\varphi = 0$) and of constant absolute value on concentric spheres. This leads to purely *radial oscillations* (or radial pulsations) during which the star remains spherically symmetric all time. For the perturbed variables at time $t$ we write

$$P(m,t) = P_0(m) + P_1(m,t) = P_0(m)\left[1 + p(m)e^{i\omega t}\right] \quad,$$

$$r(m,t) = r_0(m) + r_1(m,t) = r_0(m)\left[1 + x(m)e^{i\omega t}\right] \quad,$$

$$\varrho(m,t) = \varrho_0(m) + \varrho_1(m,t) = \varrho_0(m)\left[1 + d(m)e^{i\omega t}\right] \quad, \tag{38.1}$$

where the subscript 1 indicates the perturbations for which we have made a separation ansatz with an exponential time dependence [as in (25.17)]. The relative perturbations $p$, $x$, $d$ are assumed to be $\ll 1$.

We now insert these expressions into the equation of motion (9.2), linearize, and use the fact that $P_0$, $r_0$ obey the hydrostatic equation (9.16). Then with $g_0 = Gm/r_0^2$ we obtain

$$\frac{\partial}{\partial m}(P_0 p) = (4g_0 + r_0\omega^2)\frac{x}{4\pi r_0^2} \quad. \tag{38.2}$$

Using (9.16) again for $\partial P_0/\partial m$ and the relation

$$\frac{\partial}{\partial r_0} = 4\pi r_0^2 \varrho_0 \frac{\partial}{\partial m} \quad, \tag{38.3}$$

we find

$$\frac{P_0}{\varrho_0}\frac{\partial p}{\partial r_0} = \omega^2 r_0 x + g_0(p + 4x) \quad. \tag{38.4}$$

Quite similarly (38.1) introduced into (9.1) yields with (38.3)

398

$$r_0 \frac{\partial x}{\partial r_0} = -3x - d \quad . \tag{38.5}$$

Note that the transformation (38.3) does not mean that we go back to an Eulerian description. The partial derivative $\partial/\partial t$ describes time variations at constant $r_0$. But since $r_0 = r_0(m)$ is given by the equilibrium solution, $\partial/\partial t$ also refers to a fixed value of $m$.

We know already that perturbations of hydrostatic equilibrium proceed on a time-scale $\tau_{hydr} \ll \tau_{adj}$. We therefore assume here that the oscillations are adiabatic, which means that

$$p = \gamma_{ad} d \quad . \tag{38.6}$$

This shows again the advantage of using Lagrangian variables: the adiabatic condition has the simple form (38.6) only if $p$ and $d$ are considered functions of $m$ [or of $r_0 = r_0(m)$] and therefore give the variations in the *co-moving frame*. For the sake of simplicity we now assume that $\gamma_{ad}$ is constant in space and time. From (38.5,6) we obtain by differentiation with respect to $r_0$

$$\frac{\partial x}{\partial r_0} + r_0 \frac{\partial^2 x}{\partial r_0^2} = -3 \frac{\partial x}{\partial r_0} - \frac{1}{\gamma_{ad}} \frac{\partial p}{\partial r_0} \quad . \tag{38.7}$$

Eliminating $\partial p/\partial r_0$, $p$, and $d$ from (38.4–7) gives

$$x'' + \left( \frac{4}{r_0} - \frac{\varrho_0 g_0}{P_0} \right) x' + \frac{\varrho_0}{\gamma_{ad} P_0} \left[ \omega^2 + (4 - 3\gamma_{ad}) \frac{g_0}{r_0} \right] x = 0 \quad , \tag{38.8}$$

where a prime denotes a derivative with respect to $r_0$.

This second-order differential equation describes the relative amplitude $x(r_0)$ as function of depth for an adiabatic oscillation of frequency $\omega$. In addition one has to fulfil boundary conditions, one at the centre and one at the surface. At the centre the coefficient of $x'$ in (38.8) is singular, while the coefficient of $x$ remains regular, since $g_0 \sim m/r_0^2 \sim r_0$. Because one has to demand that $x$ is regular there, this gives the central boundary condition $x' = 0$.

With a simple expansion into powers of $r_0$ of the form $x = a_0 + a_1 r_0 + a_2 r_0^2 + \ldots$, one finds that the regular solution starts from the centre outwards with $a_1 = 0$ and

$$a_2 = -\frac{1}{10} \frac{\varrho_c}{\gamma_{ad} P_c} \left[ \omega^2 + (4 - 3\gamma_{ad}) \frac{4\pi}{3} G \varrho_c \right] a_0 \quad , \tag{38.9}$$

where the subscript c indicates central values of the unperturbed solution.

For the surface the simple condition $P_1 \equiv pP_0 = 0$ is often used. However, one can find a slightly more realistic boundary condition. We simplify the atmosphere by assuming its mass $m_a$ to be comprised in a thin layer at $r = R(t)$, which follows the changing $R$ during the oscillations and provides the outer boundary condition at each moment by its weight. We neglect, however, its inertia. Then at the bottom of the "atmosphere" we have

$$4\pi R^2 P - \frac{Gm_a M}{R^2} = 0 \quad , \tag{38.10}$$

and in the equilibrium state we have

$$4\pi R_0^2 P_0 = \frac{G m_{\mathrm{a}} M}{R_0^2} \quad .$$

(38.11)

Using this and (38.1), we find from (38.10) that after linearization

$$p + 4x = 0 \quad .$$

(38.12)

We can rewrite this condition in terms of $x$ and $x'$. If we replace $p$ in (38.12) by (38.6) and then $d$ by (38.5), the outer boundary condition at $r_0 = R_0$ becomes

$$\gamma_{\mathrm{ad}} R_0 x' - (4 - 3\gamma_{\mathrm{ad}}) x = 0 \quad .$$

(38.13)

The interior boundary condition at $r_0 = 0$ was

$$x' = 0 \quad .$$

(38.14)

If we multiply the differential equation (38.8) by $r_0^4 P_0$, we can write it in the form

$$(r_0^4 P_0 x')' + \frac{r_0^4 \varrho_0}{\gamma_{\mathrm{ad}}} \left[ \omega^2 + (4 - 3\gamma_{\mathrm{ad}}) \frac{g_0}{r_0} \right] x = 0 \quad .$$

(38.15)

Together with the (linear, homogeneous) boundary conditions (38.13,14) this defines a classical *Sturm–Liouville problem* with all its consequences.

From the theory of eigenvalue problems of the Sturm–Liouville type, a series of theorems immediately follows that we shall here list without proofs (which can be found in standard textbooks):

1. There is an infinite number of eigenvalues $\omega_n^2$.
2. The $\omega_n^2$ are real and can be placed in the order $\omega_0^2 < \omega_1^2 < \ldots$, with $\omega_n^2 \to \infty$ for $n \to \infty$.
3. The eigenfunction $x_0$ of the lowest eigenvalue $\omega_0$ has no node in the interval $0 < r_0 < R_0$ ("fundamental"). For $n > 0$, the eigenfunction $x_n$ has $n$ nodes in the above interval ("$n$th overtone").
4. The normalized eigenfunctions $x_n$ are complete and obey the orthogonality relation

$$\int_0^{R_0} r_0^4 \varrho_0 x_m x_n dr_0 = \delta_{mn} \quad ,$$

(38.16)

where $\delta_{mn}$ is the Kronecker symbol.

The eigenfunctions permit the investigation of the evolution in time of any arbitrary initial perturbation described by $x_m = x_m(r_0)$, $\dot{x}_m = \dot{x}_m(r_0)$ at $t = 0$. Indeed if one writes down the expansion of the initial perturbations in terms of the eigenfunctions,

$$x_m(r_0) = \sum_{n=0}^{\infty} c_n x_n(r_0) \quad , \quad \dot{x}_m(r_0) = \sum_{n=0}^{\infty} d_n x_n(r_0) \quad ,$$

(38.17)

where the $c_n$, $d_n$ are real, then

$$x(r_0, t) = \mathrm{Re} \left[ \sum_{n=0}^{\infty} (a_n \, e^{i\omega_n t} + b_n \, e^{-i\omega_n t}) x_n(r_0) \right] \quad ,$$

$$\dot{x}(r_0, t) = \mathrm{Re} \left[ \sum_{n=0}^{\infty} i\omega_n \, (a_n \, e^{i\omega_n t} - b_n \, e^{-i\omega_n t}) x_n(r_0) \right] \tag{38.18}$$

with complex coefficients $a_n$, $b_n$, fulfil the time-dependent equation of motion (38.15) with the initial conditions (38.17) at $t = 0$ if $a_n$, $b_n$ satisfy

$$a_n + b_n = c_n \quad , \quad \mathrm{Re}\,[i\omega_n(a_n - b_n)] = d_n \quad . \tag{38.19}$$

Now we come to the question of stability. Since the perturbations are assumed to be adiabatic, it is dynamical stability we are asking for. We have seen that $\omega_n^2$ is real, so that if $\omega_n^2 > 0$, then $\pm\omega_n$ is real, and the perturbations according to (38.1) are purely oscillatory (with constant amplitude): the equilibrium is dynamically stable. If $\omega_n^2 < 0$, then $\pm\omega_n$ is purely imaginary, say $\pm\omega_n = \pm i\chi$ with real $\chi$. The general time-dependent solution for this model is a sum of expressions of the form

$$A x_n \, e^{-\chi t} + B x_n \, e^{\chi t} \quad , \tag{38.20}$$

where $A$, $B$ are complex constants. Hence at least one of the two terms describes an amplitude growing exponentially in time. This term will necessarily show up in the expansion (38.18) of an arbitrary perturbation and dominate after sufficient time: the equilibrium is dynamically unstable.

The two regimes are separated by the case of marginal stability with $\omega_0^2 = 0$, which according to earlier considerations (§ 25.3.2) is expected to occur for $\gamma_{ad} = 4/3$. We now show that this in fact follows from the rather general formalism used here. For simplicity let us assume that $P_0 \to 0$ at the outer boundary.

Integration of (38.15) over the whole star for the fundamental mode ($n = 0$) gives

$$\left[ r_0^4 P_0 x_0' \right]_0^{R_0} + \frac{\omega_0^2}{\gamma_{ad}} \int_0^{R_0} r_0^4 \varrho_0 x_0 \, dr_0$$

$$+ \frac{4 - 3\gamma_{ad}}{\gamma_{ad}} \int_0^{R_0} r_0^3 \varrho_0 g_0 x_0 \, dr_0 = 0 \quad . \tag{38.21}$$

The boundary term on the left vanishes and we find

$$\omega_0^2 = (3\gamma_{ad} - 4) \frac{\int_0^{R_0} r_0^3 \varrho_0 g_0 x_0 \, dr_0}{\int_0^{R_0} r_0^4 \varrho_0 x_0 \, dr_0} \quad . \tag{38.22}$$

Since $x_0$, as eigenfunction of the fundamental, does not change sign in the interval, we have sign $\omega_0^2 = \mathrm{sign}(3\gamma_{ad} - 4)$. Therefore $\gamma_{ad} > 4/3$ gives $\omega_0^2 > 0$, and the equilibrium is dynamically stable, because all $\omega_n^2 > \omega_0^2$ for $n > 0$ (see above). If $\gamma_{ad} < 4/3$, then for the fundamental (and possibly for a finite number of overtones) $\omega_n^2 < 0$, and the equilibrium is dynamically unstable.

Here we have assumed that $\gamma_{ad}$ is constant throughout the stellar model, though the main result is unchanged if $\gamma_{ad}$ varies; in order to guarantee dynamical stability, then, a mean value of $\gamma_{ad}$ has to be $> 4/3$.

Of course, we could have carried through the whole procedure using $m$ as independent variable instead of $r_0$. Then (38.4, 5) would have had to be replaced by the equivalent equations (25.19, 20).

## 38.2 The Homogeneous Sphere

To illustrate the procedure of § 38.1 we apply it to the simplest, but very instructive, case of a gaseous sphere of constant density, where we have an easy analytical access to the eigenvalues and eigenfunctions.

If $\varrho$ is constant in space, then

$$r_0 = \left(\frac{3m}{4\pi\varrho_0}\right)^{1/3} \quad , \quad g_0 = \frac{Gm}{r_0^2} = \frac{4\pi}{3} Gr_0\varrho_0 \quad , \tag{38.23}$$

and from integration of the equation of hydrostatic equilibrium (2.3) we find

$$P_0(r_0) = \frac{2\pi}{3} G\varrho_0^2\left(R_0^2 - r_0^2\right) \quad , \tag{38.24}$$

where $R_0$ is the surface radius in hydrostatic equilibrium.

If we introduce the dimensionless variable $\xi = r_0/R_0$ and define

$$\tilde{A} := \frac{3\omega^2}{2\pi G\varrho_0\gamma_{ad}} + \frac{2(4 - 3\gamma_{ad})}{\gamma_{ad}} \quad , \tag{38.25}$$

then instead of (38.8) we can write

$$\frac{d^2x}{d\xi^2} + \left(\frac{4}{\xi} - \frac{2\xi}{1 - \xi^2}\right)\frac{dx}{d\xi} + \frac{\tilde{A}}{1 - \xi^2} x = 0 \quad . \tag{38.26}$$

This differential equation has singularities at the centre and at the surface and we look for solutions which are regular at both ends.

The simplest such solution of (38.26) is obvious: $x = x_0 = $ constant is an eigenfunction for $\tilde{A} = 0$. The corresponding eigenfrequency follows from (38.25):

$$\omega_0^2 = \frac{4\pi}{3} G\varrho_0 (3\gamma_{ad} - 4) \quad . \tag{38.27}$$

This represents the fundamental, since the eigenfunction $x = $ constant has no node. The expression (38.27) for the eigenvalue follows immediately from (38.22) for $x_0 = $ constant, $\varrho_0 = $ constant. Note that (38.27) shows the famous period–density relation for pulsating stars: $\omega_0^2/\varrho_0 = $ constant.

For the overtones we try polynomials in $r_0$. Indeed if for the first overtone we take $x = 1 + b\xi^2$ with constant $b$, then (38.26) can be solved with $b = -7/5$ and $\tilde{A} = 14$. The corresponding eigenvalue is obtained from (38.25,27) and we have

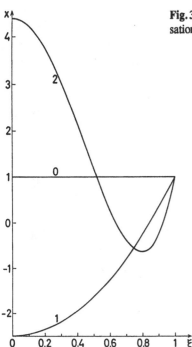

$$\omega_1^2 = \omega_0^2 \left(1 + \frac{7\gamma_{ad}}{3\gamma_{ad} - 4}\right) \quad ; \quad x_1 = 1 - \frac{7}{5}\xi^2 \quad . \tag{38.28}$$

The eigenfunction has one node at $\xi = (5/7)^{1/2}$, i.e. at $r_0 = 0.845\,R_0$. For $\gamma_{ad} = 5/3$ the ratio of the frequencies of first overtone and fundamental is $\omega_1/\omega_0 = 3.56$.

One can now try higher polynomials with free coefficients in order to find the higher overtones. But we leave this to the reader, the first three eigenfunctions being illustrated in Fig. 38.1.

### 38.3 Pulsating Polytropes

Let us now investigate the (spherically symmetric) radial oscillations of polytropic models of index $n$ as discussed in §19. We therefore express the quantities of the unperturbed model which appear in the coefficients of (38.8),

$$r_0 \quad , \quad \varrho_0 g_0/P_0 \quad , \quad \varrho_0/P_0 \quad , \quad \varrho_0 g_0/(P_0 r_0) \quad ,$$

by the Lane–Emden function $w(z)$ and by its dimensionless argument $z$. From (19.9) we have

$$g_0 = \frac{\partial \Phi_0}{\partial r_0} = A\Phi_c \frac{dw}{dz} \quad ; \quad A^2 = \frac{4\pi G}{[(n+1)K]^n}(-\Phi_c)^{n-1} \quad , \tag{38.29}$$

while (19.7) yields

$$\varrho_0 = \left[ \frac{-\Phi_c w}{(n+1)K} \right]^n \quad , \tag{38.30}$$

the subscript c denoting central values in the unperturbed model. If we use the polytropic relation (19.3), we find

$$\frac{\varrho_0}{P_0} = \frac{1}{K} \varrho^{-1/n} = -\frac{n+1}{\Phi_c w} \quad , \tag{38.31}$$

and we then have

$$\frac{g_0 \varrho_0}{P_0} = -A \frac{n+1}{w} \frac{dw}{dz} \tag{38.32}$$

and

$$\frac{g_0}{r_0} = \frac{\Phi_c A^2}{z} \frac{dw}{dz} \quad . \tag{38.33}$$

If we replace $r_0$ by $z = Ar_0$, the oscillation equation (38.8) becomes

$$\frac{d^2 x}{dz^2} + \left( \frac{4}{z} + \frac{n+1}{w} \frac{dw}{dz} \right) \frac{dx}{dz}$$

$$+ \left[ \Omega^2 - \frac{(4 - 3\gamma_{ad})(n+1)}{\gamma_{ad}} \frac{1}{z} \frac{dw}{dz} \right] \frac{x}{w} = 0 \quad . \tag{38.34}$$

Equation (38.34) is singular at the centre ($z = 0$) and at the surface ($w = 0$). $\Omega$ is a dimensionless frequency:

$$\Omega^2 = \frac{n+1}{\gamma_{ad}(-\Phi_c)A^2} \omega^2 \quad . \tag{38.35}$$

In (38.34) only $\gamma_{ad}$, the polytropic index $n$, and the Lane–Emden function for this index appear. Therefore the dimensionless eigenvalue $\Omega^2$ obtained from (38.34) depends only on $n$ and $\gamma_{ad}$, but not on other properties of the polytropic model, say $M$ or $R$. The relation (38.35) between $\Omega$ and $\omega$ can be expressed differently. Using (38.30) for the centre ($w = 1$) and (38.29) we have

$$\omega^2 = \frac{\gamma_{ad}(-\Phi_c)A^2}{n+1} \Omega^2 = \frac{4\pi G \gamma_{ad} \varrho_c}{n+1} \Omega^2 \quad . \tag{38.36}$$

Since for a given $n$ the central density $\varrho_c$ and the mean density $\bar{\varrho}$ of the whole unperturbed model differ only by a constant factor, one finds from (38.36) $\omega^2 = $ constant $\cdot \bar{\varrho}$, or with the period $\Pi = 2\pi/\omega$

$$\Pi \sqrt{\bar{\varrho}} = \left[ \frac{(n+1)\pi}{\gamma_{ad} G \Omega^2} \left( \frac{\bar{\varrho}}{\varrho_c} \right)_n \right]^{1/2} \quad . \tag{38.37}$$

For a given mode, say the fundamental, the right-hand side depends only on the polytropic index $n$ and on $\gamma_{ad}$. This is the famous *period–density relation*. It is also approximately fulfilled for more realistic stellar models.

If one assumes for a $\delta$ Cephei star that $M = 7M_\odot$ and $R = 80R_\odot$, its mean density is $\approx 2 \times 10^{-5}$ g cm$^{-3}$. If the period is $11^d$, then $\Pi(\bar\varrho)^{1/2} \approx 0.049$ ($\Pi$ in days, $\bar\varrho$ in g cm$^{-3}$). This constant gives a period of about 220 days for a supergiant with $\bar\varrho = 5 \times 10^{-8}$ g cm$^{-3}$, while for a white dwarf (with $\bar\varrho \approx 10^6$ g cm$^{-3}$) it gives a period of 4s. Indeed the supergiant period is of the order of those observed for Mira stars, while very short periods are observed for white dwarfs.

The dimensionless equation (38.34) depends on $n$ and $\gamma_{ad}$, where the polytropic index $n$ is a measure of the density concentration, say of $\varrho_c/\bar\varrho$, while $\gamma_{ad}$ is a measure of the stiffness of the configuration. If $\gamma_{ad} = 4/3$, then $\Omega = 0$ is an eigenvalue and $x =$ constant the corresponding eigenfunction, as can be seen from (38.34); the model is then marginally stable and after compression does not go back to its original size. The larger $\gamma_{ad}$, the better the stability, since the compressed model will expand more violently after being released. This can be understood with the help of the considerations in §25.3.2.

Numerical solutions of the eigenvalue problem show how variations in $n$ and $\gamma_{ad}$ modify the solutions. Because of the singularities of (38.34) at both ends of the interval $0 < z < z_n$ ($z_n$ is the value of $z$ for which the Lane–Emden function of index $n$ vanishes) the numerical solution is not straightforward. The simplest way is to choose a trial value $\Omega = \Omega^*$ and to start two integrations with power series regular at $z = 0$ and at $z = z_n$. The outward and inward integrations are continued to a common point somewhere, say at $z^* = z_n/2$. There the two solutions will have neither the same value $x(z^*)$ nor the same derivative $(dx/dz)^*$. Since the differential equation is linear and homogeneous, we can multiply one of the solutions by a constant factor such that both get the same value at $z^*$. But then they probably still disagree in $(dx/dz)^*$. Agreement in the derivatives can be achieved by gradually improving $\Omega$, carrying out new integrations, and so on. By such iterations a solution for the whole interval can be obtained.

Whether by such a procedure one arrives at the fundamental or at an overtone depends in general on the trial $\Omega^*$. If it is near the fundamental, we will end up with the fundamental eigenvalue and eigenfunction. In any case the number of nodes will reveal which mode has been found.

Since (38.34) is linear and homogeneous, the solution may be multiplied by an arbitrary constant factor, in which way we can normalize the solution such that at the surface $x(z_n) = 1$. For the polytrope $n = 3$ the eigenfunctions of different modes for $\gamma_{ad} = 5/3$ are shown in Fig. 38.2 and the eigenfunction of the fundamental for different values of $\gamma_{ad}$ is displayed in Fig. 38.3.

The variation of $\gamma_{ad}$ is indeed important. To see this, we assume an ideal monatomic gas with radiation pressure as discussed in §13. From (13.16, 21, 24) we find after some algebra that

$$\gamma_{ad} = \frac{1}{\alpha - \delta\nabla_{ad}} = \frac{32 - 24\beta - 3\beta^2}{24 - 21\beta} \quad . \tag{38.38}$$

For the limit cases $\beta = 1$ ($P_{rad} = 0$) and $\beta = 0$ ($P_{gas} = 0$) the adiabatic exponent $\gamma_{ad}$ takes the values 5/3 and 4/3 respectively. We see that our assumption $\gamma_{ad} =$ constant throughout the model holds only as long as $\beta =$ constant. Fortunately this is the case

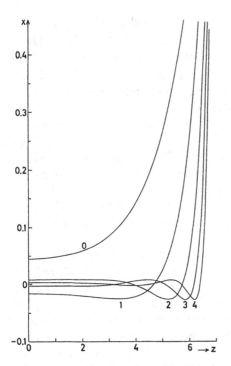

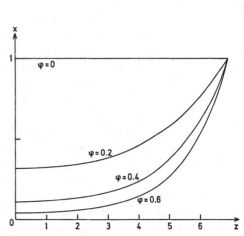

**Fig. 38.2.** Eigenfunctions for radial adiabatic pulsations of the polytrope $n = 3$ for $\varphi = 0.6$. (After SCHWARZSCHILD, 1941)

**Fig. 38.3.** The fundamental eigenfunction for radial adiabatic pulsations of the polytrope $n = 3$ for different values of $\varphi$. Radiation pressure diminishes the ratio of the amplitude at the surface to that of the centre. If the radiation pressure dominates the gas pressure completely ($\varphi = 0$) the relative amplitude $x$ is constant

for the polytrope $n = 3$, since $1 - \beta \sim T^4/P$ and $T \sim w$, $P \sim w^{n+1}$. In (38.34) the radiation pressure only appears in the quantity

$$\varphi := -\frac{4 - 3\gamma_{\mathrm{ad}}}{\gamma_{\mathrm{ad}}} = 3 - \frac{4}{\gamma_{\mathrm{ad}}} \ . \tag{38.39}$$

For vanishing and dominating radiation pressure, $\varphi$ takes the values 0.6 and 0 respectively.

Fundamental and overtone solutions of (38.34) for $n = 3$ and for different values of $\varphi$ have been found numerically by SCHWARZSCHILD (1941). For $\varphi = 0.6$ ($\gamma_{\mathrm{ad}} = 5/3$) the (dimensionless) eigenfrequency for the fundamental and the first overtones are $\Omega_0^2 = 0.1367$, $\Omega_1^2 = 0.2509$, $\Omega_2^2 = 0.4209$, $\Omega_3^2 = 0.6420$, $\Omega_4^2 = 0.9117$. The corresponding eigenfunctions are shown in Fig. 38.2.

The influence of $\beta$ on the fundamental eigenfunction can be seen in Fig. 38.3. With increasing radiation pressure ($\varphi$ decreasing) the relative amplitude $x$ drops less and less steeply from the surface to the centre. The ratio $x_{\mathrm{surface}}/x_{\mathrm{centre}}$ is 22.4 for $\varphi = 0.6$ and 9.1 for $\varphi = 0.4$. In the limit $\varphi \to 0$ (pure radiation pressure) $x$ even becomes constant. Indeed, for $\gamma_{\mathrm{ad}} = 4/3$ and for the eigenvalue $\Omega = 0$, $x = \text{constant}$ is a solution as we know already.

# § 39 Non-adiabatic Spherical Pulsations

When a star oscillates, its mass elements will generally not change their properties adiabatically. The outward-going heat flow, as well as the nuclear energy production, is modulated by the rhythm of the pulsation and both effects cause deviations from adiabaticity. However, since the pulsation takes place on the hydrostatic time-scale, which is short compared to $\tau_{KH}$, the deviations from adiabaticity should be small in most parts of the stellar interior. In order to demonstrate the main effects of the non-adiabatic terms on the equation of motion, we discuss them at first for the simple piston model.

## 39.1 Vibrational Instability of the Piston Model

We go back to the description of § 25.2.2. Equation (25.14) gives three eigenvalues $\sigma$ for non-adiabatic oscillations of the piston model. The adiabatic period $\sigma = \pm\sigma_{ad} = \pm i\omega_{ad}$ (with $\omega_{ad}^2 > 0$) would be obtained for $e_P = e_T = 0$. For small non-adiabatic terms $e_P$ and $e_T$ we now write $\sigma = \sigma_r \pm \sigma_{ad}$ as in (25.15) and assume that the real part is small, $|\sigma_r| \ll \omega_{ad}$. Then, neglecting terms of the order $\sigma_r^3$, $\sigma_r^2$, $e_P\sigma_r$, $e_T\sigma_r$ and introducing $\gamma_{ad}$ instead of 5/3, we find from (25.14) that

$$\frac{h_0 u_0}{g_0}\left(3\sigma_{ad}^2\sigma_r \pm \sigma_{ad}^3\right) - \frac{h_0}{g_0}\left(e_P + e_T\right)\sigma_{ad}^2 + \gamma_{ad}u_0\left(\sigma_r \pm \sigma_{ad}\right) - e_T = 0 . \quad (39.1)$$

Since $\sigma_{ad}$ has to obey the adiabatic equation [cf. (25.15)]

$$\frac{h_0}{g_0}\sigma_{ad}^2 + \gamma_{ad} = 0 \quad , \quad\quad\quad\quad\quad\quad\quad\quad\quad\quad\quad (39.2)$$

(39.1) becomes

$$2u_0\sigma_r = \nabla_{ad}e_T + e_P \quad , \quad\quad\quad\quad\quad\quad\quad\quad\quad\quad\quad (39.3)$$

where we have introduced $\nabla_{ad} := (\gamma_{ad} - 1)/\gamma_{ad}$.

We now assume $\varepsilon_0 = \kappa_0 = 0$, then $e_P = 0$, $e_T = -\chi T_0/m^*$ [see (25.13)], and we find that

$$2u_0\sigma_r = -\nabla_{ad}\frac{\chi T_0}{m^*} \quad . \quad\quad\quad\quad\quad\quad\quad\quad\quad\quad (39.4)$$

Therefore, since $\nabla_{ad} > 0$, one has $\sigma_r < 0$, meaning that the oscillation is damped. During each cycle heat leaves and enters the gas in the container by way of the leak, kinetic energy of the piston is lost and added to the surroundings as heat.

Similarly in a star the flow of heat modulated by the oscillation can damp the motion. Since the deviation from adiabaticity is more pronounced in the outer regions, the damping time is determined by the Kelvin–Helmholtz time-scale of the outer layers. In his classic book, EDDINGTON (1925) estimated that the damping time of $\delta$ Cephei stars would be of the order of 8000 years and concluded that there must exist a mechanism which maintains their pulsations. He actually discussed two possible mechanisms which can be easily demonstrated with the piston model.

The first is called the $\kappa$ *mechanism*, since here it is the modulated absorption of radiation which can yield vibrational instability.

If for the sake of simplicity we assume that $\chi = \varepsilon_0 = 0$, then according to (25.13) one has $e_P = \kappa_0 F \kappa_P$, $e_T = \kappa_0 F \kappa_T$, and therefore (39.3) becomes

$$2u_0 \sigma_r = \kappa_0 F \left( \nabla_{ad} \kappa_T + \kappa_P \right) \quad . \tag{39.5}$$

The model is vibrationally unstable ($\sigma_r > 0$) if $(\nabla_{ad}\kappa_T + \kappa_P) > 0$. This means that the instability occurs if during adiabatic compression ($d \ln P > 0$) the absorption coefficient increases: $d \ln \kappa = (\nabla_{ad}\kappa_T + \kappa_P)d \ln P > 0$. Then in the compressed state more energy is absorbed than in eqilibrium and the ensuing expansion is slightly enhanced. For analogous reasons the state of maximum expansion is followed by an enhanced compression.

In stars the outgoing radiative flux can similarly cause an instability if the stellar opacity increases/decreases during the phase of contraction/expansion. As we shall see (§ 39.4), this is the mechanism which indeed drives the $\delta$ Cephei stars.

In the so-called $\varepsilon$ *mechanism* the possible cause for an instability is the modulated nuclear energy generation. In order to discuss a simple case, we assume $\chi = \kappa_0 = 0$ and find from (39.3) with (25.13) that

$$2u_0 \sigma_r = \varepsilon_0 \left( \nabla_{ad}\varepsilon_T + \varepsilon_P \right) \quad . \tag{39.6}$$

This model is vibrationally unstable for any nuclear burning ($\varepsilon_0 > 0$), since all terms on the right-hand side are $> 0$. For example, the CNO cycle has typically $\varepsilon_T \gtrsim 10$, $\varepsilon_P = 1$ while $\nabla_{ad} \approx 0.4$.

In the two cases discussed above, the piston model in a certain sense mimics the stability behaviour of different layers in a star. Since $\tau_{KH} \gg 1/\omega_{ad}$, the non-adiabatic effects in a pulsating star are small, and as in the piston model one can expect that the oscillations are almost adiabatic, as described in § 38. But the non-adiabatic effects will cause a small deviation of the eigenfrequency from the adiabatic value. Indeed, since the temperature variations are different in different regions of the star, these regions exchange an additional heat which – like the heat flow through the leak – causes a damping (radiative damping). A destabilizing effect on the star is caused by those regions where the opacity increases during contraction ($\kappa$ mechanism) as well as those with a nuclear burning where $\varepsilon$ increases during contraction ($\varepsilon$ mechanism).

### 39.2 The Quasi-adiabatic Approximation

In order to determine the vibrational stability behaviour of a star, one has to solve the four ordinary differential equations (25.19–22) for the perturbations $p$, $x$, $\lambda$, $\vartheta$, together with homogeneous boundary conditions at the centre and at the surface. In addition to the "mechanical" boundary conditions (38.13, 14) one has at the centre

$$l_0 \lambda = 0 \quad \text{at} \quad m = 0 \; . \tag{39.7}$$

As a rough outer boundary condition one can assume that at the surface the relation $L = 4\pi R^2 \sigma T^4$ holds throughout the oscillation period, yielding

$$l = 2x + 4\vartheta \; . \tag{39.8}$$

This relation is not exactly true, since the photosphere (where $T = T_{\text{eff}}$) does not always belong to the same mass shell during the oscillation. With a more detailed theory of the behaviour of the atmosphere during the oscillations one can replace (39.8) by another, but also linear and homogeneous, outer boundary condition.

The homogeneous linear equations (25.19–22) and boundary conditions (38.13, 14) and (39.7, 8) define an eigenvalue problem for the eigenvalue $\omega$.

Here we will restrict ourselves to a simplified treatment, the *quasi-adiabatic approximation*. For the given unperturbed equilibrium model we first solve the adiabatic problem described in § 38, thereby obtaining a set of adiabatic eigenvalues $\omega_{\text{ad}}^{(n)}$ with the eigenfunctions $p_{\text{ad}}^{(n)}$, $x_{\text{ad}}^{(n)}$, $\vartheta_{\text{ad}}^{(n)} = \nabla_{\text{ad}} p_{\text{ad}}^{(n)}$, where the upper index $n$ labels the different eigenvalues. In the following we will drop $n$, though keeping in mind that the procedure described here and in § 39.3 can be carried out for each of the adiabatic eigenvalues. Of course, the real oscillations will not proceed exactly adiabatically, which, for example, is shown in luminosity perturbations. To determine an approximation to the relative luminosity perturbation $\lambda$ we differentiate $\vartheta_{\text{ad}}$ with respect to $m$ and find from (25.22)

$$\lambda = \frac{P_0}{\nabla_{\text{ad}} P_0'} \vartheta_{\text{ad}}' + 4x_{\text{ad}} - \kappa_P p_{\text{ad}} + (4 - \kappa_T)\vartheta_{\text{ad}} \; . \tag{39.9}$$

In this quasi-adiabatic approximation, therefore, the non-adiabatic effects determining $\lambda$ are calculated from adiabatic eigenfunctions. The correct procedure would require the use of non-adiabatic eigenfunctions on the right-hand side of (39.9), while in a strictly adiabatic case we would expect $\lambda = \lambda_{\text{ad}} = 0$. One can use the non-adiabatic variation $\lambda$ of the local luminosity in order to estimate the change of $\omega$ due to non-adiabatic effects.

For this, one assumes the star to be forced into a periodic oscillation. If non-adiabatic processes are taken into account, periodicity can only be maintained if, during each cycle, energy is added to or removed from the whole star. If energy has to be added to maintain a periodic oscillation, the star is damped; if energy has to be removed, it is excited. In order to determine the energy necessary for maintaining a periodic pulsation one defines the energy integral.

### 39.3 The Energy Integral

Suppose we want to make a star undergo periodic radial pulsations. If it is vibrationally unstable, then during each cycle a certain amount $W$ of energy has to be taken out to maintain periodicity. If the star is vibrationally stable, the energy $-W$ has to be fed into the star during each period to avoid a damping of the amplitude. In both cases $W$ is the energy to be taken out to overcome excitation or damping. Therefore, if the star is left alone, $W > 0$ *gives amplitudes increasing in time (excitation) while for $W < 0$ the oscillation is damped.*

To determine $W$ we consider a shell of mass $dm$ which gains the energy $dq/dt$ per units of mass and time. The energy gained per unit mass per cycle is the integral of $(dq/dt)dt$ taken over one cycle. Therefore the energy

$$dW = dm \oint \frac{dq}{dt} dt \qquad (39.10)$$

has to be taken out of the mass shell to maintain periodicity. If we replace $dq/dt$ by

$$\frac{dq}{dt} = -\cos \omega t \frac{d(l_0 \lambda)}{dm} \quad , \qquad (39.11)$$

and if we integrate over all mass shells, we have

$$W = -\int_0^M dm \oint \cos \omega t \frac{d(l_0 \lambda)}{dm} dt \quad . \qquad (39.12)$$

It is obvious that this integral vanishes: in the linear approximation there is neither damping nor excitation.

However, owing to a trick invented by Eddington it is still possible to determine the second-order quantity $W$ with the help of solutions of the first-order theory. Since in the adiabatic case the eigenvalues are real, the time dependence of $x$, $p$, $\vartheta$, and according to (39.9) that of $\lambda$, can be expressed by the factor $\cos \omega t$.

We first prove that

$$\oint \frac{dq}{dt} dt = \oint \vartheta \frac{dq}{dt} \cos \omega t \, dt \quad , \qquad (39.13)$$

up to second order. Indeed, since the specific entropy $s$ is a state variable, the integral of $ds$ over one cycle vanishes exactly. We now write $ds = dq/T$. Since we use only solutions of the adiabatic case, we can consider the variation of $T$ as real and can write $T = T_0(1 + \vartheta_{\text{ad}} \cos \omega t)$, which is correct in the first order. With the (real) adiabatic solutions $x_{\text{ad}}$, $p_{\text{ad}}$ and $\vartheta_{\text{ad}}$ according to (39.9) $\lambda$ also is real, and therefore $dq/dt$ is real too, as can be seen from (39.12). Therefore

$$0 = \oint \frac{ds}{dt} dt = \oint \frac{1}{T_0}(1 - \vartheta_{\text{ad}} \cos \omega t)\frac{dq}{dt} dt$$

$$= \frac{1}{T_0} \oint \frac{dq}{dt} - \frac{1}{T_0} \oint \vartheta_{\text{ad}} \cos \omega t \frac{dq}{dt} dt \quad . \qquad (39.14)$$

This equation is exact in the second order. It therefore proves (39.13). Should the integral on the left of (39.13) vanish in the first order, its value in the second order is given by the integral on the right of (39.13), which does not vanish. We can therefore write from (39.10) by using (39.11)

$$W = \int_0^M dm \oint \vartheta_{\text{ad}} \cos \omega t \frac{dq}{dt} dt = - \int_0^M dm \oint \vartheta_{\text{ad}} \frac{d(l_0 \lambda)}{dm} \cos^2 \omega t \, dt$$

$$= - \int_0^M dm \oint \left[ \vartheta_{\text{ad}} \lambda \frac{dl_0}{dm} + l_0 \vartheta_{\text{ad}} \frac{d\lambda}{dm} \right] \cos^2 \omega t \, dt \quad . \tag{39.15}$$

The time dependence of the real part is $\cos^2 \omega t$, which integrated over $2\pi$ gives $\pi/\omega$. With $dl_0/dm = \varepsilon_0$ we therefore obtain

$$W = -\frac{\pi}{\omega} \left[ \int_0^M \vartheta_{\text{ad}} \lambda \varepsilon_0 \, dm + \int_0^M l_0 \vartheta_{\text{ad}} \frac{d\lambda}{dm} \, dm \right] \quad . \tag{39.16}$$

In fact we see that only second-order terms ($\sim \vartheta_{\text{ad}} \lambda$ and $\sim \vartheta_{\text{ad}} d\lambda/dm$) appear in the expression for $W$. We can now solve the adiabatic equations, insert the resulting $\vartheta_{\text{ad}}$, differentiate $\lambda$ given in (39.9), and determine $W$ from (39.16).

### 39.3.1 The $\kappa$ Mechanism

We consider here regions of the star in which no energy generation takes place ($\varepsilon_0 = 0$) and therefore in which $l_0$ = constant. Since the adiabatic equations for the determination of $x$, $p$, $\vartheta$ are linear and homogeneous, the solutions are determined only up to a common factor. We choose it here such that $x_{\text{ad}} = 1$ at the surface. We further choose the initial point of time such that the maximal expansion of the surface is at $t = 0$. Then the first equation of (25.17) can be written $r = r_0(1 + x_{\text{ad}} \cos \omega t)$, and for $x > 0$ (expansion) the variations $\vartheta_{\text{ad}}$ and $p_{\text{ad}}$ are certainly $< 0$ there. Since, for the fundamental, $\vartheta_{\text{ad}}(< 0)$ does not change sign throughout the star, one can immediately see from (39.16) that a region where $\lambda$ increases outwards ($d\lambda/dm > 0$) gives a positive contribution to $W$: such a region has an excitational effect on the oscillation, while regions with $d\lambda/dm < 0$ have a damping influence. The last two terms on the right of (39.9) together with $\vartheta_{\text{ad}} = \nabla_{\text{ad}} p_{\text{ad}}$ can be written as

$$4\nabla_{\text{ad}} \, p_{\text{ad}} - (\kappa_P + \nabla_{\text{ad}} \kappa_T) \, p_{\text{ad}} \quad . \tag{39.17}$$

Note that the term in parenthesis is identical with a term we encountered in (39.5) for the piston model. If for the sake of simplicity we assume $\kappa_P$, $\kappa_T$, $\nabla_{\text{ad}}$ to be constant and observe that, for the fundamental, $p_{\text{ad}} < 0$ increases inwards, then for $\kappa_P + \nabla_{\text{ad}} \kappa_T > 0$ the term $-(\kappa_P + \nabla_{\text{ad}} \kappa_T) p_{\text{ad}} > 0$ gives a contribution that helps to increase $\lambda$ in an inward direction. This has a stabilizing effect. The term $4\nabla_{\text{ad}} p_{\text{ad}} < 0$ in (39.17) decreases with $p_{\text{ad}}$ in an outward direction and has a damping effect independently of $\kappa$. This damping corresponds to the effect of the leak in the piston model.

The $\kappa$ mechanism is responsible for several groups of variable stars. Before we discuss its effect on real stars we shall first deal with the other mechanism that can maintain stellar pulsations.

### 39.3.2 The $\varepsilon$ Mechanism

The terms in the energy integral discussed in §39.3.1 appear everywhere in a star where radiative energy transport occurs. However, there we have excluded nuclear energy generation, which can also be modulated by the oscillations. To investigate its influence we now concentrate on the terms which come from $\varepsilon$. If we put $l_0(d\lambda)/dm$ equal to the perturbation of the energy generation rate $\varepsilon$: $\varepsilon_0(\varepsilon_P p_{ad} + \varepsilon_T \vartheta_{ad}) = \varepsilon_0(\varepsilon_P + \nabla_{ad}\varepsilon_T)p_{ad}$, we find from (39.16) that

$$
W_\varepsilon = -\frac{\pi}{\omega}\left[ \int_0^M \vartheta_{ad}\lambda\varepsilon_0\, dm + \int_0^M \vartheta_{ad}\varepsilon_0(\varepsilon_P + \nabla_{ad}\varepsilon_T)\, p_{ad}\, dm \right]
$$

$$
= -\frac{\pi}{\omega}\int_0^M \vartheta_{ad}[\lambda + (\varepsilon_P + \nabla_{ad}\varepsilon_T)\, p_{ad}]\varepsilon_0\, dm \quad . \tag{39.18}
$$

Here we again find the excitation mechanism working if $\varepsilon_P + \nabla_{ad}\varepsilon_T > 0$, which is already known to us from the piston model of §39.1. All terms in the integral (39.18) contribute to the energy integral only in the very interior, where $\varepsilon_0 \neq 0$. Since the amplitudes of the eigenfunctions there are normally small compared to their values in the outer regions, one often ignores the contribution of the energy generation and instead of $W$ computes $W_\kappa = W - W_\varepsilon \approx W$. We come back to the case where $W_\varepsilon$ becomes important in §39.5.

## 39.4 Stars Driven by the $\kappa$ Mechanism – The Instability Strip

If one has determined the adiabatic amplitudes for a given stellar model, one can derive $\lambda$ from (39.9) and evaluate $W$ according to (39.16). We shall first describe the influence of different layers.

In the outer layers, where deviations from adiabaticity are biggest, the $\kappa$ mechanism and the damping term $4\nabla_{ad}p_{ad}$ in (39.17) become important and the sign of $(\kappa_P + \nabla_{ad}\kappa_T)$ determines whether the $\kappa$ mechanism acts to damp or to excite. To illustrate this it is useful not only to plot on a $\lg P - \lg T$ diagram lines of constant opacity, but also to indicate at each point the slope given by $\nabla_{ad} = (d\lg T/d\lg P)_{ad}$ as in Fig. 39.1. The $\kappa$ mechanism provides excitation if one comes to higher opacities when going along the slope towards higher pressure. For a monatomic gas one has $\nabla_{ad} = 0.4$. However, ionization reduces $\nabla_{ad}$ appreciably (see Fig. 14.1b), which according to Fig. 39.1 favours instability. This is easily seen for a simple Kramers opacity with $\kappa_P = 1$ and $\kappa_T = -4.5$: then the decisive term $(\kappa_P + \nabla_{ad}\kappa_T)$ is $-0.8$ for $\nabla_{ad} = 0.4$, while it is $\geq 0$ for $\nabla_{ad} \leq 0.222$.

In the near-surface layers of a star with an effective temperature of about 5000 K there are two regions where ionization, together with a suitable form of the function $\kappa = \kappa(P,T)$, acts in the direction of instability. The outer one is quite close to the surface, where hydrogen is partially ionized, followed immediately by the first ionization of helium (see Fig. 14.1, which is plotted for the sun). Below this ionization zone, $\nabla_{ad}$ goes back to its standard value of 0.4. But still deeper another region of excitation occurs caused by the second ionization of helium. This

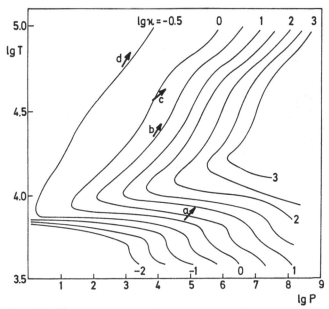

**Fig. 39.1.** Lines of constant opacity $\kappa$ in the $\lg P$–$\lg T$ plane (all values in cgs). Four arrows are shown that indicate the direction in which a mass element moves during adiabatic compression. For the arrows labelled $a$, $b$, and $d$ the direction is given by $\nabla_{\mathrm{ad}} = 0.4$. In case $a$ the arrow points in the direction of increasing $\kappa$, i.e. the $\kappa$ mechanism has a "driving" effect on pulsations. In cases $b$ and $d$ the arrows point in the direction of decreasing $\kappa$, indicating a "damping" (or almost neutral) effect on pulsation. In case $c$ the direction of the arrow is different from that of the other ones, since $\nabla_{\mathrm{ad}}$ is here reduced by the second ionization of helium. Because of this reduction, the arrow points in the direction of increasing $\kappa$ and this ionization region can contribute considerably to the excitation of pulsations in Cepheids

turns out to be the region which contributes most to instability. In still deeper layers the $\kappa$ mechanism has a damping effect, but their influence is very small, since the oscillations become more adiabatic the deeper one penetrates into the star. For an estimate of the right depth of the HeII ionization zone, see COX (1967) and Sect. 27.7 of COX, GUILI (1968).

In Fig. 39.2 the exciting and damping regions of the outer layers of a $\delta$ Cephei star of $7M_\odot$ are shown. For a star right in the middle of the Cepheid strip the "local" energy integral

$$w(m) = -\int_m^M dm \oint \cos \omega t \frac{dl_0 \lambda}{dm} dt \qquad (39.19)$$

is plotted as a function of depth in Fig. 39.3, where $\lg P$ has been used as a measure of the depth. There one can see which regions excite the oscillations ($dw/d\lg P > 0$) and which have a damping effect ($dw/d\lg P < 0$). According to (39.12) $w(0) = W$.

In order that excitation wins over damping it is necessary that the zones of ionization, which provide the excitation, contain a sufficient part of the mass of the star. This means that these zones have to be situated at suitable depths, and since ionization is mainly a function of temperature, we can conclude that it is essentially

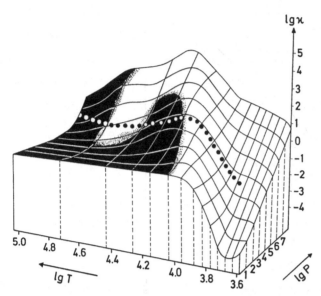

**Fig. 39.2.** An opacity surface ("$\kappa$ mountain") for the outer layers of a star as in Fig. 17.5. But this time the dependence with respect to $P$ (in dyn cm$^{-2}$) and $T$ (in K) is shown. The dotted line corresponds to the stratification inside a Cepheid of $7M_{\odot}$. The white areas of the "mountain" indicate regions which excite the pulsation, the black ones those which damp it. The excitation in the region of $\lg T \approx 4.6$ is due to the second ionization of helium

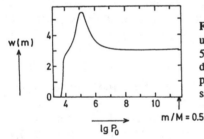

**Fig. 39.3.** The "local" energy integral $w(m)$ (in arbitrary units) as defined in (39.19) for a star of $7M_{\odot}$ and $T_{\mathrm{eff}} = 5300$ K as a function of the unperturbed pressure $P_0$ (in dyn cm$^{-2}$). $w(m)$ increases in regions which excite the pulsation, and falls in those regions which damp the pulsation. (After BAKER, KIPPENHAHN, 1965)

a question of the surface temperature that decides whether a star is vibrationally stable or unstable via the $\kappa$ mechanism.

Let us compare stellar models of the same mass (say in the range 5 to $10M_{\odot}$), of roughly the same luminosity, and consider values for the effective temperature which range from the main sequence to the Hayashi line. At the main sequence and in some range to the right of it, the outer layers of the stars are too hot: hydrogen is fully ionized far up into the atmosphere and even the second ionization of helium is almost complete up to the photosphere. Therefore the $\kappa$ mechanism due to ionization as discussed in § 39.3.1 does not provide much excitation. The main contribution to $W$ comes from the layers which are in the region of the $\lg P$–$\lg T$ plane of Fig. 39.1 where the $\kappa$ mechanism has a damping effect. Therefore the pulsation of such hot stars is damped. But the smaller the effective temperature, i.e. the further to the right in the HR diagram, the deeper inwards are the zones of partial ionization of H and He. Then a higher percentage of the stellar matter lies in the regions of

414

excitation shown in Fig. 39.2. At effective temperatures below about 6300 K the ionization zones are located such that their excitation overcomes the damping of the other layers: such stars start to pulsate with increasing amplitude. This critical temperature, which decreases slightly with increasing luminosity, defines the left ("blue") border of an instability region in which $W > 0$. This border coincides roughly with the left border of the strip in which the observed Cepheids are located.

When considering models with still lower effective temperatures, one has to keep in mind that (39.9) only holds in *radiative* regions. To determine the influence of convective layers a theory of time-dependent convection would be necessary. In particular such a theory should tell us whether in a given convective layer the energy transport is less or more efficient when the star is compressed. It may be that convective transport in a pulsating star provides so much damping that all such models are quite stable. But although several attempts have been made to extend the mixing-length theory correspondingly, there is at present no reliable time-dependent theory for convection. Therefore we can only state that the energy integral $W \approx W_\kappa$ becomes unreliable if convection becomes important in the layers where the $\kappa$ mechanism would be effective. This is the case particularly for stars close to the Hayashi line. Consequently predictions of the right ("red") border of the instability strip are not reliable.

Anyway, there is an instability strip with a probable width of a few $10^2$ K, not too far from, and roughly parallel to, the Hayashi line, extending through almost all of the HR diagram. All stellar models evolving into this strip will become vibrationally unstable via the $\kappa$ mechanism and start to pulsate. In order to predict that we can observe a corresponding pulsating star, the passage through the strip has to be slow enough.

This is fulfilled for models of typically $5 \ldots 10 M_\odot$, which during the phase of helium burning loop away from and back to the Hayashi line, thereby passing through the instability strip at least twice. These passages, in which models represent the classical Cepheids, are discussed in detail in § 31.3. Depending on $M$, the passages occur at quite different luminosities: the larger $M$, the higher $L$. Using the adiabatic approximation one can easily determine the periods of the fundamental for models of very different $L$ inside the instability strip. In this way one obtains a theoretical *period–luminosity relation* that is in satisfying agreement with the observed one. It is interesting to note that the passages through the instability strip do not follow lines of $R = $ constant. Since the radius and therefore the mean density changes, the period–density relations predict a certain amount of change (in both directions) of the period of a Cepheid, which might just be detectable with modern instruments.

Of much smaller mass are the helium-burning stars located on the horizontal branches of the HR diagrams of globular clusters. Where these branches intersect the downward continuation of the instability strip, one finds the RR Lyrae stars (§ 32.7) (VAN ALBADA, BAKER, 1971, 1973). Like the classical Cepheids these are pulsating stars driven by the $\kappa$ mechanism. It seems, however, as if some of them oscillate in the first overtone. For a review see IBEN (1974).

Even further down in the HR diagram, in the region of the main sequence, the instability strip is marked by another group of observed pulsating stars, the so-called $\delta$ Scuti stars or dwarf Cepheids.

Above the location of the RR Lyrae stars in the HR diagram of globular clusters one sometimes finds stars which lie in the instability strip and are therefore pulsating: the W Virginis stars (Fig. 32.10). In contrast to the classical Cepheids, which belong to population I, these stars are of population II. It is not surprising that they do not obey the same period–luminosity relation as Cepheids. According to the evolutionary considerations of § 32.8 they are low-mass stars in an evolutionary stage later than that of the horizontal branch. They obviously have lower masses than the Cepheids, which have travelled more or less horizontally from the main sequence into the instability strip. Let us assume that at the same point inside the instability strip there are two stars, a population I star of, say, $7 M_\odot$ and a population II star of, say, $0.8 M_\odot$. The $\kappa$ mechanism will make both of them pulsate. Being at the same point in the HR diagram, the two stars have the same radii. Therefore the population II star has the lower mean density and according to the period–density relation a longer period than the population I star, although their luminosities are the same. Since the luminosity increases with the period, it follows that pulsating population I stars have a higher luminosity than pulsating population II stars of the same period. In the history of astronomy the clarification of this difference between the two period–luminosity relations caused the revision of the cosmic distance scale by W. Baade in 1944. This increase of the cosmic distance scale amounted to no less than a factor of 2, which caused the comment "The Lord made the universe – but Baade doubled it".

Up to now we have based our considerations on a linear quasi-adiabatic approximation. In the linear theory the amplitude of the solution is not determined and the time dependence is given by almost sinusoidal oscillations with amplitudes growing or decreasing very slowly in time. In reality a vibrationally unstable star would start to oscillate with increasing amplitudes until the oscillations had grown so much that they could not be described by a linear theory any more. Once the non-linear terms in the equations have become important, they have the effect of limiting the increase of amplitudes and causing a time dependence of the solutions which differs considerably from sinusoidal behaviour. Indeed the light curves of most of the observed pulsating stars have constant amplitude and are far from being sinusoidal.

Attempts have been made to reproduce the observed light curves of Cepheids by solving the non-linear equations numerically with varied parameters. A special goal was to determine the masses of Cepheids by comparing their observed light curves with computed ones (see CHRISTY, 1975). This comparison seems to indicate lower masses for Cepheids than expected from evolution theory (compare the discussion in § 31.3). The discrepancy between pulsational and evolutionary mass is still an open question (see, for instance, SIMON, 1987).

Besides the linearization of the equations, we have additionally simplified the problem of pulsations by applying the quasi-adiabatic approximation. With some more effort, however, one can also solve the full set of linear *non-adiabatic* equations. These four equations demand four linear boundary conditions. If they are properly chosen, one obtains one complex eigenvalue $\omega$. Since the time dependence is given by $\exp(i\omega t)$, the imaginary part $\omega_I$ of $\omega$ determines vibrational stability. The energy integral (39.12), computed with the function $\lambda$ obtained from (39.9), is connected to $\omega_I$ when one is close to the adiabatic case (BAKER, KIPPENHAHN,

1962). In most cases the quasi-adiabatic approximation seems to be sufficient. If, however, pure helium stars cross the instability strip, the oscillations are far from being adiabatic, and therefore the quasi-adiabatic approximation becomes very unreliable. This can become important, for instance, if the oscillations of stars of the type R Coronae Borealis are being investigated (WEISS-RÖMER, 1987).

## 39.5 Stars Driven by the $\varepsilon$ Mechanism

In most stars the $\varepsilon$ mechanism discussed in § 39.1 and § 39.3.2 cannot overcome the damping, the reason being that it only works in the central regions of the stars where nuclear energy is released. But there the amplitudes of the oscillations are usually very small compared to the amplitudes in the near-surface regions, which – if the star is not in the instability strip – damp the oscillations by way of the $\kappa$ mechanism.

Figure 38.3 shows that for polytropes for which the radiation pressure can be neglected, the amplitude ratio $x_{\text{centre}}/x_{\text{surface}}$ is small, while it increases with decreasing $\varphi$ until the ratio becomes 1 for $\varphi = 0$ (negligible gas pressure). Since the integrand of the energy integral is quadratic in the amplitudes of the oscillations, we can expect that the $\varepsilon$ mechanism becomes more important the larger the fraction of the radiation pressure.

This is of importance at the upper end of the hydrogen main sequence (§ 22.4), because for such stars the ratio of radiation pressure to gas pressure strongly increases with $M$. Numerical calculations with realistic stellar models instead of polytropes indicate that the $\varepsilon$ mechanism makes the main-sequence stars pulsate if their mass exceeds a critical value of about $60M_\odot$ (SCHWARZSCHILD, HÄRM, 1959); this value depends slightly on the chemical composition.

Why, then, do we not see pulsating stars in the extension of the main sequence towards higher luminosities? Non-linear pulsation calculations (APPENZELLER, 1970, ZIEBARTH, 1970) indicate that the amplitudes would grow until, with each cycle, a thin mass shell is thrown into space. This would continue until the total mass is reduced to the critical mass of, say, $60M_\odot$. Then the pulsation would stop.

Correspondingly the onset of a vibrational instability due to the $\varepsilon$ mechanism limits the helium main sequence towards large $M$ (see § 23.1). The critical upper mass for helium stars depends on the content of heavier elements and lies between $7 \ldots 8M_\odot$ (BOURY, LEDOUX, 1965).

# § 40 Non-radial Stellar Oscillations

We use spherical coordinates $r, \vartheta, \varphi$ and describe the velocity of a mass element by a vector $v$ having the components $v_r$, $v_\vartheta$, $v_\varphi$. For the radial pulsations treated in the foregoing sections, the velocity has only one non-vanishing component, $v_r$, which depends only on $r$. This is so specialized a motion that one might wonder why a star should prefer to oscillate this way at all. In fact it is easier to imagine the occurrence of perturbations that are *not* spherically symmetric, for example those connected with turbulent motions or local temperature fluctuations. They can lead to non-radial oscillations, i.e. oscillatory motions having in general non-vanishing components $v_r$, $v_\vartheta$, $v_\varphi$, all of which can depend on $r$, $\vartheta$, and $\varphi$. It is obvious that the treatment of the more general non-radial oscillations is much more involved than that of the radial case, but they certainly play a role in observed phenomena (see § 40.4). We will limit ourselves to indicating a few properties of the simplest case: small (linear), adiabatic, poloidal-mode oscillations. For more details see, for instance, COX (1976, 1980), UNNO et al. (1979).

## 40.1 Perturbations of the Equilibrium Model

The unperturbed model (subscript 0) is assumed to be spherically symmetric, in hydrostatic equilibrium ($\varrho_0 \nabla \Phi_0 + \nabla P_0 = 0$) and at rest (velocity $v_0 = 0$). We now consider perturbations which shift the mass elements over very small distances. For any mass element at $r$, $\vartheta$, $\varphi$, the displacement relative to its equilibrium position is described by the vector $\boldsymbol{\xi}$ with the components $\xi_r$, $\xi_\vartheta$, $\xi_\varphi$, which, in general, depend on $r$, $\vartheta$, $\varphi$, $t$. Owing to this displacement, such variables as pressure, density or gravitational potential will change. This can be described either in a Lagrangian form (changes inside the displaced element) denoted by

$$P = P_0 + DP \quad , \quad \varrho = \varrho_0 + D\varrho \quad , \quad \Phi = \Phi_0 + D\Phi \quad , \quad v = d\boldsymbol{\xi}/dt \qquad (40.1)$$

or as Eulerian perturbations (local changes), which we write as

$$P = P_0 + P' \quad , \quad \varrho = \varrho_0 + \varrho' \quad , \quad \Phi = \Phi_0 + \Phi' \quad , \quad v = \partial\boldsymbol{\xi}/\partial t \qquad (40.2)$$

and which are preferred in the following. The linearized connection between the two types of perturbations of any quantity $q$ is

$$Dq = q' + \boldsymbol{\xi} \cdot \nabla q_0 = q' + \xi_r \frac{\partial q_0}{\partial r} \quad . \qquad (40.3)$$

(The last equality holds since $\nabla q_0$ is a purely radial vector.) Together with $\boldsymbol{\xi}$, all perturbations are functions of $r$, $\vartheta$, $\varphi$, and $t$. We have to perturb the Poisson equation and the equations of motion and continuity.

The acceleration due to gravity,

$$g = -\nabla\Phi \quad , \tag{40.4}$$

and its perturbations $Dg$ or $g'$ are given by the potential $\Phi$. Poisson's equation (2.23), together with (40.2), yields after linearization

$$\nabla^2\Phi' = 4\pi G\varrho' \quad . \tag{40.5}$$

The equation of motion for the moving mass element is

$$\varrho\frac{dv}{dt} = \varrho g - \nabla P \quad . \tag{40.6}$$

With (40.1) this gives the linearized equation

$$\varrho_0\frac{d^2\xi}{dt^2} = g_0 D\varrho + \varrho_0 Dg - \nabla(DP) \quad , \tag{40.7}$$

where the forces on the right-hand side are measured relative to equilibrium. From (40.7) and (40.3), the Eulerian equation of motion follows:

$$\varrho_0\frac{\partial^2\xi}{\partial t^2} = -\varrho_0\nabla\Phi' - \varrho'\nabla\Phi_0 - \nabla P' \quad . \tag{40.8}$$

On the right-hand side of this expression, the restoring force is represented by 3 terms, the last of which is due to pressure variations, while the others are gravitational terms. The first stems from the changed gravitational acceleration and is usually small compared with the second, which is essentially a buoyancy term.

The equation of continuity, $\partial\varrho/\partial t + \nabla(\varrho v) = 0$, after insertion of (40.1) and linearization, takes the form

$$D\varrho + \varrho_0\nabla\cdot\xi = 0 \quad , \tag{40.9}$$

which together with (40.3) is transformed to

$$\varrho' + \xi\cdot\nabla\varrho_0 + \varrho_0\nabla\cdot\xi = 0 \quad . \tag{40.10}$$

We do not have to consider the equations of energy and energy transfer, since we assume the changes to be adiabatic. The condition for adiabaticity in Lagrangian form is simply [cf. (38.6)]

$$\frac{DP}{P_0} = \gamma_{\mathrm{ad}}\frac{D\varrho}{\varrho_0} \quad , \tag{40.11}$$

which is transformed by (40.3) to the Eulerian condition

$$P' + \xi\cdot\nabla P_0 = \frac{P_0}{\varrho_0}\gamma_{\mathrm{ad}}(\varrho' + \xi\cdot\nabla\varrho_0) \quad . \tag{40.12}$$

We shall see below that the equations derived for the perturbations constitute a fourth-order system. So we need in addition 4 boundary conditions.

At the surface, we require continuity of the Lagrangian variation of $\nabla\Phi$ through the surface, and a vanishing pressure perturbation, $DP = 0$, such that no forces are transmitted to the outside. These outer boundary conditions are then written as

$$\left(\frac{\partial\Phi'}{\partial r} + \boldsymbol{\xi}\cdot\nabla\Phi_0\right)_{\text{in}} = \left(\frac{\partial\Phi'}{\partial r} + \boldsymbol{\xi}\cdot\nabla\Phi_0\right)_{\text{out}} \quad ,$$

$$P' + \boldsymbol{\xi}\cdot\nabla P_0 = 0 \quad . \tag{40.13}$$

At the centre, the perturbations are required to be regular, which also yields 2 boundary conditions, say,

$$P' = 0 \quad , \quad \Phi' = 0 \quad . \tag{40.14}$$

## 40.2 Normal Modes and Dimensionless Variables

The perturbations are to be determined from (40.5, 8, 10, 12) and (40.13, 14). Aside from the perturbations $\boldsymbol{\xi}$, $\Phi'$, $P'$, $\varrho'$, these equations contain only quantities of the unperturbed equilibrium model, for which we now drop the subscript 0.

We specify the perturbations $q(r, \vartheta, \varphi, t)$ in the usual way, assuming that all of them depend on the variables as factorized in the following separation ansatz

$$q(r, \vartheta, \varphi, t) = \tilde{q}(r) Y_l^m(\vartheta, \varphi)\, e^{\mathrm{i}\omega t} \quad . \tag{40.15}$$

The perturbations are supposed to vary on all concentric spheres like the well-known spherical harmonics $Y_l^m(\vartheta, \varphi)$ of degree $l$ and order $m$ (see, for instance, KORN, KORN, 1968). In time they vary periodically with frequency $\omega$. The dependence on $r$ is comprised in the function $\tilde{q}(r)$. The $Y_l^m$ are solutions of

$$\frac{\partial^2 Y_l^m}{\partial\vartheta^2} + \mathrm{ctg}\,\vartheta\frac{\partial Y_l^m}{\partial\vartheta} + \frac{1}{\sin^2\vartheta}\frac{\partial^2 Y_l^m}{\partial\varphi^2} + l(l+1)Y_l^m = 0 \quad , \tag{40.16}$$

and can be written as

$$Y_l^m = K(l, m) P_l^m(\cos\vartheta)\cos m\varphi \quad , \tag{40.17}$$

where $K$ is a coefficient depending on $l$, $m$, and $P_l^m(x)$ are the associated Legendre functions. Degree and order are specified by choosing the integers

$$l > 0 \quad , \quad m = -l, \ldots, +l \quad . \tag{40.18}$$

A change of $l$, $m$ changes the angular variation on concentric spheres. A few examples are illustrated in Fig. 40.1. Generally speaking, the larger $l$, the more node lines ($Y = 0$) are present, and the smaller are the enclosed areas in which the matter moves in the same radial direction (e.g. outwards). For example, $l = 2$ is a quadrupole oscillation, $l = 1$ a dipole oscillation, and $l = 0$ the special case of the earlier discussed radial pulsations.

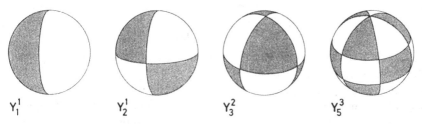

Fig. 40.1. Node lines of some spherical harmonics $Y_l^m$. Corresponding oscillations would show, for example, outward motion in the shaded areas and inward motion in the other parts of the sphere

We shall discuss here only perturbations of the form (40.15). The resulting oscillations of that form are called *poloidal modes*. It should be mentioned that there exists the additional class of *toroidal modes*, which do not have the form (40.15); they are independent of time and have purely transverse displacements (without radial components).

In order to get an overview of the problem, it is convenient to introduce dimensionless variables, for example,

$$\eta_1 = \frac{1}{r}\xi_r \quad ; \quad \eta_2 = \frac{1}{gr}\left(\frac{P'}{\varrho} + \Phi'\right) \quad ; \quad \eta_3 = \frac{1}{gr}\Phi' \quad ; \quad \eta_4 = \frac{1}{g}\frac{\partial \Phi'}{\partial r} \quad . \tag{40.19}$$

Since they are proportional to $P'$, $\varrho'$, $\Phi'$, we have according to (40.15)

$$\eta_j = \tilde{\eta}_j(r)Y_l^m(\vartheta, \varphi)\,e^{i\omega t} \quad , \quad j = 1, 2, 3, 4 \quad . \tag{40.20}$$

The density perturbation, which does not appear in (40.19), will always be replaced by terms in $P'$ (and then in $\eta_2 - \eta_3$) via (40.12).

The equation of motion (40.8), together with (40.19), becomes after some algebra:

$$-\frac{\omega^2}{g}\boldsymbol{\xi} = [W(\eta_1 - \eta_2 + \eta_3) + (1 - U)\eta_2]\,\boldsymbol{e}_r - r\boldsymbol{\nabla}\eta_2 \quad , \tag{40.21}$$

where $\boldsymbol{e}_r$ is a unit vector in the $r$ direction. The dimensionless quantities

$$
\begin{aligned}
U &:= \frac{r}{m}\frac{\partial m}{\partial r} = \frac{1}{g}\frac{\partial(gr)}{\partial r} \quad , \\
V &:= -\frac{r}{P}\frac{\partial P}{\partial r} = \frac{g\varrho r}{P} \quad , \\
W &:= \frac{r}{\varrho}\frac{\partial \varrho}{\partial r} - \frac{r}{P\gamma_{\mathrm{ad}}}\frac{\partial P}{\partial r}
\end{aligned}
\tag{40.22}
$$

are to be taken from the equilibrium model. Equation (40.21) is easily verified. Its radial component will be treated later, while the tangential components

$$\frac{\omega^2}{g}\xi_\vartheta = \frac{\partial \eta_2}{\partial \vartheta} \quad , \quad \frac{\omega^2}{g}\xi_\varphi = \frac{1}{\sin \vartheta}\frac{\partial \eta_2}{\partial \varphi} \quad , \tag{40.23}$$

are used immediately in the equation of continuity. But first we replace $\omega$ by a dimensionless frequency $\sigma$, setting

$$\frac{\omega^2 r}{g} = C\sigma^2 \quad , \quad C = \left(\frac{r}{R}\right)^3 \frac{M}{m} \quad , \quad \sigma^2 = \omega^2 \frac{R^3}{GM} \quad . \tag{40.24}$$

This frequency is scaled by a time of the order of the hydrostatic adjustment time, or of the period of the radial fundamental.

When transforming the equation of continuity (40.10), we evaluate the term $\nabla \cdot \boldsymbol{\xi}$ by using (40.23), introduce (40.20), and eliminate all derivatives of $Y_l^m$ with respect to $\vartheta$ and $\varphi$ with the help of (40.16). Then all terms are proportional to $Y_l^m \exp(\mathrm{i}\omega t)$, which can thus be dropped. One finally obtains

$$r\frac{\partial \tilde{\eta}_1}{\partial r} = \left(3 - \frac{V}{\gamma_{\mathrm{ad}}}\right)\tilde{\eta}_1 + \left[\frac{l(l+1)}{C\sigma^2} + \frac{V}{\gamma_{\mathrm{ad}}}\right]\tilde{\eta}_2 - \frac{V}{\gamma_{\mathrm{ad}}}\tilde{\eta}_3 \quad . \tag{40.25}$$

Similarly one finds from the radial component of the equation of motion (40.21)

$$r\frac{\partial \tilde{\eta}_2}{\partial r} = (W + C\sigma^2)\tilde{\eta}_1 + (1 - U - W)\tilde{\eta}_2 + W\tilde{\eta}_3 \quad . \tag{40.26}$$

The next equation is simply obtained by differentiating the definition of $\eta_3$ in (40.19) with respect to $r$, which gives

$$r\frac{\partial \tilde{\eta}_3}{\partial r} = (1 - U)\tilde{\eta}_3 + \tilde{\eta}_4 \quad . \tag{40.27}$$

In the Poisson equation (40.5), after elimination of $\varrho'$ by (40.12), we introduce (40.19) and again use (40.16), arriving at

$$r\frac{\partial \tilde{\eta}_4}{\partial r} = -UW\tilde{\eta}_1 + \frac{UV}{\gamma_{\mathrm{ad}}}\tilde{\eta}_2 + \left[l(l+1) - \frac{UV}{\gamma_{\mathrm{ad}}}\right]\tilde{\eta}_3 - U\tilde{\eta}_4 \quad . \tag{40.28}$$

With (40.25–28) we have obtained 4 ordinary, linear differential equations with real coefficients (given by the equilibrium model) for the 4 dimensionless variables $\tilde{\eta}_1, \ldots, \tilde{\eta}_4$. In addition there are 4 algebraic equations arising from the boundary conditions. This constitutes an eigenvalue problem with the eigenvalue $\sigma^2$.

Note that it is the assumption of adiabaticity which has reduced the problem to 4th order in the spatial variables. For the full non-adiabatic case one additionally has to consider the perturbations of the temperature and of the energy-flux vector. The perturbed energy equation contains *first* derivatives with respect to time, which according to (40.15) give terms multiplied by $\mathrm{i}\omega$. Therefore the equations become complex and the non-adiabatic problem is of order 12 in real variables. On the other hand, for $l = 0$ one obtains the adiabatic radial oscillations, for which the problem is reduced to second order.

## 40.3 The Eigenspectra

For adiabatic non-radial oscillations we have obtained an eigenvalue problem of 4th order in the spatial variables and non-linear in the eigenvalue $\omega^2$ (or the dimensionless $\sigma^2$). The problem can be shown to be self-adjoint, so that the eigenfunctions are orthogonal to one another. They have been found to form a complete set if complemented by the toroidal modes.

The eigenvalues obey an extremal principle. The self-adjointness assures that all eigenvalues are real. This means that the motion is either purely periodic ($\omega^2 > 0$, $\omega$ real: dynamical stability) or purely aperiodic ($\omega^2 < 0$, $\omega$ imaginary: dynamical instability).

Neither the equations (40.25–28) nor the boundary conditions contain explicitly the order $m$ of the spherical harmonics. Therefore to each eigenvalue of a given $l$ correspond $2l + 1$ solutions (for the different $m$ values $-l, \dots 0, \dots, +l$). This degeneracy can be removed, for example by centrifugal or tidal forces.

The general discussion is very much complicated by the fact that the eigenvalue $\lambda = \sigma^2$ appears non-linearly in the set (40.25–28). In order to see the typical properties of the eigenspectra, we use an approximation introduced by Cowling, assuming that the perturbation of the gravitational potential can be neglected. We then do not need (40.27, 28) and are left with a second-order problem. This approximation becomes the better, the more the oscillation is limited to the outer layers (e.g. high overtones of acoustic modes with sufficiently large $l$). The second-order problem still contains terms proportional to $\sigma^2$ [from (40.26)] and terms proportional to $1/\sigma^2$ [from (40.25)]. In order to simplify this we consider two asymptotic cases ($\sigma^2 \to \infty$ and $\sigma^2 \to 0$), in both of which the problem becomes of the classical Sturm–Liouville type.

*For large $\sigma^2$* we neglect the terms proportional to $1/\sigma^2$. The only coefficient containing $\sigma$ then is $\sigma^2/c_s^2$, with the velocity of sound given by $c_s^2 = \gamma_{ad}P/\varrho$. This problem has an infinite series of discrete eigenvalues $\lambda_k = \sigma_k^2$ with an accumulation point at infinity. Such oscillations are produced by acoustic waves propagating with $c_s$. They are dominated by pressure variations and are therefore called *p modes*. For sufficiently simple stellar models, they are easily ordered as $p_1, p_2, \dots, p_k$ where $k$ is the number of nodes of their eigenfunction $\xi_r$ between centre and surface. They are analogous to the radial oscillations ($l = 0$), except for the dynamical stability: while the radial fundamental is unstable for $\gamma_{ad} < 4/3$, the $p$ modes are all stable under reasonable conditions.

*For small $\sigma^2$* we neglect the terms proportional to $\sigma^2$. The only coefficient containing $\sigma$ is now $\omega_{ad}^2 l(l + 1)/(\sigma^2 r^2)$, where $\omega_{ad}$ is the Brunt–Väisälä frequency as introduced in §6.2. This problem has an infinite series of eigenvalues $\lambda_k = 1/\sigma_k^2$ with an accumulation point at $\lambda = \infty$, i.e. at $\sigma^2 = 0$. The motions are dominated by gravitational forces and are therefore called *g modes* (again ordered as $g_1, g_2, \dots, g_k$ according to the number $k$ of nodes).

The stability of the $g$ modes depends essentially on $W$, defined in (40.22). This quantity is connected with the problem of convective stability discussed in §6. One can easily verify from (6.18) that the Brunt–Väisälä frequency of an adiabatically oscillating mass element is given by

$$\omega_{ad}^2 = -grW \quad . \tag{40.29}$$

And $rW > 0$ is just the criterion (6.4) for convective instability against adiabatically displaced elements. If in the whole star $W < 0$ (convective stability everywhere), then all $g$ modes are stable ($\sigma^2 > 0$, $\sigma$ real). Such modes are also called $g^+$ modes and are produced by propagating gravity waves. If the star contains a region where $W > 0$ (convective instability), then unstable $g^-$ modes also exist ($\sigma^2 < 0$, $\sigma$ imaginary). So we see that convective stability (instability) coincides with dynamical stability (instability) of non-radial $g$ modes; the onset of convection appears as the manifestation of unstable $g$ modes.

The non-linearity in $\lambda = \sigma^2$ of the full set (40.25–28) implies that the eigenspectrum of stars is a combination of the above-described partial spectra: it contains high-frequency $p$ modes as well as low-frequency $g$ modes, which can be split up into the stable $g^+$ and the unstable $g^-$. Between the $p$ and $g$ modes of relatively simple stars there is another one, called the $f$ mode, since it has no node between centre and surface (like the radial fundamental).

As mentioned above, the stable modes are produced by propagating waves. From the appropriate dispersion relations with horizontal wave numbers $[l(l+1)]^{1/2}/r$ one finds that for propagating acoustic waves $\omega \geq \omega_0 := \frac{1}{2} c_s (d \ln \varrho / dr)$, and for propagating gravity waves $\omega \leq \omega_{\mathrm{ad}}$, where at any place $\omega_0 > \omega_{\mathrm{ad}}$. These conditions define two main regions ($G$ and $A$) of propagation inside a star: one in the deep interior for gravity waves, the other in the envelope for acoustic waves (see Fig. 40.2). These regions act like cavities or resonators, inside which modes can be "trapped". At certain frequencies (the eigenvalues) the propagating waves produce standing waves by reflections at the borders such that they come back in phase with themselves. The simple polytropic model demonstrated in Fig 40.2 is typical for the situation with homogeneous main-sequence stars. When during the evolution the central concentration of the model increases and a chemical inhomogeneity is built up, the maximum of the $G$ region near the core increases far above the minimum of the $A$ region in the envelope. Then the $g_1$ mode can move above the $p_1$ mode, etc.

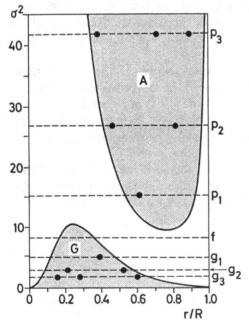

**Fig. 40.2.** Propagation diagram for oscillations with degree $l = 2$ in a polytropic star with index $n = 3$. The square of the dimensionless frequency $\sigma$ is plotted against the distance from the centre. Propagation of acoustic and gravity waves is possible in the shaded regions $A$ and $G$ respectively. For the lowest modes the eigenvalues (*broken lines*) and the positions of the nodes of the eigenfunction $\xi_r$ (*dots*) are indicated. (After SMEYERS, 1984)

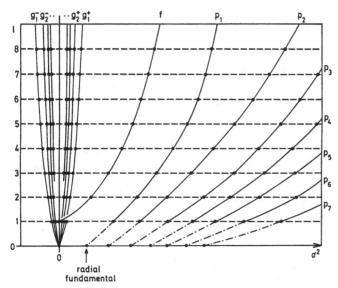

**Fig. 40.3.** In this scheme the dots indicate the eigenvalues $\sigma^2$ (plotted as abscissa) for a few modes of non-radial adiabatic oscillations with different orders $l$ of the spherical harmonics (plotted as ordinate). Eigenvalues for the same type of mode are connected by a solid line. Dot-dashed lines give the connexion to the corresponding radial modes with $l = 0$ ($p_1$ to the radial fundamental, $p_2$ to the first radial overtone, etc.). For $l = 1$ the $f$ mode has $\sigma^2 = 0$ (no oscillatory motion, see text)

When they are close to each other, resonance effects provide that they exchange their properties and avoid an exact coincidence of the eigenvalues (avoided level crossing, as known, say, from quantum mechanics). So the eigenspectra can be rather involved, particularly for evolved stars.

Fig. 40.3 illustrates the eigenspectra for different values of $l$ (degree of the spherical harmonics) for the case of a rather simple star. The radial oscillations are found at $l = 0$. For dipole oscillations ($l = 1$) the $f$ mode must have $\sigma = 0$, since otherwise it would result in an oscillatory motion of the centre of gravity, which is not possible without external forces. However, oscillations having nodes outside the centre are possible for $l = 1$, since then, for example, the core always moves in the opposite direction to the envelope such that the centre of gravity remains at rest. For higher $l$ values the eigenspectra are generally shifted to higher frequencies. The connection between the different $p$ modes and the radial modes as shown in the figure is based on physical considerations, as well as on solutions of (40.25–28) for continuously varying $l$ (where of course only those for integer $l$ have a physical meaning).

## 40.4 Stars Showing Non-radial Oscillations

When applying the above described formalism to models of real stars, a basic question is whether such oscillations in fact proceed adiabatically. Strictly speaking, one would have to test the model for its vibrational stability and look for the imaginary part of $\omega$ derived in a full non-adiabatic treatment. This is, however, so cumbersome that one usually confines oneself to a quasi-adiabatic approximation, similar to that described for the radial case in § 39: the adiabatically calculated eigenfunctions are used to determine an "energy integral", describing the growth or damping rate of the amplitude.

There is a variety of stars and stellar types that are known or suspected to undergo non-radial oscillations. We shall briefly mention a few of them.

The best established group of non-radial oscillators are certain white dwarfs (cf. VAN HORN, 1984), among them the ZZ Ceti variables, which are of type DA. They exhibit periods typically between a few $10^2$ and $10^3$s, often split up into close pairs. These periods are certainly too long for radial oscillations of white dwarfs, but can well be explained by $g^+$ modes. Rotation of the white dwarf splits them up into oscillations with different order $m$. The corresponding gravity waves are "trapped" in a superficial hydrogen layer which, according to its thickness, acts as a resonator for certain modes. They are excited by the $\kappa$ mechanism in zones of partial ionization. Other groups of oscillating white dwarfs, of type DB and very hot ones, have also been found.

The $\beta$ Cephei stars, which are situated somewhat above the upper main sequence, are widely assumed to be non-radial oscillators. Some of them also seem to show the effect of rotational mode splitting. The nature of their oscillations is not yet really understood. Suspects for non-radial oscillations are also found among the $\delta$ Scuti stars and some types of supergiants.

A very interesting example of observed non-radial oscillations is our sun (compare e.g. CHRISTENSEN-DALSGAARD, 1984; DEUBNER, GOUGH, 1984). Detailed spectral investigations of the solar surface have shown that, again and again, areas roughly $10^5$ km across start oscillating in phase for some time. The first detected and best-known oscillations have periods around 5 minutes. They represent standing acoustic waves trapped mainly in a region from somewhere below the photosphere down into the upper convective zone. Power spectra with $\omega$ plotted against the horizontal wave number show clearly that the phenomena contain mode oscillations with very many modes (many degrees $l$ and radial orders $k$). These spectra can be compared with corresponding ones calculated for standard solar models, which are thus tested. For example, the $P$-$\varrho$ stratification in the solar interior determines the variation of the velocity of sound, which is decisive for the existence of standing acoustic waves with certain wave numbers and frequencies. This important new test for the interior structure of the sun (mostly of its envelope, since only the lowest-mode oscillations are noticably affected by the central region) has been called "helioseismology", in analogy to the investigation of the earth's interior by way of seismic waves. For the solar envelope, the information to be derived concerns, for example, the depth of the convection zone, which seems to be more or less as in calculated solar models. There are indications that the central region of the models requires some modification, but not necessarily the same as suggested by the neutrino experiments. Conclusive results are yet to be expected. The excitation of the observed oscillations is unclear; it is possibly due to turbulent velocity fields.

# IX    Stellar Rotation

# §41 The Mechanics of Rotating Stellar Models

The theory of rotating bodies with constant densities (liquid bodies) has been investigated thoroughly by McLaurin, Jacobi, Poincaré, and Karl Schwarzschild. We first start with a summary of their results without deriving them. The reader who wants to go more into the details may use the book by LYTTLETON (1953).

Most of the results have been obtained for solid-body rotation, i.e. for constant angular velocity $\omega$ of the self-gravitating liquid body. In this case the centrifugal acceleration $c$ has a potential, say $c = -\nabla V$ with $V = -s^2\omega^2/2$, where $s$ is the distance from the axis of rotation. If $\Phi$ is the gravitational potential, then according to the hydrostatic equation the total potential $\Psi := \Phi + V$ must be constant on the surface. The main difficulty in determining the surface of a rotating liquid body lies with the gravitational potential, which in turn depends on the form of the surface.

## 41.1 Uniformly Rotating Liquid Bodies

For sufficiently slow rotation with constant angular velocity, the rotating liquid bodies are spheroids (i.e. axisymmetric ellipsoids) called *McLaurin Spheroids*.

In order to examine the behaviour of rotating liquid masses, we define their gravitational energy $E_g$

$$E_g := \frac{1}{2}\int \varrho\Phi \, dV \quad , \tag{41.1}$$

where $\Phi$ is the gravitational potential vanishing at infinity and $dV$ is the volume element. The expression (41.1) is the generalization for non-spherical bodies of the definition (3.3).

Indeed in the spherical case with

$$\frac{d\Phi}{dr} = \frac{Gm}{r^2} \tag{41.2}$$

we have from (3.3)

$$E_g = -G\int_0^R \frac{m\,dm}{r} = -\frac{1}{2}G\int_0^R \frac{d(m^2)}{dm}\frac{1}{r}\,dm$$

$$= -\frac{1}{2}G\frac{M^2}{R} - \frac{1}{2}G\int_0^R \frac{m^2 dr}{r^2}$$

$$= -\frac{1}{2}\frac{GM^2}{R} - \frac{1}{2}\int_0^R \frac{d\Phi}{dr}m\,dr = \frac{1}{2}\int_0^R \Phi\,dm \quad , \tag{41.3}$$

in agreement with the definition (41.1) for the more general (non-spherical) case.

The kinetic energy

$$T := \frac{1}{2} \int v^2 dm \tag{41.4}$$

is supposed to contain only the energy due to the macroscopic rotational motion, but not that due to the thermal motion of the molecules. Let us further define the dimensionless quantity

$$\chi := \frac{\omega^2}{2\pi G \varrho} \quad . \tag{41.5}$$

It is of the order of the ratio of centrifugal acceleration to gravity at the equator and is a measure of the "strength" of rotation.

We now describe some results on the equilibrium configurations and their stability. The derivations and some details of the configurations can be found in the classic book by JEANS (1928) and in that of LYTTLETON (1953).

The shape of McLaurin spheroids is described by the eccentricity $e$ of the meridional cross-section,

$$e^2 = \frac{a^2 - c^2}{a^2} \quad , \tag{41.6}$$

where $a$, $c$ are the major and the minor half axes of the meridional cross-section. A sequence of increasing $e$ leads from the sphere ($e = 0$) to the plane parallel layer ($e = 1$), and one can label each of these configurations by its value of $\chi$. But the correspondence between $e$ and $\chi$ is not unique. For each value of $\chi < 0.2247$ there exist two configurations with different values of $e$. For example, in the limit case of zero rotation with $\chi = 0$, the sphere as well as the infinite plane parallel layer are two possible equilibria, the latter of which obviously is not stable. Along the series of increasing excentricity $e$, neither $\chi$ nor $T$ are monotonic, but one can show that the angular momentum and $E_g$ vary monotonically. Furthermore, $\omega$ does not vary monotonically with the total angular momentum: if we start with a liquid self-gravitating sphere ($e = 0$) and feed in angular momentum, the angular velocity, and with it the excentricity, increases. But once the excentricity exceeds the value of 0.9299, the angular velocity decreases again, even with further increasing angular momentum. The reason for this is that the momentum of inertia increases faster than the angular momentum and therefore $\omega$ must decrease again.

But long before this, namely at $e = 0.8127$ or at $\chi = 0.1868$, the McLaurin spheroids become unstable. At this point the sequence of configurations shows a bifurcation (Fig. 41.1): another branch of stable models occurs which have a quite different shape. They are triaxial ellipsoids, the so-called *Jacobi ellipsoids*. Beyond the point of bifurcation, a McLaurin spheroid is unstable, the Jacobi ellipsoid of the same mass and angular momentum having a lower total (macroscopic kinetic plus gravitational) energy. Therefore, if there is a mechanism like friction which can use up macroscopic energy and transform it into heat, the spheroids become ellipsoids. The transition takes place on the time-scale of friction as defined in § 43.

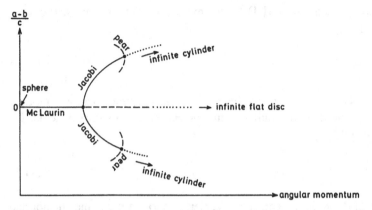

**Fig. 41.1.** Sequences of the McLaurin and Jacobian equilibrium configurations of a rotating incompressible fluid. In this schematic representation, each configuration is characterized by its angular momentum and its value of $(a-b)/c$, where $a$, $b$, $c$ are the 3 axes of an ellipsoid. Solid lines indicate dynamically and secularly stable configurations, broken lines secularly unstable and dotted lines dynamically unstable models. The branches of pear-shaped configurations are also indicated, although they cannot be plotted in a diagram with that ordinate. For more details see LEDOUX (1958)

In analogy to the case of a blob of excess molecular weight (see §6.5) in hydrostatic equilibrium with its surroundings, the motion is controlled by a dissipative process (there heat flow, here friction). One therefore calls the instability of the McLaurin spheroids also *secular*. Instead of the oblateness, one often uses the ratio $\xi := T/|E_g|$, which reaches the value 0.1376 at the point of bifurcation. Stability analysis shows that if $\xi$ exceeds another critical value (of about 0.16), the triaxial ellipsoids also become unstable and then assume a pear-shaped form (see Fig. 41.1).

It should be noted that here we have interpreted sequences of varying dimensionless parameters $e$, $\chi$, $\xi$ as sequences of models with increasing angular momentum, while mass and density were assumed to be constant. Models with the same dimensionless parameters can also be obtained by a sequence of increasing density, while mass and angular momentum are kept constant. In this way one can conclude from the foregoing discussion that a freely rotating body (mass and angular momentum constant) that contracts (density increasing) can start with slow rotation as a McLaurin spheroid, and can then become triaxial and finally pear-shaped. Indeed before the Jacobi ellipsoids become long cigars they become dynamically unstable. An ensuing fission may then split the body in two.

However, one cannot use this scenario to explain the existence of binary stars, since in stars the density increases towards the centre. Then solid-body rotation has different consequences, as we will see in §41.2. Numerical calculations, though, do show that rotating stars also become unstable against non-axisymmetric perturbations when $T/|E_g|$ comes close to 0.14.

## 41.2 The Roche Model

Since the liquid-body approximation ($\varrho$ = constant) is extremely bad for stars, one can go to the other extreme in which practically all gravitating mass is in the centre. In Roche's approximation one assumes that the gravitational potential $\Phi$ is the same as if the total mass of the star were concentrated at the centre. Then $\Phi$ is spherically symmetric:

$$\Phi = -\frac{GM}{r} \; . \tag{41.7}$$

For solid-body rotation, the centrifugal acceleration can again be derived from the potential

$$V = -\frac{1}{2}s^2\omega^2 \; , \tag{41.8}$$

where $s$ is the distance from the axis of rotation. If $z$ is the distance from the equatorial plane, then $r^2 = s^2 + z^2$, and the total potential is

$$\Psi = \Phi + V = -\frac{GM}{(s^2 + z^2)^{1/2}} - \frac{1}{2}s^2\omega^2 \; . \tag{41.9}$$

The acceleration $-\nabla\Psi$ in the co-rotating frame is the sum of gravitational and centrifugal accelerations. A set of surfaces $\Psi$ = constant is plotted in Fig. 41.2. The advantage of the Roche approximation is that the gravitational field is given independently of the rotation. Eccentricity does not affect gravity. In order to investigate the rotating Roche configurations, we consider the surfaces of constant total potential $\Psi$:

$$\frac{GM}{(s^2 + z^2)^{1/2}} + \frac{\omega^2 s^2}{2} = \text{constant} = \frac{GM}{r_{\rm p}} \; , \tag{41.10}$$

where $r_{\rm p}$, the polar radius, is the distance from the centre to the point where the

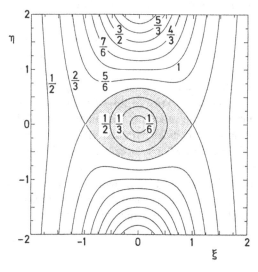

Fig. 41.2. The lines of constant total potential $\Psi$ for the Roche model in the meridional plane. They are labelled by their values of $r_{\rm p}/s_{\rm cr}$. The coordinates are $\xi = s/s_{\rm cr}$, $\eta = z/s_{\rm cr}$. The shaded area is inside the critical surface

surface intersects the axis of rotation (i.e. the value of $z$ for $s = 0$). With the abbreviations

$$a = \frac{1}{r_p} \quad , \quad b = \frac{\omega^2}{2GM} \quad , \tag{41.11}$$

we find for the equipotential surfaces

$$z^2 = \frac{1}{(a - bs^2)^2} - s^2 \quad . \tag{41.12}$$

In the equatorial plane $z = 0$, at the circle $s = s_{cr}$ with

$$s_{cr}^3 = \frac{GM}{\omega^2} \tag{41.13}$$

the gradient of $\Psi$ vanishes. The corresponding critical surface intersects the axis of rotation at $z = \pm 2/3 s_{cr}$ and separates closed surfaces from those going to infinity (Fig. 41.2). In the equatorial plane $z = 0$, gravity dominates inside the critical circle, while outside, the centrifugal acceleration dominates. Both compensate each other exactly at the critical circle. Numerical integration for the volume inside the critical surface gives

$$V_{cr} = 0.1804 \times 4\pi \, s_{cr}^3 \quad . \tag{41.14}$$

Let us now assume that a stellar model just fills its critical volume: $\bar{\varrho} = M/V_{cr}$. We redefine the dimensionless quantity $\chi$ by

$$\chi := \frac{\omega^2}{2\pi G \bar{\varrho}} \quad , \tag{41.15}$$

which is of the order of centrifugal acceleration over gravity at the equator. The model fills its critical volume if $\chi = \chi_{cr} = 0.36075$, as can be obtained from the condition of the balance of centrifugal and gravitational acceleration together with (41.14,15). Rotating models which do not fill their critical volume have $\chi < \chi_{cr}$.

In order to see the rotational behaviour of the Roche model, let us start with very slow rotation so that the stellar surface lies safely within the critical equipotential.

If we speed up the rotation, the volume of the model star will grow, since centrifugal forces "lift" the matter and therefore reduce the effective gravity. We first ignore this effect, assuming that the stellar volume remains unchanged (in spite of the speed-up). Then with increasing $\omega$, according to (41.13, 14), the critical surface will shrink and come closer to the surface of the model. Consequently the model surface becomes more and more oblate until it coincides with the critical surface. In reality the stellar volume will grow as the angular velocity speeds up and the model will reach its critical stage even earlier.

A critically rotating star cannot hold the matter at the equator. What happens if then the angular velocity increases even more? From a first glance at Fig. 41.2 one might expect that the matter can easily escape along equipotential surfaces into infinity. However, one has to keep in mind that the equipotentials plotted there only hold for solid-body rotation. If matter leaving the star at the equator were to be forced, say, by magnetic fields, to co-rotate, it would indeed be swept into space. But if there is no such mechanism, the matter would have to conserve its angular momentum and remain in the neighbourhood of the star. If $\omega$ = constant, the centrifugal acceleration ($\sim s$) dominates over gravity ($\sim s^{-2}$) for large values of $s$. But in the case of constant specific angular momentum ($\omega \sim s^{-2}$), the centrifugal acceleration ($\omega^2 s \sim s^{-3}$) drops more steeply with $s$ than gravity.

We have here considered the case of a star with increasing angular velocity and constant (or increasing) volume. A more realistic case would be that a slowly rotating star contracts. If then its radius decreases, the angular velocity increases like $R^{-2}$ while its critical surface shrinks proportionally to $s_{cr} \sim \omega^{-2/3} \sim R^{4/3}$. The critical surface therefore shrinks faster than the star, which will become more and more oblate until its surface is critical. Then the centrifugal force balances the gravitational one at the equator. With further shrinking the star loses mass, which is left behind as a rotating disk in the equatorial plane. This is similar to Laplace's scenario of the pre-planetary nebula.

### 41.3 Slowly Rotating Polytropes

In a homogeneous gaseous sphere there is no density concentration towards the centre, while for the Roche model the assumed density concentration is too extreme compared to that of real stars. Polytropes approximate real stars better, at least with respect to their density distribution. For slowly rotating polytropes (small values of $\chi$), equilibrium solutions have been found by solving ordinary differential equations for solid-body rotation.

As in the case of the non-rotating polytropes (see § 19), one has to solve the Poisson equation for the gravitational potential. But since the centrifugal acceleration according to (41.8) can be derived from the potential $V$, we combine $\Phi$ and $V$ to obtain the total potential $\Psi$ as in (41.9). Then instead of (19.7), we have in the co-rotating frame

$$\varrho = \left[ \frac{-\Psi}{(n+1)K} \right]^n \quad , \tag{41.16}$$

and since $\Delta\Phi = 4\pi G\varrho$, $\Delta V = -2\omega^2$, we find

$$\Delta\Psi = 4\pi G\varrho - 2\omega^2 \quad , \tag{41.17}$$

and with (41.16)

$$\Delta\Psi = 4\pi G \left[ \frac{-\Psi}{(n+1)K} \right]^n - 2\omega^2 \quad . \tag{41.18}$$

If we now replace $r$ in the Laplace operator by the dimensionless variable $y = Ar$, where $A$ is defined as in (19.9), we obtain for $w := \Phi/\Phi_c$, with the help of (41.16),

$$\Delta_y w = w^n - \frac{\omega^2}{2\pi G\varrho_c} \quad , \tag{41.19}$$

with $\Delta_y = A^2\Delta$, where $\Delta$ is the Laplace operator. In spherical coordinates, for the case of axial symmetry,

$$\Delta_y \equiv \frac{1}{y^2 \sin \vartheta} \left[ \frac{\partial}{\partial y} \left( y^2 \sin \vartheta \frac{\partial}{\partial y} \right) + \frac{\partial}{\partial \vartheta} \left( \sin \vartheta \frac{\partial}{\partial \vartheta} \right) \right] \quad . \tag{41.20}$$

The last term on the right-hand side of (41.19) is a measure of the strength of rotation. We therefore now define for polytropes

$$\chi := \frac{\omega^2}{2\pi G \varrho_c} \quad , \tag{41.21}$$

and we can write (41.19) in the form

$$\Delta_y w = w^n - \chi \quad . \tag{41.22}$$

This partial differential equation corresponds to the Emden equation (19.10), which indeed is obtained for $\omega \to 0$. Equation (41.22) holds in the interior of the polytrope, while outside, the solution has to obey the Laplace equation, which here is $\Delta_y w = 0$, and has to be regular at infinity. For $\chi \ll 1$ one can approximate the solution $w(y, \vartheta)$ by an expansion in Legendre polynomials $L_i(\vartheta)$ with even $i$:

$$w = w_0(y) + \chi w_1(y) + \chi w_2(y) L_2(\cos \vartheta) + \dots \quad , \tag{41.23}$$

where $w_0(y)$ is the solution of the Lane–Emden equation. The perturbation of first order in $\chi$ is split into a spherically symmetric term and a non-spherical one, which vanishes if averaged over a sphere. The terms of higher order in $\chi$ are not explicitly written down. If the expansion (41.23) is introduced into (41.22), then the terms of the same dependence on $\vartheta$ and of the same order in $\chi$ give ordinary differential equations in $y$. Similarly the Laplace equation for the outside can be reduced to a set of ordinary differential equations by the expansion (41.23).

Numerical calculations by CHANDRASEKHAR (1933) show that the oblateness of the surface defined by $(r_{\text{equ}} - r_{\text{pole}})/r_{\text{equ}}$ is $3.75\chi$, $5.79\chi$, $9.82\chi$, $41.81\chi$, $468.07\chi$ for the polytropes of index $n = 1$, 1.5, 2, 3, 4 respectively.

# §42 The Thermodynamics of Rotating Stellar Models

The theory of the structure of rotating stars becomes relatively simple if the centrifugal acceleration can be derived from a potential $V$:

$$\omega^2 s e_s = -\nabla V \quad , \tag{42.1}$$

where $e_s$ is a unit vector perpendicular to the axis of rotation (pointing outwards) and $s$ is the distance from this axis. One can easily see that a sufficient and necessary condition for the existence of such a potential is that in the system of cylindrical coordinates $s$, $\varphi$, $z$ the angular velocity depends on $s$ only: $\partial\omega/\partial z = \partial\omega/\partial\varphi = 0$, i.e. $\omega$ is constant on cylinders. We call such an angular-velocity distribution (to which the case of solid-body rotation also belongs) *conservative*.

## 42.1 Conservative Rotation

In this case the potential $V$ is

$$V = -\int_0^s \omega^2 s\, ds \quad . \tag{42.2}$$

We again combine gravitational and centrifugal potentials to form the total potential

$$\Psi := \Phi + V \quad . \tag{42.3}$$

If we now include centrifugal acceleration in the equation of hydrostatic equilibrium [compare with (2.20)], we obtain

$$\nabla P = -\varrho \nabla \Psi \quad . \tag{42.4}$$

Equation (42.4) indicates that the vectors $\nabla P$ and $-\nabla\Psi$ are parallel. In other words, the equipotential surfaces $\Psi = \text{constant}$ coincide with the surfaces of constant pressure, which means that the pressure is a function of $\Psi$: $P = P(\Psi)$. It then follows that $\varrho = -dP/d\Psi$ is also a function of $\Psi$ only. If we now have an ideal gas, then $T/\mu = P/(\varrho\mathfrak{R})$ is a function of $\Psi$. In a chemically homogeneous star, therefore, $T = T(\Psi)$, i.e. the temperature is constant on equipotential surfaces.

Since not $T$ but $T/\mu$ is constant on equipotentials, the temperature varies proportionally to $\mu$ on these surfaces if the chemical composition is not homogeneous. We have already encountered this case in §6.5, where we dealt with a blob of material with a higher molecular weight than that in the surroundings. In the blob the temperature was higher.

Note that this is a consequence of hydrostatic equilibrium: even small deviations from hydrostatic equilibrium can cause considerable temperature variations on equipotential surfaces, which can be seen in the case with negligible rotation. Then from (42.4) one can conclude that $P$, $\varrho$, and $T/\mu$ are constant on the equipotential surfaces of the gravitational field, say, of the earth. We know that if we light a match, the air on the horizontal equipotential planes intersecting the flame will not have the high temperature of the fire. The reason is that with the flame a circulation system is set up. With this motion, inertia terms disturb the equation of hydrostatic equilibrium. Although they cause only small perturbations, the inertia terms are sufficient to allow lower temperatures outside the flame.

In the following we discuss only the case of strict hydrostatic equilibrium for a chemically homogeneous ideal gas and therefore have $P = P(\Psi)$, $\varrho = \varrho(\Psi)$, $T = T(\Psi)$.

Note that the coincidence of $P$ and $\varrho$ surfaces only holds if the rotation is conservative. Otherwise they are inclined to each other (see § 43.2).

### 42.2 Von Zeipel's Theorem

We now investigate radiative energy transport in a homogeneous, hydrostatic star with conservative rotation. The equation for radiative transport (5.8) in vector form is

$$F = -\frac{4ac}{3\kappa\varrho}T^3\nabla T \quad , \tag{42.5}$$

where $F$ is the vector of the radiative energy flux. With $T = T(\Psi)$ and with $-\nabla\Psi = g_{\text{eff}}$, the effective gravitational acceleration consisting of gravitational *and* centrifugal acceleration, one finds

$$F = \frac{4ac}{3\kappa\varrho}T^3\frac{dT}{d\Psi}g_{\text{eff}} = -k(\Psi)g_{\text{eff}} \quad , \tag{42.6}$$

since also $\kappa(\varrho, T) = \kappa(\Psi)$. In the non-rotating case this equation is equivalent to (5.9). We now look for the equation of energy conservation and restrict ourselves to stationary states with complete equilibrium. Then, instead of (4.23), we have from (42.6)

$$\nabla \cdot F = -\frac{dk}{d\Psi}(\nabla\Psi)^2 - k(\Psi)\Delta\Psi$$
$$= -\frac{dk}{d\Psi}(\nabla\Psi)^2 - k(\Psi)\left(4\pi G\varrho - \frac{1}{s}\frac{d(s^2\omega^2)}{ds}\right) = \varepsilon\varrho \quad , \tag{42.7}$$

where we have made use of $\Delta\Phi = 4\pi G\varrho$ and of (42.2). ($\Delta$ is the Laplace operator.) One can easily see that this equation cannot be fulfilled. We consider a chemically homogeneous star; then $P$, $\varrho$, and $T$ are constant on the equipotential surfaces $\Psi =$ constant. Therefore the terms $\varepsilon\varrho$ as well as $4\pi G\varrho k(\Psi)$ are constant on equipotential surfaces, but in general the remaining two terms on the left are not, and they do not cancel each other. This can be easily seen in the case of solid-body rotation, for which $(s^{-1})d(s^2\omega^2)/ds$ is a constant, while $(\nabla\Psi)^2$ always varies on equipotential surfaces, the effective gravity at the equator being smaller than at the poles.

The fact that radiative transport and the simple equation of energy conservation cannot be fulfilled simultaneously was first pointed out by VON ZEIPEL (1924) and is known as von Zeipel's theorem. The solution of the problem was independently found by EDDINGTON (1925) and VOGT (1925).

## 42.3 Meridional Circulation

What is to be expected if (42.7) cannot be fulfilled? Then there must be regions in the star which would cool off, since radiation carries more energy out of a mass element than is generated by thermonuclear reactions. In other regions the mass elements would heat up. But cooling and heating cause buoyancy forces and meridional motions occur in addition to rotation. In order to maintain a stationary state as assumed, one has to demand that meridional motions contribute to the energy transport. They carry away energy from regions where radiation cannot transport all the energy generated and they bring energy to regions which otherwise would cool off.

In order to derive the velocity field of the circulation, we write the first law of thermodynamics in the co-moving frame:

$$\nabla \cdot F = \varepsilon \varrho - \varrho T \frac{d\sigma}{dt} \quad . \tag{42.8}$$

We here denote the specific entropy by $\sigma$ (instead of $s$) to avoid confusion with the distance from the axis. With $d\sigma = dq/T$, and with (4.18), one has

$$T \frac{d\sigma}{dt} = c_P \frac{dT}{dt} - \frac{\delta}{\varrho} \frac{dP}{dt} \quad . \tag{42.9}$$

If we replace the derivatives in the co-moving frame by those in a coordinate system at rest with respect to the stellar centre, i.e. $d/dt = \partial/\partial t + v \cdot \nabla$, we find

$$\nabla \cdot F = \varepsilon \varrho - c_P \varrho \frac{\partial T}{\partial t} + \delta \frac{\partial P}{\partial t} - v[c_P \varrho \nabla T - \delta \nabla P] \quad , \tag{42.10}$$

and for thermal equilibrium

$$\nabla \cdot F = \varepsilon \varrho - c_P \varrho T v \left[ \frac{1}{T} \nabla T - \frac{\delta}{c_P \varrho T} \nabla P \right] \quad . \tag{42.11}$$

With $\nabla T = \nabla \Psi (dT/d\Psi)$ and $\nabla P = \nabla \Psi (dP/d\Psi)$, the usual abbreviation $\nabla = d\ln T/d\ln P$, and (4.21), we can write

$$\nabla \cdot F = \varepsilon \varrho - \frac{c_P \varrho T}{P} (\nabla - \nabla_{\text{ad}})(v \cdot \nabla P) \quad . \tag{42.12}$$

The components of the meridional velocity field have to fulfil this equation together with the continuity equation, which in the stationary case becomes $\nabla \cdot (\varrho v) = 0$.

We can simplify (42.12) if we assume $\chi$, as defined in (41.5), to be small and ignore higher-order terms in $\chi$. Since $v$ is of first order in $\chi$, the last term in (42.12) can be replaced by $\left[c_P \varrho T(\nabla - \nabla_{ad})/P\right]_0 \nabla P_0 \boldsymbol{v}$, where the subscript 0 indicates the values of the corresponding non-rotating model. Since $\nabla P_0 = -\varrho_0 g_0$ and $g_0$ has only a radial component given by $-|g_0| = -g_0$, we have, instead of (42.12),

$$\nabla \cdot \boldsymbol{F} = \varepsilon \varrho + \left[\frac{c_P \varrho^2 T}{P}(\nabla - \nabla_{ad})g\right]_0 v_r \quad . \tag{42.13}$$

Comparing the non-rotating case, we have now introduced a new unknown variable $v_r$, which in spherical coordinates $r$, $\varphi$, $\vartheta$ together with the velocity component in the $\vartheta$ direction has to fulfil the continuity equation

$$\frac{1}{r^2}\frac{\partial(\varrho r^2 v_r)}{\partial r} + \frac{1}{r \sin \vartheta}\frac{\partial(\varrho v_\vartheta \sin \vartheta)}{\partial \vartheta} = 0 \quad . \tag{42.14}$$

Equations (42.13,14) are the necessary conditions for determining also the velocity field.

## 42.4 The Non-conservative Case

Above we have shown the existence of meridional circulation only for a conservative angular-velocity distribution. We now discuss the situation in a non-conservative case. For this we choose $\omega = \omega(r)$, but restrict ourselves to slow rotation. The equations to be solved are

$$\nabla P = -\varrho \nabla \Phi + \boldsymbol{c} \quad , \tag{42.15}$$

$$\nabla \cdot \boldsymbol{F} = \varepsilon \varrho + \left[\frac{c_P \varrho^2 T}{P}(\nabla - \nabla_{ad})g\right]_0 v_r \quad . \tag{42.16}$$

$$\boldsymbol{F} = -\frac{4ac}{3\kappa\varrho}T^3 \nabla T \quad , \tag{42.17}$$

$$\Delta \Phi = 4\pi G \varrho \quad , \tag{42.18}$$

where the functions $\varrho$, $\varepsilon$, $\kappa$ are assumed to be known functions of $P$ and $T$. Without rotation the solutions are spherically symmetric, but rotation produces deviations from that symmetry. The centrifugal acceleration $\boldsymbol{c}$ appearing in (42.15) has the components

$$c_r = \omega^2 r \sin^2 \vartheta = \frac{2}{3}\omega^2 r\,(1 - L_2) \quad , \tag{42.19}$$

$$c_\vartheta = \omega^2 r \sin \vartheta \cos \vartheta = -\frac{1}{3}\omega^2 r \frac{\partial L_2}{\partial \vartheta} \quad , \tag{42.20}$$

where we have introduced the second Legendre polynomial $L_2(\vartheta) = (3\cos^2 \vartheta - 1)/2$.

In order to solve the system (42.15–18), we split all the scalar functions into a spherically symmetric part (subscript 0) and one which is proportional to $L_2(\vartheta)$:

$$P(r,\vartheta) = P_0(r) + P_2(r)L_2(\vartheta) \quad , \quad T = T_0 + T_2 L_2 \quad , \quad \Phi = \Phi_0 + \Phi_2 L_2 \quad , \tag{42.21}$$

with $|P_2| \ll P_0$, $|T_2| \ll T_0$. For the vectors $F$ and $v$ we write

$$F_r = F_{r0}(r) + F_{r2}(r)L_2 \quad , \quad F_\vartheta = F_{\vartheta 2}(r)\frac{dL_2(\vartheta)}{d\vartheta} \quad ,$$

$$v_r = 0 + v_{r2}(r)L_2(\vartheta) \quad , \quad v_\vartheta = v_{\vartheta 2}(r)\frac{dL_2(\vartheta)}{d\vartheta} \quad , \tag{42.22}$$

with $|F_{r2}|$ and $|F_{\vartheta 2}|$ being small compared to $|F_{r0}|$. It should be noted that in this notation the quantities $P_0$, $T_0$, ... are not identical with the corresponding functions of the non-rotating star, since in the centrifugal acceleration there is also a spherically symmetric component, as can be seen from (42.19).

We now ignore second-order effects and count the number of equations for the four "spherical" functions $P_0$, $T_0$, $\Phi_0$, $F_{r0}$, and for the five "non-spherical" functions $P_2$, $T_2$, $\Phi_2$, $F_{r2}$, $F_{\vartheta 2}$. These are all variables appearing in (42.15–18) together with (42.21,22), if for the moment we ignore circulation ($v_r = 0$). It is obvious that each of the two scalar equations (42.16, 18) give two equations, a spherical one and a non-spherical one, though in the case of the vector equations (42.15, 17) it is different. We explain this in the case of (42.15). The $r$ component gives a "spherical" equation [compare (42.19)]

$$\frac{dP_0}{dr} = -\varrho_0\frac{d\Phi_0}{dr} + \frac{2}{3}\varrho_0\omega^2 r \tag{42.23}$$

and a "non-spherical" one

$$\frac{dP_2}{dr} = -\varrho_0\frac{d\Phi_2}{dr} - \varrho_2\frac{d\Phi_0}{dr} - \frac{2}{3}\varrho_0\omega^2 r \quad , \tag{42.24}$$

while the $\vartheta$ component gives [compare (42.20)]

$$P_2 = -\varrho_0\Phi_2 + \frac{1}{3}\varrho_0\omega^2 r \quad . \tag{42.25}$$

Therefore the vector equation (42.15) yields the "spherical" equation (42.23) and two "non-spherical" equations (42.24,25). The same holds for the vector equation (42.17). Altogether we have four equations for the four "spherical" functions but six equations for the five "non-spherical" functions. Obviously with $v_r = 0$ the problem is overdetermined. In general it can only be solved if meridional circulations are present; then the $v_r$ appearing in (42.16) is the sixth unknown "non-spherical" variable and the problem is no longer overdetermined. If $v_r$ is known, the continuity equation (42.14) together with (42.21,22) gives $v_\vartheta$.

## 42.5 The Eddington–Sweet Time-scale

To obtain an estimate of the velocity of the circulation, we restrict ourselves to slow rotation and to the conservative case. The estimate for the non-conservative case is more complicated, but the results are very similar. We also assume $\varepsilon = 0$, which holds for the outer layers. Therefore $l = $ constant.

We now can split each function $A(r, \vartheta)$ of the model uniquely into two terms:

$$A(r, \vartheta) = \overline{A}(\Psi) + A^*(r, \vartheta) \quad , \qquad (42.26)$$

where $\overline{A}(\Psi)$ is the mean value of $A(r, \vartheta)$ over the surface $\Psi = $ constant, while the integral of $A^*$ over each $\Psi$ surface vanishes:

$$\int_\Psi A^*(r, \vartheta) dS = 0 \quad , \qquad (42.27)$$

where $dS$ is the surface element of the $\Psi$ surface. Then according to (42.6), $k(\Psi) = \overline{F}/\overline{g}_{\text{eff}}$, where $F$ and $g_{\text{eff}}$ are the absolute values of $F$ and $g_{\text{eff}}$, and (42.7) can be written as

$$\nabla \cdot F = -\frac{d}{d\Psi} \left( \frac{\overline{F}}{\overline{g}} \right) g^2 - \frac{\overline{F}}{\overline{g}} \left[ 4\pi G\varrho - \frac{1}{s} \frac{d}{ds}(s^2\omega^2) \right] \quad , \qquad (42.28)$$

where we have omitted the subscript eff in the symbols $g$ and $\overline{g}$. We now split the terms of (42.28) according to (42.26). $\overline{\nabla \cdot F}$ has to be zero in the steady state in regions where there is no nuclear energy generation (otherwise it has to be equal to $\varepsilon\varrho$, a function which is also constant on $\Psi$ surfaces). But the term $(\nabla \cdot F)^*$ can only be compensated by circulation. Indeed the circulation term in (42.13) is $[c_P\varrho T(\nabla - \nabla_{\text{ad}})/P]\nabla P_0 v$. The integral of this term over equipotential surfaces vanishes because of mass conservation, as does $(\nabla \cdot F)^*$.

We now estimate $(\nabla \cdot F)^*$ for slow rotation and take $\overline{F}/\overline{g}$ from the non-rotating model, an approximation which introduces only errors of order $\chi^2$, since in the expression for $(\nabla \cdot F)^*$ the function $\overline{F}/\overline{g}$ appears multiplied only by terms of order $\chi$. Then

$$\frac{\overline{F}}{\overline{g}} = \frac{L}{4\pi Gm} \quad , \qquad (42.29)$$

$$\frac{d}{d\Psi} \left( \frac{\overline{F}}{\overline{g}} \right) = \frac{d}{dr} \left( \frac{\overline{F}}{\overline{g}} \right) \frac{dr}{d\Psi} = \frac{d}{dr} \left( \frac{L}{4\pi Gm} \right) \frac{1}{g} = -\frac{L\varrho}{m} \left( \frac{r^2}{Gm} \right)^2 \quad , \qquad (42.30)$$

and therefore

$$(\nabla \cdot F)^* = -\frac{L\varrho}{m} \left( \frac{r^2}{Gm} \right)^2 (g^2)^* - \frac{L}{4\pi Gm} \left[ \frac{1}{s} \frac{d(s^2\omega^2)}{dr} \right]^* \quad . \qquad (42.31)$$

Now (42.12) yields

$$\frac{\delta\varrho\overline{g}}{\nabla_{\text{ad}}}(\nabla_{\text{ad}} - \nabla)v_r = -\frac{L\varrho}{\overline{g}^2 m}(g^2)^* - \frac{L}{4\pi Gm} \left[ \frac{1}{s} \frac{d(s^2\omega^2)}{ds} \right]^* \quad , \qquad (42.32)$$

where in the circulation term we have made use of (4.21).

For angular velocities of the form $\omega^2 = c_1 + c_2/s^2$ the expression in the last bracket is constant and the last term vanishes for these special angular velocity distributions which include solid-body rotation ($c_2 = 0$). We first restrict ourselves to these special rotation laws. As a rough estimate, we can say that $(g^2)^*/\overline{g}^2$ is of the order of $\chi$. Indeed $g^*$, the variation of $g$ over an equipotential, is due to the difference of centrifugal acceleration between equator and poles, and therefore $g^*/g \approx \chi$, and also $(g^2)^*/g^2 \approx \chi$. We then find with $(\nabla_{ad} - \nabla)/\nabla_{ad}$ and $\delta$ of the order of 1,

$$v_r \approx \frac{L}{\bar{g}m}\chi \approx \frac{LR^2}{GM^2}\chi \quad , \tag{42.33}$$

where we have replaced $m$ and $g$ by their surface values $M$ and $GM/R^2$. (Replacing them by some mean values over the star would not change the order of magnitude.) The time it takes a mass element to move over the stellar radius, then, is the circulation time-scale $\tau_{circ}$, first derived by SWEET (1950):

$$\tau_{circ} \approx \frac{R}{v_r} \approx \frac{GM^2}{LR}\frac{1}{\chi} \approx \frac{\tau_{KH}}{\chi} \quad , \tag{42.34}$$

where we have made use of (3.19), ignoring a factor 2. For the sun one has $\chi \approx 10^{-5}$, $\tau_{KH} \approx 10^7$ years, and therefore $\tau_{circ} \approx 10^{12}$ years, which exceeds the lifetime of the sun.

This estimate has been made ignoring the last term in (42.32). If $\omega$ is not of the special form given above, the term in the bracket will be of the order of $\omega^2$, and since $\omega^*$ is constant on cylinders but not on equipotential surfaces, $\omega$ will be of the order of $\bar{\omega}$ and the term in question will be of the order of

$$\frac{L\omega^2}{4\pi GM} \approx \frac{L}{4\pi R^3}\chi \quad , \tag{42.35}$$

where we have replaced $\omega^2 R/g = \omega^2 R^3/(GM)$ by $\chi$. We estimated that the first term on the right of (42.32) is of the order of $L\varrho\chi/M$. Therefore as long as we are not too close to the surface we can replace $\varrho$ by the mean density $\bar{\varrho} = 3M/(4\pi R^3)$, so that the two terms on the right of (42.32) are of the same order and our estimates (42.33,34) also hold for rotation laws which are not of the special form $c_1 + c_2/s^2$. But near the surface the first term on the right of (42.32) becomes small owing to the factor $\varrho$, and the second becomes the leading term. Then near the surface, (42.33) has to be replaced by

$$v_r \approx \frac{\bar{\varrho}}{\varrho}\frac{LR^2}{GM^2}\chi \approx \frac{L}{G\varrho RM}\chi \quad , \tag{42.36}$$

where again we have neglected factors of the order of one. The circulation can therefore become rather fast near the surface.

The same is true at the interfaces between radiative and convective regions where $\nabla = \nabla_{ad}$, which we have excluded in our rough estimate of the left-hand side of (42.32). At these singularities the circulation speed would become so large that its inertia terms are important and (42.4) would no longer be valid.

Another more serious restriction of our estimates of $v_r$ is the assumption of a certain time-independent angular-velocity distribution. If one starts, say, with $\omega = $ constant, then circulation will occur and by conservation of angular momentum, it will immediately change the angular-velocity distribution, which in turn demands another circulation pattern.

The "proof" of the existence of meridional circulation in the theory of first order in $\chi$ as given in §42.4 rested on counting the number of linear equations and the number of variables. We showed that without circulation the problem is over determined. This, however, is only true if the linear equations are independent. But if $\omega$ is considered as a free function, it can be chosen in such a way that the equations become linearly dependent and in the first-order theory no circulation is necessary to fulfil the equations. In the (unrealistic) case $\varepsilon$ = constant, $\kappa$ = constant, the stellar-structure equations for radiative energy transport lead to a polytrope of index $n$ = 3. If $\varepsilon$ = constant, then $l/m$ = constant and one has a very special "standard model" as discussed in §19.5. It has been shown by SCHWARZSCHILD (1942) that, for this model, solid-body rotation does not demand circulation in the first-order theory. For other, more realistic stellar models there are also angular-velocity distributions for which there is no meridional circulation in the first-order theory (KIPPENHAHN, 1963).

The linear dependence of the equations can also be achieved if for a given rotation law $\omega$, the molecular weight is considered a free function and chosen in an appropriate way. We will come to this problem in the next section.

### 42.6 Meridional Circulation in Inhomogeneous Stars

We have already estimated that for the sun that $\tau_{\text{circ}}/\tau_{\text{nucl}} \approx 10^2$. But for more massive main-sequence stars the situation changes. According to (42.34)

$$\frac{\tau_{\text{circ}}}{\tau_{\text{nucl}}} \approx \frac{\tau_{\text{KH}}}{\tau_{\text{nucl}}}\frac{1}{\chi} \sim \frac{M^{1-\alpha}}{\chi} \approx \frac{M^{0.4}}{\chi} \ , \tag{42.37}$$

where we have assumed a mass–radius relation $R \sim M^\alpha$ and $\tau_{\text{KH}} \sim M^2/(RL)$, as can be derived from (3.19), and $\tau_{\text{nucl}} \sim M/L$, and we have put $\alpha = 0.6$ for the upper end of the main sequence (§22.1). Therefore, if we go from the sun to higher masses, say, to $20M_\odot$, then the ratio $\tau_{\text{KH}}/\tau_{\text{nucl}}$ (which for the sun is about 1/100) increases by a factor 3.3. Observations of rotating B stars show that $\chi$ is larger by a factor $10^5$ than for the sun. Therefore, $\tau_{\text{circ}}/\tau_{\text{nucl}}$ drops below unity towards the upper end of the main sequence, so that the circulation is rapid enough to mix the star. As a consequence one should expect that the fuel is not only used up in the central region and the star should remain chemically homogeneous. But then the stars, while converting hydrogen into helium, should move in the HR diagram from the main sequence straight towards the helium main sequence [compare (20.20, 21) for $M = M'$]. But we know from observation that the stars leave the main sequence moving towards the region of the red stars and not towards the region of the (blue) helium main sequence. This indicates that they do not mix, and the explanation was found by MESTEL (1953). Before the circulation can transport the material out of the burning region, the moving matter will have been enriched in helium. It therefore has a higher molecular weight than the surrounding into which it has been lifted. But then the effect discussed in connection with a blob of material of higher molecular weight $\mu$ in a gas of lower $\mu$ becomes important (§6.4). Let us assume that the

442

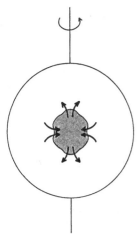

Fig. 42.1. Material of higher molecular weight in the central region of a rotating star (*grey area*) under the influence of meridional circulation

circulation lifts helium-enriched material as indicated in Fig. 42.1. Then, since in hydrostatic equilibrium $T/\mu$ must be constant on $\Psi$ surfaces, the lifted matter has a higher temperature than the matter on the same $\Psi$ surface which is not lifted. There is no buoyancy force acting on the lifted matter, since the higher molecular weight is compensated by the higher temperature. But as the lifted material adjusts thermally, it sinks back. This additional motion ("$\mu$ currents") acts against the circulation and the star can only be mixed if circulation is sufficiently fast. But even in rapidly rotating main-sequence stars, the circulation is not sufficient to mix the helium formed during hydrogen burning. Obviously layers in which the molecular weight increases in an inward direction cannot easily be penetrated by meridional circulation. One therefore often speaks of $\mu$ *barriers*.

Note that $\mu$ barriers in which no circulation occurs are not in contradiction to our "proof" of the existence of meridional circulation in rotating stars. According to our considerations in § 42.4, which also hold for inhomogeneous stars as long as $\mu$ is spherically symmetric, circulation would set in. But after a short time the circulation has modified the $\mu$ distribution and the original spherically symmetric function $\mu(r)$ has become distorted and may be of the form $\mu_0(r)+\mu_2(r)L_2(\vartheta)$. Then by counting the equations and variables as was done in § 42.4 we would not find the problem to be over determined, since $\mu_2(r)$ is an additional unknown function. It can be determined instead of $v_r$ by the "non-spherical" equations.

# §43 The Angular-Velocity Distribution in Stars

Stars formed out of an interstellar cloud contain a certain amount of angular momentum, which is distributed over the stellar mass. Suppose there were no transport of angular momentum between the mass elements during the formation and evolution of the stars; one would then have local conservation of angular momentum,

$$\frac{d(s^2\omega)}{dt} \equiv s^2\frac{\partial\omega}{\partial t} + v \cdot \nabla(s^2\omega) = 0 \quad , \tag{43.1}$$

where $v$ is the large-scale velocity in the star. Then the angular velocity $\omega(s, \vartheta)$ would be determined by the angular momenta of the mass elements in the original cloud. However, the motion of atoms, the flow of photons through matter, and instabilities that cause small-scale motions can transport angular momentum. (An example of the last of these is the convective motion in regions of dynamical instability.) We now discuss these transport mechanisms in detail.

## 43.1 Viscosity

Viscosity due to microscopic motion, like that of the molecules in a liquid, is given by the viscosity coefficient

$$\eta \approx \varrho\ell v_{\text{th}} \quad , \tag{43.2}$$

where $\ell$ is the mean free path of the particles and $v_{\text{th}}$ their mean velocity. In an ionized gas the viscosity is determined by the collisions between the ions. Therefore their mean free path and their thermal velocities have to be inserted in (43.2), and one normally obtains values for $\eta$ which in cgs units are of the order of 1.

In order to see whether viscosity is important in a star, one has to estimate the time-scale required for viscosity to influence a given angular-velocity distribution. This can be done with the $\varphi$ component of the Navier–Stokes equations of motion, which for constant viscosity can be written in the form

$$\varrho\frac{\partial\omega}{\partial t} = \eta\Delta\omega \quad , \tag{43.3}$$

where $\Delta$ is the Laplace operator. This equation is of the form of the equation of heat transfer (5.31). In analogy to (5.32), we can estimate the viscosity time-scale:

$$\tau_{\text{visc}} \approx \frac{d^2\varrho}{\eta} \quad , \tag{43.4}$$

where $d$ is the characteristic length on which $\omega$ varies. If for $d$ one takes the radius of a star, say $10^{11}$ cm, then with $\varrho \approx 1\,\mathrm{g\,cm^{-3}}$ one finds $\tau_{\mathrm{visc}} \approx 10^{22}$ s, a time-scale much longer than the cosmological time. In stars one can therefore neglect the viscosity due to the collisions between the ions.

In a star, photons can also cause viscosity, since they transport momentum. If they are absorbed after a mean free path $\ell_{\mathrm{ph}}$, they transfer their momentum to the absorbing particle. A rough estimate of this *radiative viscosity* $\eta_{\mathrm{rad}}$ is obtained if in (43.2) $\varrho$ is replaced by the mass density of the radiation field $\varrho_{\mathrm{rad}} = aT^4/c^2$, $v_{\mathrm{th}}$ is replaced by $c$, and $\ell$ by $\ell_{\mathrm{ph}} \approx 1/\kappa\varrho$, the mean free path of a photon:

$$\eta_{\mathrm{rad}} \approx \frac{aT^4}{c\kappa\varrho} \quad . \tag{43.5}$$

The characteristic time-scale according to (43.4) is

$$\tau_{\mathrm{visc}} \approx \frac{d^2\varrho}{\eta} \approx \frac{d^2\varrho^2 c\kappa}{aT^4} \quad . \tag{43.6}$$

With $d = 10^{11}\,\mathrm{cm}^2$, $\varrho = 1\,\mathrm{g\,cm^{-3}}$, $\kappa = 1\,\mathrm{cm^2\,g^{-1}}$, $T = 10^7\,\mathrm{K}$, we find the characteristic time of radiative viscosity in a star to be $10^{18}$ s, again a time-scale long compared to the lifetime of a star. One therefore can neglect the effects of viscosity not only caused by the atomic motion but also those caused by radiation: the stellar gas moves like a frictionless fluid.

It should be noted that the radiation causes a kind of viscosity similar to that of the atomic motion only in an isotropic radiation field. For a non-isotropic field the radiative viscosity is not a scalar but a tensor.

The expression (43.2) for viscosity can also be used in convective regions, where rising and falling mass elements not only transport energy as discussed in §7, but also momentum. In the picture of the mixing-length theory, one can consider the convection elements as "particles" which are created at some place, move one mixing length $\ell_{\mathrm{m}}$, and dissolve. The corresponding "turbulent viscosity" $\eta_{\mathrm{t}}$ in analogy to (43.2) is

$$\eta_{\mathrm{t}} \approx \varrho\ell_{\mathrm{m}}v_{\mathrm{t}} \quad , \tag{43.7}$$

where $v_{\mathrm{t}}$ is the convective velocity. In the case of the convective envelope of the sun, we assume $v_{\mathrm{t}}$ to be 1% of the speed of sound (as indicated in Fig. 29.3c). With $\ell_{\mathrm{m}} \approx H_P \approx 10^8$ cm, $\varrho \approx 10^{-4}\,\mathrm{g\,cm^{-3}}$, a sound velocity of $v_{\mathrm{s}} \approx 2 \times 10^6\,\mathrm{cm\,s^{-1}}$ corresponding to a temperature of $3 \times 10^4$ K, and with $v_{\mathrm{t}} \approx 0.01v_{\mathrm{s}} \approx 2 \times 10^4$ cm s$^{-1}$, we find $\eta_{\mathrm{t}} \approx 2 \times 10^8$ cgs and the corresponding time-scale $\tau_{\mathrm{visc}} \approx 5 \times 10^9$ s $\approx$ 160 years! One can therefore assume that the angular-velocity distribution in the convective zone of the sun, for instance, has reached a steady state in which the initial angular-momentum distribution is smeared out by viscosity.

However, the analogy between friction caused by molecules and that by convective blobs has its limits. While the statistical motion of molecules is isotropic to a high degree, there is no reason to suppose that convection in a stellar convective zone can be described by elements with isotropic random motion. Convection is maintained in a star by the radially outgoing energy flux. The motion is caused by buoyancy forces which are antiparallel to the (radial) gravity vector. One therefore can expect that the exchange of momentum by the turbulent elements is different in the radial direction from that in other directions. The viscosity is no longer isotropic, i.e. it is a tensor.

The macroscopic behaviour of a fluid with anisotropic viscosity is peculiar. We know that in the case of isotropic viscosity, a self-gravitating sphere which initially starts out with differential rotation approaches solid-body rotation after a viscous time-scale. This is not true any more for non-isotropic viscosity (BIERMANN, 1951). One can expect that non-isotropic turbulent viscosity causes differential rotation and should therefore not be surprised that the surface of the sun does not rotate uniformly.

In this connection it should be noted that in a large part of the solar convective zone, the layers are adiabatic (with constant $\nabla_{\mathrm{ad}}$) and surfaces of constant pressure and of constant density coincide (since $d \ln \varrho / d \ln P$ = constant). As in the barotropic case any angular-velocity distribution for which $\omega$ varies on cylinders of $s$ = constant will cause dynamically driven meridional circulation which by itself changes the angular-velocity distribution.

### 43.2 Dynamical Stability

The behaviour of incompressible homogeneous rotating fluids has been thoroughly investigated (see e.g. CHANDRASEKHAR, 1981). But in many respects compressible gases behave differently. For instance, pure rotation (without meridional motions) in the case $\varrho$ = constant can only take place if $\omega$ is constant on cylinders of $s$ = constant (compare § 42). Otherwise the curl of the centrifugal acceleration $\omega^2 s e_s$ would not vanish. But in the case of pure rotation the equation of motion in the meridional plane is

$$\frac{1}{\varrho}\nabla P + \nabla \Phi = \omega^2 s e_s \quad . \tag{43.8}$$

As long as $\varrho$ = constant, the curl of the left-hand side vanishes. For $\partial \omega / \partial z \neq 0$ one has curl $(\omega^2 s e_s) \neq 0$. Then the meridional components of the equation of motion can only be fulfilled if meridional motions occur, and with them additional terms appear in (43.8). This is also the case if the equation of state is barotropic (as for complete degeneracy), since for $P = P(\varrho)$ the curl of $(\nabla P / \varrho)$ also vanishes. The same holds if the equation of state is not barotropic, but if some other mechanism ensures that the surfaces of constant pressure and constant density coincide. One example is convection zones in their adiabatic regime. From the condition $\nabla = \nabla_{\mathrm{ad}}$ (where $\nabla_{\mathrm{ad}}$ is constant or is a function of $P$ and $T$) it follows that the surfaces of constant pressure and density coincide. If the convective region is chemically homogeneous, then the equation of state (say for an ideal gas) assures that also the pressure and density surfaces coincide. Therefore $\nabla \times (\nabla P / \varrho)$ vanishes and meridional flow occurs if $\partial \omega / \partial z \neq 0$.

But in a rotating star the pressure and density surfaces are normally inclined:

$$\nabla \times \left( \frac{1}{\varrho}\nabla P \right) = -\frac{1}{\varrho^2}\nabla \varrho \times \nabla P \neq 0 \quad . \tag{43.9}$$

Here the right-hand side is obviously proportional to the sine of the angle of inclination. The vector $\nabla P/\varrho$ is no longer a gradient; it can therefore cancel the non-conservative part of $\omega^2 s e_s$, and (43.8) can be fulfilled without any meridional velocity components.

The different behaviour of a compressible non-barotropic gas compared to that of an incompressible fluid also affects the stability behaviour.

It is well-known that the shear motion of fluids can become turbulent. Then kinetic energy of the shear flow goes into the kinetic energy of the "turbulent" elements. If friction is strong, it can prevent this transition.

In an incompressible viscous fluid, therefore, the *Reynolds number Re* decides whether the flow is turbulent or laminar (LANDAU, LIFSHITZ, vol.6, 1959):

$$Re = \frac{\varrho v d}{\eta} \quad , \tag{43.10}$$

where $v$ is a characteristic velocity difference and $d$ is a characteristic length. For high Reynolds numbers (say $Re \gg 3000$) kinetic energy of the differential motion becomes kinetic energy of the turbulent elements and the energy which is necessarily lost because of friction is small: the flow is turbulent. If, on the other hand, $Re$ is small, much more energy would have to be used up to overcome the friction of the turbulent elements than is available from the reservoir of differential motion: the flow is laminar. For a rotating star with $\varrho \approx 1 \, \mathrm{g\,cm^{-3}}$, $d \approx R \approx 10^{11} \, \mathrm{cm}$, $v \approx 10^5$ cm s$^{-1}$, and $\eta \approx 1$ cgs (molecular or radiative viscosity), we find $Re \approx 10^{16}$, which means that the flow should be highly turbulent.

But the stellar gas is not incompressible and in most cases not barotropic. Therefore for a transition from laminar to turbulent motion the energy due to the shear motion not only has to go into kinetic energy of the turbulent elements (and via friction into heat), but also into work against the buoyancy forces. Another critical dimensionless number, the *Richardson number Ri*, can be used to decide whether shear motion becomes turbulent despite the stabilizing effect of buoyancy. In the case of a plane parallel flow $v(z)$, it is defined by

$$Ri = \frac{g}{H_P} \frac{|\nabla_{\mathrm{ad}} - \nabla|}{(\partial v/\partial z)^2} \quad . \tag{43.11}$$

One can show that $Ri < 1/4$ is a sufficient condition for stability of the laminar motion. In the case of a layer in the deep interior of a star we may estimate $|\partial v/\partial z| \approx \omega R/R = \omega$, $|\nabla_{\mathrm{ad}} - \nabla| \approx 1$, $H_P \approx 10^9$ cm, $g \approx 10^5$ cm s$^{-2}$ and find that the rotation is laminar as long as $\omega < 2 \times 10^{-2}$ s$^{-1}$ or the rotation period is longer than five minutes. Only neutron stars rotate faster.

Equation (43.11) has been derived under the assumption that the turbulent elements undergo adiabatic changes during their motion. This is not necessarily always the case, not even in the very deep stellar interior. For the sake of simplicity we discuss it in the plane parallel approximation. Let us define a characteristic time-scale for a turbulent element in the case of shear instability of a plane parallel flow by $\tau_\ell = |dz/dv|$. This time-scale can be considered as the "lifetime" of the element. If its excess velocity over the mean velocity of its origin is $\Delta v = \ell |dv/dz|$, where

$\ell$ is its mean free path, then it takes the time $\tau_\ell$ to move over the distance $\ell$. The motion will only be adiabatic if $\tau_\ell \ll \tau_{adj}$, where $\tau_{adj}$ is the thermal adjustment time of the element. With (6.25) one finds as the condition for adiabatic changes of the turbulent elements of diameter $d$ (as assumed in the Richardson criterion),

$$1 \gg \frac{\tau_\ell}{\tau_{adj}} \approx \left|\frac{dz}{dv}\right| \frac{16acT^3}{\kappa \varrho^2 c_p d^2} . \tag{43.12}$$

One can see that this condition is violated for very small shear ($|dv/dz| \to 0$) as well as for small elements ($d \to 0$). Small elements always have time to adjust thermally to their surroundings while they are moving. Then the stabilizing effect of the temperature stratification disappears. The instability which then occurs for small turbulent elements can become important. But one has to keep in mind that extremely small turbulent elements cannot exist, since for them even the low molecular or radiative viscosity brakes their motion. One way of estimating the lower limit would be to assume that the smallest elements are those for which $\tau_\ell$ (which is normally short compared to the viscosity time-scale of the elements) becomes comparable to $\tau_{visc}$. This would mean that the critical size $d$ of the turbulent element is given by

$$d^2 \approx \left|\frac{dz}{dv}\right| \frac{\eta}{\varrho} , \tag{43.13}$$

while for smaller elements viscosity overcomes the instability. Since the thermal adjustment time of turbulent elements is shorter than their lifetime, however, the stabilizing effect of buoyancy is reduced and a flow can be turbulent even if $Ri < 1/4$.

There are other dynamical instabilities which are typical of rotational motion. If they occured in a star, the flows would become turbulent and the turbulent viscosity would immediately change the original angular-velocity distribution. The simplest case of such an instability can be studied by the example of an incompressible or barotropic liquid rotating, say, in a cylindrical container. The angular velocity $\omega$ may depend on $s$ only, making pure rotation possible (see § 43.3). As "mass elements" we consider the matter within two neighbouring thin tori as indicated in Fig. 43.1. Their main radii are $s_1$ and $s_2 = s_1 + ds$. Their thicknesses shall be such that

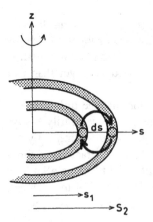

Fig. 43.1. Two tori of radii $s_1$ and $s_2$ are exchanged in order to determine the work against centrifugal forces

their mass contents $dm$ are equal. We now try to exchange the masses of the two tori by expanding the smaller one and contracting the other without changing their angular momentum and calculate the work necessary to make the exchange against the centrifugal force. The kinetic energy of a torus is $E = \omega^2 s^2 dm/2$, which for a given mass is a function of $s$. If we expand (or contract) one of the rings, then conservation of angular momentum demands $\omega \sim s^{-2}$ and therefore $E \sim s^{-2}$. At their original position ($s_1$ and $s_2$) the two tori shall have the energies $E_1$ and $E_2$ respectively. Owing to the expansion $s_1 \rightarrow s_2$, the energy of the first torus changes by an amount

$$dE_1 = \frac{E_1 s_1^2}{(s_1 + ds)^2} - E_1 = -2\frac{E_1 ds}{s_1} + 3\frac{E_1 ds^2}{s_1^2} - \ldots \quad , \tag{43.14}$$

while for the contraction $s_2 \rightarrow s_1$ of the other one we find

$$dE_2 = 2\frac{E_2 ds}{s_2} + 3\frac{E_2 ds^2}{s_2^2} + \ldots \quad . \tag{43.15}$$

Then the total energy required for the exchange of the two tori is

$$dE = dE_1 + dE_2 = 2\left[\frac{E_2}{s_2} - \frac{E_1}{s_1}\right] ds + 6\frac{E_1 ds^2}{s_1^2} + \ldots$$

$$= 2\frac{d}{ds}\left(\frac{E}{s}\right) ds^2 + 6\frac{E}{s^2} ds^2 + \ldots \quad , \tag{43.16}$$

where in the last term of (43.15), we have replaced $E_2/s_2$ by $E_1/s_1$, which only introduces third-order errors in $ds/s_1$. In the last equation (43.16) $E$ means, for instance, a value between $E_1$ and $E_2$. With $E/s = s\omega^2 dm/2$ we find

$$dE = 2\omega^2 dm \left[\frac{d\ln\omega}{d\ln s} + 2\right] ds^2 \quad . \tag{43.17}$$

Since $dE$ is the energy which has to be supplied for the exchange, $dE > 0$ indicates stability, while $dE < 0$ gives instability (energy is gained). We therefore find the condition for stability,

$$\frac{d\ln\omega}{d\ln s} > -2 \quad . \tag{43.18}$$

This is the *Rayleigh criterion*, which we have derived here in a heuristic way. It says that if the specific angular momentum $s^2\omega$ decreases with distance from the axis of rotation, the flow will be turbulent. We have to keep in mind that it has been derived by assuming axisymmetric perturbations only. Since additional non-axisymmetric instabilities may exist, (43.18) is only a necessary condition for stability. Experiments with rotating incompressible fluids between coaxial cylinders indicate that the transition from laminar to turbulent flow occurs when the left-hand side of (43.18) becomes equal to $-2$. But a liquid between a slowly rotating inner cylinder and a very rapidly rotating outer one can become turbulent even though condition (43.18) is fulfilled.

In the derivation of the Rayleigh criterion we have assumed that the gas is incompressible or at least barotropic. But in all other cases buoyancy forces become important and the work against them has to be taken into account. In the case of gas rotating with $\omega = \omega(s)$ and with gravity pointing towards the axis of rotation (as it is in the equatorial plane of a star), instead of (43.18) one has as stability condition

$$\frac{1}{s^3}\frac{\partial s^4 \omega^2}{\partial s} - g_s \frac{\partial \ln P}{\partial s}(\nabla - \nabla_{\text{ad}}) > 0 \quad , \tag{43.19}$$

where $g_s(< 0)$ is the component of gravity in the $s$ direction.

If the second term on the left is neglected, the Rayleigh criterion is recovered. Without rotation (43.19) gives the Schwarzschild criterion (6.13) for stability.

As in the case of the Rayleigh criterion the derivation of (43.19) assumes that the exchange of toroidal mass elements takes place only in the $s$ direction. If in a star the directions of gravity and of exchange do not coincide, then the Solberg–Høiland criterion decides whether the flow is stable or not. We introduce the specific entropy $\sigma$:

$$\sigma = c_P \ln \left( \varrho P^{-1/\gamma_{\text{ad}}} \right) + \text{constant} \quad . \tag{43.20}$$

As long as the equipotential surfaces are not too far from being spherical we can write approximately that

$$\boldsymbol{g} \cdot \nabla \sigma = \frac{|\boldsymbol{g}|}{H_P}(\nabla_{\text{ad}} - \nabla) \quad . \tag{43.21}$$

With the specific angular momentum $j = s^2 \omega$, the Solberg–Høiland criterion (TASSOUL, 1978; ZAHN, 1974) requires for stability

$$\frac{1}{s^2}\frac{\partial j^2}{\partial s} - \frac{|\boldsymbol{g}|}{H_P}c_P(\nabla - \nabla_{\text{ad}}) > 0 \quad , \tag{43.22}$$

$$g_z \left[ \frac{\partial j^2}{\partial s}\frac{\partial \sigma}{\partial z} - \frac{\partial j^2}{\partial z}\frac{\partial \sigma}{\partial z} \right] < 0 \quad , \tag{43.23}$$

$$g_z \frac{\partial \sigma}{\partial z} > 0 \quad . \tag{43.24}$$

All three conditions have to be fulfilled in order to obtain stability, otherwise the flow is unstable. They are necessary and sufficient for stability as long as only axisymmetric perturbations are allowed. They are also necessary for stability if non-axisymmetric perturbations are permitted.

One immediately sees that (43.22) is identical to (43.19) and gives stability for exchange in the $s$ direction. Condition (43.23) is fulfilled as long as $j$ increases on surfaces of $\sigma = $ constant on the way from the pole to the equator. Exchange on such surfaces does not imply buoyancy forces and therefore it reproduces our old condition (43.18). Condition (43.24) says that the Schwarzschild criterion has to be fulfilled for exchange in directions parallel to the axis of rotation in which there is no centrifugal acceleration.

For the problem of dynamical stability in the more general case $\omega = \omega(z, s)$, we refer to TASSOUL (1978) and ZAHN (1974).

## 43.3 Secular Stability

We have seen that buoyancy forces can stabilize angular-velocity distributions which otherwise are dynamically unstable. In the case of non-conservative rotation of a barotropic fluid, there can be no hydrostatic equilibrium between centrifugal, gravitational, and pressure accelerations. Therefore circulation currents are necessary to fulfil the equation of motion in the meridional plane. If buoyancy forces are present, equilibrium can exist for any rotation law $\omega = \omega(s, z)$ as long as gravity overcomes the centrifugal force.

However, buoyancy forces are not as reliable as, for instance, gravity. Let us consider the axisymmetric case of a fluid between two rotating cylinders, and let us assume the Rayleigh criterion (43.18) to be violated, while the Solberg–Høiland criterion (43.22–24) gives stability. We then know that if a toroidal mass element is exchanged with another one further outwards in the $s$ direction, energy is gained from centrifugal forces, but the work which goes into buoyancy is larger. Therefore, if kicked outwards, it will go back and, in the pure adiabatic case, start to oscillate around its original position. This reminds us of the oscillating blob discussed in §6. But we have seen there that a blob with an excess of molecular weight will sink while adjusting thermally. The situation is very similar in the case of a rotating star in which buoyancy forces guarantee dynamical stability.

Let us discuss the case of non-conservative rotation. It is called "baroclinic", since the $P$ and $\varrho$ are inclined against each other. Then centrifugal acceleration is not curl-free and cannot be balanced by the (conservative) gravity. We now consider a closed line in one quadrant of the meridional plane (Fig. 43.2). The vector of a line element is $dl$. Then the integral of the centrifugal acceleration taken along the line is

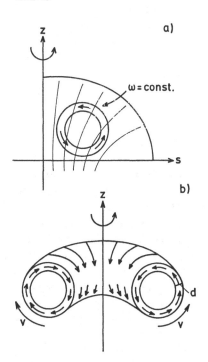

Fig. 43.2. (a) The meridional plane of a rotating star with $d\omega/dz \neq 0$. Thin lines give $\omega$ = constant. Along each closed line the integral over the centrifugal acceleration as defined in (43.25) does not vanish, giving rise to a torque which causes meridional motions as indicated in (b)

$$\oint c \cdot dl \neq 0 \quad . \tag{43.25}$$

This means that the centrifugal acceleration produces a torque on the matter along this line. In a barotropic (or incompressible) fluid this torque would cause a meridional flow. In the more general case, $\nabla P / \varrho$ can balance this torque. But the matter will follow the torque within the time-scale during which heat can leak out.

The matter will also flow if the Rayleigh criterion (43.18) is violated, but the Richardson number (43.11) gives stability. This is analogous to the case of the salt-finger experiment (see § 6.5). If we then exchange two coaxial tori adiabatically as indicated in Fig. 43.1, buoyancy will bring them back to their old position. But since it takes a finite time to return to the initial state, heat will leak out of, or go into, the two tori and they will never come back exactly to the old position. As the blobs in the salt-finger experiment exchange chemical species, here a meridional flow will exchange angular momentum. This flow is again controlled by the time during which heat can leak away from the matter.

What is the time-scale of such a thermally controlled flow? Let us go back to the baroclinic case and the example indicated in Fig. 43.2. Along each closed meridional line there is a torque. The heat exchange can take place most effectively if the thickness $d$ is small, just as the thinnest salt-finger moves fastest, as can be seen from (6.25,29). One would therefore expect that the smallest elements move fastest. Indeed, with decreasing thickness the velocity increases like $v \sim d^{-2}$. Certainly for small mass elements friction becomes important, but since the molecular (or radiative) viscosity is low, the elements slowed down by friction are rather small. Estimates indicate that they are of the order of some meters in the radiative interior of the sun.

Here we have discussed the instabilities by rather heuristic arguments. A mathematically more satisfying treatment of this problem has been carried out by GOLD-REICH, SCHUBERT (1967) and by FRICKE (1968). They find as conditions necessary for secular stability

$$\frac{\partial \ln \omega}{\partial \ln s} > -2 \quad , \quad \frac{\partial \omega}{\partial z} = 0 \quad . \tag{43.26}$$

Although the first condition is identical with (43.18) we have to keep in mind that there we discussed dynamical stability in the barotropic (or incompressible) case, while here we deal with *secular stability*. The second condition of (43.26) does not correspond to a stability condition in the barotropic case. If in this case it is violated, there is no equilibrium. Only buoyancy forces can establish equilibrium in the non-barotropic case, but this equilibrium is thermally unstable.

Several estimates have been made of the time-scale by which the thermal instabilities change the overall angular-velocity distribution, violating conditions (43.26). There is no definite answer, but it may well be that it is the Eddington–Sweet time-scale (42.34) (KIPPENHAHN et al., 1980).

What kind of angular-velocity distribution really does occur in radiative regions of stars? Let us start with a conservative angular-velocity distribution, $\omega = \omega(s)$, say with $\omega = $ constant. Then meridional motions will start. Since they are due to the thermal imbalance between polar and equatorial regions, their characteristic length-scale should be of the order of the stellar radius. They will change the angular-velocity distribution and $\omega$ will become a function of $z$ too. But then the Goldreich–Schubert–Fricke criterion (43.26) is violated and instabilities will occur, which grow fastest for small-scale perturbations. Therefore one again expects eddies of the size of metres. Although these instabilities have never been followed numerically into the non-linear regime, one can guess that on small scales no steady-state solution is possible, since the instability always creates new small-scale eddies moving in an irregular way and the circulation takes care that the $\omega$ distribution never becomes conservative. Only if the characteristic time-scale of the instability is short compared to the Eddington–Sweet time, the overall angular-velocity distribution will probably be close to a conservative one.

# References

Aizenman, M.L., Perdang, J. (1971): Astron. Astrophys. **12**, 232
Alecian, G., Vauclair, S. (1983): Fundamentals of Cosmic Phys. **8**, 369
Alexander, D.R., Johnson, H.R., Rypma, R.L. (1983): Astrophys. J. **272**, 773
Allen, C.W. (1973): *Astrophysical Quantities*, 3rd edition (Athlone Press, London)
Appenzeller, I. (1970): Astron. Astrophys. **5**, 355
Appenzeller, I., Tscharnuter, W. (1974): Astron. Astrophys. **30**, 423
Appenzeller, I., Tscharnuter, W. (1975a): Astron. Astrophys. **40**, 397
Appenzeller, I., Tscharnuter, W. (1975b): private communication
Arnett, W.D. (1967): In *High Energy Astrophysics*, Les Houches Lectures, ed. by C. DeWitt, E. Schatzman, P. Véron (Gordon and Breach, New York), Vol. 3, p. 113
Arnett, W.D. (1969): Astrophys. Space Sci. **5**, 180
Arnett, W.D., Thielemann, F.-K. (1985): Astrophys. J. **295**, 589
Arp, H., Thackeray, A.D. (1967): Astrophys. J. **149**, 73
Arponen, J. (1972): Nucl. Phys. A **191**, 257

Baade, W. (1944): Astrophys. J. **100**, 137 , see also IAU Trans. 1952, p. 397
Baade, W., Zwicky, F. (1934): Phys. Rev. **45**, 138
Bahcall, J.N., Davis, R., Jr. (1976): Science **191**, 264
Bahcall, J.N., Huebner, W.F., Lubow, S.T., Parker, P.T., Ulrich, R.K. (1982): Rev. Mod. Phys. **54**, 767
Baker, N., Kippenhahn, R. (1962): Z. Astrophys. **54**, 114
Baker, N., Kippenhahn, R. (1965): Astrophys. J. **142**, 868
Barkat, Z. (1975): Ann. Rev. Astron. Astrophys. **13**, 45
Bartenwerfer, D. (1972): Dissertation, University of Göttingen
Baym, G., Pethick, C. (1979): Ann. Rev. Astron. Astrophys. **17**, 415
Biermann, L. (1951): Z. Astrophys. **28**, 304
Boury, A., Ledoux, P. (1965): Ann. d'Astrophys. **28**, 353
Burbidge, E.M., Burbidge, G.R., Fowler, W.A., Hoyle, F. (1975): Rev. Mod. Phys. **29**, 547

Carson, T.R. (1976): Ann. Rev. Astron. Astrophys. **14**, 95
Castellani, V., Giannone, P., Renzini, A. (1971): Astrophys. Space Sci. **10**, 340
Chandrasekhar, S. (1933): Mon. Not. R. Astron. Soc. **93**, 390
Chandrasekhar, S. (1939): *An Introduction to the Study of Stellar Structure* (University of Chicago Press, Chicago)
Chandrasekhar, S. (1981): *Hydrodynamic and Hydromagnetic Stability* (Dover, Oxford, New York)
Chandrasekhar, S. (1983): *The Mathematical Theory of Black Holes* (Clarendon Press, Oxford)
Chapman, S., Cowling, T.G. (1952): *The Mathematical Theory of Non-uniform Gases*, 2nd ed. (Cambridge University Press, Cambridge)
Christensen-Dalsgaard, J. (1984): In *Theoretical Problems in Stellar Stability and Oscillations*, ed. by A. Noels, M. Gabriel, 25th Liège Intern. Astrophys. Coll., p. 155
Christy, R.F. (1964): Rev. Mod. Phys. **36**, 555
Christy, R.F. (1975): In *Problèmes d'Hydrodynamique Stellaire*, 19th Liège Intern. Astrophys. Coll., p. 173
Clayton, D.B. (1968): *Principles of Stellar Evolution and Nucleosynthesis* (McGraw-Hill, New York)

Courant, R., Friedrichs, K.O. (1976): *Supersonic Flow and Shock Waves* (Springer, New York)

Cowling, T.G. (1936): Mon. Not. R. Astron. Soc. **96**, 42 (Appendix)

Cox, A.N. (1980): Ann. Rev. Astron. Astrophys. **18**, 15

Cox, A.N., Stewart, J.N. (1965): Astrophys. J. Suppl. **11**, 22

Cox, A.N., Stewart, J.N. (1970): Astrophys. J. Suppl. **19**, 243, 261

Cox, J.P. (1967): In *Aerodynamic Phenomena in Stellar Atmospheres*, ed. by R.N. Thomas, IAU Symp. 28 (Academic Press, London), p. 3

Cox, J.P. (1976): Ann. Rev. Astron. Astrophys. **14**, 247

Cox, J.P. (1980): *Theory of Stellar Pulsation* (Princeton University Press, Princeton)

Cox, J.P., Giuli, R.T. (1968): *Principles of Stellar Structure*, Vol. I, II (Gordon and Breech, New York)


Deubner, F.L., Gough, D. (1984): Ann. Rev. Astron. Astrophys. **22**, 593


Eddington, A.S. (1925): Observatory **48**, 73

El Eid, M.F., Langer, N. (1986): Astron. Astrophys. **167**, 274

Ezer, D., Cameron, A.G.W. (1967): Canadian J. Phys. **45**, 3429


Faulkner, J. (1966): Astrophys. J. **144**, 978

Fowler, W.A., Caughlan, G.R., Zimmerman, B.A. (1967): Ann. Rev. Astron. Astrophys. **5**, 525

Fowler, W.A., Caughlan, G.R., Zimmerman, B.A. (1975): Ann. Rev. Astron. Astrophys. **13**, 69

Fowler, W.A., Caughlan, G.R., Zimmerman, B.A. (1983): Ann. Rev. Astron. Astrophys. **21**, 165

Fowler, W.A., Hoyle, F. (1964): Astrophys. J. Suppl. **9**, 201

Fricke, K.J. (1968): Z. Astrophys. **68**, 317

Fricke, K.J., Strittmatter, P.A. (1972): Mon. Not. R. Astron. Soc. **156**, 129


Gaustadt, J.E. (1963): Astrophys. J. **138**, 1050

Giannone, P., Kohl, K., Weigert, A. (1968): Z. Astrophys. **68**, 107

Gold, T. (1968): Nature **218**, 731

Goldreich, P., Schubert, G. (1967): Astrophys. J. **150**, 571

Goldreich, P., Weber, S.V. (1980): Astrophys. J. **238**, 991

Grew, K.E., Ibbs, T.L. (1952): *Thermal Diffusion in Gases*, (Cambridge University Press, Cambridge)


Hamada, T., Salpeter, E.E. (1961): Astrophys. J. **134**, 683

Hansen, C.J., Spangenberg, W.H. (1971): Astrophys. J. **168**, 71

Härm, R., Schwarzschild, M. (1972): Astrophys. J. **172**, 403

Hayashi, C. (1961): Publ. Astron. Soc. Japan **13**, 450

Hayashi, C., Hoshi, R., Sugimoto, D. (1962): Progr. Theor. Phys. Suppl. **22**, 1

Heintzmann, H., Hillebrandt, W., El Eid, M.F., Hilf, E.R. (1974): Z. Naturforsch. **29a**, 933

Henyey, L.G., Vardya, M.S., Bodenheimer, P.L. (1965): Astrophys. J. **142**, 841

Hillebrandt, W. (1986): In *Cosmological Processes*, ed. by W.D. Arnett, C.J. Hansen, J.W. Truran, S. Tsuruta (VNU Science Press, Utrecht), p. 123

Hillebrandt, W. (1987): In *High Energy Phenomena around Collapsed Stars*, ed. by F. Pacini (Reidel, Dordrecht), p. 73

Hillebrandt, W. (1989): private communication

Hofmeister, E., Kippenhahn, R., Weigert, A. (1964): Z. Astrophys. **59**, 242

Hoyle, F. (1953): Astrophys. J. **118**, 513

Hubbard, W.B., Lampe, M. (1969): Astrophys. J. Suppl. **18**, 297

Huebner, W.F. (1978): In Proc. *Informal Conf. on Status and Future of Solar Neutrino Research*, ed. by G. Friedlander, BNL Rept 50879, Vol. I, p. 107


Iben, I., Jr. (1965): Astrophys. J. **141**, 993

Iben, I., Jr. (1969): Astrophys. J. **155**, L101

Iben, I., Jr. (1974a): In *Stellar Instability and Evolution*, ed. by P. Ledoux, A. Noels, A.W. Rodgers, IAU Symp. **59** (Reidel, Dordrecht), p. 3

Iben, I., Jr. (1974b): Ann. Rev. Astron. Astrophys. **12**, 215
Iben, I., Jr. (1975): Astrophys. J. **196**, 549,
Iben, I., Jr., Renzini, A. (1983): Ann. Rev. Astron. Astrophys. **21**, 271
Iben, I., Jr., Rood, R.T. (1970): Astrophys. J. **161**, 587
Ince, E.L. (1956): *Ordinary Differential Equations* (Dover, New York)

Jeans, J. (1928): *Astronomy and Cosmogony* (Cambridge University Press, Cambridge), republished 1961 (Dover, New York)

Kähler, H. (1972): Astron. Astrophys. **20**, 105
Kähler, H. (1975): Astron. Astrophys. **43**, 443
Kähler, H. (1978): In *The HR Diagram*, ed. by A.G.D. Philip and D.S Hayes, IAU Symp. **80** (Reidel, Dortrecht), p. 303
Kähler, H., Weigert, A. (1974): Astron. Astrophys. **30**, 431
Kato, S. (1966): Publ. Astron. Soc. Japan **18**, 374
Kippenhahn, R. (1963): In *Star Evolution*, Proc. International School of Physics "Enrico Fermi", Course XXVIII, ed. by L. Gratton (Academic Press, New York), p. 330
Kippenhahn, R. (1981): Astron. Astrophys. **102**, 293
Kippenhahn, R., Weigert, A., Hofmeister, E. (1967): Meth. Comp. Phys. **7**, 129
Kippenhahn, R., Ruschenplatt, G., Thomas, H.-C. (1980a): Astron. Astrophys. **91**, 175
Kippenhahn, R., Ruschenplatt, G., Thomas, H.-C. (1980b): Astron. Astrophys. **91**, 181
Kippenhahn, R., Thomas, H.-C. (1964): Z. Astrophys. **60**, 19
Kippenhahn, R., Thomas, H.-C., Weigert, A. (1965): Z. Astrophys. **61**, 241
Kippenhahn, R., Thomas, H.-C., Weigert, A. (1968): Z. Astrophys. **69**, 265
Korn, G.A., Korn, T.M. (1968): *Mathematical Handbook for Scientists and Engineers*, 2nd ed. (McGraw-Hill, New York)
Kozlowski, M., Paczyński, B. (1975): Acta Astron. **25**, 321

Landau, L.D., Lifshitz, E.M. (1959): *Statistical Physics*, Vol. 5 of Course of Theoretical Physics (Pergamon Press, London)
Landau, L.D., Lifshitz, E.M. (1959): *Fluid Mechanics*, Vol. 6 of Course of Theoretical Physics (Pergamon Press, London)
Landau, L.D., Lifshitz, E.M. (1975): *The Classical Theory of Fields*, Vol. 2 of Course of Theoretical Physics (Pergamon, Elmsford, New York)
Langer, N., El Eid, M.F., Fricke, K.J. (1985): Astron. Astrophys. **145**, 169
Larson, R.B. (1969): Mon. Not. R. Astron. Soc. **145**, 271
La Salle, J., Lefschetz, S. (1961): *Stability by Liapunov's Direct Method with Applications* (Academic Press, New York)
Lauterborn, D., Refsdal, S., Roth, M.L. (1971): Astron. Astrophys. **13**, 119
Lauterborn, D., Refsdal, S., Weigert, A. (1971a): Astron. Astrophys. **10**, 97
Lauterborn, D., Refsdal, S., Weigert, A. (1971b): Astron. Astrophys. **13**, 119
Ledoux, P. (1958): In *Handbuch der Physik*, ed. by S. Flügge (Springer, Berlin, Heidelberg), Vol. LI, p. 605
Low, C., Lynden-Bell, D. (1976): Mon. Not. R. Astron. Soc. **176**, 367
Lyttleton, R.A. (1953): *The Stability of Rotating Liquid Masses* (Cambridge University Press, Cambridge)

Maeder, A. (1975): Astron. Astrophys. **40**, 303
Mariska, J.T., Hansen, C.J. (1972): Astrophys. J. **171**, 317
Matraka, B., Wassermann, C., Weigert, A. (1982): Astron. Astrophys. **107**, 283
McDougall, J., Stoner, E.C. (1939): Phil. Trans. R. Soc. London **237**, 67
Mestel, L. (1952): Mon. Not. R. Astron. Soc. **112**, 598
Mestel, L. (1953): Mon. Not. R. Astron. Soc. **113**, 716
Meyer-Hofmeister, E. (1967): Z. Astrophys. **65**, 164
Meyer-Hofmeister, E. (1969): Astron. Astrophys. **2**, 143

Meyer-Hofmeister, E. (1982): In Landolt-Börnstein *Numerical Data and Functional Relationships in Science and Technology*, New Series, Group VI, **2b** (Springer, Berlin, Heidelberg), p. 152
Misner, C.W., Thorne, K.S., Wheeler, J.A. (1973): *Gravitation* (Freeman, San Francisco)


Nomoto, K., Thielemann, F.-K., Miyaji, S. (1985): Astron. Astrophys. **149**, 239
Nomoto, K., Thielemann, F.-K., Yokoi, K. (1984): Astrophys. J. **286**, 644
Nomoto, K., Sugimoto, D., Neo, S. (1976): Astrophys. Space Sci. **39**, L37


Oppenheimer, J.R., Volkoff, G.M. (1939): Phys. Rev. **55**, 374


Paczyński, B. (1970): Acta Astron. **20**, 47
Paczyński, B. (1971): Acta Astron. **21**, 271
Paczyński, B. (1972): Acta Astron. **22**, 163
Paczyński, B. (1975): Astrophys. J. **202**, 558
Paczyński, B., Kozlowski, M. (1972): Acta Astron. **22**, 315
Parker, P.D., Bahcall, J.N., Fowler, W.A. (1964): Astrophys. J. **139**, 602
Petrosian, V., Beaudet, G., Salpeter. E.E. (1967): Phys. Rev. **154**, 1445
Pines, D. (1980): Journal de Physique **41**, Coll. C2, suppl. au no.3, p. C2–111
Popper, D.M. (1980): Ann. Rev. Astron. Astrophys. **18**, 115
Prandtl, L. (1925): Z. Angew. Math. Mech. **5**, 136


Rees, M.J. (1976): Mon. Not. R. Astron. Soc. **176**, 483
Refsdal, S., Weigert, A. (1970): Astron. Astrophys. **6**, 426
Renzini, A. (1987): Astron. Astrophys. **188**, 49
Richtmyer, R.D., Morton, K.W. (1967): *Difference Methods for Initial-Value Problems*, 2nd ed. (Interscience, New York)
Robertson, J.W. (1971): Astrophys. J. **164**, L105
Rood, R.T. (1973): Astrophys. J. **184**, 815
Roth, M.L. (1973): Dissertation, University of Hamburg
Roth, M.L., Weigert, A. (1972): Astron. Astrophys. **20**, 13


Salpeter, E.E. (1961): Astrophys. J. **134**, 669
Saslaw, W.C., Schwarzschild, M. (1965): Astrophys. J. **142**, 1468
Schatzman, E., Maeder, A. (1981): Astron. Astrophys. **96**, 1
Schönberg, M., Chandrasekhar, S. (1942): Astrophys. J. **96**, 161
Schönberner, D. (1979): Astron. Astrophys. **79**, 108
Schwarzschild, M. (1941): Astrophys. J. **94**, 245
Schwarzschild, M. (1942): Astrophys. J. **95**, 441
Schwarzschild, M. (1946): Astrophys. J. **104**, 203
Schwarzschild, M. (1958): *Structure and Evolution of the Stars* (Princeton University Press, Princeton)
Schwarzschild, M., Härm, R. (1959): Astrophys. J. **129**, 637
Schwarzschild, M., Härm, R. (1965): Astrophys. J. **142**, 855
Shapiro, S.L., Teukolsky, S.A. (1983): *Black Holes, White Dwarfs, and Neutron Stars. The Physics of Compact Objects* (Wiley, New York)
Shaviv, G., Salpeter, E.E. (1973): Astrophys. J. **184**, 191
Simon, N.R. (1987): In *Stellar Pulsation*, ed. by A.N. Cox, W.M. Sparks, S.G. Starrfield, Lect. Notes Phys., Vol. 274 (Springer, Berlin, Heidelberg), p. 148
Smeyers, P. (1984): In *Theoretical Problems in Stellar Stability and Oscillations*, ed. by A. Noels, M. Gabriel, Proc. 25th Liège Intern. Coll., p. 68
Spiegel, E.A. (1971): Ann. Rev. Astron. Astrophys. **9**, 323
Spiegel, E.A. (1972): Ann. Rev. Astron. Astrophys. **10**, 261
Spiegel, E.A., Zahn, J.P. (eds.) (1977): *Problems of Stellar Convection*, Lect. Notes Phys., Vol. 71 (Springer, Berlin, Heidelberg)
Spitzer, L., Jr. (1968): *Diffuse Matter in Space* (Wiley, New York)
Strom, S.E., Strom, K.M., Rood, R.T., Iben, I., Jr. (1970): Astron. Astrophys. **8**, 243

Sweet, P.A. (1950): Mon. Not. R. Astron. Soc. **110**, 548

Sweigart, A.V., Gross, P.G. (1978): Astrophys. J. Suppl. **36**, 405

Tassoul, J.-L. (1978): *Theory of Rotating Stars* (Princeton University Press, Princeton)

Thomas, H.-C. (1967): Z. Astrophys. **67**, 420

Truran, J.W., Iben, I., Jr. (1977): Astrophys. J. **216**, 797

Tscharnuter, W. (1985): In *Birth and Infancy of Stars*, ed. by R. Lucas, A. Omont, R. Stora, Les Houches, Session XLI (North Holland, Amsterdam), p. 601

Unno, W. (1967): Publ. Astron. Soc. Japan **19**, 140

Unno, W., Osaki, Y., Ando, H., Shibahashi, H. (1979): *Nonradial Oscillations of Stars* (University of Tokyo Press, Tokyo)

Van Albada, T.S., Baker, N.H. (1971): Astrophys. J. **169**, 311

Van Albada, T.S., Baker, N.H. (1973): Astrophys. J. **185**, 477

Van Horn, H.M. (1984): In *Theoretical Problems in Stellar Stability and Oscillations*, ed. by A. Noels, M. Gabriel, 25th Liège Intern. Astrophys. Coll., p. 307

Van Horn, H.M. (1986): Mitt. Astron. Ges. **67**, 63

Van Riper, K.A. (1978): Astrophys. J. **221**, 304

Vogt, H. (1925): Astron. Nachr. **223**, 229

Weaver, T.A., Zimmerman, G.B., Woosley, S.E. (1978): Astrophys. J. **225**, 1021

Weigert, A. (1966): Z. Astrophys. **64**, 395

Weinberg, S. (1972): *Gravitation and Cosmology* (Wiley, New York)

Weiss, A. (1986): private communication

Weiss, A. (1987): Astron. Astrophys. **185**, 178

Wilson, J.R. (1985): In *Numerical Astrophysics*, ed. by J.M. Centrella, J.M. LeBlanc, R.L. Bowers (Jones and Bartlett, Boston), p. 422

Woosley, S.E., Weaver, T.A. (1986): In *Nucleosynthesis and its Implications on Nuclear and Particle Physics*, ed. by J. Audouze, N. Mathieu (Reidel, Dordrecht) p. 145

Woosley, S.E. (1986): In *Nucleosynthesis and Chemical Evolution*, ed. by B. Hauck, A. Maeder, G. Meynet (Geneva Observatory), p.1

Wrubel, M.H. (1958): In *Handbuch der Physik*, ed. by S. Flügge (Springer, Berlin, Heidelberg), Vol. LI, p. 1

Zahn, J.-P. (1974): In *Stellar Instability and Evolution*, ed. by P. Ledoux, A. Noels and A.W.Rodgers, IAU Symp. 59 (Reidel, Dordrecht), p. 185

Von Zeipel, H. (1924): In *Probleme der Astronomie*, Festschrift für H. v. Seeliger, ed. by H. Kienle (Springer, Berlin) p.144

Zeldovich, Ya. B., Novikov, I.D. (1971): *Relativistic Astrophysics*, Vol. I "Stars and Relativity" (University of Chicago Press, Chicago)

Ziebarth, K. (1970): Astrophys. J. **162**, 947

# Subject Index

absorption coefficient, *see* opacity
abundances of elements,
   *see* chemical composition
accreting white dwarfs 356
accretion on protostars 258, 263, 264
— time-scale 258, 263
— luminosity 258, 261, 264
adiabatic exponent 105
— relation to Chandrasekhar's gammas 106
adiabatic temperature gradient 21
age
— sun 271
— determination for clusters 280
angular-velocity distribution in stars 444–453
— dynamical stability 446–450
— secular stability 451–453
ascending giant branch 313, 315, 322
astrophysical cross-section factor 152
asymptotic giant branch 322–324

Beta Cephei variables 426
bifurcation, rotating liquid configurations 429
binding energy per nucleon 147–148
black dwarfs 268, 366, 377
black holes 390–395
Boltzmann distribution
— excitation of atoms 107, 108
— momentum of particles 118
boundary conditions 68–72
— at the centre 68
— — series expansions 68–69
— at the surface 69–72
— — zero conditions 69
— — photospheric conditions 70
— — general formulation 70–72
— — influence on envelope 72–76
bremsstrahlung neutrinos 172
brown dwarfs 268
Brunt-Väisälä frequency 41–42, 423

carbon burning
— reactions 167
— in degenerate cores 348–349, 351–356

— in accreting white dwarfs 356
carbon flash 340, 348–349
— time-scales 351
carbon main-sequence 218–219
carbon-12 burning, pre-main-sequence 269–270
carbon–oxygen cores
— contraction and heating 344–346
— carbon flash, carbon burning 340, 348–349,
   351–356
— dynamical instability 362–364
catalyzed matter 372, 385
— $M-R$ relation for cold bodies 384
central conditions 68
central evolution
— pre-main-sequence 266–269
— through nuclear burnings 336–341
— late phases 344–347
Cepheid strip 300, 414–415
Cepheids
— evolution 300–301
— masses 301, 415–416
— change of period 300
— period–luminosity relation 300, 415–416
— excitation 411–415
Chandrasekhar's limiting mass 181–182, 368
chemical composition of stellar matter 56–62
— mass abundances 56
— change by nuclear reactions 57–58
— convective mixing 61–62
— diffusion 58–61
— mixing by circulations 442–443
— present sun 272–273
— equilibrum, catalyzed matter 372–374, 385
chlorine experiment 276
circulations, meridional 437–443
— time-scale 441
— chemical inhomogeneities 442–443
clusters of stars, *see* star clusters
CNO cycle 164–165
— in main-sequence stars 212
collapse
— interstellar clouds 254–260
— protostars 263
— evolved cores 344, 347, 356–362

461

— polytropes 187–190, 357–359
— into black holes 394
collapsing polytropes
combustion front 352–356
compact objects 365–395 (*see also* white
    dwarfs, neutron stars, black holes)
complete equilibrium (mechanical and thermal)
    25, 66
compound nucleus 150–151
Compton scattering 138
conductive opacity 142–143
conductive transport of energy 31–32
conservation of momentum 6–12
conservative rotation 435–436
continuity equation 3, 11
contraction and heating/cooling 266–268, 329,
    344–346
convection
— stability criteria (dynamical) 36–40
—— unstable $g$ modes 423–424
—— vibrational and secular stability 42–45
— mixing-length theory 49–55
—— limiting cases 52–54
—— adiabatic, superadiabatic 53–55
—— efficiency 52–55
— velocity 49–50, 55
— mixing of chemical composition 61–62
— fully convective stars 224
— in main-sequence stars 212–215
— overshooting 281–284
— semiconvection 284–285
— instability of $g$ modes 423–424
convective blocking 352
convective transport of energy 48–55
cooling of white dwarfs 377–379
core collapse
— protostars 263
— evolved cores 344, 347, 356–362
—— instabilities 347 362–362,
—— collapse time 359
—— rebounce 359–360, 362
—— neutrinos 360–362
core contraction and heating 329
core-mass–luminosity relation 312, 315, 342–343
core-mass–temperature relation 313, 315
core of neutron stars 388
Coulomb barrier of nuclei 148–149
critical rotation 432–433
crust of neutron stars 387–388
crystallization 134
— white dwarfs 370, 378–379
— neutron stars 387–388

deflagration front 352–356

degeneracy
— of electrons 118–128
—— Fermi–Dirac distribution 124
—— complete degeneracy 119–123
—— non-relativistic, extreme relativistic 123
— of ions 129
— of neutrons 381–382
degeneracy parameter 124
degenerate cores 329, 337–338, 344–349
Delta Scuti variables 426
detonation front 352–356
diffusion
— radiative energy 28–29
— chemical elements 58–61
— neutrinos in collapsed cores 361–362
dredge up of nuclear species 307, 313, 335
dynamical stability/instability 240–241,
    401–402
— critical gamma 241, 401
—— effect of general relativity 386
— protostars 263
— supermassive stars 187
— highly evolved cores 347, 357, 362–364
— configurations of catalyzed matter 385–386
— Jacobi ellipsoids 430
— local perturbations (convection) 36–40
— non-radial $g$ modes 423–424
— angular-velocity distribution 446–450
— piston model 238


Eddington's standard model 180
Eddington–Sweet time-scale 439–441
effective temperature 70
efficiency of convection 52–55
electron capture instability 357
electron scattering opacity 137–138
electron shielding of nuclear reactions 157–161
electrostatic interaction and equation of state
    370–371
energy conservation
— for stellar matter 21–23
— for the whole star 23–24
— time-dependent terms 22
— neutrino losses 23
energy integral (pulsating stars) 409–411
energy transport, see transport of energy
envelope solutions
— radiative 72–73
— convective 75
— temperature stratification 76
epsilon mechanism 408, 412
— upper-main-sequence stars 215, 218, 417
equation of motion 9–10
— non-spherical case 11

equation of state
— ideal gas and radiation 102, 104
— degenerate electron gas 123–128
— electrostatic interaction 370–371
— neutronization 135–136, 371–373
— beyond neutron drip 380–383
— for stellar matter 129–136
equations of stellar structure
— Eulerian and Lagrangian descriptions 2– 4
— equation of motion 9–10
— hydrostatic equilibrium 7
— energy equation 21–24
— transport of energy
—— radiative 27–31
—— conductive 31–32
—— convective 48–55
— change of chemical composition 57– 62
— review and summary 64– 66
—— time derivatives, time-scales 66– 67
—— hydrostatic and complete equilibrium 66– 67
—— initial values 66– 67
equilibrium
— hydrostatic 6–7, 66
— complete (hydrostatic and thermal) 66
— nuclear statistical 349–351
equilibrium composition 372–374
evolutionary mass of Cepheids 301
explosions 344, 347, 356, 359–360, 362–364

$f$ mode 424
Fermi momentum, — energy 120
Fermi–Dirac distribution 124
Fermi–Dirac integrals 125
final stages
— mass limits 340–341
fitting (shooting) method 77–78
flash 245, 331, 338, see also helium flash, carbon flash
formation of stars, see star formation
fragmentation of collapsing clouds 253–255
— stellar masses 255
fully convective stars 224

$g$ modes 423
gallium experiment 276
Gamow peak 154–157
Gamow penetration factor 149
general relativistic effects
— hydrostatic equilibrium 12–13
— neutron star masses 386–387
— dynamical stability 386
generalized main sequences 221–223

giant branch
— ascending 313
— asymptotic 322
giants, evolution to 292, 313
glitches of pulsars 388
globular cluster diagrams
— ascending giant branch 313
— horizontal branch 320–322
— asymptotic giant branch 322
Goldreich–Schubert–Fricke criterion 452
gravitational energy of stars 15
gravitational instability of interstellar clouds 248–255
— Jeans criterion 250, 253
gravitational mass of a star 12, 386–387
gravitational potential 4–5
gravothermal specific heat 17, 242–243, 331, 339

Hayashi line 224–233
— analytical approach 226–230
— "forbidden" region 224, 229–230
heat conduction 31–33
— and opacity values 142–143
heating/cooling during contraction 266–268, 344–346
helioseismology 426
helium-burning reactions 165–166
helium-burning phase
— low-mass stars 320–324
—— helium flash 316–319
—— horizontal-branch phase 320–322
—— helium-shell burning 322–323
—— asymptotic giant branch 322
—— thermal pulses 323
— massive stars 296–306
—— helium ignition 296
—— production of C, O, Ne 296–297, 306
—— time-scales 296–297
—— loops in the HR diagram 297–306
—— Cepheid phase 300–301
—— helium-shell burning 306–307
helium flash 316–319, 338, 326, 327
— mixing of composition 319–320
— time-scale 316
helium main sequence 216–217
helium-3 burning, pre-main-sequence 269
Henyey method 78–83
Henyey matrix, determinant 81–83, 96
— and stability 96
Hertzsprung–Russell diagram
— "forbidden" region 224, 229–230
— star clusters 280, 299, 320, 321
high-density (intermediate) branch 220
homologous contraction 198–199, 257, 266

— central evolution 266–269
— maximum central temperature 268
homology invariants $U, V$ 200
— for polytropic models 200–201
homology relations 191–197
— main-sequence models 194–197
— for shell-source models 309–313, 342–343
horizontal-branch stars 320–322
— zero-age models 320–321
— metal content 320
— RR Lyrae variables 321, 323
HR diagram, *see* Hertzsprung–Russell diagram
hydrodynamical methods 84
hydrogen-burning reactions 162–165
— proton–proton chain 162–164
— CNO cycle 164–165
hydrogen-burning phase,
    *see* main-sequence phase
hydrogen main sequence 207–215
— stability, upper and lower end, 215
hydrostatic equilibrium 6–7, 66
— general relativity 12–13
—— post-Newtonian approximation 13
hydrostatic time-scale 11
hyperonization 381, 382

ideal gas and radiation
— equation of state 102–104
— thermodynamic properties 104–106
ignition of nuclear burning
— minimum mass 337–338, 340–341
intermediate branch of homogeneous models
    220
initial values 66–67
instability strip 300, 321–322, 412–416
— Cepheid evolution 300–301
— RR Lyrae variables 321, 323
— W Virginis variables 322–323
— observed stars 415–416
instability, *see* stability
inverse beta decay 135, 371–372
ionization of stellar matter
— Saha equation 109–110
— pressure ionization 117
— partial ionization of H and He 110–115
—— mean molecular weight 111
—— thermodynamical properties 111–112
—— in the sun 114
isothermal-core models 288–290
— Schönberg–Chandrasekhar limit 286–288
— thermal stability 287, 290
— in the $U$–$V$ plane 206
isothermal spheres of ideal gas 183

Jacobi ellipsoids 429–430
Jeans criterion 248–253
— virial theorem 252
— Jeans mass 253, 255


$\kappa$ mechanism 408, 411
— stars driven by 412–416
Kelvin–Helmholtz time scale 18
Kramers opacity 138


Lane–Emden equation 176
— solutions 177–178
— isothermal spheres 183
— collapsing polytropes 190
later phases (post-He-burning) 328–342
— nuclear burnings 328–330, 340–342,
    347–349, 356
— central evolution 329–330, 336–341,
    344–347
— degenerate cores 329, 338–339, 344–349
— thermal pulses 333–336
— neutrino losses 339–341
— carbon flash 340, 348–349, 351
— final evolution 340–341, 344–347
Ledoux criterion 39
limiting mass (*see also* mass limit)
— white dwarfs 181–182, 368, 374
— neutron stars 384–386
linear series of models 90–91, 98
— isothermal-core models 289–290
— main sequences 219–221
— helium-core models 295–296, 305–306,
    324–327
— models of cold catalyzed matter 385–386
local uniqueness, *see* uniqueness
loops in the HR diagram 297–306
— Cepheid phase 300–301
— non-equilibrium phase 306
luminosity
— local 21–22
— surface value 21
— neutrino 23
— accretion 258, 261, 264


main-sequence models 207–215
— $M$–$R$ and $M$–$L$ relations 207–209
— central values 210–212
— radiation pressure 212
— $pp$ and CNO reactions 212
— convective regions 212–215
— instability at small and large $M$ 215

main-sequence phase 277–285
— chemical evolution 277–278
— time-scales 280
— convective overshooting 281–284
— semiconvection 284–285
main sequences
— zero-age (hydrogen) main sequence 216
— helium main sequence 216–217
— carbon main sequence 218–219
— generalized main sequences 221–223
— linear series of models 219–221
— lower end 218, 219
— vibrational stability 218
mass
— of Cepheids 301
— gravitational 12, 386–387
mass defect
— nuclei 146
— neutron stars 386–387
mass limits
— ignition of nuclear burning 337–338,
  340–341
— degenerate cores 337–338
— types of late evolution 347, 364
— final stages 340–341
mass loss
— before helium flash 320
— white-dwarf production 323
— and final stages 341
— critical rotation 432–433
mass–luminosity relation
— main sequence models 195, 207–209
— helium and carbon main sequences 216–218
mass–radius relation
— polytropic stars 181
— main-sequence models 194–195, 207–209
— white dwarfs 368–369, 373–374
— neutron stars 383–384
— models of cold catalyzed matter 385
material functions of stellar matter 101–172
maximum mass of neutron stars 384–386
McLaurin spheroids 428–430
mechanical equilibrium,
    see hydrostatic equilibrium
melting temperature 134
meridional circulations 437–443
— Eddington–Sweet time-scale 439–441
— chemical inhomogeneities 442–443
minimum mass
— ignition of nuclear burning 267–268,
  337–338, 340–341
mirror principle of radial motions 293, 306
mixing length 49
mixing-length theory of convection 49–52
mixing of chemical composition

— by convection 61–62
— by meridional circulations 442–443
molecular weight 8
— for ionized matter
— — mean value 102–104
— — per ion 103
— — per free electron 103
— partially ionized matter 111
multiple solutions
— homogeneous equilibrium models 221
— isothermal-core models 290
— helium-core models 296, 327
$\mu$ barrier, $\mu$ currents 443

neon disintegration 168, 349
neutrino losses
— energy equation 23
— thermal stability 245, 339
— temperature inversion 317–318, 340
— before helium flash 317, 318
— degenerate C–O cores 339–341
neutrino luminosity 23, 339, 341–342
neutrino sphere 362–362
neutrino trapping 361
neutrinos 169–172
— mean free path 169, 361
— from hydrogen burning 169–170
— from electron captures 170
— Urca process 170
— from leptonic interaction 170–172
— solar 275–276
— core collapse, supernovae 360–362
neutron drip 135, 372, 380
neutron stars 380–389
— formation 347, 360
— $M–R$ relation 383–384
— maximum mass 384–386
— gravitational mass 386–387
— stability 385–386
— interior models 387–388
— extension of atmosphere 387
neutronization 135–136, 360, 362
neutronization threshhold 372
neutrons
— production in thermal pulses 335
— degeneracy 381–382
— superfluid liquid 383, 388
non-radial oscillations 418–426
— eigenspectra 422–425
— dynamical stability 423–424
— observations 425–426
nuclear-burning
— reactions 161–167
— thermal stability 243–245, 339

— minimum mass for ignition 337–338
nuclear-burning phases
— pre-main-sequence burning 269–270
— hydrogen burning 277–284
— helium burning 292–321
— later burnings 328–330, 340–342, 347–349, 356
nuclear cross-sections 150–152
— resonances 151
— astrophysical cross-section factor 152
nuclear energy generation 146–149
— temperature sensitivity 156–157
— electron shielding 157–161
— burning, *see* hydrogen burning, etc
nuclear statistical equilibrium 349–351
nuclear time-scale
— shell sources 330
— late phases 342
numerical methods 77–84
— shooting (fitting) method 77–78
— Henyey method 78–83
— explicit, implicit schemes 83–84
— hydrodynamical problem 84

"onion skin model" 328–329
opacity of stellar matter 137–145
— Rosseland mean 29–31
— electron scattering 137
—— Compton scattering 138
— free–free transitions 138–139
— bound–free transitions 139–140
— bound–bound transitions 140–141
— negative hydrogen ions 141–142
— conductive opacity 142–143
oscillation of stars, *see* pulsation, non-radial oscillations oxygen burning
— reactions 167–168
— in C–O cores 351, 353–354

$p$ modes 423
$pp$ reactions 162–164
pair annihilation neutrinos 170–171
pair creation instability 362–364
partition function 107, 109
Pauli's exclusion principle 118–119
pear-shaped configurations 430
period–density relation 300, 404–405
period–luminosity relation 300, 415–416
photodisintegration 168–169, 347, 349–351, 363
photo neutrinos 171
photosphere 70
photospheric conditions 76
piston model 8–9
— mechanical properties 13–14
— thermal properties 34–35

— virial theorem 17
— stability 45–47, 235–238, 407
plasma neutrinos 171–172
Poisson equation 4
poloidal modes of non-radial oscillations 421
polytropes
— collapsing 357–359
— slowly rotating 433–434
polytropic relation,-index,-exponent 174–175
polytropic stellar models 175–190
— Lane–Emden equation 176
— radiation pressure 180
— $M$–$R$ relations 181
— isothermal ideal-gas sphere 183
— supermassive stars 186–187
— collapsing polytropes 187–190
— pulsations 403–406
post-main-sequence evolution
— massive stars 292–296
—— hydrogen-shell burning 293
—— core-contraction phase 293
—— Hertzsprung gap 295, 308
— low-mass stars 308–309
—— hydrogen-shell burning 308–309
—— ascending giant branch 313
post-Newtonian approximation 13
pre-main-sequence contraction 266–270
— central heating 266–267
— minimum mass for hydrogen ignition 268
— approach to main sequence 269–270
— time-scales 270
pre-main-sequence nuclear burning 269–270
proton–proton chain
— reactions 162–164
— in main-sequence stars 212
protostar evolution 260–265
— formation 260–262
— collapse onto condensed object 258
— collapse calculations 259–262
— $H_2$ dissociation, core collapse 263
pulsation of stars
— adiabatic spherical pulsations 398–406
—— dynamical stability 401–402
—— effect of radiation pressure 405–406
— non-adiabatic spherical pulsations 407–417
—— $\kappa$ mechanism 408, 411, 412–416
—— epsilon mechanism 408, 412, 417
—— instability strip 412–415
—— quasi-adiabatic approximation 408–409
—— non-linear effects 416
— non-radial oscillations 418–426
—— eigenspectra 422–425
—— observations 425–426
pulsational mass of Cepheids 301
pulse instability 331–332

pulses, *see* thermal pulses
pycnonuclear reactions 161, 340, 349, 371–372

quasi-adiabatic approximation 408–409, 425

radiation pressure 104
— in polytropic models 180
— in main-sequence models 212, 218
— supermassive stars 186–187
— influence on pulsations 218, 405–406, 417
radiative transport of energy 27–31
radiative viscosity 445
Rayleigh criterion 449
resonance reactions 151
Reynolds number 55, 447
Richardson number 447
Roche model 431–433
Rosseland mean of the absorption coefficient 29–31
rotation of stellar models 427–453
— rotating liquid configurations 428–430
—— stability 430
— Roche model 431–433
—— critical rotation 432–433
— polytropes 433–434
— conservative rotation 435–436
— thermodynamic properties 436–443
—— von Zeipel's theorem 436
— meridional circulations 437–443
— non-conservative rotation 438–439
— angular-velocity distribution 444–453
—— dynamical stability 446–450
—— secular stability 451–453
RR Lyrae variables 321, 323, 416

Saha equation 109–110
— limitation for high densities 115–117
salt-finger instability 44–45, 319
scale height of pressure 38
Schönberg–Chandrasekhar limit 285–292, 326–327
— in the $U$–$V$ plane 206
Schwarzschild criterion for convection 39
Schwarzschild radius,-metric 390
screening factor 160
secular (thermal) stability 241–242
— salt finger 44–45, 319
— piston model 238
— nuclear burning 243–245, 339
— neutrino losses 245, 339
— shell sources 330–332
— isothermal-core models 287, 290
— main-sequence branches 219–221
— McLaurin spheroids 430

— angular-velocity distribution 451–453
sedimentation 60–61
semiconvection 284–285, 322
— influence on loops 304
shell-source burning
— hydrogen 293, 308–309, 322–323
— helium 306–307, 323
— double-shell sources 322–323
— late phases 330
— local nuclear time-scales 330
shell-source homology 309–313, 342–343
silicon burning 169
solar neutrinos 275–276
— spectrum 275
— measurements 276
solar standard model 272–274
Solberg–Høiland criterion 450
specific heat
— radiation pressure 105
— ionization 112
— electron degeneracy 133, 377
— white-dwarf matter 377–379
— gravothermal 17, 242–243, 331, 339
stability 234–246
— and local uniqueness 95
— general considerations 234–235
— perturbation equations 239–240
— dynamical stability 240–241
— secular (thermal) stability 241–242
— vibrational stability 241–242, 407–408
— gravitational 248–255
— local perturbations 36–45
—— dynamical stability 36–40
—— vibrational stability 42–43
—— secular (salt-finger) stability 44–45
— piston model 45–47
standard model of Eddington 180
star clusters
— age determination 280
— Hertzsprung–Russell diagram 280, 299, 320, 321
star formation 248–265
— gravitational instability 248–255
— collapse of clouds 256–261
—— time-scales 253
— fragmentation 253–255
— protostar evolution 260–265
— pre-main-sequence contraction 266–270
stellar-structure equations, *see* equations of stellar structure
sun
— mass, radius 9
— luminosity 18
— time-scales 11, 18, 25
— central values 9

— age 271
— solar standard model 272–274
— solar neutrinos 275–276
— oscillations, helioseismology 426
supermassive stars 186–187
supernova explosions 344, 347, 356, 359–360, 362
— neutrinos 360–362
surface conditions 69–76
synchrotron neutrinos 172

temperature gradient
— for radiative transport 32
— in convective regions 39–43, 50–55
thermal adjustment time 33–34
thermal pulses 323, 333–336
— instability 331–332
— cycle time 336
— nuclear reactions 335
— dredge up 335
thermal stability, *see* secular stability
thermonuclear fusion 149
thermonuclear reaction rates 152–157
time-scales 25–26, 66
— hydrostatic 11, 25
— Kelvin–Helmholtz 18, 25
— nuclear 25
— thermal adjustment 33–34, 43–44, 253
— explosion 10
— free fall 10, 250, 253, 257
— hydrostatic 11
— local oscillations 41–42
— convection 352
— for the sun 11, 18, 25
— collapsing clouds 250, 253
— accretion 258, 263
— pre-main-sequence contraction 270
— main-sequence phase 280
— helium-burning phase 296–297
— meridional circulations 441
Tolman–Oppenheimer–Volkoff (TOV) equation 13
transport of energy
— radiative 27–31
— diffusion approximation 28–29
— conductive 31–32
— convective 48–55
triple alpha reaction 165–166
tunnelling probability 149
turbulence of rotational motion 447–450
turbulent viscosity 445
turning points of linear series
— main sequences 218, 219
— isothermal-core models 290
— helium-core models 306, 327
— models of cold catalyzed matter 385

$U$–$V$ plane 200–206
— convective cores 205
— radiative envelopes 203–204
— isothermal cores 206
uniqueness of solutions 85–99
— local uniqueness
— — for complete equilibrium 88–91
— — for thermal non-equilibrium 93–94
— — and stability 95–97
— non-local uniqueness 97–99
Urca process 170

variable stars
— Cepheids 300–301, 411–416
— RR Lyrae variables 321, 323, 416
— W Virginis variables 322–323, 416
— non-radial oscillators 425–426
velocity
— convection 49–50, 55
— meridional circulations 441
vibrational stability 241–242, 407–408
— excitation mechanisms 411–417
— stars in the instability strip 300, 412–416
— upper-main-sequence stars 215, 218, 417
— local mass elements 41–43
— piston model 238, 407
virial theorem 15–16
— surface terms 18
— piston model 17
viscosity of stellar matter 444–446
— radiative 445
— turbulent 445
Von Zeipel's theorem 436

W Virginis variables 322–323, 416
white dwarfs 366–379
— formation of 323–324, 341
— $M$–$R$ relation 181, 368–369
— limiting mass 181–182, 368–369, 374
— mechanical structure 366–374
— — particle interaction effects 370–374
— thermal properties 374–379
— energy reservoirs 376–377
— cooling time 378
— crystallization 370, 378–379
— initial masses of progenitors 341
— accreting 356
— non-radial oscillations 426

zero-age horizontal-branch models 320–321
zero-age main-sequence (ZAMS) 207–215
ZZ Ceti variables 426